Dimitrios G. Aggelis and Nathalie Godin (Eds.)

Acoustic and Elastic Waves: Recent Trends in Science and Engineering

MDPI

This book is a reprint of the Special Issue that appeared in the online, open access journal, *Applied Sciences* (ISSN 2076-3417) from 2015–2016, available at:

http://www.mdpi.com/journal/applsci/special_issues/acoustic_elastic_waves?view=default&listby=pubdate_published+DESC%2Cfirstpage+DESC%2Cnumber+DESC

Guest Editors
Dimitrios G. Aggelis
Dept. Mechanics of Materials and Constructions (MeMC),
Vrije Universiteit Brussel (VUB)
Belgium

Nathalie Godin
Laboratoire MATEIS, National Institute for Applied Sciences (INSA) Lyon
France

Editorial Office	*Publisher*	*Senior Assistant Editor*
MDPI AG	Shu-Kun Lin	Yurong Zhang
St. Alban-Anlage 66		
Basel, Switzerland		

1. Edition 2016

MDPI • Basel • Beijing • Wuhan • Barcelona • Belgrade

ISBN 978-3-03842-296-9 (Hbk)
ISBN 978-3-03842-297-6 (PDF)

Table of Contents

V

List of Contributors

Dimitrios G. Aggelis Department of Mechanics of Materials and Constructions, Vrije Universiteit Brussel, Pleinlaan 2, Brussels 1050, Belgium.

Wamadeva Balachandran Brunel Innovation Centre (BIC), Brunel University, Cambridge CB21 2AL, UK.

Amadeo Benavent-Climent Department of Mechanical Engineering, Polytechnic University of Madrid, Madrid 28006, Spain.

Alina Bruma CRISMAT Laboratory, Ecole Nationale Superieure d'Ingenieurs de Caen, Universite de Caen Basse Normandie, 6 Blvd Marechal Juin, Caen 14050, France; Department of Physics and Astronomy, University of Texas at San Antonio, One University of Texas at San Antonio Circle, San Antonio, TX 78249, USA.

Valerii V. Burkhovetsky Department of Physical Materials Science, Donetsk Institute for Physics and Engineering named after O.O. Galkin of National Academy of Sciences of Ukraine, 72 R. Luxemburg str., Donetsk 83114, Ukraine.

Alberto Carpinteri Department of Structural, Geotechnical and Building Engineering, Politecnico di Torino, Corso Duca degli Abruzzi 24, 10129 Torino, Italy.

María T. Casas Chemical Engineering Department, Polytechnic University of Catalonia, Av. Diagonal 647, Barcelona E-08028, Spain.

Liang Cheng Brunel Innovation Centre (BIC), Brunel University, Cambridge CB21 2AL, UK.

Alastair Clarke Cardiff School of Engineering, Cardiff University, Cardiff CF24 3AA, UK.

Mihail-Liviu Craus Nondestructive Testing Department, National Institute for Research and Development for Technical Physics, 47 D. Mangeron Blvd., Iasi 700050, Romania; Frank Laboratory for Neutron Physics, Joint Institute for Nuclear Research, 6 Joliot-Curie Avenue, Dubna 141980, Russia.

Patrizia Cutugno Department of Structural, Geotechnical and Building Engineering, Politecnico di Torino, Corso Duca degli Abruzzi 24, 10129 Torino, Italy.

Martin Dannemann Institute of Lightweight Engineering and Polymer Technology (ILK), Technische Universität Dresden, Holbeinstraße 3, 01307 Dresden, Germany.

Niels J.F. De Baerdemaeker Laboratory of Plant Ecology, Department of Applied Ecology and Environmental Biology, Faculty of Bioscience Engineering, Ghent University, Coupure links 653, 9000 Gent, Belgium.

Dieter De Baere Department of Mechanical Engineering, Vrije Universiteit Brussel, Pleinlaan 2, 1050 Brussels, Belgium.

Linus De Roo Laboratory of Plant Ecology, Department of Applied Ecology and Environmental Biology, Faculty of Bioscience Engineering, Ghent University, Coupure links 653, 9000 Gent, Belgium.

Joris Degrieck Mechanics of Materials and Structures, Department of Materials Science and Engineering, Ghent University, Technologiepark-Zwijnaarde 903, 9052 Zwijnaarde, Belgium.

Steven Delrue Wave Propagation and Signal Processing, Department of Physics, KULeuven-KULAK, Etienne-Sabbelaan 52, 8500 Kortrijk, Belgium.

Wen Deng Laboratory for Nondestructive Evaluation and Structural Health Monitoring Studies, Department of Civil and Environmental Engineering, University of Pittsburgh, 3700 O'Hara Street, 729 Benedum Hall, Pittsburgh, PA 15261, USA; School of Automation, Northwestern Polytechnical University, Xi'an 710072, China.

Angélica Díaz Chemical Engineering Department, Polytechnic University of Catalonia, Av. Diagonal 647, Barcelona E-08028, Spain.

Daniel Didie Department of Chemistry and Physics, Southeastern Louisiana University, SLU 10878, Hammond, LA 70402, USA.

David R. Didie Department of Chemistry and Physics, Southeastern Louisiana University, SLU 10878, Hammond, LA 70402, USA.

Pierre-Antoine Dubos Laboratory of Crystallography and Materials Science (CRISMAT), CNRS-ENSICAEN-Université de Caen Basse Normandie, Caen 14050, France.

Dagmar Faktorova Faculty of Electrical Engineering, University of Žilina, Univerzitná 1, Žilina 010 26, Slovakia.

Gilbert Fantozzi INSA de Lyon, MATEIS (UMR CNRS 5510), 7 avenue Jean Capelle, 69621 Villeurbanne Cedex, France.

Carol Featherston Cardiff School of Engineering, Cardiff University, Cardiff CF24 3AA, UK.

Katarzyna Gabryś Water Centre Laboratory, Faculty of Civil and Environmental Engineering, Warsaw University of Life Sciences, 02-787 Warsaw, Poland.

Antolino Gallego Department of Applied Physics, University of Granada, Granada 18071, Spain.

Tat-Hean Gan TWI Ltd., Granta Park, Great Abington, Cambridge CB21 6AL, UK; Brunel Innovation Centre (BIC), Brunel University, Cambridge CB21 2AL, UK.

Carmelo Gentile Department Architecture, Built environment and Construction engineering (DABC), Politecnico di Milano, P.za Leonardo da Vinci 32, Milan 20133, Italy.

Nathalie Godin INSA de Lyon, MATEIS (UMR CNRS 5510), 7 avenue Jean Capelle, 69621 Villeurbanne Cedex, France.

Janez Grum Faculty of Mechanical Engineering, University of Ljubljana, Aškerčeva 6, Ljubljana 1000, Slovenia.

Marco Guidobaldi Department Architecture, Built environment and Construction engineering (DABC), Politecnico di Milano, P.za Leonardo da Vinci 32, Milan 20133, Italy.

Patrick Guillaume Department of Mechanical Engineering, Vrije Universiteit Brussel, Pleinlaan 2, 1050 Brussels, Belgium.

David Gurney Department of Chemistry and Physics, Southeastern Louisiana University, SLU 10878, Hammond, LA 70402, USA.

Michaël F. Hinderdael Department of Mechanical Engineering, Vrije Universiteit Brussel, Pleinlaan 2, 1050 Brussels, Belgium.

Klaudiusz Holeczek Institute of Lightweight Engineering and Polymer Technology (ILK), Technische Universität Dresden, Holbeinstraße 3, 01307 Dresden, Germany.

Nicoleta Iftimie Nondestructive Testing Department, National Institute of R&D for Technical Physics, Iasi 700050, Romania.

Stefano Invernizzi Department of Structural, Geotechnical and Building Engineering, Politecnico di Torino, Corso Duca degli Abruzzi 24, 10129 Torino, Italy.

Vassilios Kappatos Brunel Innovation Centre (BIC), Brunel University, Cambridge CB21 2AL, UK.

Tomaž Kek Faculty of Mechanical Engineering, University of Ljubljana, Aškerčeva 6, Ljubljana 1000, Slovenia.

Mathias Kersemans Mechanics of Materials and Structures, Department of Materials Science and Engineering, Ghent University, Technologiepark-Zwijnaarde 903, 9052 Zwijnaarde, Belgium.

Maria Kogia Brunel Innovation Centre (BIC), Brunel University, Cambridge CB21 2AL, UK.

Tatiana E. Konstantinova Department of Physical Materials Science, Donetsk Institute for Physics and Engineering named after O.O. Galkin of National Academy of Sciences of Ukraine, 72 R. Luxemburg str., Donetsk 83114, Ukraine.

Dragan Kusić TECOS Slovenian Tool and Die Development Centre, Kidričeva 25, Celje 3000, Slovenia.

Giuseppe Lacidogna Department of Structural, Geotechnical and Building Engineering, Politecnico di Torino, Corso Duca degli Abruzzi 24, 10129 Torino, Italy.

Sylvie Malo Laboratory of Crystallography and Materials Science (CRISMAT), CNRS-ENSICAEN-Université de Caen Basse Normandie, Caen 14050, France.

Ryan Marks Cardiff School of Engineering, Cardiff University, Cardiff CF24 3AA, UK.

Arvid Martens Wave Propagation and Signal Processing, Department of Physics, KULeuven-KULAK, Etienne-Sabbelaan 52, 8500 Kortrijk, Belgium.

Theodore E. Matikas Department Materials Science and Engineering, University of Ioannina, Ioannina 45110, Greece.

Niels Modler Institute of Lightweight Engineering and Polymer Technology (ILK), Technische Universität Dresden, Holbeinstraße 3, 01307 Dresden, Germany.

Abbas Mohimi TWI Ltd., Granta Park, Great Abington, Cambridge CB21 6AL, UK;Brunel Innovation Centre (BIC), Brunel University, Cambridge CB21 2AL, UK.

Anastasios C. Mpalaskas Department Materials Science and Engineering, University of Ioannina, Ioannina 45110, Greece.

Amir Nasrollahi Laboratory for Nondestructive Evaluation and Structural Health Monitoring Studies, Department of Civil and Environmental Engineering, University of Pittsburgh, 3700 O'Hara Street, 729 Benedum Hall, Pittsburgh, PA 15261, USA.

Gianni Niccolini Department of Structural, Geotechnical and Building Engineering, Politecnico di Torino, Corso Duca degli Abruzzi 24, 10129 Torino, Italy.

Christophe Paget Airbus UK Ltd., Filton, Bristol BS99 7AR, UK.

Hae-Sung Park Department of Mechanical Engineering, The Graduate School, Seoul National University of Science and Technology, 232 Gongneung-ro, Nowon-gu, Seoul 01811, Korea.

Ik-Keun Park Department of Mechanical and Automotive Engineering, Seoul National University of Science and Technology, 232 Gongneung-ro, Nowon-gu, Seoul 01811, Korea.

Jordi Puiggalí Chemical Engineering Department, Polytechnic University of Catalonia, Av. Diagonal 647, Barcelona E-08028, Spain.

Rhys Pullin Cardiff School of Engineering, Cardiff University, Cardiff CF24 3AA, UK.

Lincy Pyl Department Mechanics of Materials and Constructions, Vrije Universiteit Brussel, Pleinlaan 2, 1050 Brussels, Belgium.

Mohamed R'Mili INSA de Lyon, MATEIS (UMR CNRS 5510), 7 avenue Jean Capelle, 69621 Villeurbanne Cedex, France.

Pascal Reynaud INSA de Lyon, MATEIS (UMR CNRS 5510), 7 avenue Jean Capelle, 69621 Villeurbanne Cedex, France.

Piervincenzo Rizzo Laboratory for Nondestructive Evaluation and Structural Health Monitoring Studies, Department of Civil and Environmental Engineering, University of Pittsburgh, 3700 O'Hara Street, 729 Benedum Hall, Pittsburgh, PA 15261, USA.

Andrés Roldán Department of Electronics and Computer Technology, University of Granada, Granada 18071, Spain.

Francisco Sagasta Department of Applied Physics, University of Granada, Granada 18071, Spain.

Antonella Saisi Department Architecture, Built environment and Construction engineering (DABC), Politecnico di Milano, P.za Leonardo da Vinci 32, Milan 20133, Italy.

Wojciech Sas Water Centre Laboratory, Faculty of Civil and Environmental Engineering, Warsaw University of Life Sciences, 02-787 Warsaw, Poland.

Tomohiro Sasaki Department of Mechanical Engineering, Niigata University, Ikarashi Ninocho 8050, Nishi-ku, Niigata-shi, Niigata 950-2181, Japan.

Adriana Savin Nondestructive Testing Department, National Institute of R&D for Technical Physics, Iasi 700050, Romania; Nondestructive Testing Department, National Institute for Research and Development for Technical Physics, 47 D. Mangeron Blvd., Iasi 700050, Romania.

Cem Selcuk Brunel Innovation Centre (BIC), Brunel University, Cambridge CB21 2AL, UK.

Shouzhao Sheng College of Automation Engineering, Nanjing University of Aeronautics and Astronautics, 29 YuDao St., Nanjing 210016, China.

Min Shi Institute of Mechanics, Chinese Academy of Sciences, Beijing 100190, China.

Emil Soból Department of Geotechnical Engineering, Faculty of Civil and Environmental Engineering, Warsaw University of Life Sciences, 02-787 Warsaw, Poland.

Hugo Sol Department Mechanics of Materials and Constructions, Vrije Universiteit Brussel, Pleinlaan 2, 1050 Brussels, Belgium.

Eric Starke Institute of Lightweight Engineering and Polymer Technology (ILK), Technische Universität Dresden, Holbeinstraße 3, 01307 Dresden, Germany.

Rozina Steigmann Nondestructive Testing Department, National Institute of R&D for Technical Physics, Iasi 700050, Romania; Faculty of Physics, University Al.I. Cuza, 11 Carol I Blvd, Iasi 700506, Romania.

Kathy Steppe Laboratory of Plant Ecology, Department of Applied Ecology and Environmental Biology, Faculty of Bioscience Engineering, Ghent University, Coupure links 653, 9000 Gent, Belgium.

Chenwu Sun College of Automation Engineering, Nanjing University of Aeronautics and Astronautics, 29 YuDao St., Nanjing 210016, China.

Alojzy Szymański Department of Geotechnical Engineering, Faculty of Civil and Environmental Engineering, Warsaw University of Life Sciences, 02-787 Warsaw, Poland.

Eleni Tsangouri Department of Mechanics of Materials and Constructions, Vrije Universiteit Brussel, Pleinlaan 2, Brussels 1050, Belgium.

Vitalii Turchenko Frank Laboratory for Neutron Physics, Joint Institute for Nuclear Research, 6 Joliot-Curie Avenue, Dubna 141980, Russia.

Koen Van Den Abeele Wave Propagation and Signal Processing, Department of Physics, KULeuven-KULAK, Etienne-Sabbelaan 52, 8500 Kortrijk, Belgium.

Wim Van Paepegem Mechanics of Materials and Structures, Department of Materials Science and Engineering, Ghent University, Technologiepark-Zwijnaarde 903, 9052 Zwijnaarde, Belgium.

Julie M. Vandenbossche Department of Civil and Environmental Engineering, University of Pittsburgh, 3700 O'Hara Street, 705 Benedum Hall, Pittsburgh, PA 15261, USA.

Lidewei L. Vergeynst Laboratory of Plant Ecology, Department of Applied Ecology and Environmental Biology, Faculty of Bioscience Engineering, Ghent University, Coupure links 653, 9000 Gent, Belgium.

Anja Winkler Institute of Lightweight Engineering and Polymer Technology (ILK), Technische Universität Dresden, Holbeinstraße 3, 01307 Dresden, Germany.

Sanichiro Yoshida Department of Chemistry and Physics, Southeastern Louisiana University, SLU 10878, Hammond, LA 70402, USA.

Yang Yu Collaborative Innovation Center for Advanced Ship and Deep-Sea Exploration, State Key Laboratory of Hydraulic Engineering Simulation and Safety, Tianjin University, Tianjin 300072, China.

Filip Zastavnik Department Mechanics of Materials and Constructions, Vrije Universiteit Brussel, Pleinlaan 2, 1050 Brussels, Belgium.

Xiaohui Zeng Institute of Mechanics, Chinese Academy of Sciences, Beijing 100190, China.

Liang Zhang Institute of Mechanics, Chinese Academy of Sciences, Beijing 100190, China.

Jifu Zhou Institute of Mechanics, Chinese Academy of Sciences, Beijing 100190, China.

About the Guest Editors

Dimitrios Aggelis is a Professor of the Department of Mechanics of Materials and Constructions at the Vrije (Free) University of Brussels, Belgium since October 2012. Prior to this position, he worked as an Assistant Professor in the Department of Materials Science and Engineering at the University of Ioannina, Greece (2008–2012) and as a research fellow in the Research Institute of Technology, Tobishima Corporation, Japan (2006–2008). He received his PhD degree from the Mechanical Engineering and Aeronautics Department of the University of Patras in 2004. His main area of interest includes characterization of cementitious materials, expanding also to composites and metals by use of non-destructive inspection techniques focused on elastic wave propagation. He is an active member of several technical committees of RILEM, the secretary of IAM (Damage Assessment in Consideration of Repair/Retrofit-Recovery in Concrete and Masonry Structures by Means of Innovative NDT) and was the recipient of the RILEM Robert L'Hermite medal in 2012 for outstanding contribution in the field of construction materials monitoring. He has published more than 110 papers in international journals and more than 120 papers in conference proceedings along with 10 chapters in books or stand-alone books. He is editor of the journal *Construction and Building Materials* and editorial board member of other international journals (incl., *Applied Sciences* and *NDT&E International*). He is currently, or has been involved in, teaching of Dynamics of Structures, Experimental Techniques and Non-destructive Testing of Materials, Mechanics of Materials, and Construction Materials. He was the chairman of the International Conference of Emerging Technologies in Non-Destructive Testing 6 (ETNDT6) held in Brussels, 27–29 May 2015, while he is the co-chairing the 2nd International RILEM Conference on Early Age Cracking and Serviceability in Cement-Based Materials and Structures (EAC2), 12–14 September, 2017, in Brussels, Belgium.

Nathalie Godin, Ph.D., is an Associate Professor at the National Institute of Applied Sciences (INSA) in Lyon, France since 1996. She received her PhD degree from the University of Bordeaux in 1994. She has 20 years of experience in AE and in the analysis of damage to various classes of materials. She focuses on fiber-reinforced composites as these materials have a variety of applications and they are quickly being adopted in a number of industries. She has authored over 40 articles and two book chapters and has been an invited speaker at numerous professional research conferences. She is also a board member of the French Society for Composite Materials (AMAC).

Preface to "Acoustic and Elastic Waves: Recent Trends in Science and Engineering"

Elastic waves in the active (ultrasound) and passive form (acoustic emission) provide an ideal approach for the non-destructive inspection of materials. They are used for all types of materials and structures, having many advantages over other traditional measurements. Although they are already applied in a usual basis in the structural health monitoring of structures, new challenges arise. These are related to pushing the limit of the characterization capacity further: detecting smaller cracks, providing more information on the condition of the material, predicting the useful life time, and characterizing new innovative materials that have not been tested before.

The articles contained in this issue/book include a broad range of materials from masonry and concrete to ceramics, composites, 3D printed metals and medical prosthetics and structures from cultural heritage towers and concrete beams to marine and aeronautic structures and solar panels. Characterization concerns the fracture mode, remaining fatigue life, disbonds in bonded stiffeners, localization of defects, condition of injection molding tools and many others.

We believe that this book constitutes an up-to-date collection of cutting edge applications of acoustic/elastic wave techniques, many of which were presented in the International Conference on Emerging Technologies in Non-Destructive Testing (ETNDT6) in Brussels, 27–29 May 2015. We wish to thank all the authors for submitting their work and the reviewers who contributed to the high final quality of the presented studies.

Dimitrios G. Aggelis and Nathalie Godin
Guest Editors

The Ultrasonic Polar Scan for Composite Characterization and Damage Assessment: Past, Present and Future

Mathias Kersemans, Arvid Martens, Joris Degrieck, Koen Van Den Abeele, Steven Delrue, Lincy Pyl, Filip Zastavnik, Hugo Sol and Wim Van Paepegem

Abstract: In the early 1980's, the ultrasonic polar scan (UPS) technique was developed to assess the fiber direction of composites in a nondestructive way. In spite of the recognition by several researchers as being a sophisticated and promising methodology for nondestructive testing (NDT) and materials science, little advance was made during the following 30 years. Recently however, the UPS technique experienced a strong revival and various modifications to the original UPS setup have been successfully implemented. This revival has exposed several powerful capabilities and interesting applications of the UPS technique for material characterization and damage assessment. This paper gives a short historical overview of the UPS technique for characterizing and inspecting (damaged) fiber-reinforced plastics. In addition, a few future research lines are given, which will further expand the applicability and potential of the UPS method to a broader range of (damaged) materials, bringing the UPS technique to the next level of maturity.

Reprinted from *Appl. Sci.* Cite as: Kersemans, M.; Martens, A.; Degrieck, J.; Van Den Abeele, K.; Delrue, S.; Pyl, L.; Zastavnik, F.; Sol, H.; Van Paepegem, W. The Ultrasonic Polar Scan for Composite Characterization and Damage Assessment: Past, Present and Future. *Appl. Sci.* **2016**, *6*, 58.

1. Introduction

An ultrasonic polar scan (UPS) is obtained by replacing the translational movement of a classical ultrasonic C-scan setup with a rotational movement. Hence, instead of scanning a plate surface at normal incidence, the UPS insonifies a certain material spot from as many oblique incidence angles $\psi\,(\phi, \theta)$ as possible [1]. A schematic of the UPS method is presented in Figure 1a. Depending on the employed ultrasound, we speak of a pulsed ultrasonic polar scan (P-UPS) for a broadband pulse or a harmonic ultrasonic polar scan (H-UPS) for mono-frequency ultrasound. To improve coupling of the ultrasonic wave energy in the sample, the plate sample is immersed in water. Simply recording the transmitted (or reflected) ultrasound amplitude then yields a UPS image. Figure 1b,c shows current state-of-the-art P-UPS recordings for aluminum and $[0°]_8$ carbon/epoxy (C/E) laminate, using an ultrasonic pulse with central frequency f_c = 5 MHz. The vertical incident angle θ is placed on the radial axis, the in-plane polar angle φ is represented along the angular axis, while the assigned gray (or color) scale is a measure of the transmitted (or reflected) pulse amplitude. Hence, the P-UPS image comprises a large collection of amplitudes of obliquely transmitted (or reflected) ultrasound pulses.

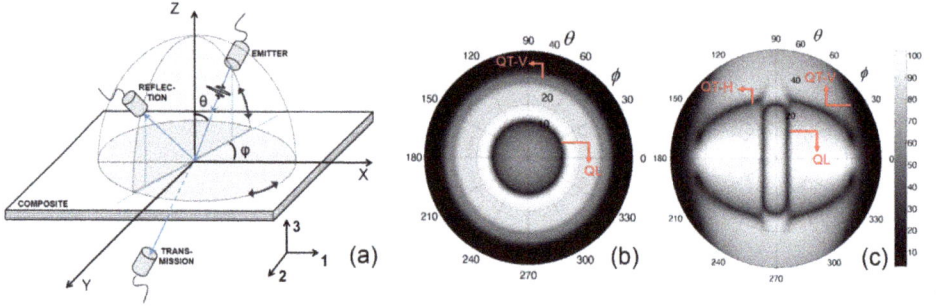

Figure 1. Schematic of ultrasonic polar scan (UPS) method (**a**). State-of-the-art pulsed ultrasonic polar scan (P-UPS) recordings at f_c = 5 MHz: aluminum with thickness d = 0.6 mm (**b**) and $[0°]_8$ C/E laminate with thickness d = 1.1 mm (**c**). "QL", "QT-H" and "QT-V" stand for quasi-longitudinal wave and quasi-transverse wave with horizontal and vertical polarization respectively.

Within the amplitude landscape of the P-UPS, characteristic contours emerge that more or less relate to the condition for critical (or in-plane) stimulation of the quasi-longitudinal, horizontally and vertically polarized quasi-transverse wave modes [2]. Note that along symmetry orientations, the prefix "quasi-" may be omitted, resulting in pure wave modes. These stimulation conditions are of course linked to the mechanical properties, like elasticity, of the insonified plate sample through Christoffel's equation and Snell's law [3,4]. As such, the P-UPS image actually represents an acoustic fingerprint of the mechanical properties. In the example of Figure 1b, the circular symmetry clearly puts on view the isotropic mechanical nature of the aluminum sample. The isotropic nature further invokes a degeneration of the quasi-transverse horizontal and quasi-transverse vertical wave, thus resulting in a P-UPS image with two characteristic contours. Figure 1c, on the other hand, displays a stretched appearance due to the unidirectional nature of the $[0°]_8$ C/E laminate. Higher stiffness values lead to smaller critical angles (Snell's law), and thus correspond to the contours being locally pressed inward in the P-UPS image. Considering the inner contour of Figure 1c, which is dominated by the quasi-longitudinally polarized wave, one can immediately determine that highest stiffness is found along $\varphi = 0°$, which obviously corresponds to the orientation of the carbon reinforcement fibers of the $[0°]_8$ C/E laminate.

In the remainder of this manuscript, a historical overview of the UPS research is given, starting from the initial results of the pioneering researchers, going over our recent advances [5], and finishing with currently investigated research lines.

2. The Past: 1981–2010

In the early 1980's, the UPS technique was first introduced by Van Dreumel and Speijer in a pulsed version in order to assess the fiber orientation of composites [1]. The beauty of the intriguing patterns made the pioneering authors state [1]: "*A library of Polar-patterns, stored as 'fingerprints', raises the possibility of laminate identification by pattern recognition*". It is unfortunate that in the following years, only one sequel study was performed by Van Dreumel and Speyer to further explore the capabilities of the ultrasonic polar scan [6].

It took fifteen years before the technique was investigated again through the work of Degrieck [7–9]. He used a modernized scanning system to obtain more accurate and detailed P-UPS experiments. The work of Degrieck has identified several practical applications of the P-UPS technique for composite materials: (i) estimation of fiber direction; (ii) determination of fiber volume fraction and porosity and (iii) detection of fatigue damage [8,9]. In addition, he and Van Leeuwen implemented a numerical procedure for simulating P-UPS images of homogeneous composites [7], in view of bringing the technique to the next level: full quantitative characterization of the elastic properties of composite materials using a mixed experimental-numerical approach. Although the gap between experiment and simulation was not yet bridged, their numerical results did contribute to the physical understanding of the formation of a P-UPS image. They found that the characteristic patterns are (more or less) a representation of critical bulk wave angles, while the global transmission amplitude exposes attenuation properties [8]. Consequently, a P-UPS image may be used for characterizing viscoelastic material properties.

The ultrasonic polar scan research has been further extended by Declercq, first as a student of Degrieck [2,10,11] and afterwards as a professor at the Georgia Institute of Technology in Metz [12]. He applied the technique for detecting tension-tension induced fatigue damage in glass fiber composites by tracing shifts in the characteristic fingerprint [10]. In addition, Declercq extended the simulation technique towards layered viscoelastic materials having arbitrary anisotropy using a global matrix method [2,11]. He experimentally implemented a time-of-flight (TOF) version of the P-UPS method and commented on the superior sensitivity (compared with amplitude recording) to the presence of damage features [12]. Here, TOF is defined as the arrival time of the peak amplitude of the transmitted ultrasound pulse. As an illustration, Figure 2 displays state-of-the-art amplitude and TOF landscape of a P-UPS for a $[0°]_8$ C/E laminate [13].

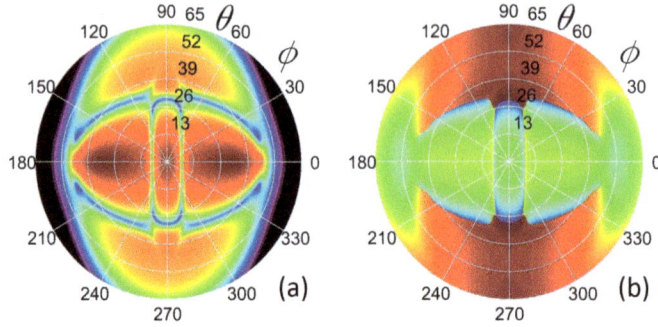

Figure 2. P-UPS for $[0°]_8$ C/E laminate: amplitude (**a**) and time-of-flight (TOF) (**b**) landscape [13].

It is clear from the above results that the UPS technique already had several applications, but it is evenly clear that many capabilities and opportunities still have to be explored.

3. The Present: 2010–2015

Building further on this background, Kersemans intensified the research on the ultrasonic polar scan method in 2010 [13]. Several barriers were identified that impede the further development of the UPS methodology. Three main barriers may be summarized as:

- lack of high quality experimental data,
- lack of a computationally efficient simulation model, and
- lack of adequate inverse modeling techniques to couple experiment to simulation.

3.1. The Ultrasonic Polar Scan Revisited

The experimental barrier has been tackled by the development of a 5-axis scanner that records ultrasonic polar scans in a fully automated way. During recording a UPS experiment, the scanner typically insonifies a material spot from more than 1,000,000 different incidence angles $\psi(\phi, \theta)$ in a timeframe of ~15 min. High-precision encoders are installed, providing accurate position feedback for the insonification direction. The USIP40 (General Electric, Hurm (Efferen), Germany), with a sampling clock of 400 MHz, is used as pulser/receiver apparatus. With this setup, we identified several pitfalls in the earlier experimental procedures, which prevented the correct and accurate recording of an ultrasonic polar scan [14]. The high quality of our experiments can already be seen in Figure 1b,c, showing P-UPS recordings for aluminum and $[0°]_8$ C/E laminate.

4

Secondly, a simulation technique has been implemented to support experimental observations [15]. The simulation model is founded on a transfer matrix technique [16–18], and allows the simulation of UPS for immersed layered viscoelastic anisotropic media. A Fourier-based approach is employed to account for the spectral nature of the considered pulsed ultrasonic wave. As all transducers show some deviation from their design conditions, the recorded output signal (in water) of the actual transducer is directly used as input for the simulation model. In Figure 3, an example is displayed for a broadband transducer, which shows a clear deviation from its designed central frequency f_c = 5 MHz.

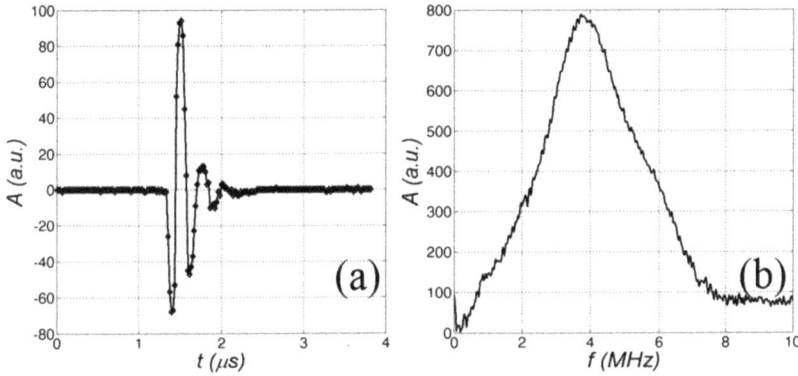

Figure 3. Excitation signal of a 5-MHz broadband transducer: temporal (**a**) and frequency (**b**) domain. The spectral content clearly reveals a central frequency around 4 MHz, instead of 5 MHz.

Compared to previous simulation models, we significantly reduced the computational time. Typically, the simulation time for a UPS is in the order of seconds. Figure 4a,b displays the simulated P-UPS of the aluminum plate and the $[0°]_8$ C/E laminate. As realistic material properties for immersion liquid and solid were considered, and the experimental broadband pulse of Figure 3 was explicitly modeled, the P-UPS simulations in Figure 4 may be straightforwardly compared with the P-UPS experiments displayed in Figure 1b,c. It is clear that a good visual comparison is obtained between experiment and simulation.

This brings us to the third barrier, which is tackled by implementing an inversion procedure to couple experiment to simulation, in view of identifying material parameters. Basically, the simulated P-UPS image is fitted to the recorded P-UPS image while updating the material properties by means of a genetic algorithm [13,19,20]. The obvious goal is to find a set of viscoelastic material parameters that minimizes the difference between both amplitude landscapes. A schematic of the optimization procedure is shown in Figure 5a. Figure 5b shows the optimized result for the $[0°]_8$ C/E laminate in terms of the viscoelastic tensor $C_{ij} = C_{ij}^R - iC_{ij}^I$, with C_{ij}^R representing the elastic properties and C_{ij}^I the attenuation characteristics.

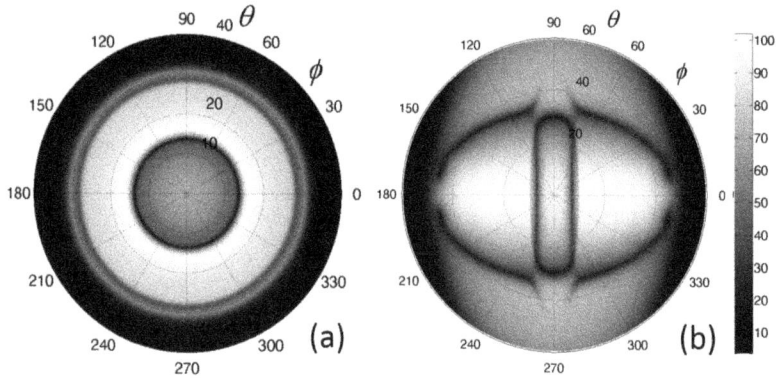

Figure 4. P-UPS simulation (ultrasound pulse of Figure 3 is used as input) for aluminum (**a**) and $[0°]_8$ C/E laminate (**b**).

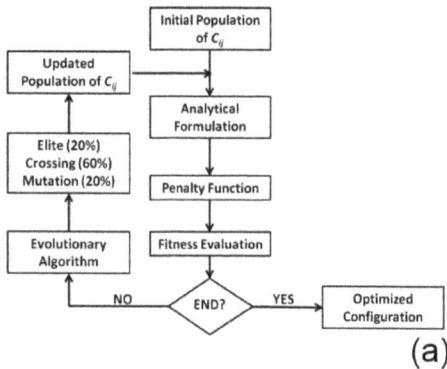

$C_{ij} = C_{ij}^R - iC_{ij}^I$	C_{ij}^R [GPa]	C_{ij}^I [GPa]	$\varepsilon = C_{ij}^I / C_{ij}^R$ [%]
C_{11}	122.716	6.901	5.62
C_{12}	6.564	1.095	16.69
C_{22}	13.468	0.859	6.38
C_{44}	3.398	0.195	5.73
C_{55}	5.856	0.444	7.58
C_{66}	6.251	0.276	4.41

(a)   (b)

Figure 5. Schematic of inversion procedure to identify composite properties (**a**) and identified properties for $[0°]_8$ C/E laminate (**b**) [13,19]. (**a**) reproduced from [19], Copyright Society for Experimental Mechanics, 2014.

In this way, the (visco)elastic properties for a range of (composite) materials have been identified, showing good correspondence with alternative identification techniques [13,19]. Apart from the obtained successes, we still encountered some drawbacks and difficulties with an inversion procedure that is based on amplitude data only. For this reason we are currently implementing a more robust inversion procedure that accounts for both the TOF and amplitude landscape of a P-UPS recording (further details in Section 4.1) [21,22].

3.2. Extensions of UPS Method

During our investigations, we found that additional information is encoded in the ultrasonic polar scan recordings. In the following, we briefly indicate a few of the various extensions that were implemented over the last few years.

3.2.1. Harmonic Ultrasonic Polar Scan

Numerical simulations suggested that the global view of a UPS image is not only a function of material parameters, but also strongly depends on the (temporal) shape of the employed ultrasound wave [2]. This has led us to the experimental and numerical investigation of a harmonic version of the ultrasonic polar scan (H-UPS) [15]. In Figure 6, a comparison is shown between P-UPS (broadband wave with central frequency $f_c = 5$ MHz) and H-UPS (harmonic wave with frequency $f = 5$ MHz) recordings of a $[0°]_8$ C/E laminate. Apart from the similar global view, imposed by the unidirectional nature of the scanned sample, one observes that the H-UPS image has a "richer" fingerprint.

Contrary to the P-UPS image, the patterns in the H-UPS image depend on the exact ultrasonic frequency. This is easily understood considering that the H-UPS image puts on view the stimulation condition of dispersive, $i.e.$, frequency dependent, Lamb waves, while the P-UPS image is mainly governed by bulk wave characteristics [2,15,23].

For the case of harmonic signals, the UPS analysis was extended to the evaluation of both the amplitude and the phase signal of the transmitted wave. In this way, knowledge can be obtained about the complex transmission coefficient. Figure 7 displays the H-UPS images for a $[0°]_8$ C/E laminate, considering the analysis of both the amplitude and the phase of the transmission signal. Note that the sharp discontinuities in the phase map are a mere consequence of phase wrapping.

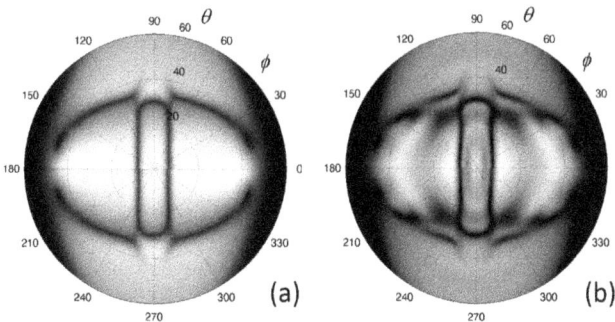

Figure 6. Experimental recordings of a $[0°]_8$ C/E laminate: P-UPS at $f_c = 5$ MHz (**a**) and harmonic ultrasonic polar scan (H-UPS) at $f = 5$ MHz (**b**).

7

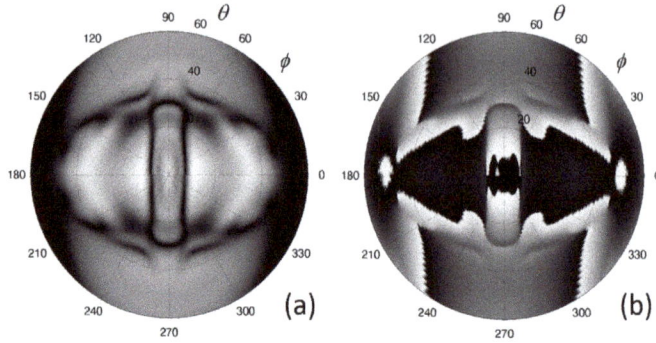

Figure 7. H-UPS recordings for $[0°]_8$ carbon/epoxy laminate at f = 5 MHz: amplitude (**a**) and phase (**b**) analysis.

3.2.2. Ultrasonic Backscatter Polar Scan (UBPS)

During our UPS investigations, we persistently observed a small amount of ultrasonic energy being backscattered to the emitter, even for large incident angles θ. Recording of the backscattered signal according to the UPS principle then results in an ultrasonic backscatter polar scan (UBPS). Depending on the employed wave, we speak of P-UBPS for ultrasonic pulses [24] and H-UBPS for quasi-harmonic ultrasound [25] (similarly as done for H-UPS and P-UPS). A schematic of the UBPS method, together with a H-UBPS recording for a $[0°]_8$ C/E laminate, is presented in Figure 8.

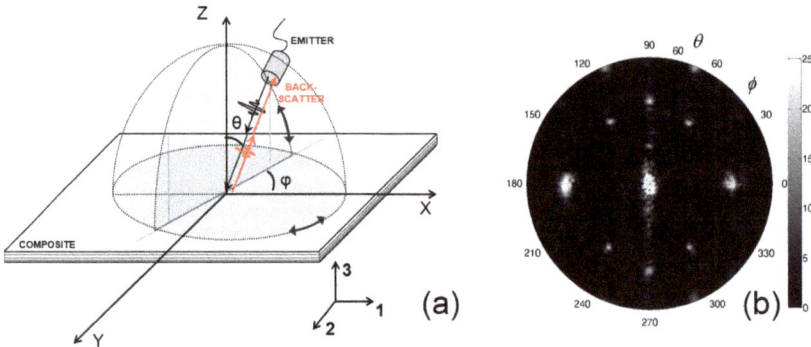

Figure 8. Schematic of ultrasonic backscatter polar scan (UBPS) method (**a**) and harmonic ultrasonic backscatter polar scan (H-UBPS) recording for a $[0°]_8$ C/E laminate (**b**).

Typically, a H-UBPS image is characterized by well-defined high amplitude backscatter spikes, which can be linked to diffraction phenomena under the

8

Bragg condition [25–28]. As such, these spikes directly expose geometrical characteristics of (sub)surface structures. For the $[0°]_8$ C/E laminate, the observed backscatter spikes originate at the imprint left by the peel-ply cloth during manufacturing of the laminate [24].

3.3. Applications of UPS and UBPS

The UPS and the UBPS, both in pulsed and harmonic version, have been applied to a range of nondestructive testing (NDT) applications involving polymeric, metallic and fiber-reinforced materials.

3.3.1. Static Damage

Various fiber-reinforced plastics have been quasi-statically loaded to induce material degradation. In Figure 9a,b, P-UPS images are shown before and after (quasi-)static shear loading of a $[−45°,+45°]_S$ C/E laminate. The indicated overlap angle ζ provides a measure of the angle between the orientation of the fiber reinforcement. Initially $\zeta = 91.5°$, indicating that the stacking of the laminate was not perfect (see Figure 9a). After loading in shear, the overlap angle reduces to $\zeta = 81.5°$ (see Figure 9b), which thus reveals a shift in fiber orientation of $10°$. Figure 9c shows the fracture area of a $[−45°,+45°]_S$ C/E laminate loaded in shear until failure, indeed displaying a very large shift in fiber orientation. The evolution of the overlap angle ζ was monitored for a range of shear load levels, and we found a clear relationship between the applied shear load level and the observed shift in fiber orientation in the P-UPS images [29].

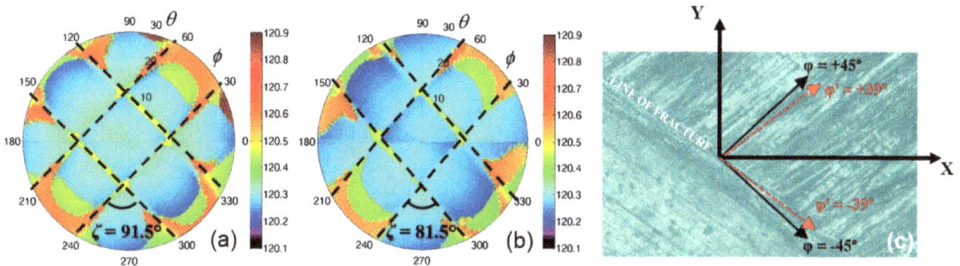

Figure 9. P-UPS images of $[−45°,+45°]_S$ C/E laminate: before (**a**) and after (**b**) applying the shear load. Microscopic image of the fracture area when shear loading a $[−45°,+45°]_S$ C/E laminate until failure (**c**).

3.3.2. Delamination

As the patterns in a H-UPS image are governed by the stimulation conditions for Lamb waves, these patterns must shift position in the presence of a delamination. This is easily understood considering that a delamination divides the original laminate in two sub-laminates, resulting in different boundary conditions. Different boundary conditions invoke different Lamb wave stimulation conditions, and as such yield different H-UPS images [15,30]. This is demonstrated in Figure 10 for a cross-ply [0°,90°]$_S$ C/E laminate (thickness d = 1.1 mm) provided with water-filled delaminations (with nominal thickness D = 50 μm) at different depth positions. Note that such types of delaminations are in general difficult to assess in thin composites using conventional normal incident ultrasonic techniques. Figure 10 clearly shows that the depth position of the delamination is well represented in the H-UPS images. Figure 10d further indicates that the H-UPS can even detect multiple overlapping water-filled delaminations.

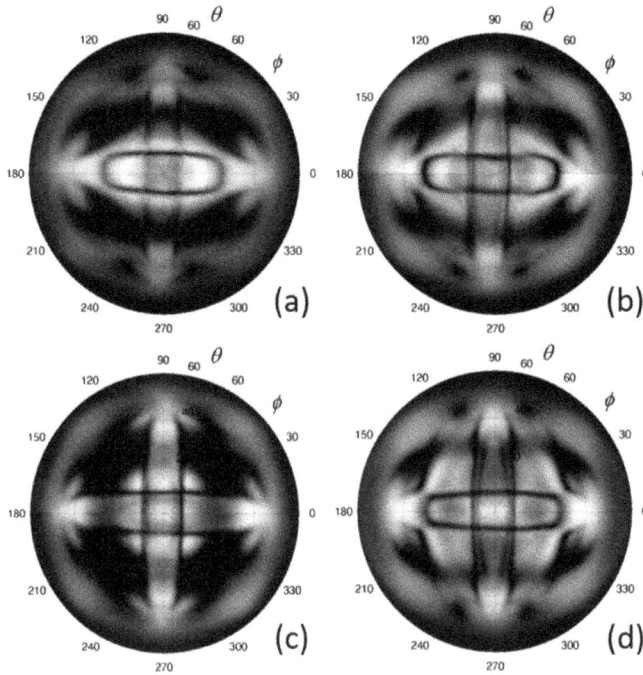

Figure 10. H-UPS recordings of a cross-ply C/E laminate at f = 5 MHz: [0°,90°,90°,0°] (**a**); [0°,W,90°,90°,0°] (**b**); [0°,90°,W,90°,0°] (**c**) and [0°,W,90°, 90°,W,0°] (**d**). "W" stands for the position of the water-filled delamination.

Excellent agreement is obtained between the results of experiment and simulation. Further, our numerical computations indicate the ability of the H-UPS to characterize delamination parameters such as depth position and delamination thickness D. Currently, an inversion model is being implemented to extract these delamination parameters in an automated way.

3.3.3. Fatigue Damage

The UPS method was also applied to fatigued composites. Fatigue loading typically leads to the initiation, progression and accumulation of micro defects. At the macroscopic level, these defects manifest themselves in a directional reduction of the stiffness properties [31,32]. Such a directional stiffness reduction is visible as a stretching (along the direction of loading) of the UPS contours. This is explicitly demonstrated in Figure 11 for a woven C/E laminate (ultimate tensile strength of 770 MPa), which was fatigued along $\varphi = 0°$: 314,111 load cycles with stress range 0–625 MPa at a frequency of 5 Hz [33]. The P-UPS experiment yielded a stiffness reduction of 12.8% along the loading direction, which is in good agreement with extensometer data yielding a stiffness reduction of 11% [33].

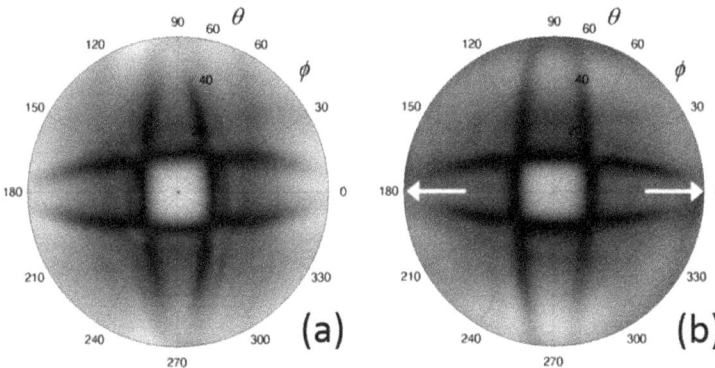

Figure 11. P-UPS recording of woven C/E laminate: before (**a**) and after (**b**) applying tension fatigue.

3.3.4. (Sub)surface Corrugation

As the characteristics of a periodic (sub)surface structure determine the position of the observed backscatter spikes in a H-UBPS experiment (see Figure 8b), we may also reverse this: start from a H-UBPS experiment and reconstruct the parameters of the periodic (sub)surface structure. A 2D subsurface corrugation has been applied to a polycarbonate sample through laser ablation. A microscopic image of the 2D corrugation is displayed in Figure 12a. The corresponding H-UBPS experiment at $f = 5$ MHz is shown in Figure 12b. Several well-defined backscatter spikes can be observed, which were used to reconstruct the in-plane parameters of the 2D corrugation. We obtained excellent agreement between the ultrasonic reconstruction, design parameters and optical measurement of the corrugation parameters (see Table 1).

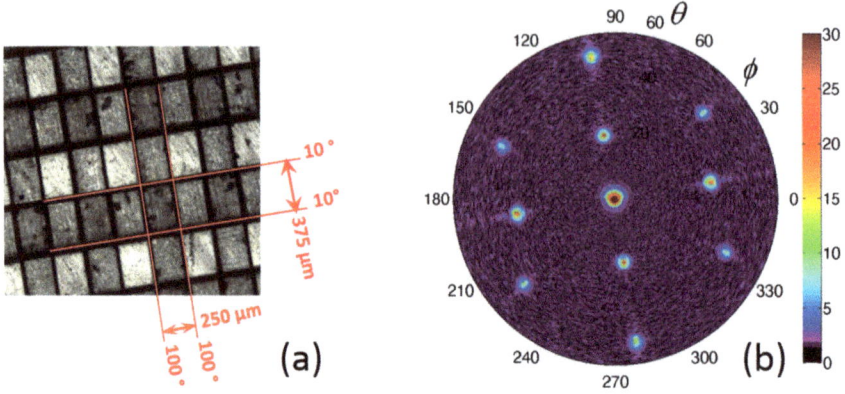

Figure 12. Microscopic image of ablated 2D subsurface corrugation (**a**) and corresponding H-UBPS recording at f = 5 MHz (**b**).

Table 1. Corrugation parameters: design (column 1), ultrasonic reconstruction (column 2) and optical measurement (column 3).

Corrugation Parameters	Design Values	Ultrasonic Reconstruction	Optical Measurement
Λ_1	250 μm	249.8 μm	250.1 μm
Φ_1	100°	99.3°	99.2°
Λ_2	375 μm	375.3 μm	374.8 μm
Φ_2	10°	9°	9.1°

The above method has also been applied to periodic subsurface structures having superimposed geometrical randomness. Although additional scattering features, such as Lyman and Rowland "ghosts" and "grass" [34,35], emerge in the H-UBPS image, we were still able to extract the basic corrugation parameters with excellent accuracy [13].

3.3.5. Strain Measurement

As the H-UBPS method is capable of reconstructing a periodic (sub)surface structure with excellent accuracy, it should also be feasible to detect any change in surface parameters due to strain. Instead of machining a periodic surface structure, which would mechanically weaken the sample, we simply exploit residual surface roughness features. The ultrasonic strain measurement technique has been demonstrated on cold-rolled DC06 steel coupons, which have a deterministic surface structure left from skin passing of the work rolls (see Figure 13). The surface structure has an average roughness Ra = 1.1 μm with standard deviation σ_{Ra} = 0.0502 μm, and has a clear degree of aperiodic features superimposed. The DC06 steel samples have been plastically strained at different levels ranging from 2% up to 35%. One can clearly observe a positional shifting of the backscatter spikes when strained (see Figure 14), from which the applied in-plane strain field was reconstructed [36].

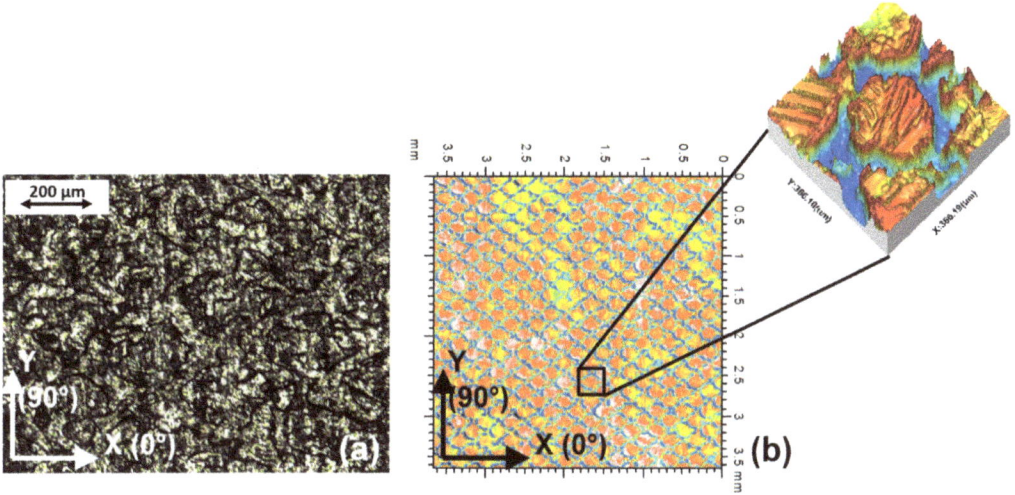

Figure 13. Optical visualization of the surface structure of DC06 steel: standard 2D microscopy (**a**) and 3D coherence correlation interferometry (after processing) (**b**).

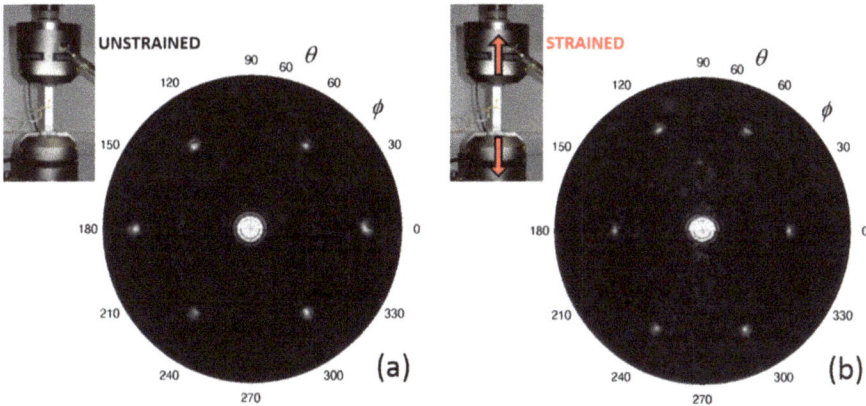

Figure 14. H-UBPS for DC06 steel coupon at f = 5 MHz: unstrained (**a**) and strained (**b**) state.

In addition, we also analyzed the response of ultrasonic broadband pulses at normal incidence, in order to detect any change in the stimulation condition of a thickness resonance (Lamb wave cut-off frequency). From this change, the out-of-plane strain component can then easily be obtained. The ultrasonically

13

obtained strain values have been confronted with the results of conventional strain measurements techniques (manual micrometer, mechanical extensometer, mono- and stereovision digital image correlation), showing excellent agreement for a wide range of applied strain values [36]. Interestingly, the developed ultrasonic strain gauge is the only method that was able to determine a 3D local strain field in a single-sided and contactless manner.

4. The (Near) Future: 2015–...

The above results illustrate we are on the right track in advancing the U(B)PS technique towards a higher level of maturity. But, of course there are still open (research) questions to be tackled. In addition, more applications and novel approaches for materials science and non-destructive testing lie ahead. Several of these opportunities are, or will further be, investigated by the present authors.

4.1. Viscoelastic Tensor

The complex C-tensor identification scheme discussed in Section 2 is a procedure solely based on the analysis of the amplitude P-UPS landscape. That system identification procedure has its merits, but still suffers from drawbacks. First, the amplitude-based inversion procedure is ill-defined for samples with a high degree of viscoelasticity. Further, for certain materials, small discrepancies between the characteristic patterns and the critical bulk wave angles were observed [19], thus requiring special care during inversion. Finally, the current amplitude-based inversion is unable to provide the complete set of orthotropic viscoelasticity parameters.

Other researchers have already tried to tackle the problem of (visco)elastic tensor determination by considering TOF (and amplitude) analysis of obliquely incident waves [37–40]. Unfortunately, their approaches also have inherent limitations. As they consider only a limited set of orientation angles φ, prior knowledge about the material symmetry axes is required [37,40,41]. Secondly, those inversion schemes often rely on bulk wave approximations, which has the consequence that the plate thickness d should match the ultrasound frequency f. Thirdly, phase effects at the liquid-solid and solid-liquid interface are often ignored, though some authors incorporated the phase effect by considering an additional iterative step [38].

One of our current research lines is trying to reconcile the above two inversion procedures, *i.e.*, combining the analysis of amplitude P-UPS landscape with TOF analysis, in order to obtain a robust, stable and widely applicable procedure for extracting viscoelastic material parameters. In order to reach this goal, we consider a two-step inversion approach [21,22]. The schematic roadmap of this approach is presented in Figure 15. First, the elastic properties are extracted on the basis of the

TOF landscape of a P-UPS recording. Second, damping parameters are obtained through analysis of the amplitude landscape of a P-UPS recording.

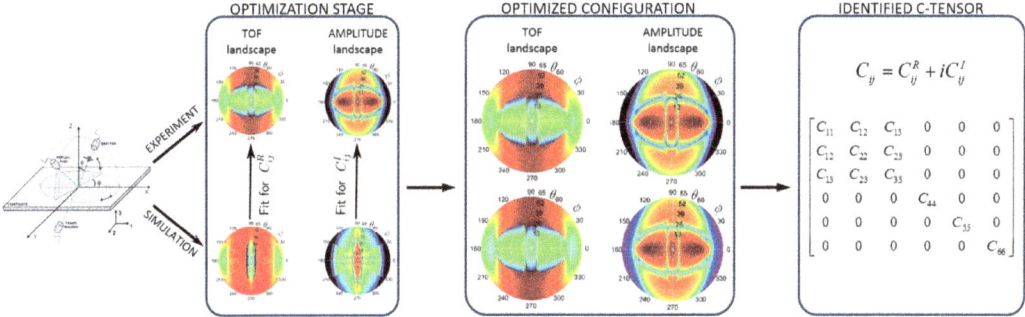

Figure 15. Roadmap for the identification of viscoelastic C-tensor using both TOF and amplitude P-UPS landscapes.

As the forward P-UPS simulation model (both amplitude and TOF) accounts for the true physical parameters, its validity is not restricted by the bulk wave approximation. Further, the optimization procedure includes a wide range of incidence angles $\Psi(\varphi,\theta)$, which makes prior knowledge about material symmetry ubiquitous. Finally, the large amount of redundant data will without doubt lead to robust and stable inversion results, even for noisy data. As such, we believe that our current effort may lead to a comprehensive approach in obtaining viscoelastic parameters for a wide range of (orthotropic) materials. As a matter of fact, we already successfully tested this system identification procedure for synthetic data [21,22], and we are now in the stage of applying this procedure to real experimental data.

4.2. Nonlinear U(B)PS

Until now, linear wave phenomena were used in the U(B)PS methodology for detecting, assessing and analyzing a range of damage features [24,29,30,33]. However, it is well-known from literature that nonlinear wave phenomena show an increased sensitivity to early stage damage features as well as to certain classes of contact defects [42–45]. For this reason, the present authors are currently upgrading the U(B)PS methodology towards the nonlinear regime. One approach is to perform two experiments with different amplitude levels or with inverted wave amplitudes, and to investigate the lack of scalability which is typical for nonlinear features. Subtraction of both experiments (after scaling to the input excitation) then provides a measure of the presence of nonlinearity. Another approach is to combine U(B)PS with spectroscopic analysis. In that way, we can easily evaluate and analyze the U(B)PS images at various harmonics of the input frequency. Our first experimental

15

results of the nonlinear version of the U(B)PS are very promising, and indicate a great potential for inspection of damaged (fiber-reinforced) materials. To validate the experimental results, we will also develop the corresponding simulation models that account for these nonlinear wave phenomena.

4.3. Redesign of UPS Scanner

As we realize that our current laboratory scanner does not meet in-field requirements, we are in the process of designing a portable UPS scanner. The current status of our portable scanner is presented in Figure 16. This device is made from mechanical parts and will only serve as an intermediate step as our final goal is to exclude any mechanical movement by using advanced phased matrix technology [46].

Figure 16. Current design of portable UPS scanner.

4.4. Comprehensive Inspection Approach

Most industrial-scale composite components are subject to a range of (accidental) loading conditions (fatigue, impact . . .), which lead to specific macroscopic damage phenomena (stiffness reduction, delamination . . .). Further considering the complex geometry (bends, stiffeners . . .) of typical engineering structures, it is clear that not a single non-destructive technique provides an objective and complete screening of the structural health of the investigated component. For this reason, the present authors are working towards a stepwise procedure in which various complementary inspection methods are applied. The first stage will involve a global testing of the component using for example, high-frequency modal analysis, in order to check for any deviations from the reference state [47,48]. Assuming there is an indication of anomalous response, the component will be further interrogated and inspected by more advanced and thorough techniques such as (lock-in) thermography and/or thermoelastic stress analysis [49–51], active shearography [51] and laser Doppler vibrometry [52]. These techniques actually provide a semi-quantitative view of the structural health of the component, and will identify "hotspots" that may deserve further attention. Finally, these identified "hotspots" will be investigated with local methods, like the UPS, in order to gain quantitative insight into the local damage state and/or local material properties.

5. Conclusions

This paper gives a short historical overview of the ultrasonic (backscatter) polar scan research for composite characterization and NDT. After introducing the pioneering results of the initiators, state-of-the-art research results are given and discussed. Various applications have been explicitly illustrated for a range of (fiber-reinforced) materials, including

○ (Visco)elastic C-tensor identification through the use of inverse methods.
○ Assessment of (quasi-)static shear damage in cross-ply C/E laminate.
○ Detection of (overlapping) delamination in thin cross-ply C/E laminate.
○ Assessment of tension-tension fatigue damage in woven C/E laminate.
○ Identification of the in-plane parameters of 2D (sub)surface corrugation.
○ Single-sided identification of a local 3D plastic strain field in DC06 steel coupon.

Final notes are given about our ongoing exploration of several novel and high-potential research lines, which will undoubtedly further advance the ultrasonic (backscatter) polar scan technique, bringing the methodology to its next level of maturity.

Acknowledgments: M.K. acknowledges the financial support of Bijzonder OnderzoeksFonds BOF (special research fund grant BOF.PDO.2015.0028.01). The authors further acknowledge the financial support of Fonds voor Wetenschappelijk Onderzoek FWO—Vlaanderen (grants FWO.G012010N and FWO.G0B9515N).

Author Contributions: M.K. conceived, designed and performed the experiments; M.K. and A.M. implemented the numerical models; L.P., F.Z., H.S. and S.D. provided samples; W.V.P., J.D. and K.V.D.A. supervised the presented research; M.K. wrote the paper.

Conflicts of Interest: The authors declare no conflict of interest.

References

1. Van Dreumel, W.H.M.; Speijer, J.L. Non-destructive composite laminate characterization by means of ultrasonic polar-scan. *Mater. Eval.* **1981**, *39*, 922–925.
2. Declercq, N.F.; Degrieck, J.; Leroy, O. Simulations of harmonic and pulsed ultrasonic polar scans. *NDT E Int.* **2006**, *39*, 205–216.
3. Rokhlin, S.I.; Chimenti, D.E.; Nagy, P.B. *Physical Ultrasonics of Composites*; Oxford University Press: Oxford, UK, 2011.
4. Rose, J.L. *Ultrasonic Waves in Solid Media*; Cambridge University Press: Oxford, UK, 1999.

5. Kersemans, M.; Martens, A.; Delrue, S.; van den Abeele, K.; Pyl, L.; Zastavnik, F.; Sol, H.; Degrieck, J.; van Paepegem, W. The ultrasonic polar scan: Past, Present and future. In Proceedings of the Emerging Technologies in NonDestructive Testing (ETNDT6), Brussels, Belgium, 27–29 May 2015; pp. 133–139.

6. Van Dreumel, W.H.M.; Speijer, J.L. Polar-scan, a non-destructive test method for the inspection of layer orientation and stacking order in advanced fiber composites. *Mater. Eval.* **1983**, *41*, 1060–1062.

7. Degrieck, J.; van Leeuwen, D. Simulatie van een Ultrasone Polaire Scan van een Orthotrope Plaat (in Dutch). In Proceedings of the 3rd Belgian National Congress on Theoretical and Aplied Mechanics, Liege, Belgium, 30–31 May 1994.

8. Degrieck, J. Some possibilities of nondestructive characterisation of composite plates by means of ultrasonic polar scans. In Proceedings of the Emerging technologies in nondestructive testing (ETNDT), Patras, Greece, 22–23 May 1995; pp. 225–235.

9. Degrieck, J.; Declercq, N.F.; Leroy, O. Ultrasonic polar scans as a possible means of non-destructive testing and characterisation of composite plates. *Insight* **2003**, *45*, 196–201.

10. Declercq, N.F.; Degrieck, J.; Leroy, O. On the influence of fatigue on ultrasonic polar scans of fiber reinforced composites. *Ultrasonics* **2004**, *42*, 173–177.

11. Declercq, N.F.; Degrieck, J.; Leroy, O. Ultrasonic polar scans: Numerical simulation on generally anisotropic media. *Ultrasonics* **2006**, *45*, 32–39.

12. Satyanarayan, L.; Weide, J.M.V.; Declercq, N.F. Ultrasonic polar scan imaging of damaged fiber reinforced composites. *Mater. Eval.* **2010**, *68*, 733–739.

13. Kersemans, M. Combined experimental-numerical study to the ultrasonic polar scan for inspection and characterization of (damaged) anisotropic materials. Ph.D. Thesis, Ghent University, Ghent, Belgium, 31 October 2014.

14. Kersemans, M.; van Paepegem, W.; van den Abeele, K.; Pyl, L.; Zastavnik, F.; Sol, H.; Degrieck, J. Pitfalls in the experimental recording of ultrasonic (backscatter) polar scans for material characterization. *Ultrasonics* **2014**, *54*, 1509–1521.

15. Kersemans, M.; Martens, A.; van den Abeele, K.; Degrieck, J.; Pyl, L.; Zastavnik, F.; Sol, H.; van Paepegem, W. The quasi-harmonic ultrasonic polar scan for material characterization: Experiment and numerical modeling. *Ultrasonics* **2015**, *58*, 111–122.

16. Rokhlin, S.I.; Wang, L. Stable recursive algorithm for elastic wave propagation in layered anisotropic media: Stiffness matrix method. *J. Acoust. Soc. Am.* **2002**, *112*, 822–834.

17. Tan, E.L. Hybrid compliance-stiffness matrix method for stable analysis of elastic wave propagation in multilayered anisotropic media. *J. Acoust. Soc. Am.* **2006**, *119*, 45–53.

18. Tan, E.L. Stiffness matrix method with improved efficiency for elastic wave propagation in layered anisotropic media. *J. Acoust. Soc. Am.* **2005**, *118*, 3400–3403.

19. Kersemans, M.; Martens, A.; Lammens, N.; van den Abeele, K.; Degrieck, J.; Zastavnik, F.; Pyl, L.; Sol, H.; van Paepegem, W. Identification of the elastic properties of isotropic and orthotropic thin-plate materials with the pulsed ultrasonic polar scan. *Exp. Mech.* **2014**, *54*, 1121–1132.

20. Kersemans, M.; Lammens, N.; Luyckx, G.; Degrieck, J.; van Paepegem, W. Quantitative measurement of the elastic properties of orthotropic composites by means of the ultrasonic polar scan method. *JEC Compos.* **2012**, *75*, 48–52.

21. Martens, A.; Kersemans, M.; Degrieck, J.; van Paepegem, W.; Delrue, S.; van den Abeele, K. Time-of-flight recorded pulsed ultrasonic polar scan for elasticity characterization of composites. In Proceedings of the Emerging Technologies in NonDestructive Testing (ETNDT6), Brussels, Belgium, 27–29 May 2015.

22. Van den Abeele, K.; Martens, A.; Kersemans, M.; Degrieck, J.; Delrue, S.; van Paepegem, W. Characterization of visco-elastic material parameters by means of the ultrasonic polar scan method. In Proceedings of the Acoustical Society of America—Fall 2015 Meeting, Jacksonville, FL, USA, 2–6 November 2015.

23. Kersemans, M.; Lammens, N.; Degrieck, J.; van den Abeele, K.; Pyl, L.; Zastavnik, F.; Sol, H.; van Paepegema, W. Extraction of bulk wave characteristics from a pulsed ultrasonic polar scan. *Wave Motion* **2014**, *51*, 1071–1081.

24. Kersemans, M.; van Paepegem, W.; Lemmens, B.; van den Abeele, K.; Pyl, L.; Zastavnik, F.; Sol, H.; Degrieck, J. The pulsed ultrasonic backscatter polar scan and its applications for NDT and material characterization. *Exp. Mech.* **2014**, *54*, 1059–1071.

25. Kersemans, M.; van Paepegem, W.; van den Abeele, K.; Pyl, L.; Zastavnik, F.; Sol, H.; Degrieck, J. Ultrasonic characterizaion of subsurface 2D corrugation. *J. Nondestruct. Eval.* **2014**, *33*, 438–442.

26. De Billy, M.; Cohentenoudji, F.; Jungman, A.; Quentin, G.J. Possibility of assigning a signature to rough surfaces using ultrasonic backscattering diagrams. *IEEE Trans. Sonics Ultrason.* **1976**, *23*, 356–363.

27. De Billy, M.; Quentin, G. Measurement of the periodicity of internal surfaces by ultrasonic testing. *J. Phys. D Appl. Phys.* **1982**, *15*, 1835–1841.

28. Herbison, S.W. Ultrasonic diffraction effects on periodic surfaces. Ph.D. Thesis, Georgia Institute of Technology, Atlanta, GA, USA, August 2011.

29. Kersemans, M.; de Baere, I.; Degrieck, J.; van den Abeele, K.; Pyl, L.; Zastavnik, F.; Sol, H.; van Paepegem, W. Nondestructive damage assessment in fiber reinforced composites with the pulsed ultrasonic polar scan. *Polym. Test.* **2014**, *34*, 85–96.

30. Kersemans, M.; Martens, A.; van den Abeele, K.; Degrieck, J.; Pyl, L.; Zastavnik, F.; Sol, H.; van Paepegem, W. Detection and localization of delaminations in thin carbon fiber reinforced composites with the ultrasonic polar scan. *J. Nondestruct. Eval.* **2014**, *33*, 522–534.

31. Giancane, S.; Panella, F.W.; Dattoma, V. Characterization of fatigue damage in long fiber epoxy composite laminates. *Int. J. Fatigue* **2010**, *32*, 46–53.

32. Taheri-Behrooz, F.; Shokrieh, M.M.; Lessard, L.B. Residual stiffness in cross-ply laminates subjected to cyclic loading. *Compos. Struct.* **2008**, *85*, 205–212.

33. Kersemans, M.; de Baere, I.; Degrieck, J.; van den Abeele, K.; Pyl, L.; Zastavnik, F.; Sol, H.; van Paepegem, W. Damage signature of fatigued fabric reinforced plastic in the pulsed ultrasonic polar scan. *Exp. Mech.* **2014**, *54*, 1467–1477.

34. Lyman, T. An explanation of the false spectra from diffraction gratings. *Proc. Am. Acad. Arts Sci.* **1903**, *39*, 39–47.

35. Gale, H.G. Rowland ghosts. *Astrophys. J.* **1937**, *85*, 49–61.

36. Kersemans, M.; Allaer, K.; van Paepegem, W.; van den Abeele, K.; Pyl, L.; Zastavnik, F.; Sol, H.; Degrieck, J. A novel ultrasonic strain gauge for single-sided measurement of a local 3D strain field. *Exp. Mech.* **2014**, *54*, 1673–1685.

37. Castaings, M.; Hosten, B.; Kundu, T. Inversion of ultrasonic, plane-wave transmission data in composite plates to infer viscoelastic material properties. *NDT E Int.* **2000**, *33*, 377–392.

38. Wang, L.; Lavrentyev, A.I.; Rokhlin, S.I. Beam and phase effects in angle-beam-through-transmission method of ultrasonic velocity measurement. *J. Acoust. Soc. Am.* **2003**, *113*, 1551–1559.

39. Lavrentyev, A.I.; Rokhlin, S.I. Determination of elastic moduli, density, attenuation, and thickness of a layer using ultrasonic spectroscopy at two angles. *J. Acoust. Soc. Am.* **1997**, *102*, 3467–3477.

40. Cawley, P.; Hosten, B. The use of large ultrasonic transducers to improve transmission coefficient measurements on viscoelastic anisotropic plates. *J. Acoust. Soc. Am.* **1997**, *101*, 1373–1379.

41. Castellano, A.; Foti, P.; Fraddosio, A.; Marzano, S.; Piccioni, M.D. Mechanical characterization of CFRP composites by ultrasonic immersion tests: Experimental and numerical approaches. *Compos. B Eng.* **2014**, *66*, 299–310.

42. Aymerich, F.; Staszewski, W.J. Impact damage detection in composite laminates using nonlinear acoustics. *Compos. A Appl. Sci. Manuf.* **2010**, *41*, 1084–1092.

43. Scalerandi, M.; Gliozzi, A.S.; Bruno, C.L.E.; van den Abeele, K. Nonlinear acoustic time reversal imaging using the scaling subtraction method. *J. Phys. D Appl. Phys.* **2008**, *41*, 1–10.

44. Meo, M.; Polimeno, U.; Zumpano, G. Detecting damage in composite material using nonlinear elastic wave spectroscopy methods. *Appl. Compos. Mater.* **2008**, *15*, 115–126.

45. Hirsekorn, S.; Rabe, U.; Arnold, W. Characterisation and evaluation of composite materials by nonlinear ultrasonic transmission measurements. In Proceedings of the International Congress on Ultrasonics (ICU 2007), Wenen, Oostenrijk, 10–12 April 2007.

46. Drinkwater, B.W.; Wilcox, P.D. Ultrasonic arrays for non-destructive evaluation: A review. *NDT E Int.* **2006**, *39*, 525–541.

47. Rucka, M. Damage detection in beams using wavelet transform on higher vibration modes. *J. Theor. Appl. Mech.* **2011**, *49*, 399–417.

48. Baba, B.O.; Thoppul, S.; Gibson, R.F. Experimental and numerical investigation of free vibrations of composite sandwich beams with curvature and debonds. *Exp. Mech.* **2010**, *51*, 857–868.

49. Pitarresi, G. Lock-in signal post-processing techniques in infra-red thermography for materials structural evaluation. *Exp. Mech.* **2013**, *55*, 667–680.

50. Wong, A.K.; Rajic, N.; Nguyen, Q. 50th anniversary article: Seeing stresses through the thermoelastic lens—A retrospective and prospective from an australian viewpoint. *Strain* **2015**, *51*, 1–15.

51. Hung, Y.Y.; Chen, Y.S.; Ng, S.P.; Liu, L.; Huang, Y.H.; Luk, B.L.; Ip, R.W.L.; Wu, C.M.L.; Chung, P.S. Review and comparison of shearography and active thermography for nondestructive evaluation. *Mater. Sci. Eng. R Rep.* **2009**, *64*, 73–112.

52. Klepka, A.; Pieczonka, L.; Staszewski, W.J.; Aymerich, F. Impact damage detection in laminated composites by non-linear vibro-acoustic wave modulations. *Compos. B Eng.* **2014**, *65*, 99–108.

Dispersion of Functionalized Silica Micro- and Nanoparticles into Poly(nonamethylene Azelate) by Ultrasonic Micro-Molding

Angélica Díaz, María T. Casas and Jordi Puiggalí

Abstract: Ultrasound micro-molding technology has proved useful in processing biodegradable polymers with minimum material loss. This makes this technology particularly suitable for the production of biomedical microdevices. The use of silica (SiO_2) nanoparticles is also interesting because of advantages like low cost and enhancement of final properties. Evaluation of the capacity to create a homogeneous dispersion of particles is crucial. Specifically, this feature was explored taking into account micro- and nano-sized silica particles and a biodegradable polyester derived from 1,9-nonanodiol and azelaic acid as a matrix. Results demonstrated that composites could be obtained with up to 6 wt. % of silica and that no degradation occurred even if particles were functionalized with a compatibilizer like (3-aminopropyl) triethoxysilane. Incorporation of nanoparticles should have a great influence on properties. Specifically, the effect on crystallization was evaluated by calorimetric and optical microscopy analyses. The overall crystallization rate was enhanced upon addition of functionalized silica nanospheres, even at the low percentage of 3 wt. %. This increase was mainly due to the ability of nanoparticles to act as heterogeneous nuclei during crystallization. However, the enhancement of the secondary nucleation process also played a significant role, as demonstrated by Lauritzen and Hoffmann analysis.

Reprinted from *Appl. Sci.* Cite as: Díaz, A.; Casas, M.T.; Puiggalí, J. Dispersion of Functionalized Silica Micro- and Nanoparticles into Poly(nonamethylene Azelate) by Ultrasonic Micro-Molding. *Appl. Sci.* **2015**, 5, 1252–1271.

1. Introduction

Ultrasonic waves are an energy source that has been employed as a plastic welding procedure for over 40 years because it is clean, efficient, and fast. In fact, absorption of vibration energy leads to polymer friction, which may result in local melting of the sample. The first descriptions of the potential use of ultrasonic waves in plastic powder molding were given by Fairbanks [1] and Crawford et al. [2]. Since then different efforts have been made to develop the great potential of this technology. Thus, ultrasonic hot embossing appears as a new, interesting process for fast and low-cost production of microsystems from polymeric materials [3–5].

A wide variety of microdevices (e.g., micro mixers, flow sensors or micro whistles) have been prepared by this technology demonstrating its feasibility.

The use of an ultrasonic source has also been proposed to achieve both the plasticization of the material and the direct injection of the molten material [6–8]. This provides a new micro-molding technology that can be a serious alternative to conventional micro-injection techniques. Micro-molding equipment (Figure 1) is mainly composed of a plasticizing chamber, a controller, a mold, an ultrasonic generator that produces high frequency (30 kHz) from line voltage, and a resonance stack or acoustic unit connected to the generator. This unit consists of: (a) a converter (piezoelectric transducer), where high frequency signals from the generator are transmitted through piezoelectric crystals that expand and contract at the same rate as electrical oscillation; (b) a booster, which amplifies or reduces mechanical oscillation (from 0 to 137.5 μm); and (c) a sonotrode, which transfers this oscillation to the polymer sample in the plasticizing chamber by applying a force (from 100 to 500 N).

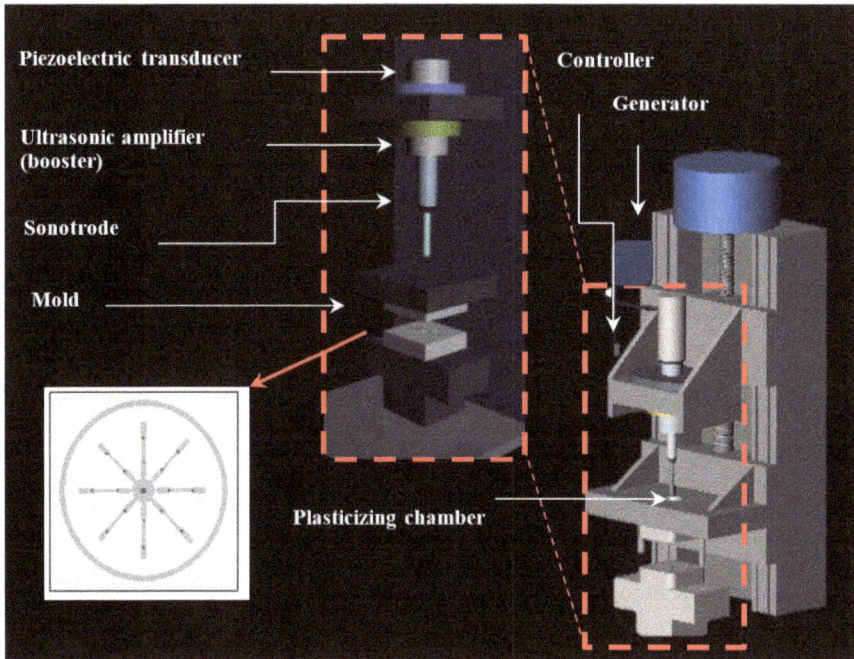

Figure 1. Scheme of the main parts of the ultrasound micro-molding machine.

The heat build-up caused by the resulting friction heat melts the polymer, which flows into the mold cavities through their feeding channels under the sonotrode pressure. A controller regulates the main processing parameters: oscillation time (e.g., from 0.5 to 10 s), molding force, and amplitude of the ultrasonic wave.

Despite the potential advantages of using high-intensity ultrasonic waves as an energy source, it should be pointed out that these waves have not only physical effects on the melt rheology of the polymer but also chemical effects on the polymer chain as a result of cavitation and the high temperature that can be achieved inside the plasticizing chamber [9–12]. Therefore, optimization of processing parameters to obtain minimally degraded samples is an essential step in applying the new micro-molding technology. In this sense, we have recently found that it is possible to establish appropriate conditions for biodegradable polymers that could be interesting for the production of biomedical devices such as polylactide (PLA) [13], poly(butylene succinate) (PBS) [14], and poly(nonamethylene azelate) (PE99) [15]. Degradation logically increases with irradiation time and amplitude, but a minimum period is necessary to ensure complete injection into the mold, as well as a minimum amount of vibration energy. Severe degradation can also occur at low ultrasonic amplitude upon application of a high molding force because of chain scissions caused by mechanical shear stress. Nevertheless, a minimum force must be applied to make melt polymer flow and mold cavity filling feasible.

The ability to provide good dispersion is another important reason to evaluate the potential of micro-molding technology. One possible application could involve the incorporation of compounds with added value (e.g., drugs with pharmacological activity) or even of nanoparticles that lead to a large surface area to volume ratio. It is well known that this feature may considerably improve the properties of the material (e.g., physical, chemical, and mechanical) with respect to the properties attained with conventionally filled composites [16–18].

Ultrasound micro-molding was recently found to be effective in obtaining nanocomposites with the final form required for a selected application, a homogeneous clay distribution up to a load of 6 wt. % and, more interestingly, an exfoliated structure without the need for a compatibilizer between the organic polymer and the inorganic silicate clay [17,18]. Furthermore, polymer degradation was minimized by adding pristine clay. These promising results were achieved using PLA, PBS, and PE99 as biodegradable matrices and montmorillonites like neat N757 and C20A, C25A, and N848 organo-modified clays. Incorporation of nanoparticles also had a remarkable influence on the crystallization kinetics of the polymer matrix. Interestingly, significant differences were found depending on the polymer type (e.g., PLA or PE99) despite obtaining similar exfoliated structures [18].

The use of silica (SiO_2) nanoparticles to enhance material properties is also of high interest due to their intrinsic advantages (low cost, nontoxicity, high modulus, and ability to modify chemical surface characteristics) [19,20]. Thus, properties can be considerably improved when particles with appropriate surface functionalization to enhance interfacial interactions are well dispersed in the polymer matrix [21–23].

Silica particles with sizes suitable for biomedical applications can be easily prepared from base-catalyzed sol-gel processes [24]. They use organosilane precursors (e.g., tetraethoxysilane), which lead to the formation of a new phase (sol) by hydrolysis and condensation reactions. Subsequently, the condensation of colloidal particles leads to the gel phase. Silica particles can be easily functionalized, giving rise to a high versatility. In fact, functional groups can be attached through covalent bonds during condensation or even after a later grafting process [25].

The present work evaluates the applicability of ultrasound micro-molding technology to prepare dispersion of silica micro- and nanoparticles in a biodegradable polymer matrix that was recently considered for preparation of clay nanocomposites (*i.e.*, PE99). The effect of silica nanoparticle addition on crystallization kinetics is also studied since they should affect nucleation and crystal growth, as recently determined for nanocomposites based on poly(ethylene oxide) [26].

2. Experimental Section

2.1. Materials

Poly(nonamethylene azelate) (PE99) was synthesized by thermal polycondensation of azelaic acid with an excess of 1,9-nonanediol (2.2:1 molar ratio), as shown in Figure 2a. Titanium tetrabutoxyde was used as a catalyst and the reaction was first performed in a nitrogen atmosphere at 150 °C for 6 h and then in a vacuum at 180 °C for 18 h. The polymer was purified by precipitation with ethanol of a chloroform solution (10 wt. %). The average molecular weight and polydispersity index determined by GPC and using poly(methyl methacrylate) standards were 13,200 g/mol and 3.1, respectively.

Preparation of silica micro/nanoparticles (Figure 2b): Deionized water (0.6 mL) and 1.2 mL of a 28% aqueous ammonium hydroxide (1.2 mL) (Aldrich, Madrid, Spain) were added to a flask of 100 mL containing absolute ethanol (50 mL). The mixture was stirred vigorously for 20 min at room temperature using a magnetic agitator. Next 1.2 mL of tetraethoxysilane (Aldrich, Madrid, Spain) were quickly added and the resultant solution was purged with dry nitrogen and stirred for another 16 h. Spheres with a homogeneous diameter close to 600 nm were obtained after filtration. The same protocol except for increasing the stirring speed to 250–1000 rpm was employed to prepare nanoparticles with diameters close to 25 nm.

Functionalization of silica micro/nanoparticles (Figure 2c): 1 g of micro/nanospheres previously dried under a vacuum was introduced into a vessel together with 66 μL of pure ethanol. The mixture was sonicated for 15 min to enhance dispersion using a VWR Ultrasonic cleaner bath (UWR International, New York, NY, USA) at 100 watts, and subsequently 66 µL of ammonium hydroxide was added under stirring for 15 min. Finally, 297 µL of (3-aminopropyl) triethoxysilane (AMPS) was quickly added and the reaction was allowed to progress under stirring for 24 h at room temperature. The resulting solid was centrifuged, repeatedly washed with ethanol, and vacuum dried.

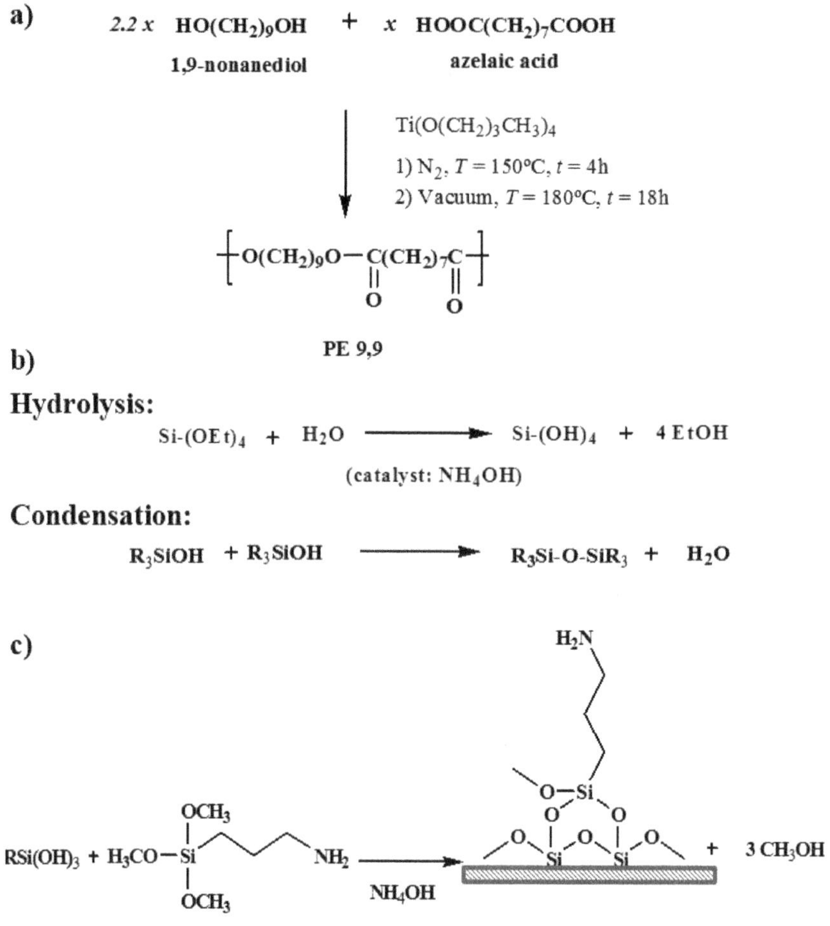

a)

$$2.2\,x \quad HO(CH_2)_9OH \quad + \quad x \quad HOOC(CH_2)_7COOH$$

1,9-nonanediol   **azelaic acid**

$Ti(O(CH_2)_3CH_3)_4$

1) N_2, $T = 150°C$, $t = 4h$
2) Vacuum, $T = 180°C$, $t = 18h$

$$\left[O(CH_2)_9O - \underset{O}{\underset{\|}{C}}(CH_2)_7\underset{O}{\underset{\|}{C}} \right]$$

PE 9,9

b)

Hydrolysis:

$$Si\text{-}(OEt)_4 \quad + \quad H_2O \quad \longrightarrow \quad Si\text{-}(OH)_4 \quad + \quad 4\,EtOH$$

(catalyst: NH_4OH)

Condensation:

$$R_3SiOH \quad + \quad R_3SiOH \quad \longrightarrow \quad R_3Si\text{-}O\text{-}SiR_3 \quad + \quad H_2O$$

c)

Figure 2. Synthesis schemes for PE99 (**a**); silica spheres (**b**); and functionalization process (**c**).

Nanocomposites will be denoted by polymer abbreviation, size (micro or nano), and wt. % of added silica (e.g., PE99-M 6 and PE99-N 6 indicate composites having 6 wt. % of micro- and nanospheres, respectively).

2.2. Micro-Molding Equipment

A prototype Ultrasound Molding Machine (Sonorus®, Ultrasion S.L., Barcelona, Spain) was employed. The apparatus was equipped with a digital ultrasound generator from Mecasonic (1000 W–30 kHz, Barcelona, Spain) a controller (3010 DG digital system, Mecasonic, Barcelona, Spain), a converter, an acoustic unit, and an electric servomotor control (Berneker and Rainer, Barcelona, Spain) dotted with software from Ultrasion S.L. Mold was thermally controlled and designed to prepare eight test specimens. Dimensions of these specimens were $1.5 \times 0.1 \times 0.1$ cm^3 and followed IRAM-IAS-U500-102/3 standards.

2.3. Measurements

Molecular weight was estimated by gel permeation chromatography (GPC) using a liquid chromatograph (Shimadzu, model LC-8A, Tokyo, Japan) equipped with an Empower computer program (Waters, Massachusetts, MA, USA). A PL HFIP gel column (Polymer Lab, Agilent Technologies Deutschland GmbH, Böblingen, Germany) and a refractive index detector (Shimadzu RID-10A) were employed. The polymer was dissolved and eluted in 1,1,1,3,3,3-hexafluoroisopropanol containing CF_3COONa (0.05 M) was employed as solvent and elution medium. Flow rate was 0.5 mL/min, the injected volume 100 μL, and the sample concentration 2 mg/mL. Polymethyl methacrylate standards were employed to determine the number and weight average molecular weights and molar-mass dispersities.

A Focused Ion Beam Zeiss Neon40 microscope (Oberkochen, Germany) operating at 5 kV was employed to get SEM micrographs of micro-molded specimens. Carbon coating was accomplished with a Mitec K950 Sputter Coater (Oberkochem, Germany) was used to coat all samples, which were then viewed at an accelerating voltage of 5 kV.

A FTIR 4100 Jasco spectrophotometer dotted with an attenuated total reflectance accessory (Specac MKII Golden Gate Heated Single Reflection Diamond ATR, Jasco International Co. Ltd., Tokyo, Japan) and a thermal controller was employed to get the FTIR spectra. Samples were placed in an attenuated total reflectance accessory with thermal control and a diamond crystal.

X-ray photoelectron spectroscopy (XPS) was performed with a SPECS system (Berlin, Germany) dotted with an XR50 source of Mg/Al (1253 eV/1487 eV) operating at 150 W. Analyses were performed in a SPECS system equipped with a high intensity twin-anode X-ray source XR50 of Mg/Al (1253 eV/1487 eV) operating at 150 W. A Phoibos 150 MCD-9 XP detector (Berlin, Germany) was employed. The overview spectra were taken with an X-ray spot size of 650 and pass energy of 25 eV in 0.1 eV steps at a pressure below 6×10^{-9} mbar. Surface composition was determined through the N 1s and Si 2p peaks at binding energies of 399.2 and 103.2 eV, respectively.

Distribution of micro/nanoparticles in the composites was evaluated by means of a Philips TECNAI 10 electron microscope (Philips Electron Optics, Eindhoven, Holland) at an accelerating voltage of 80 kV. Samples were prepared by embedding the nanocomposite specimens. A low-viscosity, modified Spurr epoxy resin was employed to embed the specimens before curing and cutting in small sections. A Sorvall Porter-Blum microtome (New York, NY, USA) equipped with a diamond knife was employed in this case. The thin sections were collected in a trough filled with water and lifted onto carbon-coated copper grids.

Thermal degradation was performed in a Q50 thermogravimetric analyzer of TA Instruments (TA Instruments, New Castle, DE, USA). Experiments were carried out at a heating rate of 10 °C/min with 5 mg samples and under a flow of dry nitrogen.

Growth rates of spherulites were measured by optical microscopy using a Zeiss Axioscop 40 Pol light polarizing microscope (Oberkochen, Germany). This was equipped with a Linkam temperature control system configured by a THMS 600 heating and freezing stage connected to an LNP 94 liquid nitrogen system. Spherulites were isothermally crystallized at the selected temperature from homogeneous melt thin films. These were prepared by melting 1 mg of the polymer between microscope slides. Subsequently, small sections of the obtained thin films were pressed between two cover slides for microscopy observations. Samples were kept at the hot stage at 90 °C for 5 min to eliminate sample history effects, and then rapidly cooled to the selected temperatures. A Zeiss AxiosCam MRC5 digital camera (Munich, Germany) was employed to follow the diameter evolution of spherulites by taking micrographs at different times. Nucleation was evaluated by counting the number of active nuclei that appeared in the micrographs. The sign of spherulite birefringence was determined by means of a first-order red tint plate placed under crossed polarizers.

Differential scanning calorimetry was performed with a TA instrument Q100 series (TA Instruments, New Castle, DE, USA) with T_{zero} technology and equipped with a refrigerated cooling system (RCS). Crystallization kinetics experiments were conducted under a flow of dry nitrogen with a sample weight around 5 mg. Calibration was performed with indium. Studies were carried out according the following protocol: Samples were firstly heated (20 °C/min) up to 25 °C above their melting temperature, subsequently held at this temperature for 5 min to eliminate the thermal history, and finally cooled to the selected temperature at a rate of 50 °C/min. Samples were kept at the isothermal temperature until baseline was reached. Finally, a new heating run (20 °C) was performed in order to determine the equilibrium melting temperature of samples.

3. Results and Discussion

3.1. Characterization of Functionalized Silica Particles

Silica particles were effectively prepared using tetraethoxysilane as a precursor. TEM micrographs clearly revealed a highly regular spherical form (Figure 3a) for silica particles obtained at a low stirring speed. These particles showed a highly homogeneous diameter that varied in a narrow range (*i.e.*, between 570 and 650 nm). Particles were slightly more irregular in form (Figure 3b) when prepared at a higher stirring speed. Diameter size was considerably reduced but a homogeneous distribution could still be observed (*i.e.*, values were always within the 20–30 nm interval). The two kinds of silica preparations will be designated as micro (M) and nano (N) particles.

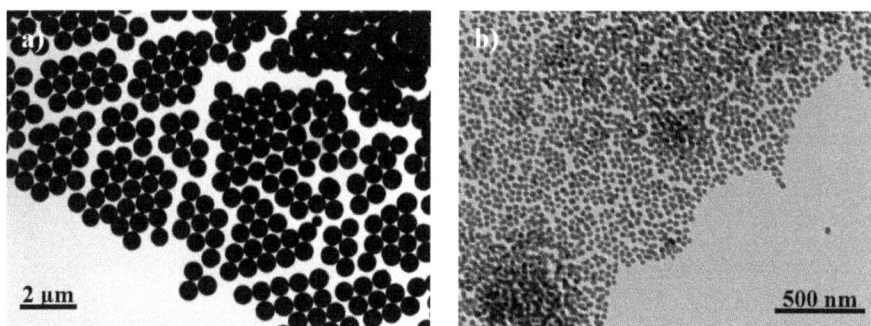

Figure 3. TEM micrographs of functionalized microspheres (**a**); and nanospheres (**b**).

Surface functionalization by reaction with (3-aminopropyl) triethoxysilane was verified by analysis of XPS spectra (Figure 4), which allowed for determining a ratio between N and Si atoms of close to 8% (*i.e.*, 8.8% and 7.3% for micro- and nanoparticles, respectively).

Thermogravimetric analyses (Figure 5) also showed a significant weight loss of functionalized silica particles because of decomposition of grafted AMPS groups, which took place at around 500 °C. This decomposition step corresponded to an approximate weight loss of 9% and, logically, was not detected in non-functionalized particles. Gradual weight loss leading to a value of only 8%–9% at 590 °C was observed for all particles. Thus, at this temperature the total loss was 17% and 8% for functionalized and non-functionalized particles, respectively. Practically no differences were found in the TGA (Figure 5a) and DTGA (Figure 5b) curves of micro- and nanoparticles.

29

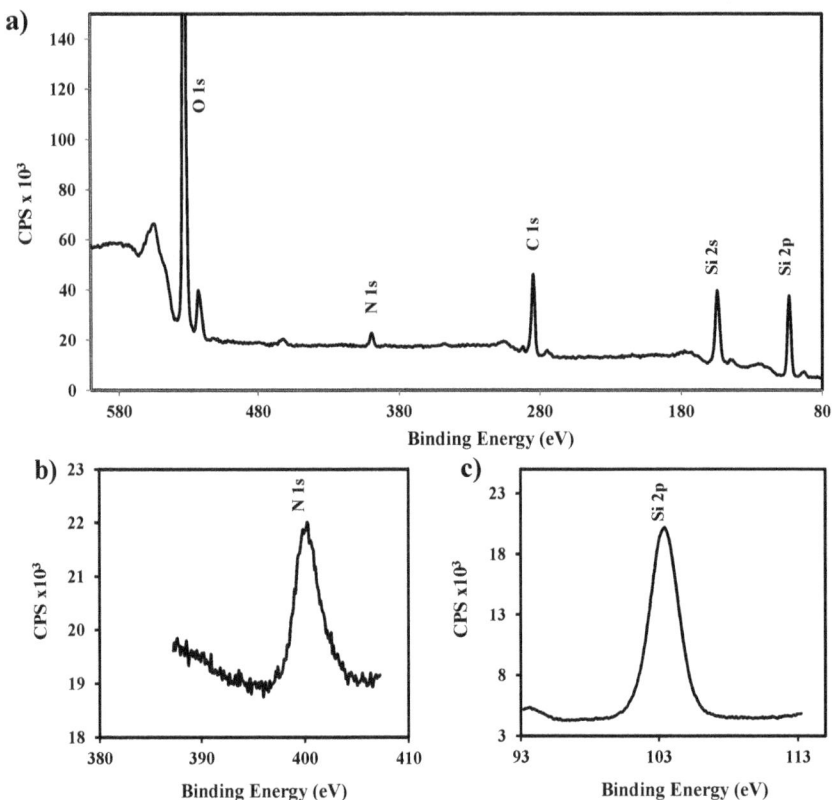

Figure 4. Full scale XPS spectra (**a**) and details corresponding to silicon (**b**) and nitrogen (**c**) XPS signals detected in the functionalized silica nanoparticles.

3.2. Dispersion of Functionalized Silica Micro- and Nanoparticles by Ultrasound Micro-Molding Technology

PE99 samples could be processed under relatively mild conditions, as previously established [15]. Thus, a minimum irradiation time of 1.2 s, low amplitude of 24 μm, and a moderate molding force of 300 N were sufficient to guarantee a 100% molding efficiency. Experimental conditions could be maintained for processing mixtures with functionalized micro- and nanospheres up to the maximum test load of 6 wt. %.

Molecular weights of raw PE99 and specimens processed under the above optimized conditions are summarized in Table 1, whereas Figure 6 compares the GPC traces of PE99, PE99-N 3, and PE99-M 6 specimens.

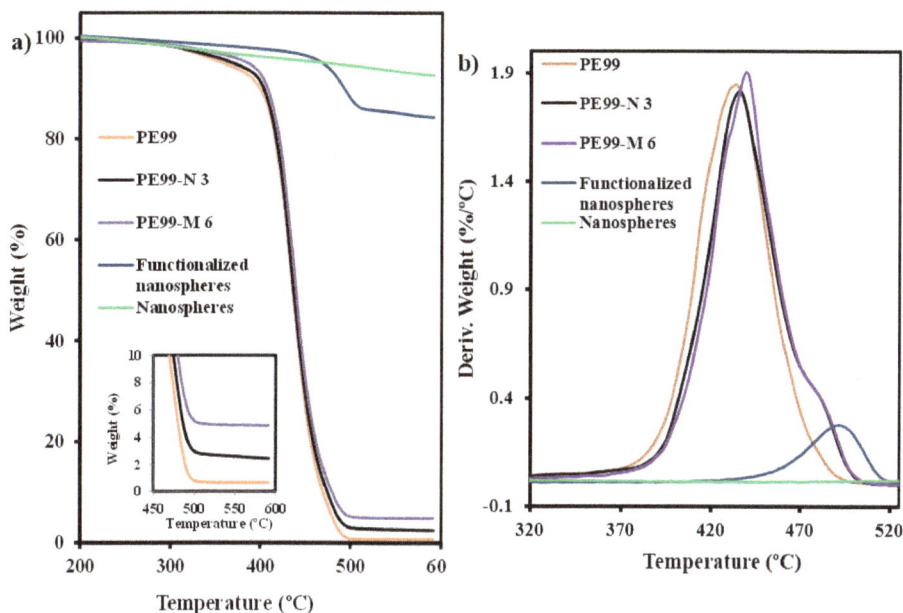

Figure 5. TGA (**a**) and DTGA (**b**) curves of silica nanospheres, functionalized silica nanospheres, and micro-molded PE99, PE99-N 3, and PE99-M 6 specimens.

Table 1. Molecular weights of processed PE 99 and their mixtures with functionalized micro- and nanoparticles [a].

Sample	Mn (g/mol)	Mw (g/mol)	Mw/Mn
PE99 (raw)	13,300	35,900	2.7
PE99	14,700	37,200	2.5
PE99-M 3	14,800	36,800	2.5
PE99-M 6	12,600	33,800	2.7
PE99-N 3	15,800	39,200	2.5
PE99-N 6	14,200	37,900	2.7

[a] Micro-molding conditions: Time (s), amplitude (μm) and force (N): 1.2, 10 and 300.

No statistically significant differences were found between samples before and after processing or upon addition of functionalized nanospheres. Only a not highly significant decrease was detected for samples loaded with the maximum percentage (6 wt. %) of microspheres. The new technology appears fully adequate to obtain micropieces with negligible degradation of composites constituted by PE99 and functionalized silica micro/nanoparticles.

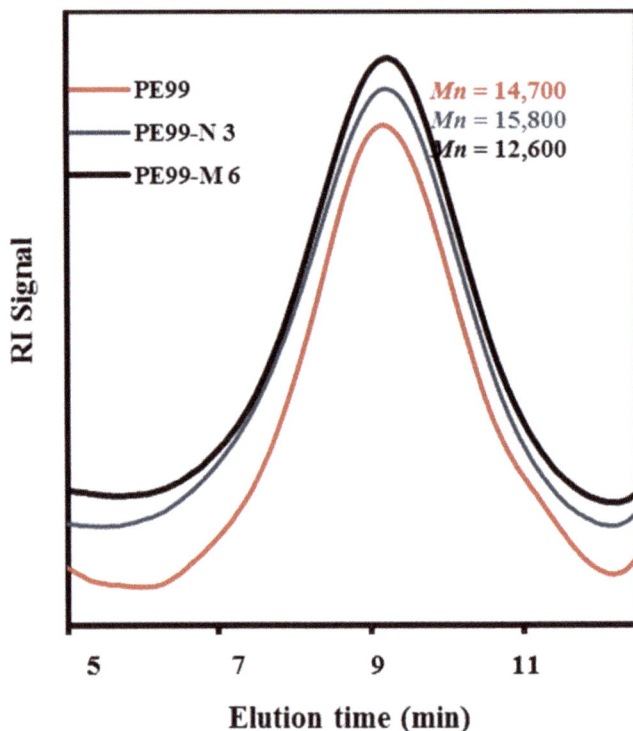

Figure 6. GPC molecular weight distribution curves determined for micro-molded PE99, PE-N 3, and PE99-M 6 specimens.

Figure 7a,b show optical and SEM images of a representative silica loaded specimen for which a regular texture is detected in the longitudinal section micrographs. All specimens were highly homogeneous, without the presence of cavities that could affect the final properties of the material.

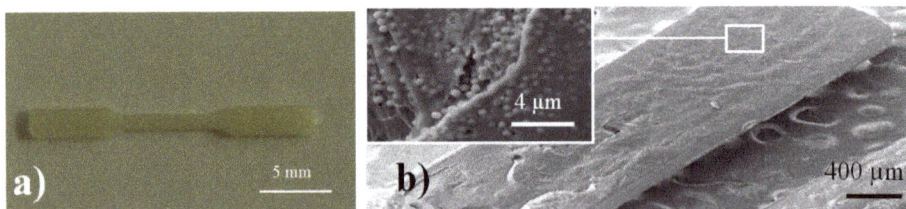

Figure 7. Image of a processed PE99 specimen (**a**); SEM micrograph of details of a micro-molded PE99-M 3 specimen (**b**). Microparticles can be observed in the inset.

FTIR spectra were also useful to verify the uniform incorporation of silica particles into micro-molded specimens and to discard concentration in the sprue. Thus, the typical Si–O stretching band at 1075 cm^{-1} was observed in every part of all loaded specimens. Hence, the spectra of specimen zones close to (proximal part) and distant from (distal part) the feeding channel showed a similar ratio between the intensity of the band associated with the C=O stretching band of the polyester and the band associated with the silica particles (see Figure 8 for representative samples containing micro- and nanoparticles).

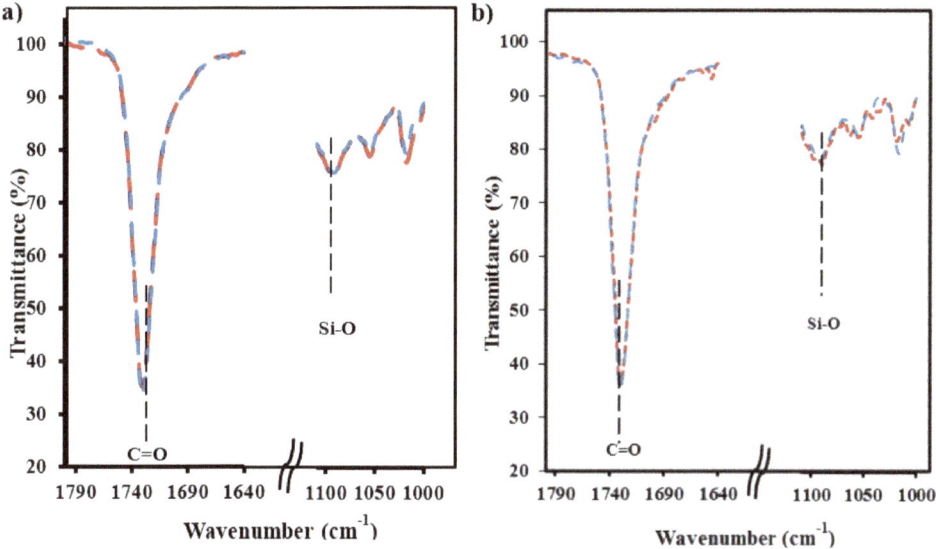

Figure 8. FTIR spectra of the characteristic C=O and Si–O stretching bands of micro-molded PE-N 3 (**a**) and PE-M 3 (**b**) specimens. Blue and red lines correspond to the proximal and distal parts of the specimens, respectively.

Analysis of particle dispersion in the processed composites was also carried out by transmission electron microscopy observation of ultrathin sections of micro-molded specimens. Flawless ultrasections could be obtained, as can be seen in the images for representative samples in Figure 9. These micrographs show the distribution of micro- and nanoparticles in the specimen, which is not completely homogeneous since density appears to be locally variable when observed at this scale.

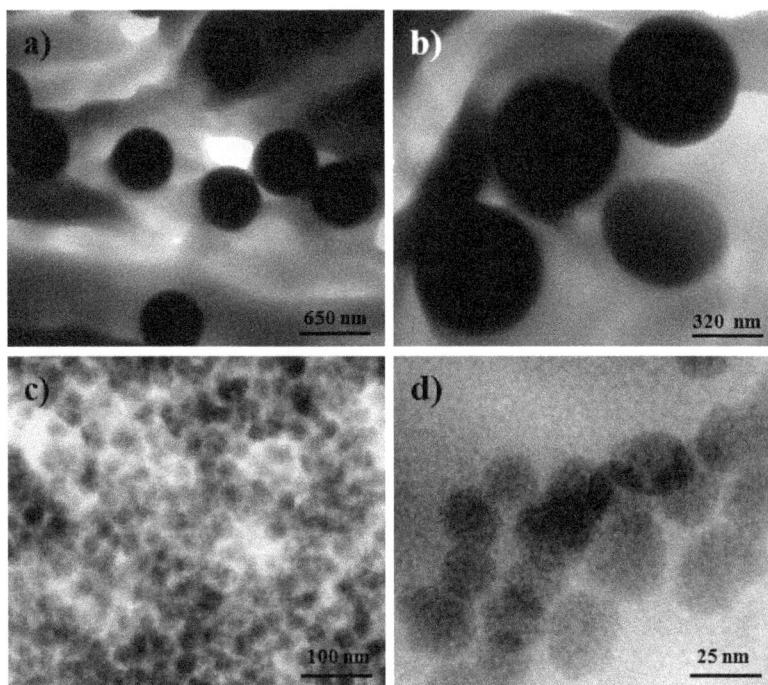

Figure 9. TEM micrographs of the dispersion of functionalized silica microspheres (**a,b**) and nanospheres (**c,d**) in the micro-molded PE99-M 3 and PE99-N 3 specimens.

Thermogravimetric analysis also gave information about the percentage of silica particles incorporated into the polymer matrix and their effect on thermal stability. In all cases, a constant char yield was attained at high temperatures, as shown by representative PE99-N 3 and PE99-M 6 specimens in Figure 5a. Furthermore, the remaining weight percentage was always in full agreement with the expected silica content (*i.e.*, close to 3 and 6 wt. %). The result is meaningful since it has demonstrated again that particles were not generally retained in the sprue. Silica was effectively led through the feeding channels by the molten polymer, giving rise to well-dispersed specimens. A close resemblance between the TGA and DTGA traces of the micro-molded PE99 sample and the silica-loaded composites was also found during the first stages of degradation (probably also as a consequence of the low content of added particles). The added particles even seem to slightly stabilize the sample (Table 2 and Figure 5). In any case, functionalization of silica particles did not have a negative impact on the ultrasound micro-molding process, as was also observed by GPC measurement. Differences were noticeable only at the end of the degradation process; the presence of shoulders around 483 °C in the DTGA plots (Figure 5b) seems to be due to decomposition of grafted AMPS groups.

Table 2. Characteristic TGA temperatures and remaining weight percentages for the decomposition of the studied micro-molded specimens.

Polymer	T_{onset} (°C)	$T_{10\%}$ (°C)	$T_{20\%}$ (°C)	$T_{40\%}$ (°C)	T_{max} (°C)	Remaining Weight (%)
PE 99	367	397	411	426	433	0
PE 99-N 3	377	405	415	431	436	2.7
PE 99-N 6	377	405	415	432	438	5.7
PE 99-M 3	377	405	417	431	437	2.9
PE 99-M 6	377	404	419	432	439	5.8

3.3. Calorimetric Studies on the Influence of Functionalized Silica Nanoparticles on the Isothermal Crystallization of Poly(nonamethylene Azelate)

Kinetic analysis was only performed for melt crystallization processes because of the impossibility of obtaining amorphous samples by cooling the melted nanocomposite at the maximum rate allowed by the equipment.

Crystallization experiments were therefore carried out in a narrow temperature interval (*i.e.*, between 56 and 59 °C) due to experimental limitations. Figure 10a shows the crystallization exotherms of the neat polyester and the PE99-N 3 nanocomposite, which allowed for determining the time evolution of the relative degree of crystallinity, $\chi(t)$. The last was calculated according to Equation (1):

$$\chi(t) = \int_{t0}^{t} (dH/dt)dt / \int_{t0}^{\infty} (dH/dt)dt \qquad (1)$$

where t_0 is the induction time and dH/dt corresponds to the heat flow rate. The evolution of crystallinity always showed a sigmoidal dependence on time for the five melt crystallization experiments performed for the different samples (Figure 10b). Experimental data were analyzed considering the typical Avrami equation [27,28]:

$$1 - \chi (t - t_0) = \exp(-Z \cdot (t - t_0)^n) \qquad (2)$$

where Z is a temperature-dependent rate constant and n the Avrami exponent whose value depends on the mechanism and geometry of the crystallization process. A normalized rate constant, $k = Z^{1/n}$, can also be calculated for comparison purposes since corresponding units (time^{-1}) are independent of the specific value of the Avrami exponent.

Plots of $\log(-\ln(1 - \chi(t - t_0)))$ against $\log(t - t_0)$ (Figure 11) allowed for determining the indicated crystallization parameters, which are summarized in Table 3. Avrami exponents for the neat polyester remain in a narrow range (*i.e.*, 2.60–2.94) and have an average value of 2.78. The determined Avrami exponent indicates a predetermined (heterogeneous) nucleation and a spherical growth under geometric constraints. Note that a slight deviation is observed with respect to the theoretical value of 3 and that a value close to 4 should be expected for a sporadic (heterogeneous) and homogeneous nucleations. It should also be pointed out than homogeneous nucleation usually requires high undercooling, which does not correspond with the performed experiments. The exponent slightly decreased for the nanocomposite (*i.e.*, 2.54–2.14, with 2.33 being the average value), suggesting an increase in geometric constraints upon incorporation of the well-dispersed nanospheres. Exponents determined for both samples were found to vary without a well-defined trend within the selected narrow temperature interval.

Reciprocal crystallization half-times ($1/\tau_{1/2}$) are also summarized in Table 3. This parameter is directly determined from DSC isotherms (*i.e.*, it corresponds to the inverse of the difference between thr crystallization start time and half-crystallization time) and can be useful to test the accuracy of the Avrami parameters since an estimated value can be obtained from them (*i.e.*, $1/\tau_{1/2} = (Z/\ln 2)^{1/n}$).

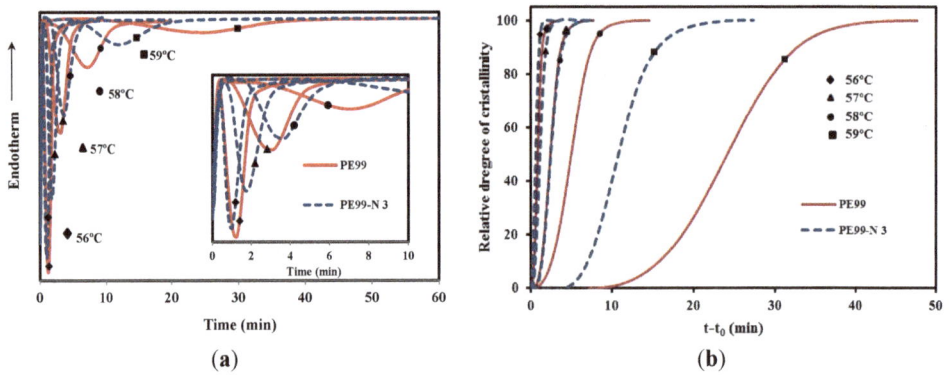

Figure 10. (a) Exothermic DSC peaks of isothermal crystallizations of PE99 (**garnet**) and PE99-N 3 (**blue**) samples at temperatures between 56 and 59 °C; (b) development of the relative degree of crystallinity of PE99 (**garnet**) and PE99-N 3 (**blue**) samples at different crystallization temperatures.

36

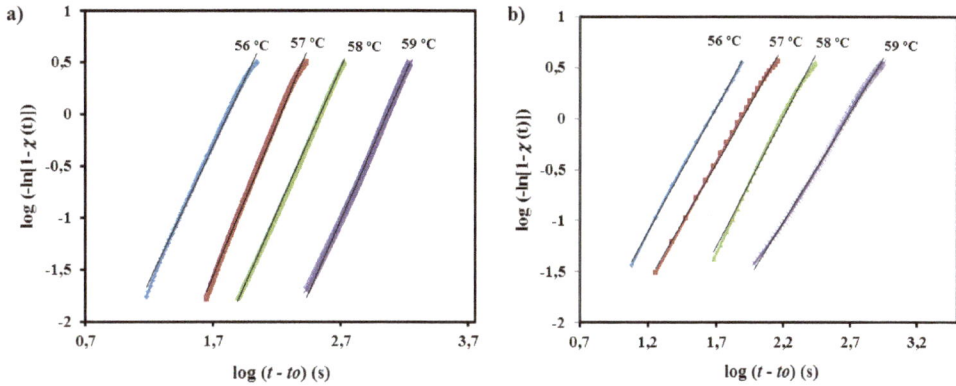

Figure 11. Avrami plots of isothermal crystallization of PE99 (**a**) and PE99-N 3 (**b**) at the indicated temperatures.

Table 3. Main crystallization kinetic parameters determined by DSC for the neat polyester and its nanocomposite with 3 wt. % of functionalized silica nanospheres.

Sample	T (°C)	$Z \times 10^6$ (s^{-n})	n	$k \times 10^3$ (s^{-1})	$1/\tau_{1/2} \times 10^3$ (s^{-1})	$(Z/\ln 2)^{1/n} \times 10^3$ (s^{-1})
PE99	56	18.48	2.60	15.23	17.93	17.5
	57	0.28	2.94	5.89	6.82	6.67
	58	0.077	2.81	2.92	3.31	3.32
	59	0.0028	2.79	0.86	0.97	0.99
PE99-N 3	56	98.56	2.42	21.98	25.97	25.58
	57	42.03	2.31	12.84	15.12	15.05
	58	2.69	2.54	6.37	7.62	7.36
	59	1.89	2.14	2.12	2.54	2.52

Variation of the overall rate constant with temperature for the neat polymer and its nanocomposites was also evaluated (Table 3). The rate for the nanocomposite increased (*i.e.*, from 2.12×10^{-3} s^{-1} to 21.98×10^{-3} s^{-1}) when crystallization temperature decreased (*i.e.*, from 59 °C to 56 °C), a trend that was also logically observed for PE99. More interestingly, the nanocomposite showed a remarkably higher crystallization rate than the neat polyester at all test crystallization temperatures (e.g., 21.98×10^{-3} s^{-1} and 15.23×10^{-3} s^{-1} were determined at 56 °C for PE99-N 3 and PE99, respectively). Therefore, incorporation of functionalized silica nanoparticles had a significant influence on the crystallization process and, logically, on the final material properties. It should be pointed out that the ratio between the two kinetic constants decreased (*i.e.*, between 2.5 and 1.4) as crystallization temperature decreased, which justifies further studies on nucleation and crystal growth processes.

3.4. Optical Microscopy Studies on the Influence of Functionalized Silica Nanoparticles on the Isothermal Crystallization of Poly(nonamethylene Azelate)

Crystallization kinetics from the melt state was also studied for micro-molded samples with and without functionalized silica nanospheres by optical microscopy. Spherulite radii grew linearly with time until impingement in both cases, as shown in Figure 12. Crystal growth rates were clearly higher for the nanocomposite at high crystallization temperatures, whereas differences were minimal when this temperature decreased. The relatively high growth rate allowed for collecting experimental data only over a narrow temperature range where crystallization was mainly governed by secondary nucleation (*i.e.*, the typical bell curve of crystal growth rate *versus* temperature could not be obtained). The increase in the crystal growth rate for the nanocomposite is peculiar and suggests favored deposition of molecules onto existing crystal surfaces, as will be discussed. At lower temperatures, this effect seems to be counterbalanced by reduced chain mobility in the presence of silica nanoparticles and spatial constraints imposed by confinement [29,30].

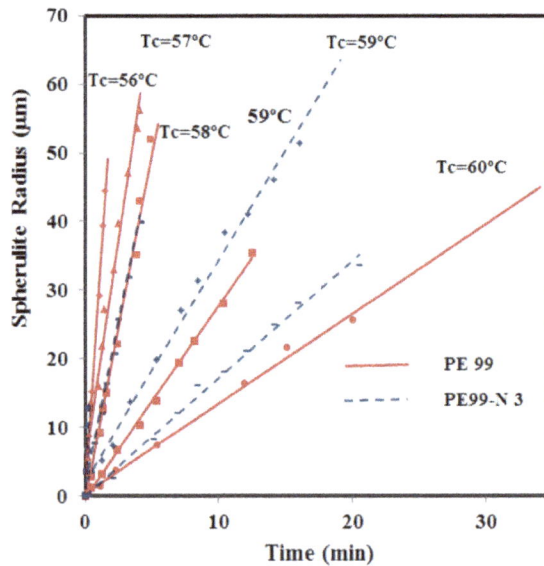

Figure 12. Variation of PE99 (**garnet**) and PE99-N 3 (**blue**) spherulite radii with time for isothermal crystallizations performed at the indicated temperatures.

Pristine polyester and nanocomposite spherulites with similar morphological features were formed (Figure 13). Thus, both samples crystallized from the melt into ringed spherulites with negative birefringence. The spacing between rings decreased significantly with decreasing crystallization temperatures (*i.e.*, 13 μm at 60 °C and

38

5 μm at 56 °C) and decreased slightly upon incorporation of nanospheres (*i.e.*, 5 μm as opposed to 4 μm for crystallizations at 56 °C). A more confusing texture of less defined rings was detected at an intermediate temperature (*i.e.*, 59 °C) for both samples. Note that pristine polyester and nanocomposite spherulites fill the field of view and have a relatively uniform size, suggesting athermal nucleation (*i.e.*, the number of nuclei remains constant during crystallization). More interestingly, this result indicates good dispersion of silica nanoparticles in the polyester matrix. In fact, adsorption of PE99 molecules onto the functionalized surface of silica nanoparticles may hinder particle-particle agglomeration and enhance colloidal stability.

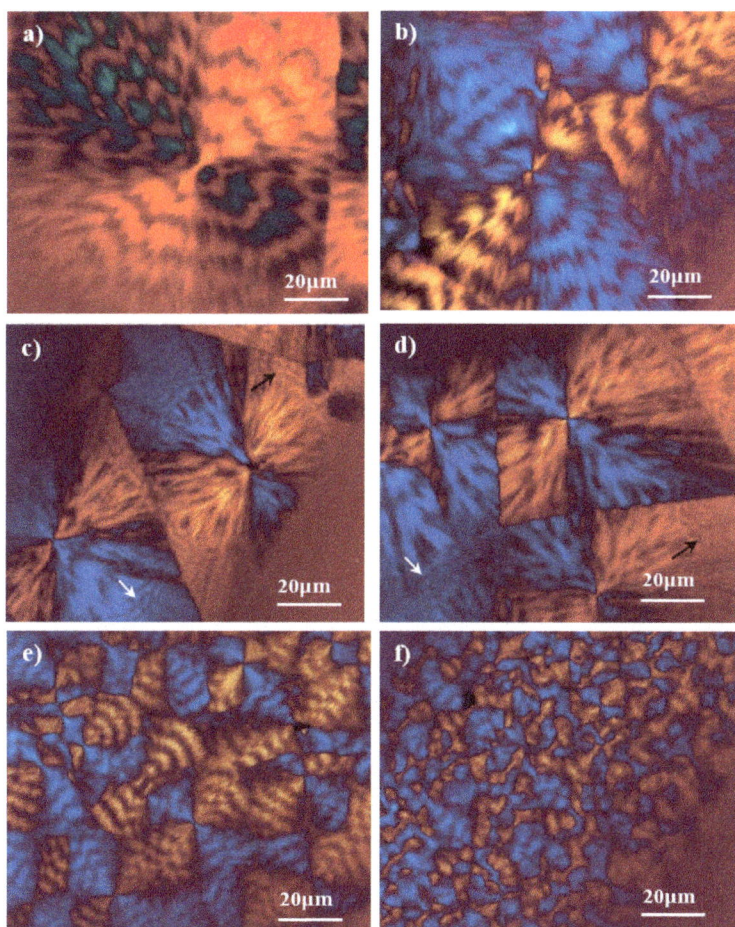

Figure 13. Optical micrographs of PE99 (**a,c,e**) and PE99-N 3 (**b,d,f**) spherulites isothermally grown at 60 °C (**a,b**); first step at 59 °C and second step at 54 °C (**c,d**); and 56 °C (**e,f**). Well-defined rings with interspacing between 3 and 2 μm were detected (see arrows) in the outer part of spherulites grown at 54 °C.

Despite the morphological similarities, great differences were detected in the primary nucleation (Figure 14). They were more remarkable at low crystallization temperatures; for example, nucleation densities of 60 and 20 nucleus/mm^2 were determined from the optical micrographs taken at 57 °C. Note that crystal growth rates were similar at this temperature; consequently, the differences determined for the overall crystallization rate (e.g., 0.01284 and 0.00589 s^{-1} for PE99 and PE99-N 3 at 57 °C, respectively) were mainly attributed to a nucleation effect. On the contrary, nucleation densities were similar at high temperatures, while crystal growth rates were clearly different. Therefore, the incorporation of nanoparticles had a strong impact on the overall crystallization rate due to differences in crystal growth rate and primary nucleation densities, which became more significant at high and low crystallization temperatures, respectively. Figure 14 also shows that the nucleation density increased exponentially for the two samples at lower crystallization temperatures. These changes in nucleation logically affected the final spherulite size. Thus, a diameter decrease from 120 μm to 35 μm and from 80 μm to 15 μm was observed for PE99 and PE99-N 3 samples, respectively, when the temperature decreased from 60 °C to 56 °C.

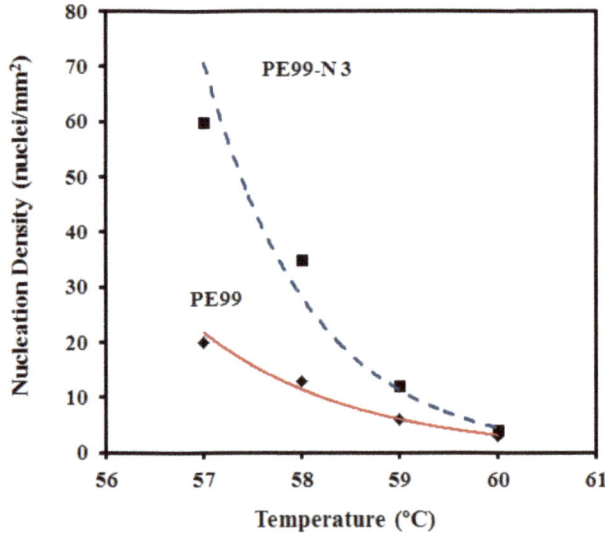

Figure 14. Change in the nucleation density with isothermal crystallization temperature for PE99 (**garnet**) and PE99-N 3 (**blue**) samples.

The logarithmic form of the Lauritzen and Hoffman equation [31] was employed to estimate the secondary nucleation constant (K_g):

$$\ln G + U^*/R(T_c - T_\infty) = \ln G_0 - K_g/(T_c(\Delta T)f) \tag{3}$$

where G is the radial growth rate, T_c is the crystallization temperature, T_∞ is the temperature below molecular motion ceases, ΔT is the degree of supercooling, f is a correction factor calculated as $2T_c/(T_m + T_c)$, U^* is the activation energy, G_0 is a constant preexponential factor, R is the gas constant, and K_g is the secondary nucleation constant.

The Lauritzen-Hoffman plot was fitted with straight lines ($r^2 > 0.97$) for micro-molded PE99 and PE99-N 3 samples when the "universal" values reported by Suzuki and Kovacs [32] (*i.e.*, $U^* = 1500$ cal/mol and $T_\infty = T_g - 30$ K) and the experimental T_g of PE99 (*i.e.*, $-45\ ^\circ$C) were used in the calculation (Figure 15).

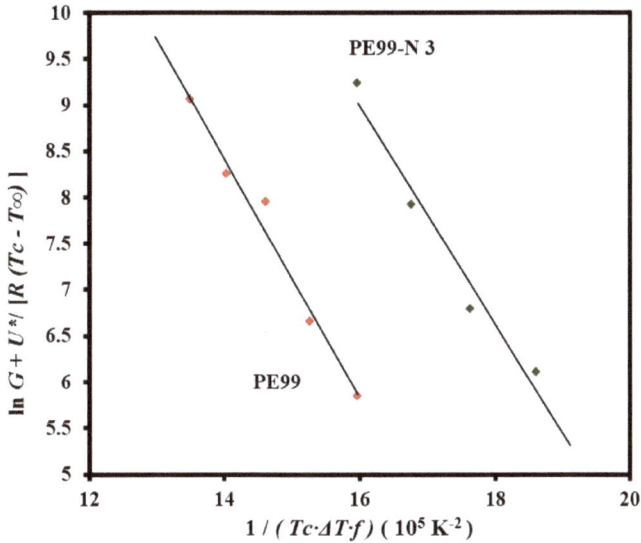

Figure 15. Plot of $\ln G + U^*/R(T_c - T_\infty)$ *versus* $1/T_c(\Delta T)f$ to determine the K_g secondary nucleation parameter of PE99 (**garnet**) and PE99-N 3 (**blue**) samples.

The nucleation term (deduced from the slope of the plot) mainly influenced the kinetic features at low supercoolings in such a way that crystallization rates could become relatively insensitive to the U^* and T_∞ parameters. Therefore, the equilibrium melting temperature was determined since it influenced the degree of supercooling, and consequently the nucleation term. Typical Hoffman-Weeks plots [33] were made with samples crystallized at different temperatures (not shown), leading to equilibrium temperatures of 79.2 $^\circ$C and 76.4 $^\circ$C for PE99 and PE99-N 3 samples, respectively. The slight change suggests less perfect lamellae upon addition of functionalized silica nanoparticles. The Lauritzen-Hoffman plot allowed for estimating secondary nucleation constants of 1.30×10^5 and 1.18×10^5 K^2 for PE99 and PE99-N 3 samples, respectively. These values indicate that the presence of functionalized nanoparticles favored the crystallization process. Thus, enhanced PE99/SiO$_2$ interfacial interaction may decrease the energy involved in the folding of polyester chains and promote their deposition on existing crystal surfaces during the secondary nucleation process [26,34].

4. Conclusions

Poly(nonamethylene azelate) and its mixtures with functionalized silica micro/nanoparticles could be micro-molded by means of ultrasonic energy and using similar time, amplitude, and force processing parameters. Minimal polymer degradation was detected in the molded specimens, as well as a dispersion of added particles. Furthermore, thermal stability was slightly improved by the addition of silica particles. Decomposition of grafted functional groups was only detected at the end of the thermal degradation process.

Silica nanoparticles had a significant influence on the crystallization process even at a low content of 3 wt. %. This point is important because crystallinity of samples needs to be controlled and specifically gives information about the mold temperature of the ultrasonic equipment that should be established to achieve a determined degree of crystallinity. Avrami exponents slightly decreased compared to the neat polymer as evidence of geometric constraints caused by incorporation of nanospheres. The overall crystallization rate of nanocomposites was always greater than that determined for the neat polymer, with the ratio of the respective constants ranging between 1.4 and 2.5.

Negative and ringed spherulites were obtained at all test temperatures, according to an athermal nucleation process. Primary nucleation was significantly enhanced by the addition of silica nanoparticles, whose functionalized surface was expected to favor the adsorption of polyester molecules. In addition, enhanced PE99/SiO$_2$ interfacial interaction favored molecular deposition on existing crystal surfaces, causing a decrease in the secondary nucleation constant and an increase in the crystal growth rate.

Acknowledgments: This research was performed in the framework of an INNPACTO project "IPT-2011-0876-420000" and was also supported by grants from MINECO/FEDER and AGAUR (MAT2012-36205, 2014SGR188). We are grateful to Xavier Planta and Jordi Romero from Ultrasion S.L. and the Fundació ASCAMM, respectively, for their technical support. Gonzalo López has also contributed to experimental work concerning the preparation and functionalization of silica particles.

Author Contributions: Angélica Diáz performed the experiments; María Teresa Casas and Jordi Puiggalí directedthe research; Jordi Puiggalí wrote the manuscript. The manuscript was finalized through contributions from all authors, and all authors also approved the final manuscript.

Conflicts of Interest: The authors declare no conflict of interest.

References

1. Fairbanks, H.V. Applying ultrasonics to the moulding of plastic powders. *Ultrasonics* **1974**, *12*, 22–24.

2. Paul, D.W.; Crawford, R.J. Ultrasonic moulding of plastic powders. *Ultrasonics* **1981**, *19*, 23–27.

3. Werner, K.S.; Burlage, K.; Gerhardy, C. Ultrasonic Hot Embossing. *Micromachines* **2011**, *2*, 157–166.

4. Khuntontong, P.; Blaser, T.; Maas, D.; Schomburg, W.K. Fabrication of a polymer micro mixer by ultrasonic hot embossing. In Proceedings of the 19th Micro-Mechanics Europe Workshop, MME 2008, Aachen, Germany, 28–30 September 2008.

5. Khuntontong, P.; Blaser, T.; Schomburg, W.K. Fabrication of molded interconnection devices by ultrasonic hot embossing of thin films. *IEEE Trans. Electron. Packag. Manuf.* **2009**, *32*, 152–156.

6. Michaeli, W.; Spennemann, A.; Gartner, R. New plastification concepts for micro injection moulding. *Microsyst. Technol.* **2002**, *8*, 55–57.

7. Michaeli, W.; Starke, C. Ultrasonic investigations of the thermoplastics injection moulding process. *Polym. Test.* **2005**, *24*, 205–209.

8. Michaeli, W.; Kamps, T.; Hopmann, C. Manufacturing of polymer micro parts by ultrasonic plasticization and direct injection. *Microsyst. Technol.* **2011**, *17*, 243–249.

9. Chen, J.; Chen, Y.; Li, H.; Lai, S.Y.; Jow, J. Physical and chemical effects of ultrasound vibration on polymer melt in extrusion. *Ultrason. Sonochem.* **2010**, *17*, 66–71.

10. Chen, G.; Guo, S.; Li, H. Ultrasonic improvement of rheological behavior of polystyrene. *J. Appl. Polym. Sci.* **2002**, *84*, 2451–2460.

11. Kang, J.; Chen, J.; Cao, Y.; Li, H. Effects of ultrasound on the conformation and crystallization behavior of isotactic polypropylene and β-isotactic polypropylene. *Polymer* **2010**, *51*, 249–256.

12. Cao, Y.; Li, H. Influence of ultrasound on the processing and structure of polypropylene during extrusion. *Polym. Eng. Sci.* **2002**, *42*, 1534–1540.

13. Sacristán, M.; Plantá, X.; Morell, M.; Puiggalí, J. Effects of ultrasonic vibration on the micro-molding processing of polylactide. *Ultrason. Sonochem.* **2014**, *21*, 376–386.

14. Planellas, M.; Sacristán, M.; Rey, L.; Olmo, C.; Aymamí, J.; Casas, M.T.; del Valle, L.J.; Franco, L.; Puiggalí, J. Micro-molding with ultrasonic vibration energy: New method to disperse nanoclays in polymer matrices. *Ultrason. Sonochem.* **2014**, *21*, 1557–1569.

15. Díaz, A.; Casas, M.T.; del Valle, L.J.; Aymamí, J.; Olmo, C.; Puiggalí, J. Preparation of micro-molded exfoliated clay nanocomposites by means of ultrasonic technology. *J. Polym. Res.* **2014**, *21*, 584–596.

16. Usuki, A.; Kojima, Y.; Kawasumi, M.; Okada, A.; Fukushima, Y.; Kurauchi, T.; Kamigaito, O. Mechanical properties of nylon 6-clay hybrid. *J. Mater. Res.* **1993**, *8*, 1185–1189.

17. Kojima, Y.; Usuki, A.; Kawasumi, M.; Okada, A.; Kurauchi, T.; Kamigaito, O. Synthesis of nylon 6-clay hybrid by montmorillonite intercalated with ε-caprolactam. *J. Polym. Sci. Part A* **1993**, *31*, 983–986.

18. Kojima, Y.; Usuki, A.; Kawasumi, M.; Okada, A.; Kurauchi, T.; Kamigaito, O. One-pot synthesis of nylon 6-clay hybrid. *J. Polym. Sci. Part A* **1993**, *31*, 1755–1758.

19. Burgaz, E. Poly(ethylene-oxide)/clay/silica nanocomposites: Morphology and thermomechanical properties. *Polymer* **2011**, *52*, 5118–5126.

20. Choi, M.; Kim, C.; Jeon, S.O.; Yook, K.S.; Lee, J.Y.; Jang, J. Synthesis of titania embedded silica hollow nanospheres *via* sonication mediated etching and re-deposition. *Chem. Commun.* **2011**, *47*, 7092–7094.

21. Yang, F.; Ou, Y.; Yu, Z. Polyamide 6/silica nanocomposites prepared by *in situ* polymerization. *J. Appl. Polym. Sci.* **1998**, *69*, 355–361.

22. Xu, X.; Li, B.; Lu, H.; Zhang, Z.; Wang, H. The effect of the interface structure of different surface-modified nano-SiO$_2$ on the mechanical properties of nylon 66 composites. *J. Appl. Polym. Sci.* **2008**, *107*, 2007–2014.

23. Vassiliou, A.A.; Papageorgiou, G.Z.; Achilias, D.S.; Bikiaris, D.N. Non-isothermal crystallisation kinetics of *in situ* prepared poly(ε-caprolactone)/surface-treated SiO$_2$ nanocomposites. *Macromol. Chem. Phys.* **2007**, *208*, 364–376.

24. Busbee, J.D.; Juhl, A.T.; Natarajan, L.V.; Tongdilia, V.P.; Bunning, T.J.; Vaia, R.A.; Braun, P.V. SiO$_2$ Nanoparticle Sequestration via Reactive Functionalization in Holographic Polymer-Dispersed Liquid Crystals. *Adv. Mater.* **2009**, *21*, 3659–3662.

25. Li, Z.; Barnes, J.C.; Bosoy, A.; Stoddart, J.F.; Zink, J.I. Mesoporous silica nanoparticles in biomedical applications. *Chem. Soc. Rev.* **2012**, *41*, 2590–2605.

26. Lee, E.; Hong, J.Y.; Ungar, G.; Jang, J. Crystallization of poly(ethylene oxide) embedded with surface-modified SiO$_2$ nanoparticles. *Polym. Int.* **2013**, *62*, 1112–1122.

27. Avrami, M. Kinetics of phase change. I General Theory. *J. Chem. Phys.* **1939**, *7*, 1103–1112.

28. Avrami, M. Kinetics if phase change. II Transformation time relations for random distribution of nuclei. *J. Chem. Phys.* **1940**, *8*, 212–224.

29. Kennedy, M.; Turturro, G.; Brown, G.R.; St-Pierre, L.E. Silica retards radial growth of spherulites in isotactic polystyrene. *Nature* **1980**, *287*, 316–317.

30. Nitta, K.H.; Asuka, K.; Liu, B.; Terano, M. The effect of the addition of silica particles on linear spherulite growth rate of isotactic polypropylene and its explanation by lamellar cluster model. *Polymer* **2006**, *47*, 6457–6463.

31. Lauritzen, J.I.; Hoffman, J.D. Extension of theory of growth of chain folded polymer crystals to large undercoolings. *J. Appl. Phys.* **1973**, *44*, 4340–4352.

32. Suzuki, T.; Kovacs, A.J. Temperature dependence of spherulitic growth rate of isotactic polystyrene. A critical comparison with the kinetic theory of surface nucleation. *Polym. J.* **1970**, *1*, 82–100.

33. Hoffman, J.D.; Weeks, J.J. Melting process and the equilibrium melting temperature of polychlorotrifluoroethylene. *J. Res. Natl. Bur. Stand.* **1962**, *66*, 13–28.

34. Wang, K.; Wu, J.; Zeng, H. Radial growth rate of spherulites in polypropylene/barium sulfate composites. *Eur. Polym. J.* **2003**, *39*, 1647–1652.

Enhancement of Spatial Resolution Using a Metamaterial Sensor in Nondestructive Evaluation

Adriana Savin, Alina Bruma, Rozina Steigmann, Nicoleta Iftimie and Dagmar Faktorova

Abstract: The current stage of non-destructive evaluation techniques imposes the development of new electromagnetic methods that are based on high spatial resolution and increased sensitivity. Printed circuit boards, integrated circuit boards, composite materials with polymeric matrix containing conductive fibers, as well as some types of biosensors are devices of interest in using such evaluation methods. In order to achieve high performance, the work frequencies must be either radiofrequencies or microwaves. At these frequencies, at the dielectric/conductor interface, plasmon polaritons can appear, propagating between conductive regions as evanescent waves. Detection of these waves, containing required information, can be done using sensors with metamaterial lenses. We propose in this paper the enhancement of the spatial resolution using electromagnetic methods, which can be accomplished in this case using evanescent waves that appear in the current study in slits of materials such as the spaces between carbon fibers in Carbon Fibers Reinforced Plastics or in materials of interest in the nondestructive evaluation field with industrial applications, where microscopic cracks are present. We propose herein a unique design of the metamaterials for use in nondestructive evaluation based on Conical Swiss Rolls configurations, which assure the robust concentration/focusing of the incident electromagnetic waves (practically impossible to be focused using classical materials), as well as the robust manipulation of evanescent waves. Applying this testing method, spatial resolution of approximately $\lambda/2000$ can be achieved. This testing method can be successfully applied in a variety of applications of paramount importance such as defect/damage detection in materials used in a variety of industrial applications, such as automotive and aviation technologies.

Reprinted from *Appl. Sci.* Cite as: Savin, A.; Bruma, A.; Steigmann, R.; Iftimie, N.; Faktorova, D. Enhancement of Spatial Resolution Using a Metamaterial Sensor in Nondestructive Evaluation. *Appl. Sci.* **2015**, *5*, 1412–1430.

1. Introduction

In recent years, several nondestructive evaluation (NDE) techniques have been developed for detecting the effect of damages/embedded objects in homogeneous media.

The electromagnetic nondestructive evaluation (eNDE) of materials consists in the application of an electromagnetic (EM) field with frequencies ranging from tens of Hz to tens of GHz, to the examined object and evaluating the interaction between the field and the eventually material discontinuities. A generic NDE system is presented in Figure 1.

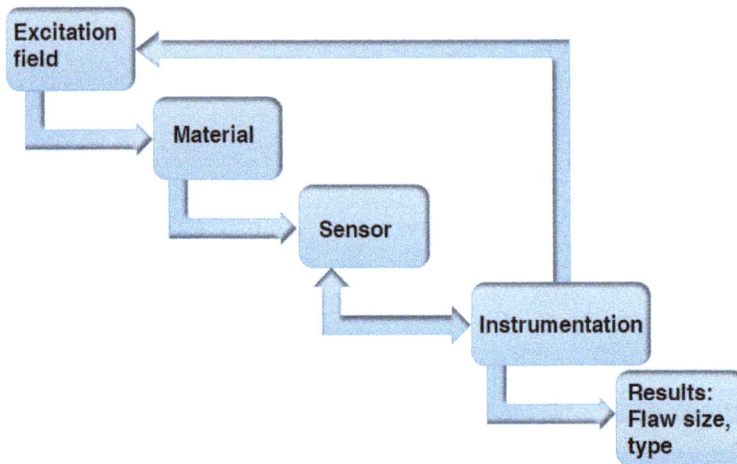

Figure 1. Principles of nondestructive evaluation operation.

Effective data acquisition and display capabilities have led to developments in extraction and recording information about discontinuities and material properties. Fundamentally, these methods were involved with the evolution of reflected and/or transmitted waves after interacting with the test part. If the examined object is electrically conductive, under the action of the incident EM field, eddy currents will be induced in the material, according to Faraday's Law [1].

If the incident EM field has low frequency, usually created by coils circulated by alternating current [2–9] or current impulses [10–13], different shapes of coils [14] assure the detection of the secondary EM field. In order to obtain a better signal/noise ratio [15–19], it is necessary to use the smallest possible lift-off (distance between the EM transducer and the controlled piece). This requires working in the near field because the generated and scattered EM waves are evanescent waves (waves that are rapidly attenuated with distance [20] and are difficult to be focalized using classical materials).

Recently, a new approach to the design of EM structures has been introduced, where the paths of EM waves are controlled by modifying constitutive parameters through a variation of space features. The interaction of the EM field with periodical metallic structures constitutes an interesting domain, both from a theoretical point of view as well as experimentally [21–24]. The metallic strip gratings (MSG) can act as filters and polarizers [25], also representing the basic elements in rigid and flexible printed circuits boards, serving to supply and to the transmission of the signals. MSGs present special properties when they are excited with a transversal electric/magnetic along the z axis (TEz/TMz) polarized EM field. These structures are intensively studied from a theoretical point of view [26,27], for obtaining complex information about their behavior in electronic applications [28] and the design of new types of metamaterials (MMs) starting from the existence of surface plasmons polaritons (SPPs) [29]. These applications impose a rigorous and rapid quality control.

Extending the study to composite materials, carbon fiber reinforced plastics (CFRP) with uniaxial symmetry [30], as well in the case of reinforcement with carbon fibers woven [7,31], the possibility of manipulating the evanescent waves that appear in the space between fibers indicates the necessity to improve the spatial resolution of the sensors, to emphasize microscopic flaws.

The composite materials found wide usage in modern technologies due to the development of industries [32], ranging from medicine and sporting goods to aeronautics components. These composite components can have various dimensions, from very small panels used in satellite structures to large ones used in miniature naval vessels as hulls, 30 m long and 25 mm thick.

Fiber-reinforced polymer composites (FRPC) have superior mechanical properties to metallic structures, assuring simple manufacturing of layered products [33]. FRPCs are classified by the type of reinforcement fibers (carbon, glass, or aramids) or the type of matrix (thermoset or thermoplastic). Despite the fact that thermoplastic composites are more expensive, they have been preferred in the construction of complex structures, due to the advantages of thermoplastic matrix such as recyclability, aesthetic finishing, high impact resistance, chemical resistance, hard crystalline or rubbery surface options, and eco-friendly manufacturing, making the entire process cost less [34]. Woven carbon fibers are recommended due to their high strength-to-weight ratio. Polyphenylene sulfide (PPS) has excellent properties [35], tying into the advantages described above. Woven carbon fibers/PPS laminates are characterized by reduced damages but are susceptible to impacts with low energies, leading to delaminations, desbonding of carbon fibers, and/or matrix cracking [36]. The aerospace industry requires the highest quality control and product release specifications [37,38]. The raw material standard, prepreg materials, require mechanical property testing (*i.e.*, interlaminar strength and tensile strength) from specimens. Final parts also require NDE [39] and structural health monitoring (SHM) [40,41]. The presence of different types of defects such as voids, inclusions, desbondings, improper cure, and delaminations are common during the manufacture and use of composite materials.

Alternative techniques based on phenomenological changes in the composite materials were developed to measure damage due to impacts: acoustic emission [42,43], infrared imaging [44,45], electrical resistance [46,47], or non-contact techniques such as digital image correlation [48] and X-ray tomography [49,50], but these methods have their limitations, being complementary. Nowadays, NDE methods are developed for post-damage inspection and assessment such as thermography [51,52], ultrasonic C-scan [53,54], eddy current [55,56], and optical fiber [57,58], offering quantitative results. Other techniques can be used for online health monitoring of CFRP structures, embedding external sensors or additional fiber input in CFRPs.

This paper proposes the possibility to enhance the spatial resolution of eNDE methods applied to MSG and to CFRP, using a sensor with MM lenses. For this, special attention is granted to the sensor based on a lens [59,60] with conical Swiss rolls, which allows for the manipulation of evanescent waves created in slits and in the dielectric insulating the carbon fibers, respectively, and can reach a spatial resolution for visualization of carbon fibers' layout and eventually of flaws such as delamination created by impact.

2. Metamaterial Sensor for eNDE and Theory

MMs, EM structures with distinguished properties, have started to be studied especially in the last few years. EM MMs belong to the class of artificially engineered materials, which can provide an engineered response to EM radiation, not available from naturally occurring materials. These are often defined as the structures of metallic and/or dielectric elements, periodically arranged in two or three dimensions [61]. The size of the structure is typically smaller than the free space wavelength of incoming EM waves. Nowadays, multitudes of MM structural elements type are known, conferring special EM properties. Depending on the frequency of the incident EM field, the type and geometrical shape of the MM may have a high relative magnetic permeability, either positive or negative [61]. Also, MM lenses allow the amplification of evanescent waves [62]. These properties strongly depend on the geometry of MM rather than their composition [63], and were experimentally demonstrated [64]. MMs have started to interest engineers and physicists due to their wide application in perfect lens [65], slow light [66], data storage [67], *etc.*

In order to find the effective permittivity and permeability of a slab of MM, the material has to be homogeneous. The permittivity and permeability can be found from the S parameters data. For a MM slab characterized by effective permittivity ε_{eff} and effective magnetic permeability μ_{eff}, the refractive index is:

$$n = \sqrt{\varepsilon_{eff}\mu_{eff}} \tag{1}$$

and the impedance is given by:

$$Z = \sqrt{\frac{\mu_{\text{eff}}}{\varepsilon_{\text{eff}}}} \qquad (2)$$

The relationship between ε and μ for a medium as well as the wave propagation through it can categorize the media into the following classes [68,69]:

(a) Double positive medium (DSP) when $\varepsilon > 0$ and $\mu > 0$; Only propagating waves;
(b) Single negative medium-electric negative (ENG), when $\varepsilon < 0$ and $\mu > 0$; only evanescent waves;
(c) Double negative medium (DNG) when $\varepsilon < 0$ and $\mu < 0$; propagating waves and evanescent waves;
(d) Single negative medium-magnetic negative (MNG), when $\varepsilon > 0$ and $\mu < 0$; only propagating waves.

When the effective electrical permittivity ε_{eff}, and the effective magnetic permeability μ_{eff} of a MM slab are simultaneously -1, the refractive index of the slab is $n = -1$ [70]. Therefore, the surface impedance of such an MM is $Z = 1$, there is no mismatch and consequently no reflection at the slab-air interface [64]. This MM slab forms a perfect lens [62] and is not only focusing the EM field, but also focuses the evanescent waves [62].

Due to experimental difficulties in obtaining a perfect lens, the manipulation of the evanescent modes can be made with an EM sensor with MM lenses that have, at the operation frequency, either $\varepsilon_{\text{eff}} = -1$ when electric evanescent modes can be manipulated, or $\mu_{\text{eff}} = -1$ when the lens can focus magnetic evanescent modes [61]. Moreover, working at a frequency that ensures $\mu_{\text{eff}} = -1$ for the same lens, the magnetic evanescent modes can be manipulated [20,71].

According to the above classes, as shown in Ref. [30], the electric evanescent modes can be manipulated with a sensor made from a special MM, named Conical Swiss Roll (CSR) [31], functioning in a frequency range that ensures the maximum μ_{eff} [60]; it did not accomplish the conditions imposed for a perfect lens but will lead to the substantial enhancement of spatial resolution. The spatial resolution of the system (the distance between two distinctively visible points) was verified according to [59] and the analysis of data obtained shows that the realization of MM lenses in the RF range is possible using the CSR, whose distortions are minimal and whose calculation is made based on Fourier optic principles.

EM sensors with MM lenses have been made using two CSRs, the operation frequencies depending both on the constitutive parameters of MM and the polarization of the incident EM field (TE_z or TM_z). Figure 2 shows the developed sensor with MM [7,59,60]. A CSR consists of a number of spiral-wound layers of an insulated conductor on a conical mandrel [31]. The EM sensor is absolute send-receive type; it has the emission part made from one-turn rectangular coil with

35 × 70 mm dimensions, using Cu wire with a 1 mm diameter. When an EM TM_z polarized waves acts, at normal incidence, the magnetic field being parallel with the y axis such that $H_x = H_z = 0$ and $H_y \neq 0$, in very near field, between carbon fibers similar with MSG, evanescent waves can appear. In the focal image point a reception coil with MM lens is placed, having one turn with 1 mm average diameter made of Cu wire with 0.1 mm diameter, to convert localized energy in electromotive force (e.m.f.). An optimized work frequency of 476 MHz assures magnetic effective permeability of 22. At this frequency, the property of CSR to act as an alternative magnetic flux concentrator has been verified.

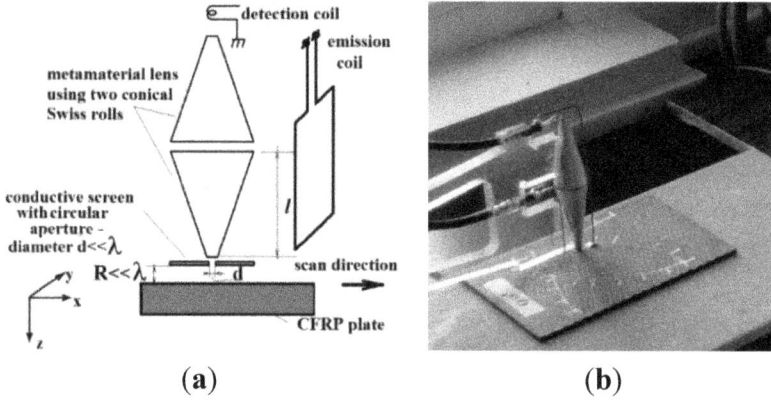

(a)   **(b)**

Figure 2. Sensor with MM lens: (**a**) schematic representation; (**b**) photograph.

The calculus of the MM lens based on CSR was presented in Ref. [31]; the field in the focal plane of the lens is given by

$$H\left(x,y,z_0+2l\right) = \frac{H_0}{\pi^2}e^{j\left(\frac{k_b+k_a}{2}x+\frac{k_b+k_a}{2}y\right)}\frac{\sin\left(\frac{k_b-k_a}{2}x\right)}{x}\frac{\sin\left(\frac{k_b-k_a}{2}y\right)}{y}, \tag{3}$$

where $z_0 = R$ (Figure 2a) with $R \ll \lambda$, and H_0 is the amplitude of incident magnetic field.

The principle scheme of sensor for the evanescent wave's detection is shown in Figure 2a. In order to enhance the spatial resolution of the sensor, a conductive screen having a circular aperture made from perfect electric conductor (PEC) material with a very small diameter is placed in front of the lens. The circular aperture serves for the diffraction of the evanescent waves that can occur on slits. This ensures paraxial incident beam. The diameter of focal spot provided by MM lens is given by [72]:

$$D = \frac{4\pi}{k_b - k_a} \tag{4}$$

and is equal with the diameter of the small basis of the conical Swiss roll, *i.e.*, 3.2 mm. The MM lens with CSR will be displaced along the *x*-axis (Figure 2a). When $k_a = 0$ and this value is inserted into Equation (4), k_b is obtained and the field in the focal plane is calculated with Equation (3). The detection principle is similar with near-field EM scanning microscopy (NFESM). NFSEM imaging is a sampling technique, *i.e.*, the sample (in our case MSG or carbon fibers) is probed point by point by raster scanning with the sensor over the sample surface and recording for energy image pixel a corresponding EM signature.

Selecting a region with dimensions $(x_c + x_d) \cdot (x_c + x_d)$, (where x_c and x_d are coordinates for conducting/dielectric material), using the Fourier optics methods [72,73], an object $O(x,y)$ that can represent the eigenmodes in function of the polarization of an incident EM field, has, while passing through the circular aperture and the lens, an image $I(x', y')$, given by [68]:

$$I(x',y') = \frac{1}{\lambda^2 d_1 d_2} \int\limits_{-\infty}^{\infty} \int\limits_{-\infty}^{\infty} \exp\left[i\frac{k\left((x'-x_1)^2+(y'-y_1)^2\right)}{2d_2}\right] P(x,y) \exp\left[i\frac{k\left(x_1^2+y_1^2\right)}{2f}\right] \times$$

$$\left(\int\limits_{-\infty}^{\infty} \int\limits_{-\infty}^{\infty} O(x,y) \exp\left[i\frac{k\left((x_1-x)^2+(y_1-y)^2\right)}{2d_1}\right] dxdy\right) dx_1 dy_1 \tag{5}$$

where $P(x,y)$ is the pupil function defined as [73]:

$$P(x,y) = \begin{cases} 1 & x^2 + y^2 \le d^2 \\ 0 & \text{otherwise} \end{cases} \tag{6}$$

$O(x,y)$ is the object defined as:

$$O(x,y) = \begin{cases} e_v(x,y) & \text{for} \quad \text{TEz} \quad \text{polarized} \quad \text{incident} \quad \text{waves} \\ h_v(x,y) & \text{for} \quad \text{TMz} \quad \text{polarized} \quad \text{incident} \quad \text{waves} \end{cases} \tag{7}$$

f is the focal distance of the lens equal with the height of CSR; λ is the wavelength in a vacuum; $k = 2\pi/\lambda$ is the wave number; $d_1 = R + l$ is the distance from the object to the center of the lens; and $d_2 = l$ is the distance from the center of the lens to the detecting coil.

The most convenient method from an experimental point of view is to measure S parameters for the MM that fill a waveguide or in free space, using an emission and reception antennas. The relation between S_{11} and S_{21}, applying the effective method [74] using a 4395A Network/Spectrum/Impedance Analyzer Agilent

(Agilent Technologies, Santa Clara, CA, USA) coupled with an Agilent 87511A S Parameters Test kit and effective refractive index n, is given by [75,76]:

$$S_{11} = \frac{R_{0_1}\left(1 - e^{j2nk_0d}\right)}{1 - R_{0_1}^2 e^{j2nk_0d}} \cdot S_{21} = \frac{\left(1 - R_{0_1}^2\right)e^{j2nk_0d}}{1 - R_{0_1}^2 e^{j2nk_0d}} \tag{8}$$

where $R_{0_1} = \frac{Z-1}{Z+1}$ and the impedance Z is obtained by inverting Equation (8) yielding:

$$Z = \pm\sqrt{\frac{\left(1 + S_{11}\right)^2 - S_{21}^2}{\left(1 - S_{11}\right)^2 - S_{21}^2}} \cdot e^{jnk_0d} = X \pm j\sqrt{1 - X^2} \tag{9}$$

where $X = \frac{1}{2S_{21}(1 - S_{11}^2 + S_{21}^2)}$.

The focal distance of the lens using an MM is given in Ref. [31] with $f \cong 1$, where l is the height of a CSR. Assuming that the passive MM slab has an effective refractive index n and impedance Z, according to Ref. [74] the effective permittivity ε_{eff} and permeability μ_{eff} are directly $\mu_{eff} = nZ$ and $\varepsilon_{eff} = \frac{n}{Z}$.

3. Studied Samples and Experimental Setup

The functioning of the MM sensor has been verified using two types of materials, MSGs and FRPC.

3.1. Metallic Strip Gratings (MSGs)

Metallic films on flexible substrates (polyimide or plastic) currently have superior mechanical properties compared to ones deposited on glass, in many aspects. Although the glass substrate is hard, the polyimide or plastic substrate is lighter, less expensive, flexible, and more suitable for use in small devices [77]. Other applications include protective coatings, EM shielding, and electric current conductors for microelectronic applications.

Two types of MSG were taken into consideration:

Flexible printed circuit with transparent polyimide support, 80 μm thickness, with silver conductive strip of 10 μm thickness, the width of the strips being $x_c = 0.6$ mm and the width of the slits being $x_d = 0.4$ mm;

Silver strips realized with polyimide support of 65 μm thickness made from a silver strip having 14 μm thickness, the width of the strips being $x_c = 1.2$ mm and the width of the slits being $x_d = 0.8$ mm.

For strips having $x_c = 1.2$ mm, we have taken into account interruptions as well as non-alignments. Silver strips were realized by successive deposition of silver paste, using an adequate stencil by screen printed method. The adhesion of silver on polyimide has been done with a thin film of resin. The conductive silver paste made from microparticles with concentration >80%, density 10.49 g/cm^3, resistivity $1 \div 3 \times 10^{-5}$ Ω cm was used [78]. The silver paste has good adhesion and fast drying speed at room temperature. At frequencies around the value of 500 MHz, the permittivity of silver [79] is $\varepsilon_m = -48.8 + j \cdot 3.16$.

The polyimide as flexible substrate can be easily embedded in 3D structures, also ensuring retention of bioactive species in the developed structure. They satisfy just about all the requirements for electronics applications, being lightweight, flexible, and resistant to heat and chemicals [80]. The flexible MSG structures are presented in Figure 3a,b.

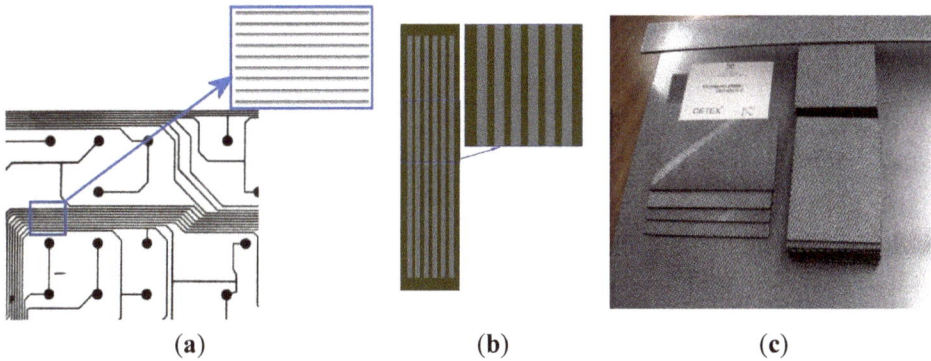

(a)   **(b)**   **(c)**

Figure 3. Studied samples: (**a**) MSGs from flexible printed circuit; (**b**) MSGs realized with polyimide support; (**c**) Plates from FRPC.

For the MSG having the width of strips x_c = 0.6 mm, the study is focused on the appearance of abnormal and/or evanescent modes for the cases where various dielectric fluids fill the gaps between the strips, and also improvement of the EM images to obtain a better spatial resolution in order to exceed the limit imposed by diffraction.

3.2. Plates from FRPC Composite Materials

The study involved quasi-isotropic FRPC plates produced by Tencate [35] (Almelo, The Netherlands), having 150 × 100 × 4.2 mm^3, containing 12 layers of 5 harness satin (5HS) carbon fibers woven with balanced woven fabric [81] (Figure 3c). The matrix is made of PPS, a thermoplastic polymer consisting of aromatic rings linked with sulfide moieties. It is resistant to chemical and thermal attack, and the amount of gas released due to matrix ignition is low. The carbon fibers are T300JB type (TORAYCA, Santa Ana, CA, USA); their volume ratio is 0.5 ± 0.03 and the density is 1460 kg/m^3.

Carbon fiber woven embedded into PPS offers strength of composite to impact. The FRPC has the transverse electric conductivity between 10 S/m and 100 S/m and longitudinal conductivity ranging between 5×10^3 S/m and 5×10^4 S/m and is paramagnetic, allowing eNDE. The samples were impacted with an energy of 8 J [59], in order to induce delaminations. The impacts are induced with FRACTOVIS PLUS 9350 CEAST (Instron, Nordwood, MA, USA), which allows the modification of impact energies as well as the temperature during impact. The conditions of the impact are designed according to ASTM D7136 [82] at a temperature of 20 °C. Under impact, the FRPCs suffer delamination, usually accompanied by a dent, deviation, and/or breaking of the carbon fibers. In all cases, a reduction in the space between fibers in the thickness direction appears and this causes an increase in fiber contact, leading to a decrease of electrical resistance in the thickness direction and modifying the local electrical conductivity both in the plane of the fibers and perpendicularly on fibers [83]. The energy absorbed by the composite serves as the plastic deformation of the composite in the contact zone, being dissipated through internal friction between the matrix's molecules, carbon fibers, and matrix-carbon fibers, as well as at the creation of delaminations. Typical records of force *vs.* time during impact can give information about the FRPC status (delaminated or not) [84].

The generation and detection of evanescent waves from slits/fibers has been made using a sensor with MM lens and the equipment presented in Figure 4. The rectangular frame used for the generation of the TM_z polarized EM field has 35×70 mm dimensions, from a 1 mm diameter Cu wire. This represents the excitation part of the transducer. The lens is constructed from two CSRs with the large bases being placed face to face. The insulated conductor is a copper foil with 18 μm thickness laminated adhesiveless with a polyimide foil (LONGLITE™200, produced by Rogers Corporation (Connecticut, CT, USA)), in order to reduce the losses at high frequencies, with 12 μm thickness (Figure 2b). Each CSR has 1.25 turns, 20 mm diameter large base, 3.2 mm diameter small base, aperture angle of 20°, and 50 mm height. The reception coil with 1 mm diameter and one turn from 0.1 mm diameter Cu wire was placed in the focal plane of the lens. A grounded screen made from the same insulated conductor as the CSR, having a circular aperture with diameter $d = 100$ μm, is placed in front of the lens (Figure 2a). The distance between screen aperture and the surface to be examined has been maintained at 20 ± 1 μm.

The EM sensor with MM lens, presented above, is connected to a Network/ Spectrum/Impedance Analyzer type 4395A Agilent USA Analyzer (Agilent Technologies, Santa Clara, CA, USA). During the measurements, the transducer was maintained in fixed position and samples were displaced with a XY displacement system, type Newmark—Newmark Systems Inc. (Santa Margarita, CA, USA) that ensures the raster scanning in plane with ± 10 μm precision and rotation with $\pm 2''$.

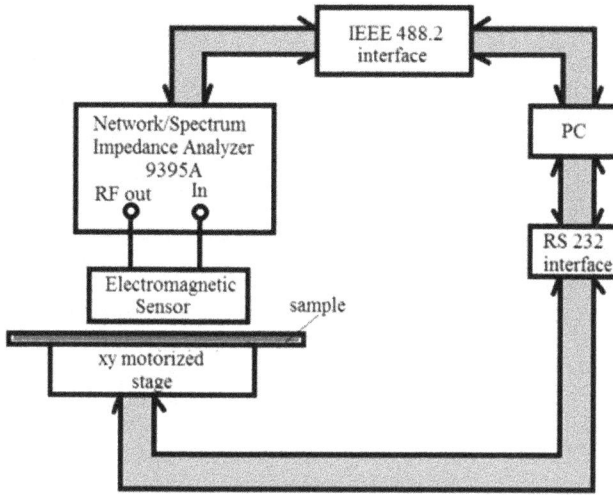

Figure 4. Experimental setup.

The measurement system is commanded via PC through an RS232 interface (National Instruments, Mopac Expwy, Austin, TX, USA) for displacement system and IEEE 488.2 for Network/Spectrum/Impedance Analyzer (National Instruments, Mopac Expwy, Austin, TX, USA). The data acquisition and storage are made by software developed in Matlab 2012b (The MathWorks, Inc., Natick, MA, USA). The e.m.f. induced in the reception coil of the measurement system is the average of 10 measurements at the same point in order to reduce the effects of the white noise, the bandwidth of the analyzer being set to 10 Hz, also to diminish the noise level.

The measurements of S parameters were carried out with Agilent S Parameters Test kit 87511A (Agilent Technologies, Santa Clara, CA, USA) coupled to Agilent 4395A Analyzer.

4. Results and Discussion

The EM sensor used in eNDE must accomplish two roles:
Induce eddy current into the conductive material to be examined, and
Emphasize their flow modifications due to material degradation.

The simplest method to create time-variable magnetic fluxes that can induce eddy current into the material to be examined is represented by the coils circulated by alternative currents, by current impulses, or more alternative currents with different frequencies. The emission part of the EM sensor has this role.

To emphasize the induced eddy current and effect of material degradation over their propagation, sensors sensitive to variation of magnetic field can be used, in our case a sensor with MM lens. The theoretical and experimental study of eigenmodes [71] that appear in the studied MSGs open new domains of applications in the eNDE of stratified structures.

The method and the developed sensor can serve as the eNDE of conductive strips, in order to eventually detect interruptions as well as non-alignments of silver strips or short-circuits between the traces.

According to [59], in the case of MSG with silver strips and sub-wavelength features, excited with TM_z polarized incident EM field, causing a single evanescent mode to appear in the space between strips but disappearing when water is inserted (ε_{water} = 81). This mode could be detected and visualized using the sensor described above.

For MSG with features compatible with the λ value of incident EM field, TM_z polarized; the structure presents known selective properties of transmission and reflection [85]. When an EM TM_z polarized waves acts, at normal incidence, for MSG, according to Ref. [27], the reflection coefficient for a strip grating with $x_0 \ll \lambda_0$ is practically 1.

When MSG represents a layered structure having the features x_c = 0.6 mm, x_d = 0.4 mm, and thickness of 10 μm deposited on polyimide having a relative permittivity of 4.8, the study focuses on the appearance of abnormal and/or evanescent modes for the cases where dielectric fluids fill the gaps between the strip.

Using the sensors with MM lenses, it is experimentally confirmed that, in the space between the strips, evanescent and abnormal modes appear in the case of modes excited with a TM_z polarized EM wave with λ = 0.6 m. The images present an increasing of the amplitude of the signal induced in the reception coil of the evanescent and abnormal modes created in slits at the scanning of a 1×1 mm^2 region with a 10 μm step in both directions. When the slits are filled with air, only the evanescent mode will be generated; the amplitude of the signal induced in the reception coil has the shape presented in Figure 5a.

In the central zone of the slits the amplitude is maximum, followed by an accentuated decreasing towards the flanks of the metallic strips. The existence of a single evanescent mode, foreseen theoretically [7], is experimentally confirmed by the existence of a local maximum in the middle zone of the slits, with maximum amplitude on the middle of the slits, followed by an accentuated decreasing, symmetrically on the flanks of the strip. The two secondary maxima with smaller amplitude are at ±42 μm from the vertical strip wall.

In slits abnormal modes are generated, in the case of excitation with a TM_z polarized wave, for large values of the liquid dielectric constants larger than 10, when ε_d = 17.9, (isopropyl alcohol) (Figure 5b). Because the real component of the propagation constant β_v for isopropyl alcohol is smaller (β_v = 4.2 + i·104.321) than the imaginary component, abnormal modes will be generated in slits; the EM image of these modes shows a similar behavior to the case of air in the slits. The amplitude of the signal is smaller and has a central maximum that is more flat (Figure 5b). This gives new ways for MSGs in sub-subwavelength regime to be used as sensors (as biosensors using evanescent modes generated space between strips and extremely low frequency plasmons).

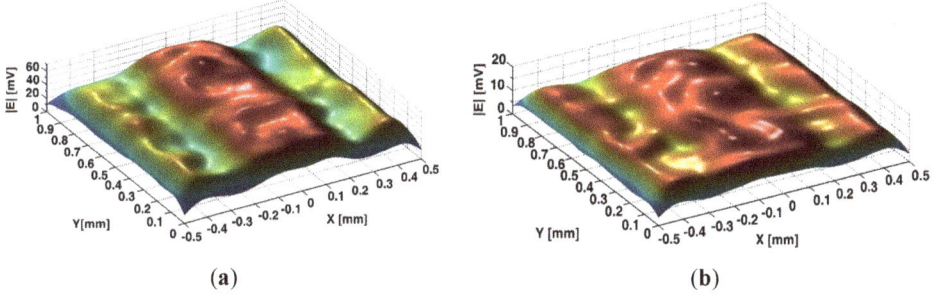

(a)   (b)

Figure 5. Image of evanescent and abnormal modes generated in slits for TMz polarized excitation at frequency of 476 MHz: (**a**) slits filled with air; (**b**) slits filled with isopropyl alcohol.

In order to obtain a silver strip with different widths and thickness in the range of micrometers, different masks/stencils with different microstrip widths were used. Once a metal coating has been applied to a surface, it is critical to determine the adhesion properties of the deposit.

In the case of an MSG made from silver strip having 14 μm thickness, the width of strips where $x_c = 1.2$ mm and the width of slits is $x_d = 0.8$ mm is analyzed, considering that the wavelength of incident field is $\lambda = 0.6$ m (corresponding to a 500 MHz frequency).

A region of 16×16 mm^2 from MSG has been scanned with 0.1-mm steps in both directions. The scanning along the x direction was done to correspond with a few periods of gratings, $x_0 = x_c + x_d$. The working frequency was 476 MHz.

For MSG excitation with EM field TM$_z$ polarized, the simulation was performed using XFDTD 6.3 software produced by REMCOM (State College, PA, USA) [86]. According to Ref. [79], the value of dielectric permittivity of silver is $\varepsilon_m = -48.8 + j\cdot3.16$. Figure 6 shows the dependency of e.m.f. amplitude induced in the reception coil of a sensor on the scanning of the MSG taken into study, the image showing that this type of sensor correctly relies on conductive strips with 14 μm thickness and eventual non-adherence to support and/or interruptions.

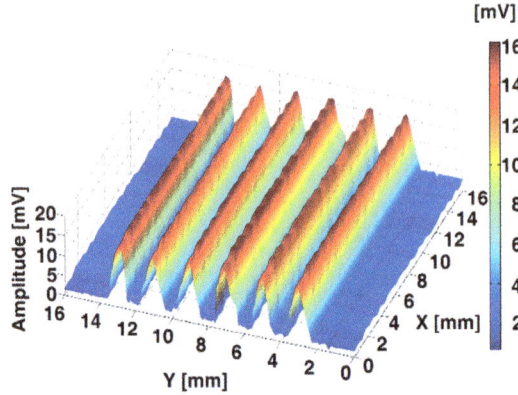

Figure 6. Amplitude of e.m.f. induced in the reception coil at the scanning of silver strip grating.

These results are in good concordance with theoretical estimations. They confirm a good adhesion of the silver paste on polyimide, as well as good alignment of the strips. It is known that the biosensing characteristic is strongly dependent on the deposition condition, which affects the physical properties of the thin film. Using the procedure and sensor described earlier for eNDE of MSG, interruption of a strip stopped the propagation of evanescent waves in the nearest slit so that the amplitude of e.m.f induced in the reception coil practically decreased to zero when the aperture's sensor was in the corresponding region of the slit. The roughness feature is emphasized by the propagation of surface polaritons [29].

For the second type of material taken into study (FRPC), the EM behavior of the composite was simulated by FDTD software; the samples were CAD designed following textiles features, and compared with eNDE tests. One cell of woven carbon fiber has been designed in the TexGenTextile Geometric modeler software (University of Nottingham, Nottingham, UK) (Figure 7a) and exported in CAD format in order to be used in FDTD software (XFDTD REMCOM, State College, PA, USA). The 5 × 5 tows were represented in different colors in order to easily follow their intersection when woven. The cell dimension in FDTD simulation is 0.04 mm, and the grid size is 146 × 146 × 45 cells. The perfectly matched layer boundary condition was applied at the grid boundaries.

The result of the simulation presented in Figure 7b is a snapshot from a field sequence, showing the H_y field progress at a particular slice of the geometry. In this case, the role of conductive strips is taken by carbon fibers, which act as MSG [71]; the apparition of evanescent waves can also be emphasized.

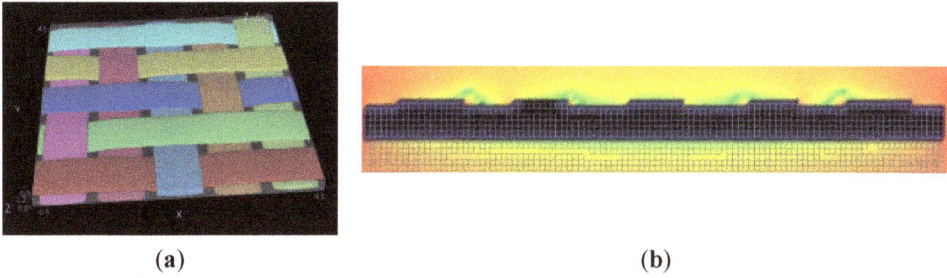

(a)   (b)

Figure 7. Simulation of one woven carbon cell: (a) TexGen—different colors show the intersections in the woven 5HS; (b) XFDTD—H_y propagation in a plane orthogonal at composite.

The detection of eventual delamination or the characterization of carbon fibers' woven structure using the eNDE method and the sensors with MM lens is an emerging nondestructive technique combining the advantages of conventional eddy current testing and evanescent wave detection, giving a higher resolution than a classical eddy current.

The results are shown in Figure 8, which presents a scan with 1-mm step of a 60×60 mm^2 region from the FRPC sample impacted with 8 J, scanned in both directions.

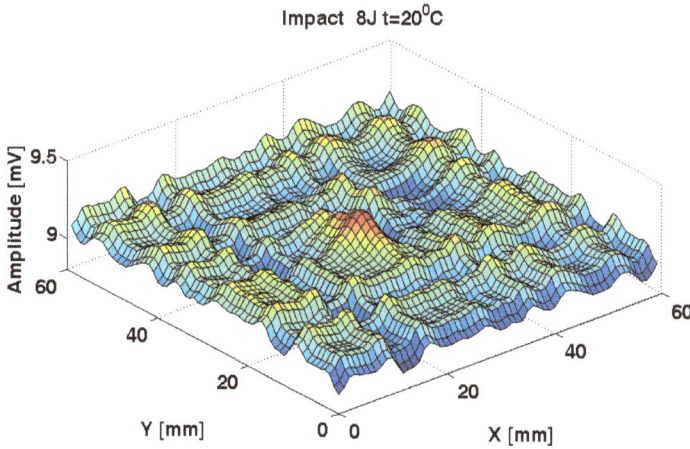

Figure 8. Sensor data with MM lens for 8 J impact.

It can be shown that the MM lens allow the enhancement of spatial resolution, with the layout of the woven being emphasized. The proposed method can thus be extended not only for the evaluation of MSG but also for the eNDE of FRPC in order to evaluate delaminations as well as the woven layout.

59

5. Conclusions

The structures' conductive grating allows, by extension, the estimation of results from eNDE of few real situations such as evaluation of metallic strips in printed circuit, MSG in the sub-wavelength regime as a biosensor, and FRPC.

In order to significantly enhance the spatial resolution of the EM method, the use of evanescent waves that can appear in slits, in the space between FRPC and on the edge of open microscopic cracks, is proposed. Special attention is granted to the MM lens based on CSR configurations, with an optimal frequency that assures the concentration of the incident electromagnetic field as well as the effective manipulation of the evanescent waves. The EM field TMz polarized can be created with a rectangular frame, having the plane perpendicular on the plane of MSG/FRPC and fed with an alternating current.

The use of evanescent waves and sensors with MM lens allows for the manipulation of evanescent waves to reach a spatial resolution of approximately $\lambda/2000$.

The performance of the EM sensors with MM lens is improved regarding sensitivity and spatial resolution by using the evanescent wave that can appear in the space between slits for structures excited with a polarized TMz plane wave.

Acknowledgments: This paper is partially supported by a grant from the Romanian Ministry of Education CNCS-UEFISCDI, project number PN-II-ID-PCE-2012-4-0437.

Author Contributions: Adriana Savin has analyzed the experimental data and wrote of the paper. Adriana Savin, Alina Bruma have contributed to development of the theoretical aspects of the EM sensor with MM lens and functioning principle. Adriana Savin and Rozina Steigmann have designed and realized the EM sensor with MM lens. Nicoleta Iftimie and Dagmar Faktorova performed the experiment and used the EM sensor with MM lens on different MSG in the sub-wavelength regime to involve them as biosensors, and on FRPC to evaluate delamination as well as the woven layout. All of the authors read and approved the final manuscript.

Conflicts of Interest: The authors declare no conflict of interest.

References

1. Bladel, V. *Electromagnetic Fields*, 2nd ed.; IEEE Press: Piscataway, NJ, USA, 2007.
2. Theodoulidis, T.P.; Kriezis, E.E. Impedance evaluation of rectangular coils for eddy current testing of planar media. *NDT E Int.* **2002**, *35*, 407–414.
3. Sablik, M.J.; Burkhardt, G.L.; Kwun, H.; Jiles, D.C. A model for the effect of stress on the low frequency harmonic content of the magnetic induction in ferromagnetic materials. *J. Appl. Phys.* **1988**.
4. Harfield, N.; Yoshida, Y.; Bowler, J.R. Low-frequency perturbation theory in eddy-current non-destructive evaluation. *J. Appl. Phys.* **1996**, *80*, 4090–4100.
5. Theodoulidis, T.P.; Bowler, J.R. Eddy-current interaction of a long coil with a slot in a conductive plate. *IEEE Trans. Magn.* **2005**, *41*, 1238–1247.

6. Grimberg, R.; Udpa, L.; Savin, A.; Steigmann, R.; Palihovici, V.; Udpa, S.S. 2D eddy current sensor array. *NDT E Int.* **2006**, *39*, 264–271.

7. Grimberg, R.; Savin, A.; Steigmann, R. Electromagnetic imaging using evanescent waves. *NDT E Int.* **2012**, *46*, 70–76.

8. Auld, B.A.; Moulder, J.C. Review of advances in quantitative eddy current non-destructive evaluation. *J. Nondestruct. Eval.* **1999**, *18*, 3–36.

9. Li, Y.; Theodoulidis, T.; Tian, G.Y. Magnetic field based multi-frequency eddy current for multilayered specimen characterization. *IEEE Trans. Magn.* **2007**, *43*, 4010–4015.

10. Wang, L.; Xie, S.; Chen, Z.; Li, Y.; Wang, X.; Takagi, T. Reconstruction of stress corrosion cracks using signals of pulsed eddy current testing. *Nondestruct. Test Eva.* **2013**, *28*, 145–154.

11. He, Y.; Pan, M.; Luo, F.; Tian, G. Pulsed eddy current imaging and frequency spectrum analysis for hidden defect nondestructive testing and evaluation. *NDT E Int.* **2011**, *44*, 344–352.

12. Tai, C.C.; Yang, H.C.; Liang, D.S. Pulsed eddy current for metal surface cracks inspection: Theory and experiment. *AIP Conf. Proc.* **2002**, *21*, 388–395.

13. Wilson, J.W.; Tian, G.Y. Pulsed electromagnetic methods for defect detection and characterization. *NDT E Int.* **2007**, *40*, 275–283.

14. Mook, G.; Hesse, O.; Uchanin, V. Deep penetrating eddy currents and probes. *Mater. Test* **2007**, *49*, 258–264.

15. Wong, B.S. *Non-Destructive Testing-Theory, Practice and Industrial Applications*; Lap-Lambert Publishing GmbH KG: Saarbrücken, Germany, 2014.

16. Dodd, C.V.; Deeds, W.E. Analytical Solutions to Eddy-Current Probe-Coil Problems. *J. Appl. Phys.* **1968**, *39*, 2829–2838.

17. Grimberg, R.; Savin, A.; Radu, E.; Mihalache, O. Nondestructive evaluation of the severity of discontinuities in flat conductive materials by an eddy-current transducer with orthogonal coils. *IEEE Trans. Magn.* **2000**, *36*, 299–307.

18. Bowler, J.R.; Jenkins, S.A.; Sabbagh, L.D.; Sabbagh, H.A. Eddy-current probe impedance due to a volumetric flaw. *J. Appl. Phys.* **1991**, *70*, 1107–1114.

19. Bihan, Y.L. Study on the transformer equivalent circuit of eddy current nondestructive evaluation. *NDT E Int.* **2003**, *36*, 297–302.

20. Grbic, A.; Eleftheriades, G.V. Growing evanescent waves in negative-refractive-index transmission-line media. *Appl. Phys. Lett.* **2003**, *82*, 1815–1817.

21. Petit, R.L.; Botten, L.C. *Electromagnetic Theory of Gratings*; Springer-Verlag: Berlin, Germany, 1980.

22. Collin, R.E. *Field Theory of Guided Waves*; IEEE Press: New York, NY, USA, 1991.

23. Munk, B. *Frequency Selective Surfaces: Theory and Design*; John Wiley & Sons, Inc.: New York, NY, USA, 2000.

24. Balanis, C.A. *Antenna Theory: Analysis and Design*, 3rd ed.; John Wiley & Sons, Inc.: Hoboken, NJ, USA, 2005.

25. Brand, G.I. The strip grating as a circular polarizer. *Am. J. Phys.* **2003**, *71*, 452–457.

26. Arnold, M.D. An efficient solution for scattering by a perfectly conducting strip grating. *J. Electromagnet. Wave* **2006**, *20*, 891–900.

27. Peterson, A.F.; Ray, S.L.; Mittra, R. *Computational Methods for Electromagnetics*; IEEE Press: New York, NY, USA, 1998.

28. Porto, J.A.; Garcia-Vidal, F.J.; Pendry, J.B. Transmission resonances on metallic gratings with very narrow slits. *Phys. Rev. Lett.* **1999**, *83*, 2845–2849.

29. Kolomenski, A.; Kolomenskii, A.; Noel, J.; Peng, S.; Schuessler, H. Propagation length of surface plasmons in a metal film with roughness. *Appl. Opt.* **2009**, *48*, 5683–5691.

30. Grimberg, R. Electromagnetic metamaterials. *Mater. Sci. Eng. B* **2013**.

31. Grimberg, R.; Savin, A.; Steigmann, R.; Serghiac, B.; Bruma, A. Electromagnetic non-destructive evaluation using metamaterials. *Insight* **2011**, *53*, 132–137.

32. Pilato, L.A.; Michno, M.J. *Advanced Composite*; Springer Verlag: Berlin, Germany, 1994.

33. Morgan, P. *Carbon Fibers and Their Composites*; CRC Press: Boca Raton, FL, USA, 2005.

34. Vieille, B.; Taleb, L. About the influence of temperature and matrix ductility on the behavior of carbon woven-ply PPS or epoxy laminates: Notched and unnotched laminates. *Compos. Sci. Technol.* **2011**, *71*, 998–1007.

35. TenCate Advanced Composites. Available online: http://www.tencate.com/ (accessed on 10 September 2015).

36. Kaw, A.K. *Mechanics of Composite Materials*, 2nd ed.; CRC Press: Boca Raton, FL, USA, 2006.

37. Standard AS 9100/2009. In *Quality Systems-Aerospace-Model for Quality Assurance in Design, Development, Production, Installation and Servicing*; SAE: Warrendale, PA, USA, 1999.

38. Boller, C.; Meyendorf, N. State-of-the-art in Structural Health monitoring for aeronautics. In Proceedings of the International Symposium on NDT in Aerospace, Fürth/Bavaria, Germany, 3–5 December 2008; pp. 1–8.

39. Elmarakbi, A. Advanced Composite Materials for Automotive Applications: Structural Integrity and Crashworthiness. John Wiley & Sons Ltd.: Chichester, West Sussex, UK, 2013.

40. Boller, C. Next generation structural health monitoring and its integration aircraft design. *Int. J. Syst. Sci.* **2000**, *31*, 1333–1349.

41. Salvado, R.; Lopes, C.; Szojda, L.; Araújo, P.; Gorski, M.; Velez, F.J.; Krzywon, R. Carbon Fiber Epoxy Composites for Both Strengthening and Health Monitoring of Structures. *Sensors* **2015**, *15*, 10753–10770.

42. Kordatos, E.Z.; Aggelis, D.G.; Matikas, T.E. Monitoring mechanical damage in structural materials using complimentary NDE techniques based on thermography and acoustic emission. *Compos. Part B Eng.* **2012**, *43*, 2676–2686.

43. Adden, S.; Pfleiderer, K.; Solodov, I.; Horst, P.; Busse, G. Characterization of stiffness degradation caused by fatigue damage in textile composites using circumferential plate acoustic waves. *Compos. Sci. Technol.* **2008**, *68*, 1616–1623.

44. He, Y.; Tian, G.Y.; Pan, M.; Chen, D. Impact evaluation in carbon fiber reinforced plastic (CFRP) laminates using eddy current pulsed thermography. *Compos. Struct.* **2014**, *109*, 1–7.

45. Mian, A.; Han, X.; Islam, S.; Newaz, G. Fatigue damage detection in graphite/epoxy composites using sonic infrared imaging technique. *Compos. Sci. Technol.* **2004**, *64*, 657–666.

46. Vavouliotis, A.; Paipetis, A.; Kostopoulos, V. On the fatigue life prediction of CFRP laminates using the electrical resistance change method. *Compos. Sci. Technol.* **2011**, *71*, 630–642.

47. Park, J.M.; Kwon, D.J.; Wang, Z.J.; DeVries, K. Nondestructive sensing evaluation of surface modified single-carbon fiber reinforced epoxy composites by electrical resistivity measurement. *Compos. Part B Eng.* **2006**, *37*, 612–626.

48. Seon, G.; Makeev, A.; Cline, J.; Shonkwiler, B. Assessing 3D shear stress-strain properties of composites using Digital Image Correlation and finite element analysis based optimization. *Compos. Sci. Technol.* **2015**, *117*, 371–378.

49. Nikishkov, Y.; Airoldi, L.; Makeev, A. Measurement of voids in composites by X-ray Computed Tomography. *Compos. Sci. Technol.* **2013**, *89*, 89–97.

50. Schilling, P.J.; Karedla, B.R.; Tatiparthi, A.K.; Verges, M.A.; Herrington, P.D. X-ray computed microtomography of internal damage in fiber reinforced polymer matrix composites. *Compos. Sci. Technol.* **2015**, *65*, 2071–2078.

51. Ren, W.; Liu, J.; Tian, G.Y.; Gao, B.; Cheng, L.; Yang, H. Quantitative non-destructive evaluation method for impact damage using eddy current pulsed thermography. *Compos. Part B Eng.* **2013**, *54*, 169–179.

52. Hung, Y.Y.; Chen, Y.S.; Ng, S.P.; Liu, L.; Huang, Y.H.; Luk, B.L.; Ip, R.W.L.; Wu, C.M.L.; Chung, P.S. Review and comparison of shearography and active thermography for nondestructive evaluation. *Mater. Sci. Eng. R Rep.* **2009**, *64*, 73–112.

53. Solodov, I.; Döring, D.; Rheinfurth, M.; Busse, G. New Opportunities in Ultrasonic Characterization of Stiffness Anisotropy in Composite Materials. In *Nondestructive Testing of Materials and Structures*; Springer: Dordrecht, The Netherlands, 2013; Volume 6, pp. 599–604.

54. Kolkoori, S.; Hoehne, C.; Prager, J.; Rethmeier, M.; Kreutzbruck, M. Quantitative evaluation of ultrasonic C-scan image in acoustically homogeneous and layered anisotropic materials using three dimensional ray tracing method. *Ultrasonics* **2014**, *54*, 551–562.

55. Cacciola, M.; Calcagno, S.; Megali, G.; Pellicano, D.; Versaci, M.; Morabito, F.C. Eddy current modeling in composite materials. *PIERS Online* **2009**, *5*, 591–595.

56. Grimberg, R.; Savin, A.; Steigmann, R.; Bruma, A.; Barsanescu, P. Ultrasound and eddy current data fusion for evaluation of carbon-epoxy composite delaminations. *Insight* **2009**, *51*, 1–25.

57. Lamberti, A.; Chiesura, G.; Luyckx, G.; Degrieck, J.; Kaufmann, M.; Vanlanduit, S. Dynamic Strain Measurements on Automotive and Aeronautic Composite Components by Means of Embedded Fiber Bragg Grating Sensors. *Sensors* **2015**, *15*, 27174–27200.

58. Takeda, S.; Okabe, Y.; Takeda, N. Delamination detection in CFRP laminates with embedded small-diameter fiber Bragg grating sensors. *Compos. Part A* **2002**, *33*, 971–980.

59. Savin, A.; Steigmann, R.; Bruma, A.; Šturm, R. An Electromagnetic Sensor with a Metamaterial Lens for Nondestructive Evaluation of Composite Materials. *Sensors* **2015**, *15*, 15903–15920.

60. Grimberg, R.; Savin, A. Electromagnetic Transducer for Evaluation of Structure and Integrity of the Composite Materials with Polymer Matrix Reinforced with Carbon Fibers. Patent RO126245-A0, 2011.

61. Pendry, J.; Holden, A.J.; Robbins, D.J.; Stewart, W.J. Magnetism from conductors and enhanced non-linear phenomena. *IEEE Trans. Microw Theory Tech.* **1999**, *47*, 47–58.

62. Pendry, J.B. Negative Refraction makes a perfect lens. *Phys. Rev. Lett.* **2000**, *85*, 3966–3969.

63. Cai, W.; Chettiar, U.K.; Kildishev, A.V.; Shalaev, V.M. Optical cloaking with metamaterials. *Nat. Photonics* **2007**, *1*, 224–227.

64. Smith, D.R.; Pendry, J.B.; Wiltshire, M.C.K. Metamaterials and negative refractive index. *Science* **2004**, *305*, 788–792.

65. Shelby, R.A.; Smith, D.R.; Schultz, S. Experimental Verification of a Negative Index of Refraction. *Science* **2001**, *6*, 77–79.

66. Bai, Q.; Liu, C.; Chen, J.; Cheng, C.; Kang, M. Tunable slow light in semiconductor metamaterial in a broad terahertz regime. *J. Appl. Phys.* **2010**.

67. Wuttig, M.; Yamada, N. Phase-change materials for rewriteable data storage. *Nat. Mater.* **2007**, *6*, 824–832.

68. Engheta, N.; Ziolkowski, R.W. *Electromagnetic Metamaterials: Physics and Engineering Explorations*; John Wiley & Sons, Inc.: New York, NY, USA, 2006.

69. Zouhdi, S.; Sihvola, A.; Vinogradov, A.P. *Metamaterials and Plasmonics: Fundamentals, Modelling, Applications*; Springer: New York, NY, USA, 2008.

70. Veselago, V.G. The electrodynamics of substances with simultaneously negative values of ϵ and μ. *Phys. Uspekhi* **1968**, *10*, 509–514.

71. Grimberg, R.; Tian, G.Y. High Frequency Electromagnetic Non-destructive Evaluation for High Spatial Resolution using Metamaterial. *Proc. R. Soc. A* **2012**, *468*, 3080–3099.

72. Born, M.; Wolf, E. *Principle of Optics*, 5th ed.; Pergamon Press: Oxford, UK, 1975.

73. Goodman, J.W. *Introduction to Fourier Optics*, 3rd ed.; Roberts and Company: Englewood, CO, USA, 2005.

74. Kong, J.A. *Electromagnetic Wave Theory*; EMW Publishing: Cambridge, MA, USA, 2000.

75. Shelby, R.A.; Smith, D.R.; Nemat-Nasser, S.C.; Schultz, S. Microwave transmission through a two-dimensional, isotropic, left-handed metamaterial. *Appl. Phys. Lett.* **2001**, *78*, 489–491.

76. Chen, X.; Grzegorczyk, T.M.; Wu, B.I.; Pacheco, J., Jr.; Kong, J.A. Robust method to retrieve the constitutive effective parameters of metamaterials. *Phys. Rev. E* **2004**.

77. Prentice, G.A.; Chen, K.S. Effects of current density on adhesion of copper electrodeposits to polyimide substrates. *J. Appl. Electrochem.* **1998**, *28*, 971–977.

78. SPI Supplies. Available online: http://www.2spi.com/ (accessed on 10 October 2015).

79. Palik, E.D. *Handbook of Optical Constants of Solids*; Academic Press: London, UK, 1985.

80. Schlesinger, M. *Deposition on Nonconductors, Chapter 15 on Modern Electroplating*, 5th ed.; John Wiley & Sons: Hoboken, NJ, USA, 2010; pp. 413–420.

81. Akkerman, R. Laminate mechanics for balanced woven fabrics. *Compos. Part B Eng.* **2006**, *37*, 108–116.

82. Standard Test Method for Measuring the Damage Resistance of a Fiber-Reinforced Polymer Matrix Composite to a Drop-Weight Event. ASTM International: West Conshohocken, PA, USA, 2005.

83. Menana, H.; Féliachi, M. Electromagnetic characterization of the CFRPs anisotropic conductivity: Modeling and measurements. *Eur. Phys. J. Appl. Phys.* **2011**.

84. Ullah, H.; Abdel-Wahab, A.A.; Harland, A.R.; Silberschmidt, V.V. Damage in woven CFRP laminates subjected to low velocity impacts. *J. Phys. Conf. Ser.* **2012**.

85. Balanis, C. *Advanced Engineering Electromagnetics*; John Wiley & Sons: Hoboken, NJ, USA, 1989.

86. Kunz, K.S.; Luebbers, R.J. *The Finite Difference Time Domain Method for Electromagnetics*; CRC Press: Boca Raton, FL, USA, 1993.

Monitoring Techniques of Cerium Stabilized Zirconia for Medical Prosthesis

Adriana Savin, Mihail-Liviu Craus, Vitalii Turchenko, Alina Bruma,
Pierre-Antoine Dubos, Sylvie Malo, Tatiana E. Konstantinova and
Valerii V. Burkhovetsky

Abstract: The purpose of this paper is to emphasize the improvement of Zr-based ceramics properties as a function of addition of Ce ions in the structure of the original ceramics. The structural investigations proposed in this paper cover X-ray, and neutron diffraction offered the first indication of the variation of the phase composition and the structural parameters, micro-hardness measurements as well as non-destructive evaluations in order to analyze the structural properties of these materials with utmost importance in fields such as medicine, where these composite materials are used in hip-implants or dental implants/coatings. In combination of Resonant Ultrasound Spectroscopy, which makes use of the resonance frequencies corresponding to the normal vibrational modes of a solid in order to evaluate the elastic constants of the materials, we emphasize a unique approach on evaluating the physical properties of these ceramics, which could help in advancing the understanding of properties and applications in medical fields.

Reprinted from *Appl. Sci.* Cite as: Savin, A.; Craus, M.-L.; Turchenko, V.; Bruma, A.; Dubos, P.-A.; Malo, S.; Konstantinova, T.E.; Burkhovetsky, V.V. Monitoring Techniques of Cerium Stabilized Zirconia for Medical Prosthesis. *Appl. Sci.* **2015**, *5*, 1665–1682.

1. Introduction

Recent developments in advanced dental materials [1,2], solid-oxide fuel-cell design to oxygen detection [3], nuclear waste confinement [4], optics [5], medical prosthesis [6,7], and catalytic [8] technologies have drawn attention towards the remarkable structural properties of zirconia (ZrO_2)-based ceramics. In all the current technological applications, ZrO_2-based ceramics are preferred due to their advanced mechanical properties such as high-fracture toughness and bulk modulus, corrosion resistance, chemical inertness, low chemical conductivity and biocompatibility [9]. It is well known that three polymorphic forms of pure ZrO_2 can be found: the monoclinic state, $P2_1/c$, stable at temperatures below 1170 °C; the tetragonal phase, $P4_2/nmc$ stable in the temperature range between 1170 and 2370 °C; and the cubic, Fm-3m phase, appearing at a temperature above 2370 °C [10]. These ceramics have mechanical properties, promoting them to special applications. Under external stress, as grinding or impact, transition from the tetragonal (t) to monoclinic (m) phase can appear at normal temperatures, being followed by an increase of volume of at least 4%, causing compressive stress. This can cause material failure. Ceria (Ce_2O_3) or yttria (Y_2O_3) are used to stabilize these ceramics at room temperature and allowing t → m transformation to prevent crack propagation.

The main method used for the stabilization of the ZrO_2 tetragonal phase is the introduction of stabilization components in the zirconia lattice, such as Mg, Ce, Fe, Y, *etc.* [11–14]. At nanoscale level, the metastable phase formation in ZrO_2 can be induced by including in the oxide structure some vacancy defects [11]. Although, the stabilization effect of the oxygen vacancies in tetragonal ZrO_2 is not yet well understood (contrary to cubic ZrO_2) [14], the concentration of oxygen vacancies in the lattice required to stabilize the tetragonal phase is found in phases like $ZrO_{1.97}$ and $ZrO_{1.98}$ for tetragonal zirconia doped by rare-earth elements, mainly cerium (Ce) and yttrium (Y).

Our aim is to investigate the influence of the phase change of ZrO_2 ceramics following doping with Ce on the mechanical properties. We investigated the structure of these ceramics and the phase stabilization using X-ray and neutron diffraction, Scanning Electron Microscopy (SEM) as well as microstructure characterization methods, including micro-hardness measurements. Moreover, we employed Resonant Ultrasound Spectroscopy (RUS) as a non-destructive evaluation method in order to estimate the presence of low-density regions, state of sintering, and the presence and development of small cracks in the structure, by evaluating the complete elasticity matrix. RUS appears to be a feasible method for the inspection of zirconia ceramics based on the changes of the resonance frequencies, being easily applied for the quality control procedure of such ceramic samples; essentially monitoring of a single resonance peak within the frequency band included in experiment would be sufficient for this purpose. We emphasize the properties of zirconium-based ceramics for applications in the biomedical field, such as ceramic femoral heads used in hip implant procedures, which are very resistant to scratches resulting from debris caused by accumulation of bone parts, cement, or metal that occasionally fall between artificial joint surfaces, but are extremely fragile.

2. Materials and Methods

The Ce doped zirconia samples have been obtained through the standard ceramic technology [15], a mixture of ceria and zirconia oxides ($Zr_{1-x}Ce_xO_2$ (x = 0–0.17)) being used in proportions established *a priori*. After grinding and pressing, the samples were shaped in form of cylinders. Finally we treated cylindrical samples at a temperature of 1500 °C for six hours in air.

The phase composition, the lattice constants, space group, average size of coherent blocks, microstrains, and positions of cations and anions in the unit cell have been determined using XRD and ND data.

For XRD analysis the samples surface were prepared by a standard metallographic technique [16]. XRD data were acquired with a X'pert Pro MPD PANalytical diffractometer (PANalytical Inc., Westborough, MA, USA) and a BRUKER AXS D8—Advance diffractometer (Bragg—Brentano geometry, CuKα, 2θ range of 20°–80°) (BRUKER AXS Inc., Madison, WI, USA). A powder diffraction software package, which includes the standards of the Crystallography Open Database (COD) [17], was used in order to identify the phase composition of the samples. The microstrains and average size of crystalline blocks of the thin layer from the surface of the bulk samples were obtained using the XRD data, processed by the Rietveld method, PowderCell [18]. LaB_6 was used as standard for XRD measurements (NIST SRM-660—a lanthanum hexaboride powder LaB_6, for line position and line shape).

The structural analysis of the $Zr_{1-x}Ce_xO_2$ (x = 0.0, 0.09, 0.13, 0.17) ceramic samples was performed also using the time-of-flight (TOF) High Resolution Fourier Diffractometer (HRFD) at the IBR-2 pulsed reactor in Joint Institute for Nuclear Research, Dubna, Russia [19]. At the HFRD diffractometer, the correlation technique of data acquisition is used, providing a very high resolution of $\Delta d/d$ ~0.001, practically constant in a wide interval of d_{hkl} spacings [19]. We collected the high resolution diffraction patterns by using a detector placed at backscattering angles 2θ = ±152°, d_{hkl} = 0.6–3.6 Å. It is important to mention that via X-ray diffraction, a layer of typically 10 μm thickness has been investigated, whereas in the case of neutron diffraction, a thickness of tens of centimeters of a sample has been investigated.

Information about microstructural parameters, such as crystalline size (average size of crystalline blocks) and microstrains were obtained from the lines broadening. From the dependence between the square of Full Width at Half Maximum (W) of neutron peaks and the square of the interplanar spacing, d^2, described by the Equation (1) [20,21]

$$W^2 = C_1 + (C_2 + C_3)d^2 + C_4d^4 \tag{1}$$

we extracted the microstrain values (ε).

Here, C_1, C_2 are the refining constants of the resolution function of HRFD obtained from a standard Al_2O_3 (SRM-676 of NIST, USA); C_3 ~ε^2, where ε is the microstrain; C_4 ~$1/L$ where L is the size of the coherently scattering blocks. The values of ε and L can be determined from the true physical characteristics of line broadening, whose value is defined as the difference between broadening of peaks of the experimental sample and the standard Al_2O_3.

RUS is a complex method that allows the determination of elastic constant and the elements of elasticity matrix for samples with certain shapes [22–26]. Sample geometry affects data acquisition [26]. For cylinders with a high ratio of length to diameter we have few excitation modes and the spectra are simple [26,27]. In the case of the studied samples the ratio is around the unit and the spectrum requires more analysis than a long bar. The increasing of spectral complexity supposes adjacent analyses before concluding about these samples. In order to determine a parameter, more resonance frequencies must be searched [25]. This requires a lot of computational calculation time and repeated tests. The same number of normal modes for a short cylinder compared with a long cylinder requires a narrower frequency range.

In the case of cylindrical samples, having diameter approximately equal with height, the first mode is the fundamental torsional mode allowing the direct determination of shear modulus. RUS involves scanning the resonance structure of a compact specimen (in our case a ceramic cylinder) with the aim of determining its mechanical properties [28,29]. In principle, RUS is composed of three parts. Successful use of this method implies the obtaining of a resonant ultrasound spectrum, meaning obtaining the frequency answer for a certain excitation for the analyzed sample. It is necessary to develop a computing method which can help in predicting the resonant ultrasound spectrum for the estimated parameters. The third component represents the iterative refinement of the estimated parameters, in agreement with the measured spectrum. Each of the steps mentioned above have multiple solutions. In the classical RUS experiment, a sample is placed between two US transducers. The primary advantages of this contact technique are its relative simplicity and low cost.

For the RUS method, the samples have been supported by two identical piezoelectric ultrasonic transducers, for emission and reception respectively, placed at the opposite ends of the cylindrical sample (see Figure 1). The measurements are carried out via transducers with contact. The coupling of the transducer with the specimen influences which modes are measured. When the sample is pinned on its edge, more modes are excited and the modes are better defined than when the transducers are placed on the ends of the cylinder. The probe is fixed between the emission and reception transducers in order to accomplish the condition of a stress free surface. The equipment allows the setting so that for the established position of the cylindrical sample, the contact on the edge assures the excitation of a maximum number of possible resonances for the fixed geometry.

A Network/Spectrum/Impedance Analyzer 4395A (Agilent Technologies, Santa Clara, CA, USA), generates a sweep frequency ranging between 80 and 250 kHz with a 1 kHz step. The signal is amplified using a Power amplifier AG 1012 (T&C Power Conversion Inc., Rochester, NY, USA) and applied to the US emission transducer is applied to the B port of the 4395 A Agilent, the spectrum being acquired by a PC used to program the functioning of the equipment as well with a numerical code developed in Matlab 2012b (The MathWorks Inc., Natick, MA, USA, 2012) via a PCIB interface (National Instruments, Mopac Expwy, Austin, TX, USA).

The command of the power amplifier is made using the same PC via a RS232 interface (National Instruments). For the determination of the elasticity and shear moduli of the samples, the propagation speed of the longitudinal and transversal ultrasound waves were measured using a transmission procedure. The emission transducer is applied at one base of the sample via a delay line from Perspex® (Perspex Acrylic Brand, Darwen, Lancashire, UK) 20 mm in length, while the reception transducer is applied on the other base.

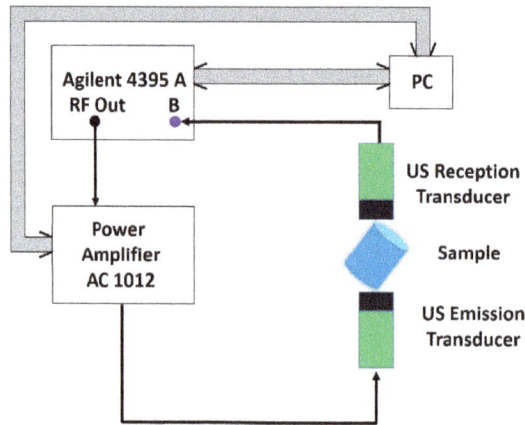

Figure 1. RUS Experimental setup schematic block diagram. Thin lines correspond to coaxial cable for sinusoidal wave transmission; thick lines are GPIB cable for digitalized data.

A G5KB GE transducer (General Electric Measurement & Control, Boston, MA, USA) with a central frequency of 5 MHz has been used for the measurement of the propagation speed of the longitudinal waves, the coupling being made using a ZG-F Krautkramer (General Electric Measurement & Control, Boston, MA, USA), whereas for the transversal waves, a MB4Y GE transducer (General Electric Measurement & Control, Boston, MA, USA) was used, with a central frequency of 4 MHz, the coupling being made with honey. The generation of emission impulses and the reception of signals delivered by the reception transducer has been made using a PR 5073 Pulser Receiver former Panametrics NDT USA (Olympus Corporation Waltham, MA, USA).

The densities of the samples were obtained by means of Archimedes' method, by using water as fluid. The measurements of microhardness were performed by Vickers method, with a load of 50 g and a dwell time of 20 s.

3. Results and Discussion

3.1. Structural Parameters

The experimental results obtained from X-ray and neutron data emphasize that the $Zr_{1-x}Ce_xO_2$ ($x = 0$–0.17) samples do not have a homogeneous structure (see Figure 2). The sample containing pure ZrO_2 has a monoclinic ($P2_1/c$) crystalline structure [10]. Once the concentration of Ce cations increases, the crystal structure changes, from monoclinic to tetragonal ($P4_2/nmc$) (see Figure 2) with two formula units ($Z = 2$) per unit cell. The ND diffractograms indicated that the sintered $Zr_{0.91}Ce_{0.09}O_2$ ($x = 0.09$) contains both monoclinic and tetragonal phases, whereas when $x > 0.09$, the samples only contain tetragonal phases, with a unit cell volume increasing from 69.2 $Å^3$ ($x = 0.13$) to 69.7 $Å^3$ ($x = 0.17$) (see Figure 2 and Table 1).

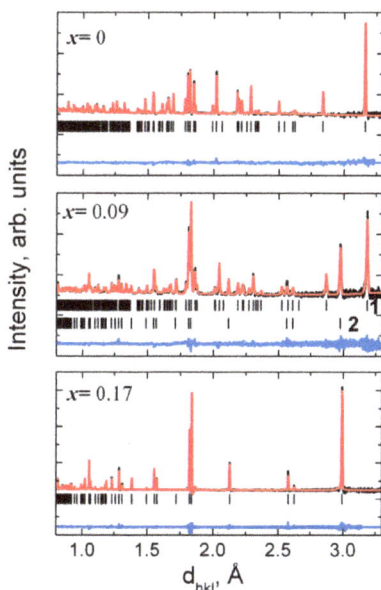

Figure 2. Diffraction patterns for the $Zr_{1-x}Ce_xO_2$ samples measured using HRFD at room temperature, the data were treated using a Fullprof program. The experimental and calculated diffractograms and the difference between calculated and observed diffractograms are shown. The difference is weighted by the mean-squares deviation for each point. Below each graph, ticks indicate the position of the calculate peaks for each phase. 1—monoclinic structure ($P2_1/c$ (SG 14)); 2—tetragonal structure ($P4_2/nmc$ (SG 137)).

Data were acquired at room temperature and refined using the Fullprof program [30,31], and shown in Figure 2. From the analysis of the neutron diffraction data, we were able to define with precision the phase composition, crystal lattice, and the atomic structure of each phase.

Lattice parameters and atomic fractional positions, as well as background parameters, scale factors, occupancy, and isotropic thermal parameters were refined in this case (see Table 1).

Table 1. Refined structural parameters of $Zr_{1-x}Ce_xO_2$ (x = 0.0–0.17) at room temperature, obtained by processing the data collected using the High Resolution Fourier Diffractometer at the Joint Institute for Nuclear Research, Dubna, Russia (x represents the nominal Ce concentration).

Parameters	x				
	$x = 0$	$x = 0.09$		$x = 0.13$	$x = 0.17$
	Monoclinic P2$_1$/c	Monoclinic P2$_1$/c	Tetragonal P4$_2$/nmc	Tetragonal P4$_2$/nmc	Tetragonal P4$_2$/nmc
a, (Å)	5.1453 ± 0.0001	5.2046 ± 0.0003	3.6236 ± 0.0002	3.6358 ± 0.0001	3.6452 ± 0.0002
b, (Å)	5.2091 ± 0.0001	5.2155 ± 0.0003	3.6236 ± 0.0002	3.6358 ± 0.0001	3.6452 ± 0.0002
c, (Å)	5.3116 ± 0.0001	5.3776 ± 0.0002	5.2163 ± 0.0003	5.2378 ± 0.0001	5.2471 ± 0.0002
β, (°)	99.225 ± 0.001	98.938 ± 0.0002	90.0	90.0	90.0
V_{cell}/Z, (Å^3)	35.130	36.05	34.247	34.619	34.860
Volume fraction	1.0	0.378 ± 0.004	0.622 ± 0.006	1.00	1.00
R_{wp}, %	9.5	14.3		10.4	10.4

The R_{wp} values indicate a good fit between the calculated and the experimental data (see Table 1). The refined atomic positions are indicated in Table 2. Temperature corresponding to the ions oscillations have been taken into account in the isotropic approximation.

The neutron peaks broadening (W), related to the sample of ZrO_2, decreases as the concentration of Ce ions is increasing (see Figure 3). It is observable that the experimental points can be fitted with linear functions and consequently, the size effect is absent ($L > 3000$ Å).

As it can be clearly observed, the volume of the unit cell increases as the concentration of Ce cations is increased (see Figure 4). For $x = 0.09$ a change of the unit cell symmetry, from monoclinic to tetragonal, takes place. The change of the symmetry is associated with a sudden decrease of the volume associated (see Figure 4). The slope of the linear functions decreases as the concentration of Ce ions is increasing. This behavior signifies that microstrains in the investigated samples are decreasing with the concentration of Ce ions, as shown in Figure 5.

Table 2. Atomic positions and isotropic thermal parameters of $Zr_{1-x}Ce_xO_2$ ($x = 0$–0.17) calculated using the Rietveld method from neutron diffraction data. $P2_1/c$: Zr (4e) (xyz); O1 (4e) (x,y,z); O2 (4e) (x,y,z); $P4_2$/nmc: Zr (2a) (0.75, 0.25, 0.75); O (4d) (0.25, 0.25, z) (ND data).

Atomic Parameters		Ce Concentration				
		$x = 0$	$x = 0.09$		$x = 0.13$	$x = 0.17$
		Monoclinic $P2_1/c$	Monoclinic $P2_1/c$	Tetragonal $P4_2$/nmc	Tetragonal $P4_2$/nmc	Tetragonal $P4_2$/nmc
Zr/Ce	x	0.2752 ± 0.0001	0.2683 ± 0.0004	0.7500	0.75000	0.75000
	y	0.0405 ± 0.0002	0.0364 ± 0.0003	0.2500	0.25000	0.25000
	z	0.2084 ± 0.0001	0.2084 ± 0.0004	0.7500	0.75000	0.75000
	Biso (Å^2)	0.116 ± 0.000	0.287 ± 0.000	0.283 ± 0.00	0.143 ± 0.00	0.169 ± 0.00
	Occupation	1.0	0.903/0.097	0.89/0.09	0.87/0.13	0.83/0.17
O1	x	0.0710 ± 0.0001	0.0533 ± 0.0005	0.2500	0.25000	0.25000
	y	0.3341 ± 0.0002	0.3146 ± 0.0005	0.2500	0.25000	0.25000
	z	0.3452 ± 0.0003	0.3604 ± 0.0004	0.0476 ± 0.0002	0.0474 ± 0.0001	0.0467 ± 0.0001
	Biso (Å^2)	0.096 ± 0.003	0.59 ± 0.04	0.37 ± 0.03	0.419 ± 0.001	0.521 ± 0.002
	Occupation	1.0	1.0	1.0	1.0	1.0
O2	x	0.4488 (1)	0.4500 ± 0.0004	-	-	-
	y	0.7570 (3)	0.7513 ± 0.0005	-	-	-
	z	0.4799 (2)	0.4735 ± 0.0004	-	-	-
	Biso (Å^2)	0.137 (3)	0.59 ± 0.04	-	-	-
	Occupation	1.00	1.00	-	-	-

First sample, for which $x = 0.0$, have the baddeleyite structure [32]. The modification of the crystal structure due to the substitution of Zr ions with Ce ions is related to the variation of the observed distance between Zr–O ions (see Table 2 and Figure 6). Cerium cations have two possible valence: +3 and +4. On the same type of places Ce^{3+}/Ce^{4+} have a larger ionic radius as Zr^{4+} cations ($r_{Ce3+} = 1.143$ Å; $r_{Ce4+} = 0.97$ Å; $r_{Zr4+} = 0.84$ Å, for C.N. = 8, after Shannon [31]). The substitution of Zr^{4+} with Ce^{4+}/Ce^{3+} should lead to an increase of the lattice constants, of the unit cell volume and of the microstrains. On other hand, the increase of the lattice energy can cause a change in crystal structure and can sit at the basis of the explanation related to the observed transition from monoclinic to tetragonal structure at room temperature. The transition is accompanied by a relaxation of energy part due to the microstrain. In the same time, the substitution of Zr^{4+} with Ce^{3+} leads to a corresponding decrease of oxygen concentration, implicitly to the appearance of anionic voids and/or other types of crystalline defaults, for example dislocations.

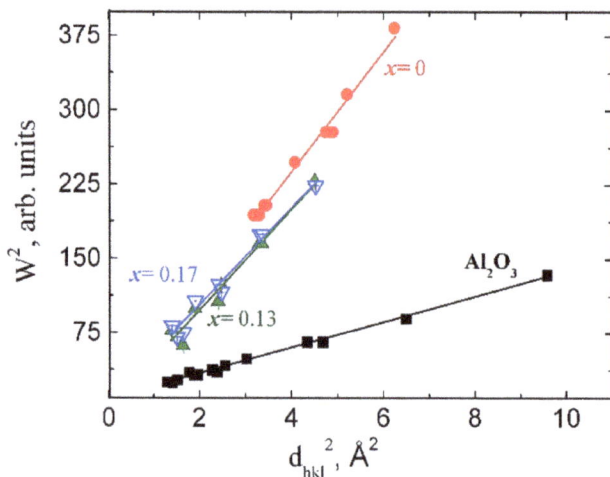

Figure 3. The dependency of the square of FWHM of diffraction peaks *vs.* the square of inter-planar spacing.

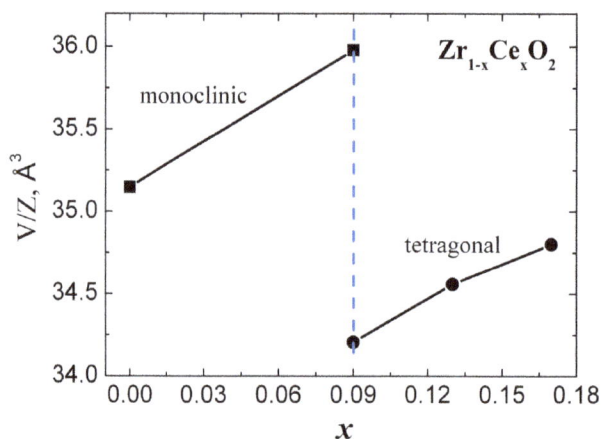

Figure 4. Variation of the unit cell volume of $Zr_{1-x}Ce_xO_2$ *vs.* Ce concentration (x). Z—the number of molecules in the unit cell.

XRD data were obtained from the surface of the bulk samples and correspond to the structure of a thin layer (see Table 3). The common phase for these samples has a tetragonal structure, $P4_2/nmc$. For a concentration $x = 0.09$, we observed a mixture of tetragonal ($P4_2/nmc$, SG 137) and monoclinic ($P2_1/c$, SG 14) phases (see Table 1).

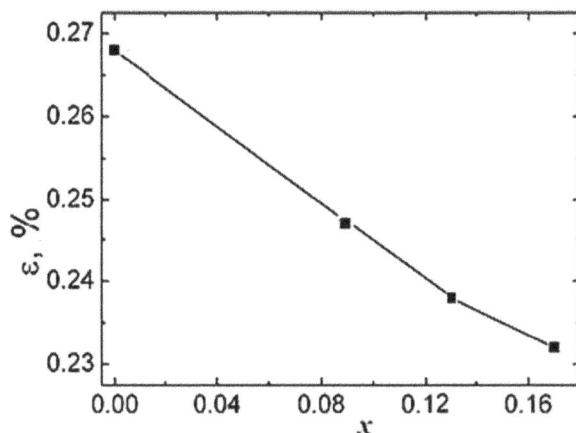

Figure 5. The dependency of microstrain, ε, on the concentration of Ce ions (x).

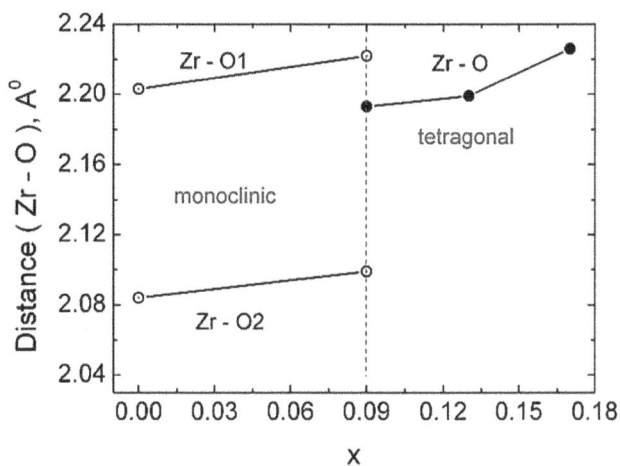

Figure 6. The dependence of Zr-O distances *vs.* the Ce concentration (x).

The distances Zr-O for each phase and Ce concentration were calculated by using the position of cations and anions in the unit cells (see Table 4). The average size of crystalline blocks and the microstrains were obtained by using the Williamson-Hall plot. For the calculated density of the samples (see Table 3) we take account of the monoclinic and tetragonal phase concentrations.

75

Table 3. Lattice constants (a, b, c, β), unit cell volume (V_{cell}), average size of crystalline blocks (*L*), microstrains (ε), and calculated densities ($\rho_{cal,phase}$) and observed densities ($\rho_{cal,sample}$) of each phase and each $Zr_{1-x}Ce_xO_2$ (*x* = 0.9–0.17) sample at room temperature, obtained by processing the data collected with the Bruker Advance diffractometer (XRD data).

Parameters	Ce Concentration *x*				
	x = 0.09		*x* = 0.13		*x* = 0.17
	Monoclinic P2₁/c	Tetragonal P4₂/nmc	Tetragonal P2₁/c	Tetragonal P4₂/nmc	Tetragonal P4₂/nmc
a, (Å)	5.1959 ± 0.0001	3.6218 ± 0.0001	5.1933 ± 0.0001	3.6269 ± 0.0001	3.6442 ± 0.0001
b, (Å)	5.2161 ± 0.0001	3.6218 ± 0.0001	5.2247 ± 0.0001	3.6269 ± 0.0001	3.6442 ± 0.0001
c, (Å)	5.3698 ± 0.0001	5.2170 ± 0.0001	5.3803 ± 0.0001	5.2431 ± 0.0001	5.2440 ± 0.0001
β, °	98.93 ± 0.01	90.0	98.93 ± 0.01	90.0	90.0
V_{cell}/Z, (Å³)	35.940	34.217	36.054	34.417	34.824
L, (Å)	1320	354	1137	605	1606
ε, (%)	0.07	0.086	0.027	0.051	0.036
R_{wp}, %	10.5		11.4		9.2
$\rho_{cal,phase}$ (kg/m³)	5693	7747	5675	7795	7798
Volume fraction	0.90 ± 0.03	0.10 ± 0.03	0.84 ± 0.02	0.16 ± 0.02	1.00
$\rho_{cal,sample}$ (kg/m³)	5898		6014		7798

As observed, including for large Ce ion concentration, in the layer from the surface of the samples corresponding to *x* = 0.09 and *x* = 0.13 a tetragonal phase appears concomitantly with the monoclinic phase (see Table 3 and Figure 7). As shown in Figure 7, we can observe that the samples contain a small amount of unknown phases.

The variation of the lattice constants and the unit cell volume, corresponding to the thin layer from the samples surface as a function of the substitution of Zr with Ce, is similar to the variation of the lattice constant observed for the entire sample volume (see Tables 1 and 3). An investigation related to the difference between the chemical composition at the surface and inside the sample is ongoing.

Table 4. Atomic positions, isotropic thermal parameters, and Zr-O distances of $Zr_{1-x}Ce_xO_2$ (x = 0.9–0.17) calculated using the Rietveld method from XRD data. (P 2_1/c: Zr (4e) (xyz); O1 (4e) (x,y,z); O2 (4e) (x,y,z); P 4_2/nmc: Zr (2a) (0.75, 0.25, 0.75); O (4d) (0.25, 0.25, z).

Atomic Parameters		Ce Concentration x				
		x = 0.09		x = 0.13		x = 0.17
		P2$_1$/c	P4$_2$/nmc	P2$_1$/c	P4$_2$/nmc	P4$_2$/nmc
Zr/Ce	x	0.2953 ± 0.0002	0.75000	0.2885 ± 0.0001	0.75000	0.75000
	y	0.0287 ± 0.0001	0.25000	0.0227 ± 0.0001	0.25000	0.25000
	z	0.2277 ± 0.0003	0.75000	0.2463 ± 0.0003	0.75000	0.25000
	B_{iso} (Å^2)	0.400 ± 0.000	0.407 ± 0.000	0.412 ± 0.000	0.410(0)	0.433(0)
O1	x	0.0640 ± 0.0001	0.25000	0.0630 ± 0.0001	0.25000	0.25000
	y	0.3291 ± 0.0003	0.25000	0.3301 ± 0.0003	0.25000	0.25000
	z	0.3291 ± 0.0001	0.04654 ± 0.0001	0.3281 ± 0.0001	0.0467 ± 0.0001	0.0203 ± 0.0001
	B_{iso} (Å^2)	0.704 ± 0.001	0.246 ± 0.003	0.860 ± 0.002)	0.419 ± 0.001	0.519 ± 0.001
O2	x	0.4251 ± 0.0001	-	0.4291 ± 0.0001	-	-
	y	0.7561 ± 0.0002	-	0.7610 ± 0.0001	-	-
	z	0.4844 ± 0.0003	-	0.4801 ± 0.0001	-	-
	B_{iso} (Å^2)	0.800 ± 0.000	-	0.920 ± 0.000	-	-
d_{Zr-O1}(Å)		2.1946	-	2.2210	-	-
d_{Zr-O2}(Å)		2.2135	-	2.2387	-	-
d_{Zr-O}(Å)		-	2.1933	-	2.1987	2.2257

3.2. Resonant Ultrasound Spectroscopy

In order to determine mechanical properties of a compact specimen (in our case a ceramic cylinder), the resonance structure is scanned by RUS [22,23,29,33]. This analysis comes in completion and adds more information related to the structural analysis performed above. In comparison to other ultrasound methods, resonant techniques are particularly interesting because they allow for easy and inexpensive detection of both internal and surface defects with a single test and it has a suite of advantages, among them is its applicability to small volume specimens. RUS is based on the principle that the mechanical resonant response of solids depend strongly on its elastic moduli, shape, and density. Resonant (or natural) frequencies of a system can be either measured or calculated by solving equations of motion for the known shape [34]. The reverse is also true: if resonant frequencies of an object are known, its elastic properties can be determined [23,34,35].

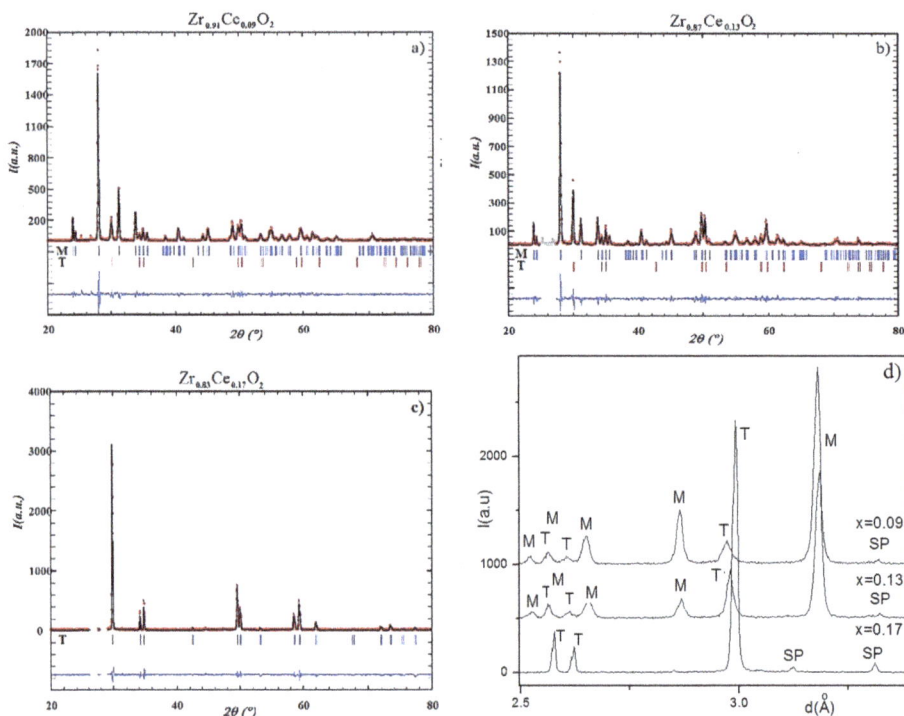

Figure 7. X-ray diffractograms corresponding to (**a**) $x = 0.09$; (**b**) $x = 0.13$; (**c**) $x = 0.17$; (**d**) difractograms presenting the dependence of intensity (I) on interplanar distances (d) and Ce concentration (x). The vertical bars correspond to the monoclinic (M) and tetragonal (T) phases (FullProf program was used for determination of parameters). SP represents the foreign phase.

Inhomogeneity in an object may be identified from a resonant frequency spectrum by resonant frequency shifts, peak splitting, increases in peak width, and changes in amplitude. The method is based on the estimation of resonant eigenfrequencies [27], based on an eigenvalue and eigenfunction method [26]. For these, we have used an equipment configuration as the one shown in Figure 1. The ceramic sample is supported by two piezoelectric ultrasound transducers, for emission and reception respectively, placed at opposite edges of the ceramic cylinders. The resonance spectra were traced for the cylindrical samples noted corresponding to $x = 0.09$, 0.13 and 0.17 Ce concentrations in the samples ($\pi 2$, $\pi 3$, $\pi 4$), whose properties have been presented in Table 4. Due to the fact that the samples $\pi 2$–$\pi 4$ are axisymmetric, isotropic, and homogeneous [25], we can conclude that every mode observed on samples must fall into the following three categories [33]:

- Torsional axisymmetric pure share motion consisting of rigid rotation of rings of materials around the sample axis. The frequency of these modes depend strongly on the sample's shear velocity
- Extensional axisymmetric mixtures of compression and shear modulus
- Flexural modes along paths that are tilted with respect to the cylinder axis. These modes occur in pairs named doublets, with the same resonance frequency.

Figure 8 shows the resonance spectra for the samples $\pi2$–$\pi4$ in the frequency range 80 kHz to 250 kHz. The RUS spectrum describes a large amplitude response detected when the frequency corresponds to one of the samples eigenfrequencies.

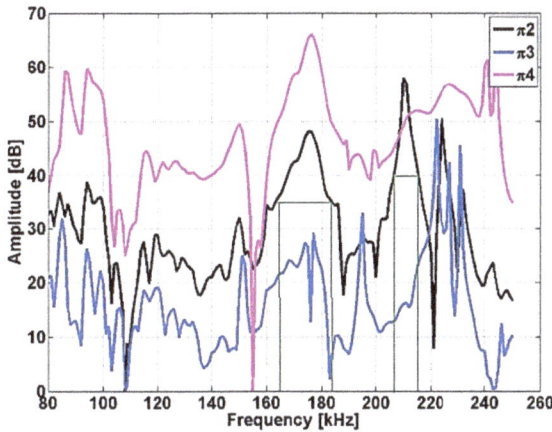

Figure 8. Resonance ultrasound spectra for samples $\pi2$–$\pi4$.

Figure 9 describes the typical response of one from the three tested samples, namely $\pi2$, in a frequency range comprised between 80 and 250 kHz chosen based on our preliminary analysis performed on similar materials.

It is immediately observed that for the three samples, there is a pronounced amplitude response in intervals of frequency corresponding to the samples eigenfrequency. The amplitude responses in the swept frequency range are in tight connection to the sample's properties, especially the ones related to the density. The maximum displacements appear as red patches in the above representations, the intermediate ones appear in yellow and green and the minimum ones appear in blue. It is noticeable that for the analyzed frequencies, the maximum displacements take place at the edges of the cylinders in opposite directions, with a minima in the middle of the samples.

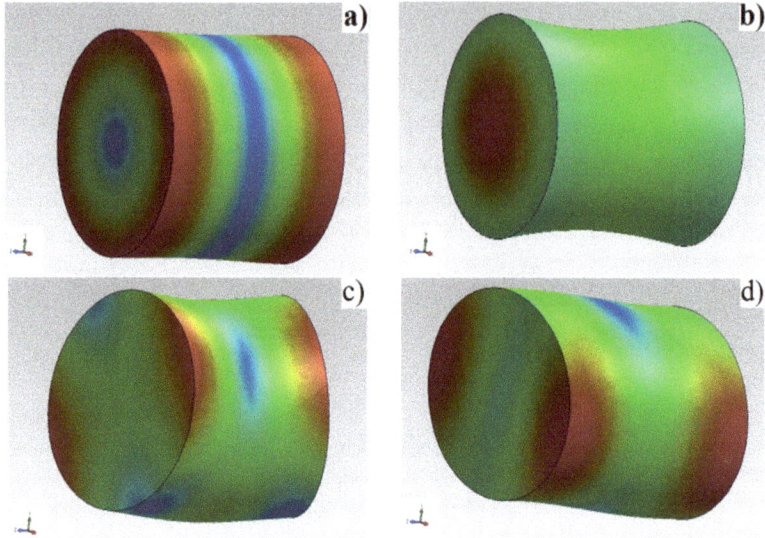

Figure 9. Resonant modes for sample $\pi 2$: (**a**) extensional mode, f = 163 kHz; (**b**) extensional mode, f = 185 kHz; (**c**) flexural mode f = 208 kHz; (**d**) flexural mode f = 214 kHz.

The eigenfrequency intervals exhibit a slight shift towards smaller values as the density of the samples decrease (see Table 3 and Figure 8). Resonant ultrasound spectroscopy is a reliable technique for monitoring structural modifications related to samples density.

In Figure 9, we are showing the simulation of deformations for two extensional modes (a,b) and two flexural modes (c,d) for sample $\pi 2$. The simulations have been made using the finite element method using SolidWorks 2011 (Dassault Systèmes SolidWorks Corporation, Waltham, MA, USA), Simulation-Frequency toolbox [36] using a solid mesh type, curvature based mesher, with a total number of 13145 nodes and 8777 total elements.

The resonance frequencies obtained by simulations could give us information similarly to the ones experimentally obtained, that could be chosen at full width at half maximum of the amplitude [33]. It is important to specify that the simulated information is useful in order to determine which of the resonances are observables for an investigated spectrum. The inversion of data has been used in order to determine the elastic properties from the measured resonance spectra, using the conjugate gradient method with minimizing the objective function [33]:

$$F = \sum_i w_i \big(f_i^{(p)} - f_i^{(m)} \big)^2 \tag{2}$$

here, $f^{(p)}$ and $f^{(m)}$ are the computed and respective measured frequencies, respectively; w_i are the measured weights, which describe the confidence in the performed measurements.

The optimization problem has been numerically computed in Matlab 2013a. Due to the fact that the number of peaks and the corresponding frequencies is relatively small, the inversion was only applied to determine the Elastic (E) and shear (G) moduli and not the geometrical dimensions and densities of the cylindrical samples made from zirconia.

An increase of the relative density with Ce concentration takes places (see Table 5). We considered that a decrease of the pores concentration takes place with the increase of the Ce concentration in the samples.

Table 5. Some crystallographic and mechanical characteristics of $Zr_{1-x}Ce_xO_2$ samples.

Ce concentration (x)	Molecular Mass	Crystallographic Structure	Relative Density (%)	Elasticity Modulus (GPa)	Shear Modulus (GPa)	Poisson Ratio
0.09	127.62	Monoclinic + tetragonal	88.9	145.85	56.49	0.291
0.13	129.58	Monoclinic + tetragonal	94.7	168.53	64.47	0.307
0.17	131.53	tetragonal	99.9	193.43	73.16	0.322

3.3. Microhardness Measurements

Microhardness measurements were performed on both natural (as sintered) faces of the samples and the resulting face after cutting with a diamond wire. The Vickers microhardnesses increase with the increase of the Ce concentration in the samples. On other hand, the micro-hardness of the sample corresponding to the natural face is smaller than those obtained from the cutting face, except those corresponding to the sample with $x = 0.17$ (see Table 6). The results are in agreement with the literature data.

For the samples with $x > 0.09$ we observe a larger microhardness on the face after cutting as comparing with those corresponding to the natural face. It is observable that the results related to the hardness measurements from the sample having a concentration $x = 0.17$ are slightly different compared to the other two concentrations, before and after the mechanical cutting is made. A possible explanation can be that, once the cutting of the natural (original) phase is made, a micro-crack could have appeared on the surface of this sample, the measured hardness being lower on a damaged surface, a result in agreement with literature observations [37].

Table 6. Results of microhard ness measurements (x—nominal Ce concentration in the samples; S—the surface area of the resulting indentation; h—indentation depth; HV—microhardness of the sample).

	Results Corresponding to the Natural Face				Results Corresponding to the Face after Cutting with Diamond Wire		
x	S (μm^2)	H (μm)	HV (GPa)	x	S (μm^2)	H (μm)	HV (GPa)
0.09	54.47	1.4	9.0 ± 0.7	0.09	53.39	1.4	9.2 ± 0.7
0.13	37.60	1.2	13.0 ± 0.1	0.13	34.51	1.1	14.2 ± 0.7
0.17	22.43	0.9	21.9 ± 0.6	0.17	27.18	1.0	18.0 ± 0.7

We performed the chemical analysis on the surface of the samples and on the surface of fractured samples (see Table 7). The results confirmed that the increase of the Ce concentration will produce an increase of unit cell volumes of monoclinic and the tetragonal phases, respectively (see Tables 1 and 3), in agreement with the literature. On other side, we observed that for the same sample the unit cell volumes will increase from the surface to the inner part (see Tables 1 and 3). It is not clear if we have a continuous decrease of Ce concentration from inner side to the surface of the sample or only the surface layer has another chemical composition as compared with the inner side.

Table 7. The chemical composition from EDX measurements (see Figure S1).

Nominal Composition	Chemical Composition at the Surface of the Samples	Chemical Composition at the Surface Obtained after Fracture
$Zr_{0.91}Ce_{0.09}O_2$	$Zr_{0.88\pm0.03}Ce_{0.12\pm0.03}O_{1.82\pm0.03}$	$Zr_{0.84\pm0.03}Ce_{0.16\pm0.03}O_{2.00\pm0.03}$
$Zr_{0.87}Ce_{0.13}O_2$	$Zr_{0.85\pm0.03}Ce_{0.15\pm0.03}O_{2.00\pm0.03}$	$Zr_{0.79\pm0.03}Ce_{0.21\pm0.03}O_{2.00\pm0.03}$
$Zr_{0.83}Ce_{0.17}O_2$	$Zr_{0.79\pm0.03}Ce_{0.21\pm0.03}O_{2.00\pm0.03}$	$Zr_{0.75\pm0.03}Ce_{0.25\pm0.03}O_{2.00\pm0.03}$

4. Conclusions

The substitution of the Zr with Ce in $Zr_{1-x}Ce_xO_2$ leads to a change of the phase composition, a gradual transition from the monoclinic to tetragonal structure. Concerning the substitution of Zr with Ce we observed a difference of the phase composition between the surface layer of the sample and the phase composition of the bulk samples. We attributed this difference to the various oxygen concentration in the surface layer and in the bulk sample. The investigations performed by means of neutron diffractometry, which "viewed" a large volume of the samples, indicated a transition from the monoclinic to tetragonal phase for $x = 0.09$. On other hand, XRD investigations, which "viewed" only a thin layer have shown that the phase composition of the samples corresponding to $x = 0.09$ and 0.13 represents a mixture between a monoclinic phase and a tetragonal phase, the tetragonal phase concentration increasing with the increase of Ce concentration. The unit cell volumes of the tetragonal and monoclinic phases and Zr-O distances increase with the increase of Ce concentration, for the inner part of the samples. The calculated and measured densities increase also with the increase of Ce concentration. A maximum of the average size of coherent blocks and of the microstrain appear in the tetragonal phase for the sample with $x = 0.13$. The increase of Ce concentration leads also to an increase of the volumes of both monoclinic and tetragonal unit cells. From the variation of unit cell volume of the surface layer, compared with those corresponding to the unit cell volume of the sample core, we conclude that a small but systematic decrease of the large radii cation concentration takes place at the surface layer. An increase of the relative density and of the mechanical parameters (elasticity and shear moduli) was obtained with the increase of Ce concentration.

RUS is a reliable technique, which emphasizes the eigenfrequency intervals which exhibit a slight change as a function of samples composition, for monitoring structural modifications related to sample density. Slight material anisotropy leads to splitting of the higher modes but not of the fundamental torsion mode. In the case of the studied samples, whose ratio is around the unit, the interpretation is favorable because the torsional mode is the lowest one, well separated from the others for $v > 0$, allowing immediate extraction of the shear modulus and its damping. The initiation of fracture of ceramic elements can be due to the presence of low density zone-containing dispersed high density agglomerates in the volume and any deformation which will be immediately apparent through changes in the resonance modes, with deviations from the normal spectrum. RUS can be used for quality control of certain ceramic elements of hip prosthesis, such as femoral heads. If the elements are incorrectly sintered, with a density smaller than the prescribed value and the elastic and shear moduli smaller, important modifications appear in the shape of the spectrum and the resonance frequencies.

Future research in this area should concentrate on the SEM analysis concerning the ceramic samples, whose aspects change dramatically with increasing concentrations of Ce ions, an effect which needs to be further analyzed in a future study.

Acknowledgments: This work was supported by a grant of the Ministry of National Education, National Research Council (CNCS)—Executive Unit for Financing Higher Education, Research, Development and Innovation (UEFISCDI), project number PN-II-ID-PCE-2012-4-0437 and bilateral cooperation with JINR Dubna by the grant "Zirconia doped with transition metals. Synthesis, electric and magnetic properties", protocol 04-4-1121-2015/2017.

Author Contributions: Adriana Savin wrote of the paper, except the parts which concern the results corresponding to the ND experiments (written by Vitalii Turchenko) and XRD experiments (written by Mihail-Liviu Craus). Valerii V.Burkhovetsky has made EDX measurements. Tatiana E. Konstantinova had obtained the samples. Alina Bruma, Pierre-Antoine Dubos, and Sylvie Malo have made hardness measurements and analyzed the experimental data. All authors have participated in analysis and comments of the results. All of the authors read and approved the final manuscript.

Conflicts of Interest: The authors declare no conflict of interest.

References

1. Guazzato, M.; Albakry, M.; Ringer, S.P.; Swain, M. Strength, fracture toughness and microstructure of a selection of all-ceramic materials. Part II. Zirconia-based dental ceramics. *Dent. Mater.* **2004**, *20*, 449–456.
2. Panadero, R.A.; Roman-Rodriguez, J.L.; Ferreiroa, A.; Sola-Ruiz, M.F.; Fons-Font, A. Zirconia in fixed prosthesis. A literature review. *J. Clin. Exp. Dent.* **2014**, *6*, 66–73.
3. Badwal, S.P.S.; Giddey, S.; Munnings, C.; Kulkarni, A. Review of Progress in High Temperature Solid Oxide Fuel Cells. *J. Aust. Ceram. Soc.* **2014**, *50*, 23–37.
4. Ashton Acton, Q. *Electrolytes: Advances in Research and Application*; Scholarly Editions: Atlanta, GA, USA, 2011.
5. Juškevičius, K.; Audronis, M.; Subačius, A.; Drazdys, R.; Juškėnas, R.; Matthews, A.; Leyland, A. High-rate reactive magnetron sputtering of zirconia films for laser optics applications. *Appl. Phys. A* **2014**, *116*, 1229–1240.
6. D'Antonio, J.A.; Capello, W.N.; Naughton, M. Ceramic bearings for total hip arthroplasty have high survivorship at 10 years. *Clin. Orthop. Relat. Res.* **2012**, *470*, 373–381.
7. Bal, B.S.; Garino, J.; Ries, M.; Rahman, M.N. A review of ceramic bearing materials in total joint arthroplasty. *Hip Int.* **2007**, *17*, 21–30.
8. Mercer, P.D.L.; van Ommen, J.G.; Doesburg, E.B.M.; Burgraff, A.J.; Ross, J.R.H. Zirconia as a support for catalysts Influence of additives on the thermal stability of the porous texture of monoclinic zirconia. *Appl. Catal.* **1991**, *71*, 363–391.
9. Depprich, R.; Zipprich, H.; Ommerborn, M.; Naujoks, C.; Wiesmann, H.P.; Kiattavorncharoen, S.; Lauer, H.C.; Meyer, U.; Kubler, N.; Handschel, J. Osseointegration of zirconia implants: An SEM observation of the bone-implant interface. *Head Face Med.* **2008**, *4*, 1–8.
10. French, R.H.; Glass, S.J.; Ohuchi, F.S.; Xu, Y.-N.; Ching, W.Y. Experimental and theoretical determination of the electronic structure and optical properties of three phases of ZrO_2. *Phys. Rev. B* **1994**, *49*, 5133.
11. Okabayashi, S.; Kono, S.; Yamada, Y.; Nomura, K. Fabrication and magnetic properties of Fe and Co co-doped ZrO_2. *AIP Adv.* **2011**, *1*.
12. Bechepeche, A.P.; Treu, O.; Longo, E. Experimental and theoretical aspects of the stabilization of zirconia. *J. Mater. Sci.* **1999**, *34*, 2751–2756.
13. Zhu, W.Z. Effect of cubic phase on the kinetics of the isothermal tetragonal to monoclinic transformation in ZrO_2 (3 mol% Y_2O_3) ceramics. *Ceram. Int.* **1998**, *24*, 35–43.
14. Li, P.; Chen, I.; Penner-Hahn, J.E. Effect of Dopants on Zirconia Stabilization—An X-ray Absorption Study: I, Trivalent Dopants. *J. Am. Ceram. Soc.* **1994**, *77*, 1289–1295.

15. ISO 13356–2015—Implants for surgery—Ceramic materials based on yttria-stabilized tetragonal zirconia. Available online: http://www.iso.org/iso/home/store/catalogue_ics/catalogue_detail_ics.htm?ics1=11&ics2=040&ics3=40&csnumber=62373 (accessed on 18 September 2015).

16. Samuels, L.E. *Metallographic Polishing by Mechanical Methods*, 4th ed.; ASM International: Novelty, OH, USA, 2003.

17. Crystallography Open Database. Available online: http://www.crystallography.net (accessed on 15 September 2015).

18. Kraus, W.; Noltze, G. *Noltze, Powder Cell for Windows, 2.4*; Federal Institute for Materials Research and Testing: Berlin, Germany, 2000.

19. Balagurov, A.M. Scientific Reviews: High-Resolution Fourier Diffraction at the IBR-2 Reactor. *Neutron News* **2005**, *16*, 8–12.

20. Bruno, G.; Efremov, A.M.; Clausen, B.; Balagurov, A.M.; Simkin, V.N.; Wheaton, B.R.; Webb, E.J.; Brown, D.W. On the stress-free lattice expansion of porous cordierite. *Acta Mater.* **2010**, *58*, 1994–2003.

21. Hauk, V. *Structural and Residual Stress Analysis by Non-Destructive Methods*; Elsevier: Amsterdam, the Netherlands, 1997.

22. Visscher, W.M.; Migliori, A.; Bell, T.M.; Reinert, R.A. On the normal modes of free vibration of inhomogeneous and anisotropic elastic objects. *J. Acoust. Soc. Am.* **1991**, *90*, 2154–2162.

23. Migliori, A.; Sarrao, J.; Visscher, W.M.; Bell, T.; Lei, M.; Fisk, Z.; Leisure, R. Resonant ultrasound spectroscopic techniques for measurement of the elastic moduli of solids. *Phys. B Condens. Matter* **1993**, *183*, 1–24.

24. Migliori, A.; Maynard, J.D. Implementation of a modern resonant ultrasound spectroscopy system for the measurement of the elastic moduli of small solid specimens. *Rev. Sci. Instrum.* **2005**, *76*.

25. Demarest, H.H., Jr. Cube-Resonance Method to Determine the Elastic Constants of Solids. *J. Acoust. Soc. Am.* **1971**, *49*, 768–775.

26. Jaglinski, T.R.; Lakes, S. Resonant ultrasound spectroscopy of cylinders over the full range of Poisson's ratio. *Rev. Sci. Instrum.* **2011**, *82*.

27. Myasnikov, D.V.; Konyashkin, A.V.; Ryabushkin, O.A. Identification of eigenmodes of volume piezoelectric resonators in resonant ultrasound spectroscopy. *Tech. Phys. Lett.* **2010**, *36*, 632–635.

28. Carter, C.B.; Norton, M.G. *Ceramic Materials*, 2nd ed.; Springer: New York, NY, USA, 2013.

29. Migliori, A.; Sarro, J.L. *Resonant Ultrasound Spectroscopy: Applications to Physics, Materials Measurement and Nondestructive Evaluation*; Wiley: New York, NY, USA, 1997.

30. Rodriguez-Carvajal, J. Recent advances in magnetic structure determination by neutron powder diffraction. *Phys. B Condens. Matter* **1993**, *192*, 55–69.

31. Shannon, R.T. Revised effective ionic radii and systematic studies of interatomic distances in halides and chalcogenides. *Acta. Crystallogr. Sect. A* **1976**, *32*, 751–767.

32. Howard, C.J.; Hill, R.J.; Reichert, B.E. Structures of ZrO_2 polymorphs at room temperature by high-resolution neutron powder diffraction. *Acta. Crystallogr. Sect. B Struct. Sci.* **1988**, *44*, 116–120.

33. Zadler, B.J.; le Rousseau, J.H.; Scales, J.A.; Smith, M.L. Resonant ultrasound spectroscopy: Theory and application. *Geophys. J. Int.* **2004**, *156*, 154–169.

34. De Silva, C.W. *Vibration: Fundamentals and Practice*; CRC Press: Boca Raton, FL, USA, 2000.

35. Ren, F.; Case, E.D.; Morrison, A.; Tafesse, M.; Baumann, M.J. Resonant ultrasound spectroscopy measurement of Young's modulus, shear modulus and Poisson's ratio as a function of porosity for alumina and hydroxyapatite. *Philos. Mag.* **2009**, *89*, 1163–1182.

36. SolidWorks 2011—Simulation-Frequency Toolbox User Guide. Available online: www.solidworks.com (accessed on 18 September 2015).

37. Wachtman, J.B.; Cannon, W.R.; Matthewson, M.J. *Mechanical Properties of Ceramics*; Springer International Publishing: Cham, Switzerland, 2014.

Correlation between Earthquakes and AE Monitoring of Historical Buildings in Seismic Areas

Giuseppe Lacidogna, Patrizia Cutugno, Gianni Niccolini, Stefano Invernizzi and Alberto Carpinteri

Abstract: In this contribution a new method for evaluating seismic risk in regional areas based on the acoustic emission (AE) technique is proposed. Most earthquakes have precursors, *i.e.*, phenomena of changes in the Earth's physical-chemical properties that take place prior to an earthquake. Acoustic emissions in materials and earthquakes in the Earth's crust, despite the fact that they take place on very different scales, are very similar phenomena; both are caused by a release of elastic energy from a source located in a medium. For the AE monitoring, two important constructions of Italian cultural heritage are considered: the chapel of the "Sacred Mountain of Varallo" and the "Asinelli Tower" of Bologna. They were monitored during earthquake sequences in their relative areas. By using the Grassberger-Procaccia algorithm, a statistical method of analysis was developed that detects AEs as earthquake precursors or aftershocks. Under certain conditions it was observed that AEs precede earthquakes. These considerations reinforce the idea that the AE monitoring can be considered an effective tool for earthquake risk evaluation.

Reprinted from *Appl. Sci.* Cite as: Lacidogna, G.; Cutugno, P.; Niccolini, G.; Invernizzi, S.; Carpinteri, A. Correlation between Earthquakes and AE Monitoring of Historical Buildings in Seismic Areas. *Appl. Sci.* **2015**, *5*, 1683–1698.

1. Introduction

A new method to evaluate seismic risk in regional areas is proposed as an attempt to preserve Italian historical and architectural cultural heritage. To this purpose, the spatial and temporal correlations between the acoustic emission (AE) data obtained from the monitoring sites and the earthquakes that have occurred in specific ranges of time and space are examined.

The predictive power and the non-invasive procedure of the AE technique (no external excitation is provided, since the source of energy is the damage process itself) can be exploited to preserve Italian cultural heritage, as historic buildings and monuments are exposed to seismic risk and, in general, to severe, long-term, cyclic loading conditions or harsh environmental conditions.

Buildings and structures exposed to the action of earthquakes with moderate magnitude, which are rather frequent in the central and southern regions of Italy, may undergo accelerated aging and deterioration. Such damage processes, often inaccessible to visual inspection, eventually lead to an increased vulnerability to the actions of major earthquakes, with catastrophic results [1].

The two regions considered in this paper surround the Italian cities of Varallo and Bologna, where the authors have recently carried out AE monitoring of important historical monuments for structural stability assessment. In particular, in the Italian Renaissance Architectural Complex of "The Sacred Mountain of Varallo" the structure of the Chapel XVII was analyzed, while in the City of Bologna the stability of the "Asinelli Tower", known as the highest leaning tower in Italy, was evaluated [2,3].

According to seismic analysis, several studies investigated spatial and temporal correlations of earthquake epicenters, involving, for example, the concepts of Omori's Law and fractal dimensions [4–6]. Other authors tried to study the complex phenomenon of seismicity using an approach that is able to analyze the spatial location and time occurrence in a combined way, without subjective a priori choices [7]. The present approach leads to a self-consistent analysis and visualization of both spatial and temporal correlations based on the definition of correlation integral [8].

Based on these considerations, we have tried not only to analyze the seismic activity in the considered regions, but also to correlate it with the AE activity detected during the structural monitoring. By adopting a modified Grassberger-Procaccia algorithm, we give the cumulative probability of the events' occurrence in a specific area, considering the AE records and the seismic events during the same period of time. In the modified integral the cumulative probabilities $C(r, \tau)$ are the function of the regional radius of interest, r, and of the time interval, τ, both considering the peak of AEs as earthquake "precursors" or "aftershocks" [9].

2. Description of Chapel XVII and Asinelli Tower

In Varallo, placed in Piedmont in the province of Vercelli, there is the relevant Italian Renaissance Architectural Complex named "The Sacred Mountain of Varallo". It was built in the 15th century on a cliff above the town of Varallo and is composed of a basilica and 45 chapels.

Because of its high level of damage due to regional earthquakes that have occurred [2], we chose to consider Chapel XVII, known as the chapel of the "Transfiguration of the Christ on the Mount Tabor", for the AE analysis. This structure, having a circular plan, was built with stone masonry and mortar. The interior walls of the chapel are also equipped with some valuable frescoes (Figure 1).

In Bologna, Emilia-Romagna, there is another important masterpiece of Italian architectural and historical heritage, the "Asinelli Tower", which is also the symbol of the city. To study the damage evolution of the structure, we have recently analyzed the effect of repetitive and impulsive natural and anthropic events, such as earthquakes wind, or vehicle traffic, with the AE technique [3].

Figure 1. Sacred Mountain of Varallo, Mount Tabor Chapel XVII.

The Asinelli Tower was built in the early 12th century, and it rises to a height of 97.30 m above the ground. From the ground level, up to a height of 8.00 m, the tower is surrounded by an arcade built at the end of the 15th century. Studies conducted in the early 20th century revealed that the Asinelli Tower leaned westward by 2.25 m, and other recent studies have confirmed that its leaning is of 2.38 m, and it remains the tallest leaning tower in Italy (Figure 2).

3. Acoustic Emission Monitoring

The AE monitoring is performed by analyzing the signals received from the transducers through a threshold detection device that counts the burst signals, which are greater than a certain voltage.

The adopted USAM (Unit for Synchronous Multichannel Acquisition) acquisition system consists of six pre-amplified AE sensors, equipped with six units of data storage, a central unit for synchronization, and a trigger threshold. The output voltage signals of the piezoelectric transducers (PZT) were filtered with a pass-band from 50 kHz to 800 kHz and a detection threshold of 100 μV. The obtained data from this device are the cumulative counting of each mechanical wave, the acquisition time, the measured amplitude in volts, the duration, and the number of oscillations across the threshold value for each AE signal [2,3]. Then, at the post-processing stage, we discarded signals with a duration shorter than 3 μs and containing less than three oscillations across the detection threshold in order to filter out electrical noise spikes.

Figure 2. The Asinelli and the adjacent Garisenda Towers in the city center of Bologna. The Asinelli tower is the tallest one on the right.

3.1. Chapel XVII

Firstly, we consider the analysis carried out in Chapel XVII of Varallo. The AE monitoring was led on the frescoed masonry on the north wall of the Chapel. The wall showed some damage: a vertical crack of about 3.00 m in length and fresco detachment. Four AE sensors were located around the vertical crack, while two were placed near the fresco detachment (Figure 3). The AE monitoring was conducted in two phases for a total duration of about 14 weeks. The first phase started on 9 May 2011 and finished on 16 June 2011; the second one was from 5 July 2011 to 5 September 2011. The monitoring results related to the chapel's structural integrity are reported in [2]. We interpreted the AE data considering the amplitude and time distribution of AE signals during the cracking phenomena. From this analysis, we found that the vertical crack, monitored on the north wall of the chapel, evolved during the acquisition period, while the process of detachment of the frescos was mainly related to the diffusion of moisture in the mortar substrate [2].

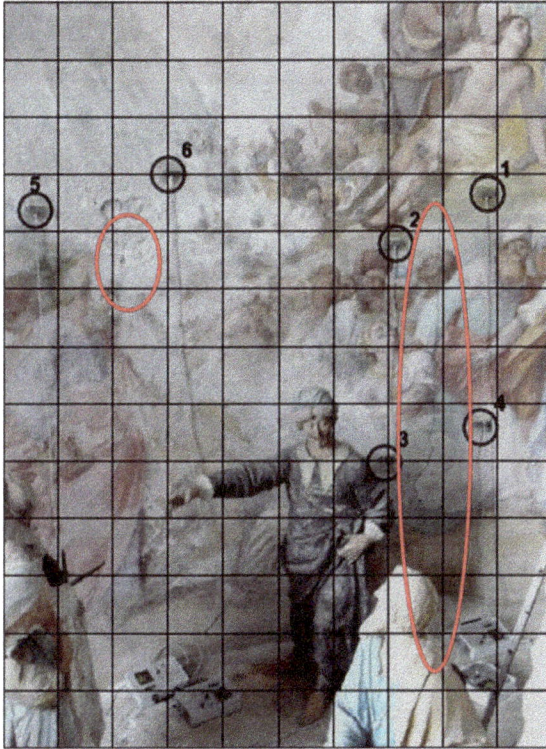

Figure 3. Chapel XVII. View of the monitored areas. Left side: sensors 5, 6, and the fresco detachment. Right side: sensors 1–4 and the vertical crack.

During the monitoring period, among all regional seismic events we considered 21 earthquakes with Richter magnitudes ≥1.2, having an epicenter within a radius from 60 to 100 km from Varallo. The strongest earthquake was a 4.3 magnitude event that occurred on 25 July 2011 at 12:31 p.m. in the Giaveno area (epicenter about 80 km from Varallo). Figure 4 displays the AE event rate, which counts simultaneous signal detection by multiple sensors as one event and the sequence of the earthquakes as functions of time, obtained during the monitoring period.

3.2. Asinelli Tower

The AE activity on Asinelli Tower was examined in a significant zone for monitoring purposes. This was developed by attaching six piezoelectric sensors to the northeast corner of the tower at an average height of *ca.* 9.00 m above ground level, immediately above the terrace atop the arcade. In this area, the double-wall masonry has an average thickness of *ca.* 2.45 m (Figure 5). AE monitoring began on 23 September 2010 and ended on 10 January 2011, for a total duration of about 16 weeks. By monitoring a significant part of this tower, the incidence of vehicle traffic, seismic activity and wind action on the damage evolution within the structure were assessed [3].

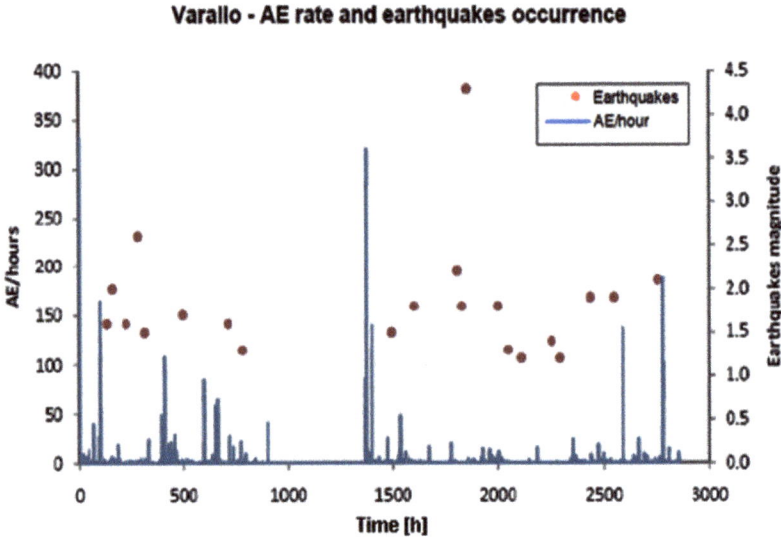

Varallo - AE rate and earthquakes occurrence

Figure 4. Chapel XVII: AE rate (blue chart) and nearby earthquake (red dots) occurrences as functions of time. The AE rate chart illustrates the number of AE events (averaged over 1 h), while seismic events are marked by points indicating occurrence time and Richter magnitude.

In this case, among all regional seismic events, we considered 43 earthquakes with a Richter magnitude ≥1.2, having an epicenter within a radius from 25 to 100 km from Bologna, as the most likely to affect the tower's stability. The strongest earthquakes were the 4.1 magnitude event that occurred on 13 October 2010 at 11:43 p.m. in the Rimini area (epicenter about 100 km from Bologna) and the 3.4 magnitude event on 21 November 2010 at 4:10 p.m. in Modena Apennines (epicenter about 50 km from Bologna). The AE instantaneous rate (averaged over 1 h) and the sequence of the earthquakes obtained during the monitoring are displayed in Figure 6. Also, in this case a correlation between peaks of AE activity in the structure and regional seismicity can be observed (Figure 6).

The tower, in fact, as in another case investigated by the authors regarding the Medieval Towers of Alba in Italy, is very sensitive to earthquake motions [9,10].

Figure 5. Front views and axonometric view of Asinelli Tower. Faces (**1**) South; (**2**) East; (**3**) North; (**4**) West; (**5**) Axonometric view. The AE transducers were applied to the northeast corner of the tower, in the zones marked with a circle.

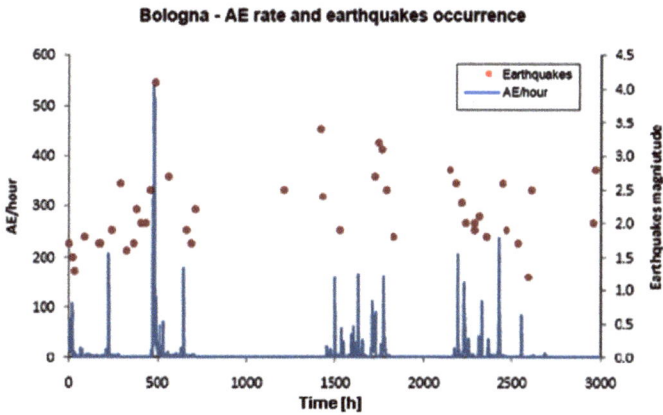

Figure 6. Asinelli Tower: AE rate (blue chart) and nearby earthquake (red dots) occurrences as functions of time. The AE rate chart illustrates the number of AE events (averaged over 1 h), while seismic events are marked by points indicating occurrence time and Richter magnitude.

4. Grassberger-Procaccia Algorithm

Acoustic emissions in materials and earthquakes in the crust are very similar phenomena though on very different scales. They both involve a sudden release of elastic energy from a source located in the medium: respectively, the tip of opening micro-cracks and the seismic hypocenter [11]. The aging and the damage of historical buildings due to the action of small and intermediate earthquakes, a situation very common in central and southern Italy, implies triggering of AE activity on these structures.

However, there appear to be some seismic events which follow AE bursts and, then, do not trigger structural AE activity. In fact, intense crises of crustal stress apparently cross large areas, as revealed by increased AE activity in several case histories from Italy [12], and precede the eventual occurrence of some earthquake within them.

This experimental evidence suggests that part of the AE activity from the structures might derive from precursive microseismic activity propagating across the ground-building foundation interface.

In this sense, AE monitoring provides twofold information: one concerning the structural damage and the other the amount of stress affecting the crust, in which the building foundation represents a sort of extended underground probe.

With this spirit we introduce a statistical approach to analyze spatial and temporal correlations between AE bursts from the structures and surrounding earthquakes, along the lines of similar investigations of the spatial and temporal correlations of earthquakes [2,3,9].

Seismicity levels of the Varallo and Bologna areas can be consistently compared by investigating the earthquakes' spatial distribution over equally sized regions (*i.e.*, circular regions with a radius of 100 km) centered at the two monitoring sites, during analogous monitoring periods (3000 h, as shown in Figures 4 and 6).

Quantification of this feature is possible through the notion of the fractal dimension of hypocenters, which is frequently estimated by the correlation integral because of its great reliability and sensitivity to small changes in clustering properties [13–15]. The correlation integral for hypocenter points embedded into a three-dimensional space is defined as follows:

$$C\left(r\right) = \frac{2}{N\left(N-1\right)} \sum_{k=1}^{N-1} \sum_{j=k+1}^{N} \theta\left(r - \left|x_k - x_j\right|\right) \tag{1}$$

where N is the total number of seismic events, x is the hypocentral coordinate vector, and Θ is the Heaviside step function ($\Theta(x) = 0$ if $x \leq 0$, $\Theta(x) = 1$ if $x > 0$). The correlation integral $C(r)$ gives the probability of finding another point in the sphere of radius r centered at an arbitrary point. For a fractal population of points, $C(r)$ scales with r as a power law, D being the correlation dimension:

$$C\left(r\right) \sim r^{D} \tag{2}$$

We apply the correlation integral to the earthquake sequences under consideration, *i.e.*, the earthquakes with epicentral distance in the range of 60–100 km from the Varallo Chapel and 25–100 km from the Asinelli Tower, which occurred during the AE monitoring periods. The choice of the minimum magnitude threshold, $m_L \geq 1.2$, was driven by completeness criteria.

The seismic events were taken from the Italian Seismological Instrumental and Parametric Data-Base ISIDE [16].

Two patterns are evidenced in Figure 7a,b: shorter ranges, 3 km $< r <$ 10–20 km, are defined by a unique slope giving a relatively high value of the correlation dimension, $D = 1.6$ (Varallo) and $D = 2.2$ (Bologna); for distances over 10–20 km, the slope gives a value of $D = 0.7$ (Varallo) and $D = 0.9$ (Bologna). For short ranges, $D \approx 2$ corresponds to a distribution of events over a fault plane, such as the active faults in the region surrounding Bologna; the highest ranges, corresponding to a value of $D < 1$, suggest a dust-like setting of seismicity as in the region surrounding Varallo (see Figure 8).The higher dimension for Bologna apparently reflects the recent and active tectonics of the Emilia-Romagna in the northeastern Italy, with respect to the seismically quiet zone of Varallo [17].

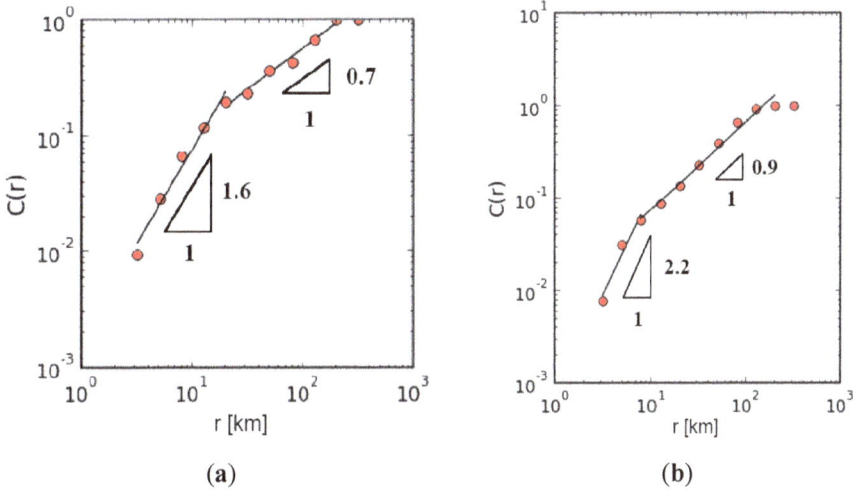

Figure 7. Correlation integral $C(r)$ *vs.* r (km) and its slope D in bi-logarithmic scale for Varallo regional seismicity (**a**) and Bologna regional seismicity (**b**).

5. Space-Time Correlation between AE and Seismic Events

Along the lines of studies on the space-time correlation among earthquakes based on a generalization of the Grassberger-Procaccia correlation integral [8], we calculate the degree of correlation between the AE time series and the sequence of nearby earthquakes recorded during the same period. No a priori assumptions on causal relationships between AE and seismic activities drive this approach, which is merely statistical. Only the choice of the time window and the region size is arbitrary.

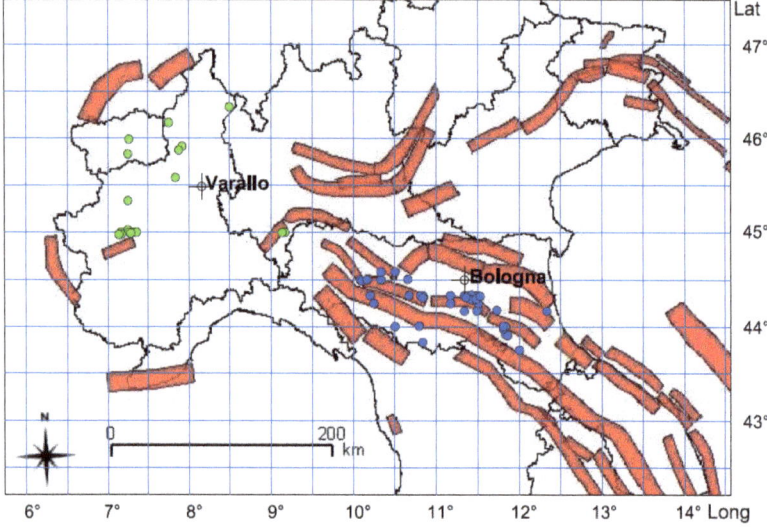

Figure 8. Map of Northern Italy depicting faults and epicenters.

The space-time combined correlation integral is defined as follows [7,9]:

$$C(r,\tau) = \frac{1}{N_{EQ}N_{AE}} \sum_{k=1}^{N_{EQ}} \sum_{j=1}^{N_{AE}} \theta\left(r - |x_k - x_0|\right) \cdot \theta\left(\tau - |t_k - t_j|\right) \quad (3)$$

where the index j runs over all the N_{AE} event bursts $\{x_0, t_j\}$, with x_0 being the coordinate vector of the monitoring site, while k runs over all the N_{EQ} seismic events $\{x_k, t_k\}$, with x_k being the epicentral coordinate vector.

Since the double-sum in Equation (3) counts pairs formed by an AE event (x_0, t_j) and a seismic event (x_k, t_k) with mutual epicentral distance $|x_0 - x_k| \leq r$ and time intervals $|t_j - t_k| \leq \tau$ among all N_{AE} and N_{EQ} possible pairs, $C(r,\tau)$ can easily be interpreted as the probability of an AE burst and an earthquake occurring with an inter-distance $\leq r$ and a time interval $\leq \tau$.

If $C(r, \tau,)$ exhibits power-law behavior both in space and time variables, time and space fractal dimensions D_t and D_s can be defined similarly to Equation (2):

$$D_t(r,\tau) = \frac{\partial \log C(r,\tau)}{\partial \log \tau} \quad (4)$$

and

$$D_s(r, \tau) = \frac{\partial \log C(r, \tau)}{\partial \log r} \tag{5}$$

The time correlation dimension $D_t(r, \tau)$ characterizes the time-coupling of AE bursts to the earthquakes occurring up to a given distance r from the AE monitoring site, whereas the space correlation dimension $D_s(r, \tau)$ characterizes the spatial distribution of nearby earthquakes with separation in time from AE bursts not exceeding a given τ.

It is worth noting that Equation (5) identifies with Equation (2) if τ equals the whole time data span, as all earthquakes would be taken into account in the Grassberger-Procaccia integral $C(r, \tau)$.

Actually, the imposed condition $|t_j - t_k| \leq \tau$ does not specify the chronological order between the two types of event. On the other hand, the AE time series and the sequence of nearby earthquakes are two very closely interrelated sets in the time domain. Therefore, a given AE burst might be either due to structural damage triggered by a seismic event or due to precursive microseismic activity. In spite of intrinsic difficulties in high-frequency propagation across disjointed media (in particular at the ground-building foundation interface), AE bursts apparently indicate widespread crustal stress crises during the preparation of a seismic event [9,12], when part of the related deformation energy stored in the Earth's crust might be transferred to the building foundations.

It is interesting to carry out a probabilistic analysis of the available data considering AE events once as preceding an earthquake, *i.e.*, as potential seismic precursors, and next as following an earthquake, *i.e.*, as structural aftershocks. In this way, the obtained conditioned probability distributions can be compared in order to discover the prevailing trend. This analysis is performed applying the modified correlation integral [9]:

$$C^{\pm}(r, \tau) = \frac{1}{N_{EQ}N_{AE}} \sum_{k=1}^{N_{EQ}} \sum_{j=1}^{N_{AE}} \theta(r - |x_k - x_0|) \cdot \theta(\tau - |t_k - t_j|) \tag{6}$$

where "+" and "−" are used to account for AE events, respectively, as seismic precursors and as aftershocks. $C^+(r, \tau)$, for example, gives the occurrence probability of an AE burst followed by an earthquake in the next interval τ and within a radius r of the monitoring site.

We applied the modified integral (Equation 6) to investigate the space-time correlation between the two above-mentioned regional seismic sequences and the AE time series from the Asinelli Tower and the Varallo Chapel (see Figures 4 and 6).

The AE signals originating from damage sources definitely localized in the building's masonry are filtered out. In this way, we can preliminarily distinguish between environmental contributions due to crustal trembling (external source) and structural damage contributions (inner sources localized by means of triangulation techniques [18]). Then, by applying the modified Grassberger-Procaccia correlation integral to the remaining AE data series and to the earthquake sequence, we obtained the cumulative probabilities $C^{\pm}(r, \tau)$ as functions of time $C^{\pm}{}_r(\tau)$ for different values of range r. This provides two-dimensional plots, which are easier to read than a three-dimensional representation of $C^{\pm}(r, \tau)$.

Figure 9a,b show that $C^{+}{}_r > C^{-}{}_r$ for all considered values of range r, suggesting that AE bursts are more likely to precede earthquakes than to follow them. The interpretation of this evidence is that the monitored structures behave as receptors of microseismic precursive activity during the preparation of a seismic event, *i.e.*, as sensitive earthquake receptors.

6. AE Clustering in Time as a Seismic Precursor or an Aftershock

Since the cumulative probabilities $C^{\pm}{}_r(\tau)$ are represented in a time domain ranging from 1 to 14–16 weeks, shorter values of τ are further investigated.

In particular, time coupling of regional seismicity to AE activity from Chapel XVII and the Asinelli Tower is analyzed in terms of D_t for τ ranging from 3 to 24 h after an earthquake occurrence, and from 10 min to 24 h before an earthquake occurrence.

The varying local slope D_t of Log C^{-} vs. Log τ reveals that the AE activity following seismic events is not equally probable over the time (see Figure 10a,b). For short time delays, $\tau = 3$–10 min, high values of the time fractal dimension, $D_t = 1.38$ and 1.24, indicate tight coupling of the AE activity to the earthquakes. In other words, AE bursts following an earthquake are more likely to occur within $\tau = 3$–10 min after the seismic event. Then, such short time delays suggest a triggering action exerted by nearby earthquakes on damage processes of Chapel XVII and the Asinelli Tower. Contrarily, for τ ranging from 1.0×10^3 to 1.440×10^3 min, the obtained low values of D_t, 0.48 and 0.45, suggest that the effects of nearby earthquakes on the structural damage evolution disappear after 24 h.

In regards to the AE events preceding earthquakes, *i.e.*, considering AE as "seismic precursors", the linear bi-logarithmic plots of C^{+} ($D_t = 1.06$ and 0.94) shown in Figure 11 describe a uniform probability density of finding AE events prior to an earthquake for a wide range of time intervals, up to $\tau = 1.440 \times 10^3$ min. In other words, we observe a constant AE activity in the 24 h preceding an earthquake. The absence of accelerating precursive activity in the presented case studies confirms the need for further investigation, possibly with the aid of electromagnetic seismic precursors.

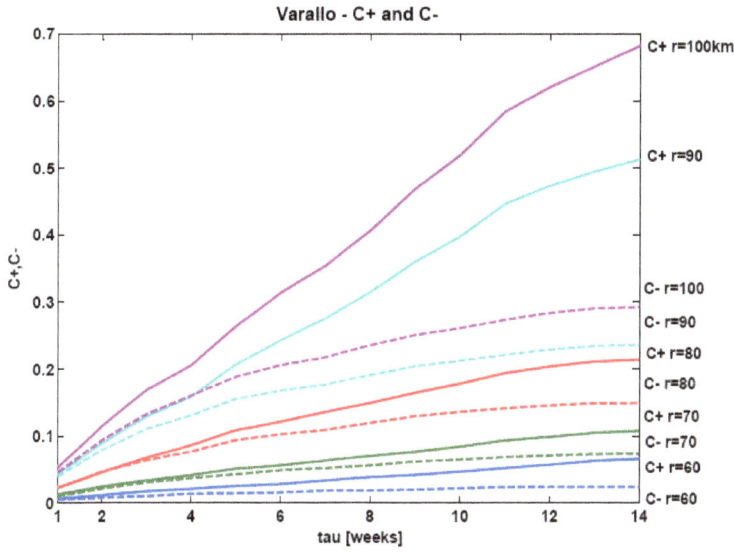

(a)

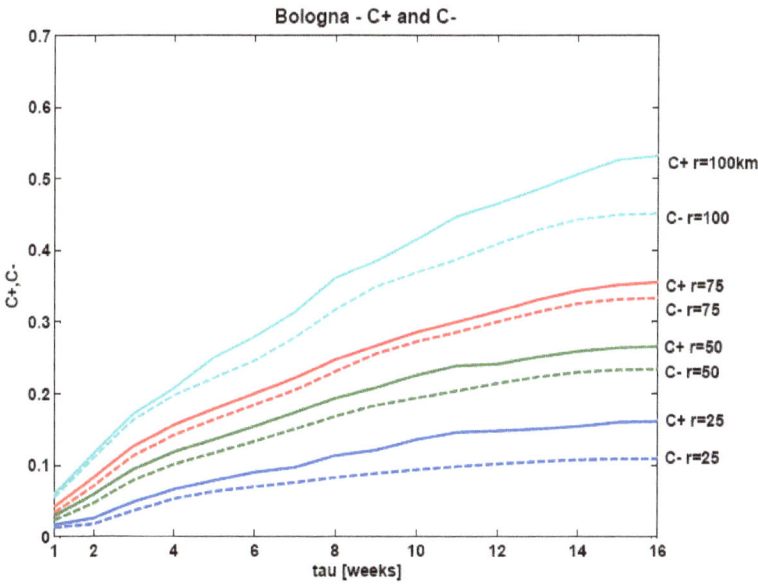

(b)

Figure 9. Modified correlation integrals $C^{\pm}_r(\tau)$, both considering AEs as earthquake "precursors" (+) and as "damage aftershocks" (−), plotted as functions of the time separation τ for different values of the spatial range r for: (**a**) Chapel XVII and area around the city of Varallo; (**b**) Asinelli Tower and area around the city of Bologna.

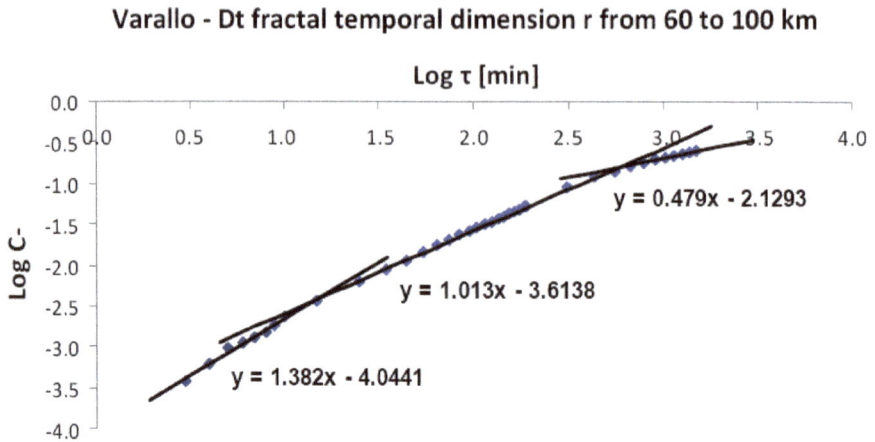

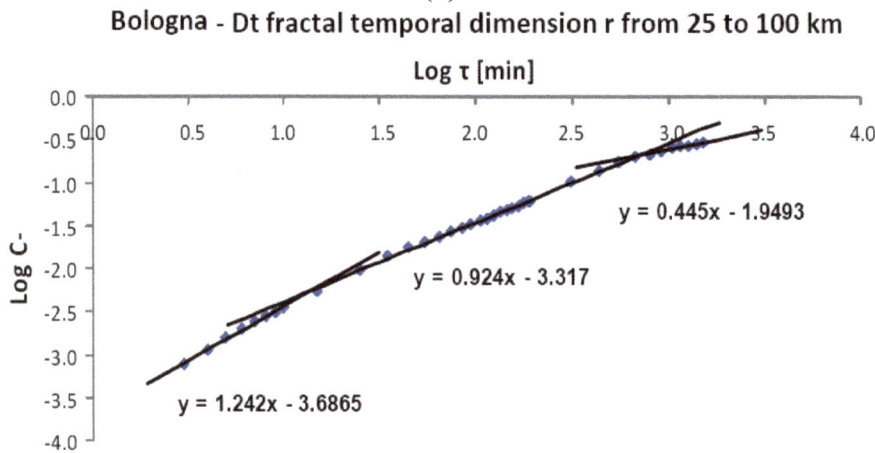

Figure 10. Analysis of time-clustering features of AE events considered as "seismic aftershocks" by the time correlation dimension D_t, as a local slope of Log C^- vs. Log τ. The time range for τ is 3 min to 24 h. The selected seismic events are all those considered during the defined monitoring periods. (**a**) Chapel XVII (Varallo); (**b**) Asinelli Tower (Bologna).

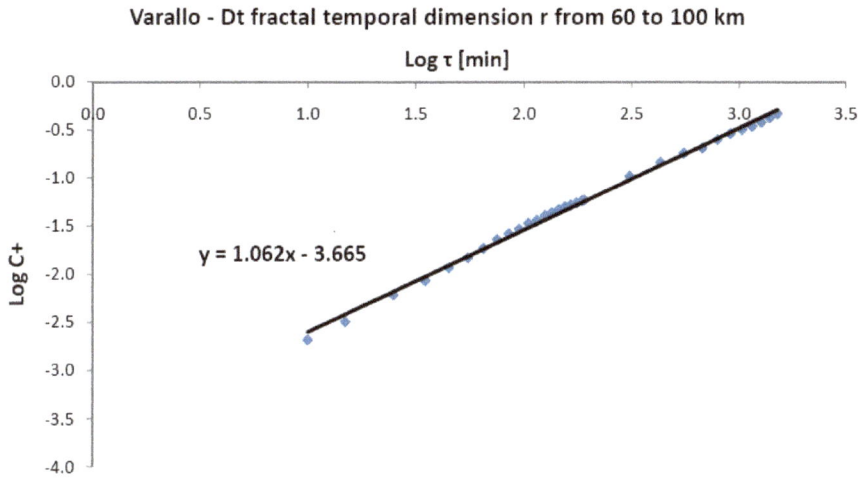

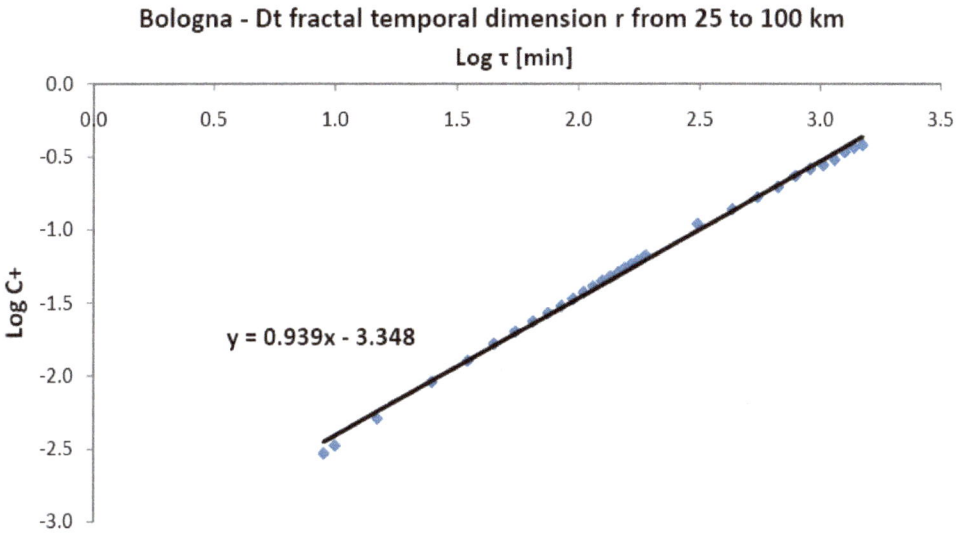

Figure 11. Analysis of time-clustering features of AE events considered "seismic precursors" by the time correlation dimension D_t, as the local slope of Log C^+ vs. Log τ. The time range for τ is 10 min to 24 h. The selected seismic events are all those considered during the defined monitoring periods. (**a**) Chapel XVII (Varallo); (**b**) Asinelli Tower (Bologna).

7. Conclusions

In order to assess the seismic hazard in different areas, it is suggested to observe the spatial and temporal correlation of seismic events and monitored AEs. It emerges that AEs behave as earthquake precursors. A statistical method of analysis is proposed, based on the Grassberger-Procaccia integral. It gives the cumulative probability of the events' occurrence in a specific area, considering the AE records and the seismic events during the same period of time. Two important constructions of Italian cultural heritage were considered: the "Sacred Mountain of Varallo" and "Asinelli Tower". Given the interesting premises that emerged from this study, the Grassberger-Procaccia integral could also be applied more extensively in several monitoring sites belonging to different seismic regions, not only for defining the seismic hazard, but also to handle it as an effective tool for earthquake risk mitigation.

However, the authors are aware that a rigorous investigation to ascertain the existence of crustal stress crises crossing areas of a few hundred kilometers' radius during the preparation of a seismic event would require simultaneous and numerous operations of AE monitoring sites, adequately placed in the territory.

Acknowledgments: E. de Filippis, Director of the Piedmont Sacred Mountains Institute, is gratefully acknowledged. The authors thank the Municipality of Bologna and Eng. R. Pisani for having allowed the study on the Asinelli Tower.

Author Contributions: A.C. supervised the research; P.C. and S.I. performed the numerical simulations; G.N. carried out AE and NDT tests; G.L. wrote the paper with discussions from all the authors.

Conflicts of Interest: The authors declare no conflict of interest.

References

1. Niccolini, G.; Carpinteri, A.; Lacidogna, G.; Manuello, A. Acoustic emission monitoring of the Syracuse Athena Temple: Scale invariance in the timing of ruptures. *Phys. Rev. Lett.* **2011**.
2. Carpinteri, A.; Lacidogna, G.; Invernizzi, S.; Accornero, F. The Sacred Mountain of Varallo in Italy: Seismic risk assessment by Acoustic Emission and structural numerical models. *Sci. World J.* **2013**, *2013*, 1–10.
3. Carpinteri, A.; Lacidogna, G.; Manuello, A.; Niccolini, G. A study on the structural stability of the Asinelli Tower in Bologna. *Struct. Control HLTH* **2015**.
4. Bak, P.; Christensen, K.; Danon, L.; Scanlon, T. Unified scaling laws for earthquakes. *Phys. Rev. Lett.* **2002**, *88*, 178501–178504.
5. Parson, T. Global Omori law decay of triggered earthquakes: Large aftershocks outside the classical aftershock zone. *J. Geophys. Res.* **2002**, *107*, 2199–2218.
6. Corral, A. Long-term clustering, scaling and universality in the temporal occurrence of earthquakes. *Phys. Rev. Lett.* **2004**.

7. Tosi, P.; de Rubeis, V.; Loreto, V.; Pietronero, L. Space-time combined correlation integral and earthquake interactions. *Ann. Geophys.* **2004**, *47*, 1–6.

8. Grassberger, P.; Procaccia, I. Characterization of strange attractors. *Phys. Rev. Lett.* **1983**, *50*, 346–349.

9. Carpinteri, A.; Lacidogna, G.; Niccolini, G. Acoustic emission monitoring of medieval towers considered as sensitive earthquake receptors. *Nat. Hazard Earth Syst. Sci.* **2007**, *7*, 251–261.

10. Carpinteri, A.; Lacidogna, G. Structural monitoring and integrity assessment of medieval towers. *J. Struct. Eng.* **2006**, *132*, 1681–1690.

11. Scholz, C.H. The frequency-magnitude relation of microfracturing in rock and its relation to earthquakes. *Bull. Seismol. Soc. Am.* **1968**, *58*, 399–415.

12. Gregori, G.P.; Paparo, G. Acoustic emission (AE). A diagnostic tool for environmental sciences and for non destructive tests (with a potential application to gravitational antennas). In *Meteorological and Geophysical Fluid Dynamics*; Schroeder, W., Ed.; Science Edition: Bremen, Germany, 2004; pp. 166–204.

13. Kagan, Y.Y.; Knopoff, L. Spatial distribution of earthquakes: The two-point correlation function. *Geoph. J. Roy. Astron. Soc.* **1980**, *62*, 303–320.

14. Hirata, T. A correlation between the *b*-value and the fractal dimension of earthquakes. *J. Geophys. Res.* **1986**, *94*, 7507–7514.

15. Carpinteri, A.; Lacidogna, G.; Niccolini, G.; Puzzi, S. Morphological fractal dimension *versus* power-law exponent in the scaling of damaged media. *Int. J. Damage Mech.* **2009**, *18*, 259–282.

16. Italian Seismological Instrumental and Parametric Data-Base (ISIDE). Available online: http://iside.rm.ingv.it/iside/standard/result.jsp?rst=1&page=EVENTS#result (accessed on 5 December 2015).

17. Burrato, P.; Vannoli, P.; Fracassi, U.; Basili, R.; Valensise, G. Is blind faulting truly invisible? Tectonic-controlled drainage evolution in the epicentral area of the May 2012, Emilia-Romagna earthquake sequence (northern Italy). *Ann. Geophys. Italy* **2012**, *55*, 525–531.

18. Carpinteri, A.; Xu, J.; Lacidogna, G.; Manuello, A. Reliable onset time determination and source location of acoustic emissions in concrete structures. *Cem. Concr. Compos.* **2012**, *34*, 529–537.

Electromagnetic Acoustic Transducers Applied to High Temperature Plates for Potential Use in the Solar Thermal Industry

Maria Kogia, Liang Cheng, Abbas Mohimi, Vassilios Kappatos, Tat-Hean Gan, Wamadeva Balachandran and Cem Selcuk

Abstract: Concentrated Solar Plants (CSPs) are used in solar thermal industry for collecting and converting sunlight into electricity. Parabolic trough CSPs are the most widely used type of CSP and an absorber tube is an essential part of them. The hostile operating environment of the absorber tubes, such as high temperatures (400–550 °C), contraction/expansion, and vibrations, may lead them to suffer from creep, thermo-mechanical fatigue, and hot corrosion. Hence, their condition monitoring is of crucial importance and a very challenging task as well. Electromagnetic Acoustic Transducers (EMATs) are a promising, non-contact technology of transducers that has the potential to be used for the inspection of large structures at high temperatures by exciting Guided Waves. In this paper, a study regarding the potential use of EMATs in this application and their performance at high temperature is presented. A Periodic Permanent Magnet (PPM) EMAT with a racetrack coil, designed to excite Shear Horizontal waves (SH_0), has been theoretically and experimentally evaluated at both room and high temperatures.

Reprinted from *Appl. Sci.* Cite as: Kogia, M.; Cheng, L.; Mohimi, A.; Kappatos, V.; Gan, T.-H.; Balachandran, W.; Selcuk, C. Electromagnetic Acoustic Transducers Applied to High Temperature Plates for Potential Use in the Solar Thermal Industry. *Appl. Sci.* **2015**, *5*, 1715–1734.

1. Introduction

Solar thermal industry is an environmentally friendly means of power generation, converting solar energy into electricity. Concentrated Solar Plants (CSPs) are employed in the solar thermal industry for collecting sunlight and converting it firstly into heat and later into electricity. There are several types of CSPs such as solar towers, dish concentrators, and parabolic trough CSPs; the latter is the most widely used [1] and is shown in Figure 1a.

Figure 1. (a) Parabolic Trough Concentrated Plants [2]; (b) reflector and absorber tube [3]; (c) absorber Tube [4].

The two essential components of parabolic trough CSPs are the long, parabolic trough shaped reflectors and the absorber tubes. The reflector collects and focuses the sunlight to the absorber tube, which is located at the focal point of the reflector and runs its whole length. The absorber tubes are composed of long, thin, stainless steel tubes covered by a glass envelope under vacuum, as shown in Figure 1b,c. Their entire surface should be exposed to sunlight and therefore there are few access points for any extra hardware to be attached to them. Inside the tubes, working fluid, either water/molten salt or synthetic oil, is flowing, absorbing the heat of the sunlight and transmitting it into heat/steam engines for the final stages of the power generation procedure.

The high temperatures (400–500 °C), the contraction/expansion endured due to the fast cooling of the tube at high temperatures, and the vibrations can result in the appearance of defects either on the surface or inside the absorber tube. Problems with absorber tubes have been reported [5–7], such as creep, thermo-mechanical fatigue, and hot corrosion. The turbulent mixing of hot and cold flow streams of the working fluid may result in temperature variations along the pipe and consequently in thermomechanical fatigue [8–11]. Corrosion is another common problem absorber tubes may suffer from [12–14]; local pitting corrosion can cause the initiation of stress corrosion cracking or result in small-scale leaks. Most stainless steel pipes are vulnerable to pitting corrosion and stress corrosion cracking as well [15]. The low flow of the working fluid may also lead the absorber tube to overheat; consequently, the absorber tube may be subject to creep damage, thermal oxidation, softening, and/or stress rupture [16].

Non Destructive Testing (NDT) techniques are suggested to be applied for the structural health monitoring or inspection of the absorber tubes. Several NDT techniques can be used for either the inspection or the monitoring of high temperature structures; Acoustic Emission (AE), Eddy Current (EC), Holographic Interferometry, Laser Ultrasonic, Guided Wave Testing (GWT), and Infrared Thermography (IR) are some of those. These techniques have been reported for operating at temperatures up to 300 °C, though with shortcomings. Some of the drawbacks are qualitative results, sensitivity to noise, laboratorial utilization, and the need for coupling media [17–20]. AE is a passive NDT technique that has been widely deployed in Structural Health Monitoring (SHM); it monitors the elastic waves generated after the initiation or propagation of a crack [21]. Nevertheless, AE is sensitive to noise and gives qualitative results. EC is a non-contact technique that can be employed for the inspection of any electrically conductive material; however, it is subject to the skin effect, leading it to be mainly used for the detection of surface and subsurface defects [22]. Holographic Interferometry can give detailed results and be used for the detection of small defects, but it is mainly used for laboratorial tests as its setup is complicating and it is sensitive to vibrations [23]. Laser Ultrasonic has a small and adjustable footprint; therefore, it can be used for the inspection of irregular surfaces and samples of small and complex geometry. It induces high frequency ultrasound and thus very small defects can be detected as well. However, its setup is complicated and it is mainly used for laboratorial tests. GWT is used for the inspection or monitoring of large structures; mainly piezoelectric transducers are employed upon the structure being tested exciting/receiving guided waves [24]. Piezoelectric transducers require direct access to the specimen and a coupling medium (usually a water-based gel), and their response cannot travel through a vacuum. IR can be used in this application mainly for overheating identification; however, the length of the pipes makes this technique practically inefficient. The camera needs to scan the whole length of the pipe, which is time-consuming and may not be possible while the absorber tube is operating [25].

The monitoring of absorber tubes requires a non-contact NDT technique that can be applied at high temperatures without the need for a couplant and can inspect the whole length of the tube from a single point. Therefore, GWT can be applied with the use of non-contact transducers that can withstand high temperatures and excite/receive guided waves and more particularly a T(0,1) wave mode (or SH_0) [26,27]. The EMAT can be used for this application, since it is a non-contact technology that has been used in GWT and can be applied for the inspection of structures under hostile conditions such as moving specimen and elevated temperatures.

In the following section the main operating principles of EMATs are described and emphasis is also placed on the potentials and limitations of EMATs on guided waves and high temperatures. A brief description of this study follows. The results obtained from the theoretical study are shown in Section 4, where the dispersion curves of the absorber tubes were calculated and processed and a PPM EMAT with racetrack coil has been evaluated mainly regarding its guided wave purity characteristics at both room and high temperatures. In Sections 5 and 6 the experimental setup and the results of the experimental evaluation of a pair of PPM

EMATs are presented. The EMATs were tested as far as their efficiency to excite SH_0 is concerned and the parameters that may affect their performance at both room and high temperatures. The EMATs were validated for their wave purity, their sensitivity to noise, their lift-off limitations, their power requirements, and their performance at high temperatures while they were exciting/receiving guided waves. Hence, a comparison between the simulated and the experimental results is finally presented. A high temperature EMAT exciting SH_0 wave and operating up to 400–550 °C for either monitoring or long-term inspection is also to be accomplished in the near future.

2. EMAT Technology

EMAT is a non-contact technology of transducers that can be used for the inspection of moving structures or a specimen that operates at elevated temperatures. Their response can travel through a vacuum as well, making them even more suitable for this application compared to piezoelectric transducers. EMATs can excite/receive guided waves and thus they can be employed in GWT for either the inspection or the monitoring of large structures [27–30]. Hence, high-temperature EMATs may be used for the inspection or monitoring of absorber tubes. Nevertheless, the high-temperature EMATs that have been reported so far have been designed for thickness measurements [31–33]. Therefore, a high-temperature EMAT that excites/receives guided waves for the long-term inspection of high-temperature structures is still required.

A typical EMAT transducer is composed of either a permanent magnet or an electromagnet for the generation of a static magnetic field, and a coil. The coil is driven by an alternating current that generates a dynamic magnetic field. This dynamic magnetic field induces an eddy current in the specimen placed below the coil. If the specimen being tested is an electrical conductor and non-ferromagnetic, mainly Lorentz force is exerted upon the material particles; Equation (1) shows that Lorentz force is equal to the product of eddy current density and the overall magnetic field:

$$FL = Je \times (Bst + Bdyn) \tag{1}$$

where F_L is Lorentz force, J_e is the eddy current density, B_{st} stands for the static magnetic field, and B_{dyn} refers to the dynamic magnetic field. In this application, the absorber tubes are made of 316 L stainless steel, which is paramagnetic; therefore the main force generated in them would be Lorentz force.

The EMAT configuration and the material properties of its main components affect EMAT performance at both room and high temperatures. The material the coil is made of influences the EMAT performance. The impedance of the coil affects the energy transmitted to the specimen and the SNR/quality of the signal received; Equation (2) shows how the resistance increases with temperature rise and Equation (3) demonstrates the relationship between the temperature and the noise level of the signal received:

$$R(T) = R0\left[1 + \alpha\left(T - T0\right)\right] \tag{2}$$

where R_0 is the resistance of a single turn coil, α is the temperature coefficient of resistance, and T_0 is the room temperature.

$$V\mathrm{Noise} = (4K\beta TR\mathrm{EMAT})^2 \tag{3}$$

where K is the Boltzmann constant, T is the temperature measured in Kelvin, β is the bandwidth, and R_{EMAT} is the resistance of the EMAT coil per turn. Consequently, the coil should be preferably made of a low-resistance material such as copper. However, copper is known to oxidize at high temperatures and, therefore, other materials could be used for this application such as silver, platinum, nickel, or constantan. Hernandez-Valle (2011) has already made an electromagnet EMAT that can operate at up to 600 °C without any cooling system for thickness measurements and apart from designing a high temperature electromagnet he has also tested several coil designs at high temperatures [32].

The permanent magnets also have limitations regarding their maximum operating temperature. The Maximum Operating Temperature (MOT) of a magnet is equal to half of its Curie Point; beyond Curie Temperature, the strength of the magnet decreases rapidly. Consequently, high Curie Point magnets are preferable for high-temperature applications. Nevertheless, the magnetic strength of the main two types of high-temperature magnets, Alnico (with a maximum operating temperature of 500 °C) and SmCo (with a maximum operating temperature of 300 °C), is smaller than the magnetic strength of Neo (NdFeB) magnets. However, NdFeB magnets cannot be used at temperatures higher than 200 °C.

A cooling system may also be required so that both the magnets and the coil will operate efficiently. Idris *et al.* have designed and tested a water-cooled EMAT that can obtain signals up to 1000 °C for thickness measurements. The EMAT was exposed to the heat source for as much time as it needed for the signal to be recorded and then it was removed [33]. Oil- and air-cooled EMATs have also been designed. Generally, the reported high-temperature EMATs seem to operate at high temperatures for short periods of time, which are suitable for inspection, but they cannot be used in long-term condition monitoring. Consequently, an EMAT operating at high temperatures (500 °C) for long periods of time is required and could be used for the structural health monitoring of high-temperature structures such as the absorber tubes.

3. Our Methodology

This study is divided into two main parts, the theoretical validation of the PPM EMAT design and the experimental evaluation of a pair of PPM EMATs. Figure 2 shows the schematic of the methodology followed in this study. In the theoretical part, the dispersion curves of a 3 mm thick, 316 L stainless steel plate were calculated for both room and high temperatures. The resonant frequency of EMAT and the wave velocity of SH_0 at the resonant frequency of EMAT for room temperature, 60 °C, 100 °C and 180 °C are also calculated. Electromagnetic simulations in COMSOL were also carried out for the theoretical validation of this EMAT design regarding its wave purity characteristics at both room and high temperatures. Experiments were conducted with a pair of PPM EMATs with racetrack coil on a stainless steel plate. During the experiments the EMATs were tested regarding their wave purity potentials, their lift-off limitations, their power requirements, and their performance at high temperatures (up to 180 °C). The results obtained from the experimental procedure were compared with the results from the theoretical validation of the EMAT and the conclusions of this study are summarized and presented in the last part of this paper.

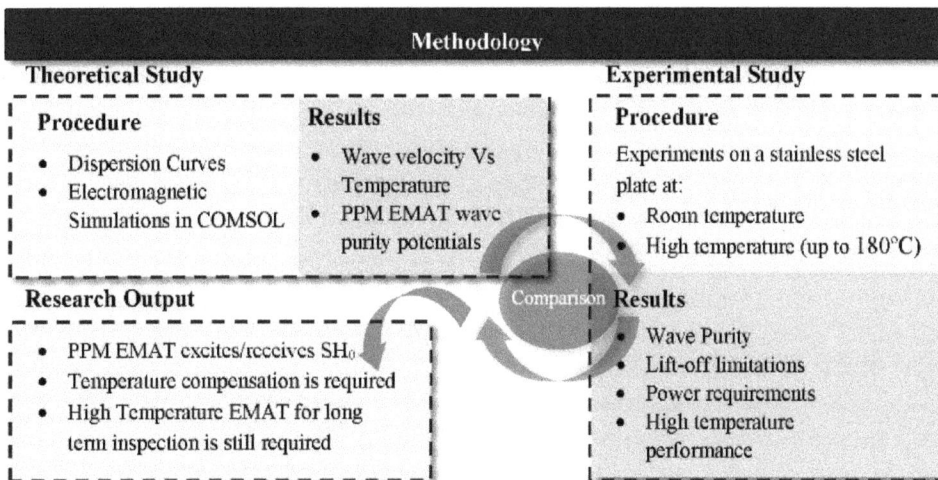

Figure 2. Methodology schematic.

4. Theoretical Study

4.1. Dispersion Curves

GWT is an NDT technique that can be employed in this applicaton. Low-frequency waves can travel big distances without being significantly attenuated;

however, their relatively large wavelength limits the size of defect that can be detected. A 10 mm length of defect was set as the minimum size defect that should be detected; considering that the wavelength should be smaller than the double of the minimum size defect, in this case the wavelength should be smaller than 20 mm. A SH_0 wave of 12 mm wavelength was introduced to this structure.

As the wave propagates inside the specimen, it strikes at the boundaries of the medium, resulting in the change of its waveform. Consequently, there is an infinite number of possible wave modes that may appear in the material; their velocity changes in respect not only to the material properties of the specimen but to its geometry as well. Hence, the same mode at a different frequency may propagate with a different mode shape and velocity. This phenomenon is called dispersion; most of the wave modes in guided waves are dispersive Nevertheless, $T(0,1)$ (or, alternatively, SH_0 on plates) is not dispersive, making the interpretation of the signal received less complicated. Its displacement is also in-plane, making it more suitable for this application, since the wave will propagate all along the pipe without it being affected by the working fluid that flows inside the pipe; $T(0,1)$ cannot propagate in liquids. The number of wave modes that may appear in a plate is also smaller compared to a pipe and therefore the interpretation of the signal received from a plate may be less complicated as well.

In this preliminary study, the EMAT is simulated and designed for exciting SH_0 waves, which will propagate axially on a 316 L stainless steel plate. The dispersion curves of a 3 mm thick, 316 L stainless steel plate of 8000 kg/m^3 density, 195 GPa Young's modulus, and 0.285 Poisson ratio have been calculated and demonstrated in Figure 3. This figure shows that the wavelength curve crosses the SH_0 curve at 256 kHz frequency and thus this should be the resonant frequency of the EMAT. The wavelength curve also crosses the S_0, A_0, and A_1 curves; Table 1 shows at which frequency the wavelength curve crosses each wave mode curve and their wave velocity at these frequencies. Hence, an EMAT designed to excite SH_0 of 12 mm wavelength should be driven with an AC current of 256 kHz frequency. Otherwise if the EMAT is tuned to any other frequency, it will be likely for it to excite Lamp waves instead of SH_0. This is more possible if the coil design is meander instead of racetrack, for meander coils are used for Lamp waves as well. At 256 kHz, the SH_0 and the A_0 have the same group velocity. As a result, both wave modes can be excited/received from the EMAT, simultaneously resulting in a more complicated signal, since A_0 is dispersive at this frequency. Nevertheless, the orientation of the displacement of each wave mode is different; the SH_0 has an in-plane displacement while the A_0 has an out-of-plane displacement. Thus, a further study should be conducted regarding the wave mode, and more particularly the displacement to which the PPM EMAT is sensitive. Consequently, an electromagnetic simulation calculating and showing the amplitude and the direction of the wave modes generated by a PPM EMAT is needed.

Material properties such as the Young's Modulus, Poisson ratio, and density of a specimen change with temperature rise as well, resulting in variations in the velocity of the propagating wave mode. The Young's modulus and density of the specimen decrease with an increase in temperature, while the Poisson ratio increases. These changes result in a decrease in the velocity of the SH_0. Table 1 summarizes the wave velocity of SH_0, S_0, A_0, and A_1 at room temperature, 60 °C, 100 °C, and 180 °C. Hence, as the temperature rises and the wave velocity decreases, any reflections received from the EMAT should shift in time. Hence, the temperature should be kept stable during the inspection or temperature compensation should take place during the signal interpretation.

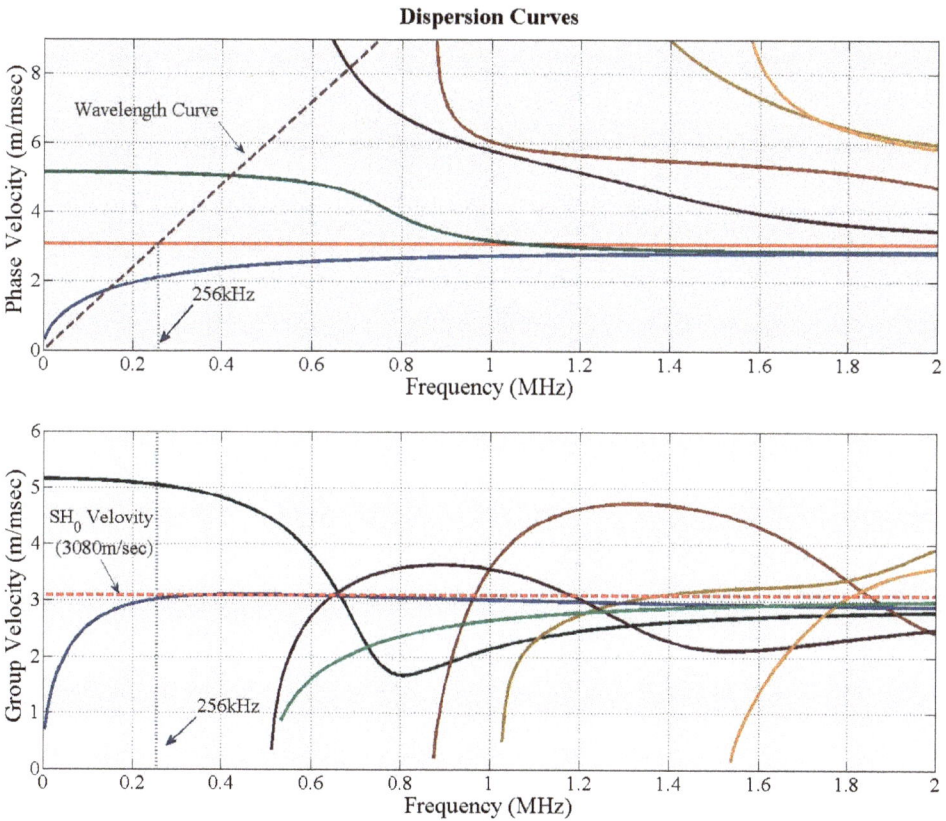

Figure 3. Dispersion curves of a 3 mm thick 316 L stainless steel plate.

Table 1. Dispersion Curves, wave velocity, and frequency.

Dispersion Curves—SH_0/S_0/A_0/A_1 Frequency & Wave Velocity								
Value	Frequency (kHz)				Velocity (m/s)			
Temperature (°C)	22	60	100	180	22	60	100	180
SH_0	256	255	254.2	250.6	3080	3067	3051	3008
S_0	420	419	415.5	405	5044	5022	5003	4944
A_0	143	140	139	136	1736	1720	1711	1685
A_1	676	673	668	657	8243	8217	8203	8129

4.2. Electromagnetic Simulations

Both the coil design and the magnet configuration affect the distribution of Lorentz force in space, its direction, and its amplitude. As a result, the wave mode the EMAT is to generate and/or receive gets affected by the configuration of its two main components. An EMAT configuration that can be used for the SH_0 is the PPM EMAT with racetrack coil [26–40].

In a PPM EMAT, the distance between two adjacent magnets whose magnetic field has the same direction (pitch) is equal to the wavelength; the arrangement of the magnets is illustrated in Figure 4. Racetrack coil is also used in this EMAT configuration. In this coil design there are no gaps within its turns, resulting in its broadband response in frequency. Hence, only the magnets' arrangement and the frequency of the AC current can affect the frequency response of EMAT.

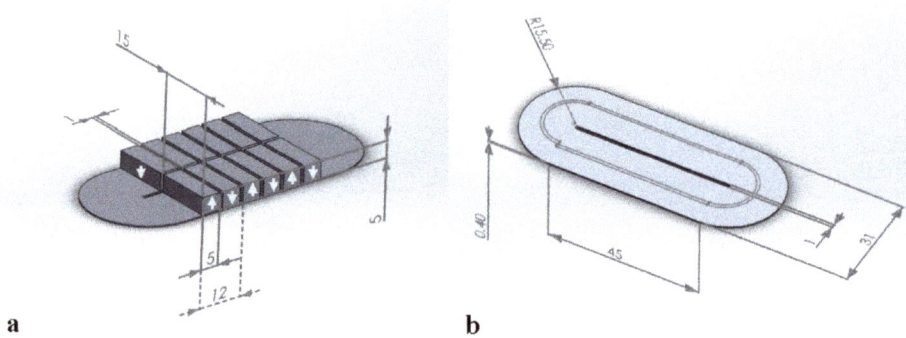

Figure 4. (a) Magnet arrangement in a PPM EMAT; (b) racetrack coil.

Finite Element Analysis (FEA) was used for the theoretical validation of this EMAT design regarding its guided wave purity characteristics at room temperature and at 180 °C. A coupled electromagnetic and mechanical analysis was carried out in COMSOL. A 3D model was created, in which a 12-magnet PPM EMAT was evaluated regarding eddy current, magnetic flux, and Lorentz force distribution. In this model two arrays of six Nd-Fe-B magnets each was simulated; each magnet has a 15 mm width, 5 mm depth, 5 mm height, and magnetization of 750 kA/m, and the direction of their magnetic flux is on the z axis. Their arrangement in space is the same as Figure 4 and, thus, the distance between one another is 1 mm. Due to the high requirements in time and computational power these complicated models have, the copper coil has been simplified and designed as two rectangular blocks of 35 mm width, 15 mm depth, and 0.4 mm height. The AC current driven to the coil is a two-cycle, Hanning windowed sinusoidal wave of 20 A amplitude and 256 kHz central frequency with direction on the x axis. The orientation of the excitation current flowing inside one rectangular coil is opposite to the orientation of the current inside the other coil. The EMAT has a lift-off of 0.6 mm from the specimen, which is a 316 L stainless steel plate of 3 mm thickness, 750 mm width, and 750 mm depth and of the same material properties as the plate simulated in the previous section at room temperature and at 180 °C. The material, magnetic, and electrical properties of the EMAT did not change with temperature.

Figure 5a,b show the excitation/eddy current at 4 μs and the static magnetic flux distribution, respectively. Both the electric and the magnetic field are uniformly distributed. The orientation of the eddy current alters between the two sides of the coil and it is on the x–y plane. The orientation of the magnetic field alters as it is depicted in Figure 5b; its orientation is mainly on the z axis and its maximum strength is observed at the center of each magnet separately, as is expected. Thus, the Lorentz force generated should mainly result in an in-plane displacement. A probe was also placed 30 cm away from the EMAT for obtaining the in-plane (x–y plane) and out-of-plane (x–z plane) displacement. Figure 5c shows that the in-plane displacement maximizes at 98 μs while the out-of-plane displacement maximizes at 100 μs; the Time of Flight (ToF) of the in-plane displacement matches with the wave velocity of the SH_0, as was calculated from the dispersion curves. while the out-of-plane displacement can be the A_0. Hence, this EMAT configuration may excite the A_0 wave mode as well. However, the maximum value of the out-of-plane displacement is significantly smaller than the in-plane displacement and therefore experimental validation of this EMAT design regarding its guided wave purity characteristics is still required.

Figure 5d shows the in-plane and out-of-plane displacement 30 cm away from the EMAT at 180 °C; both displacements are shifted in time, as was expected since the wave velocity changes with temperature rise and the ToF matches with the wave velocity as it was calculated from the dispersion curves. The amplitude of the in-plane displacement decreased; however, the amplitude of the out-of-plane displacement did not decrease with the temperature rise. As was mentioned, in the FEA model only the material properties of the specimen changed so that the effect of the temperature rise would be simulated. Nevertheless, the temperature rise affects the components of the EMAT as well, as was mentioned in the introduction, and thus a further decrease in the amplitude of both displacements may be observed in the experimental results, which the current model does not take into account.

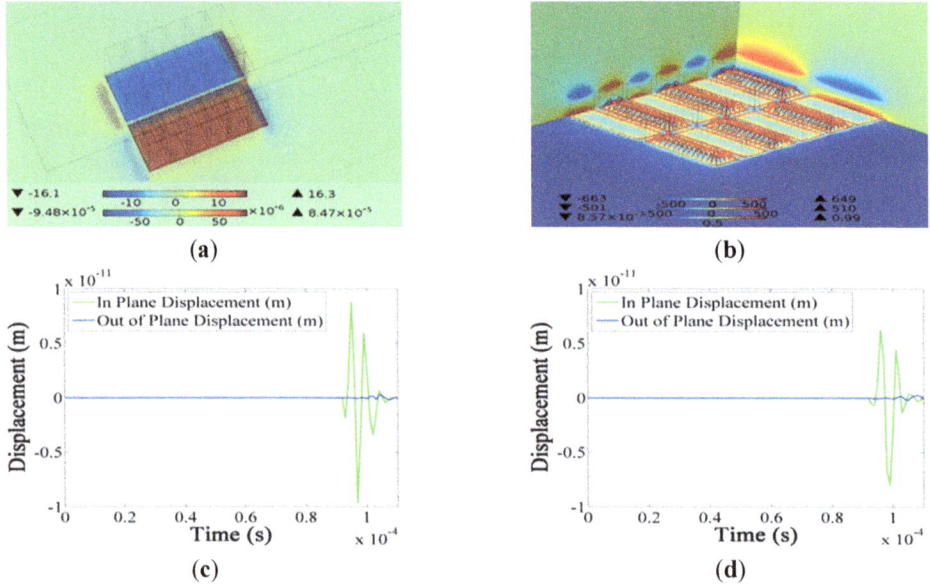

Figure 5. (a) Excitation/eddy current distribution at 12 μs; (b) magnetic flux distribution of the 12 Nd-Fe-B; (c) in-plane/out-of-plane displacement 30 cm away from the EMAT at room temperature; (d) in-plane/out-of-plane displacement 30 cm away from the EMAT at 180 °C.

Consequently, this EMAT configuration can mainly generate in-plane displacement (SH_0); however, it may also generate a small out-of-plane displacement (A_0) at this specific frequency. This can result in complicating signal analysis, since the out-of-plane displacement can be dispersive and mode conversion can also occur, which can lead to incorrect conclusions regarding the structural integrity of the specimen. In GWT, EMATs should excite a single wave mode, so that the signal received from the specimen can provide valid information regarding the structural integrity of the specimen. Also, the temperature rise affected the ultrasonic response of the EMAT, as was expected; amplitude attenuation and shifting in time were the main two changes in the signal received as the temperature increased. Hence, the wave purity characteristics of the PPM EMAT for GWT are of great importance for this application as well as the effect of temperature on the ultrasonic response of EMATs and thus an experimental validation of this EMAT design regarding its guided wave characteristics and its high-temperature performance is still required. The simulation results will be compared with the experimental results.

5. Experimental Setup

A pair of PPM EMATs with racetrack coil, manufactured by Sonemat Limited (Warwick, UK), was experimentally evaluated at both room and high temperatures. The pitch of the magnets is 12 mm and equal to the wavelength of the SH_0 [41]. During these experiments, the EMATs were tested regarding defect detection using guided waves at both room and high temperatures. Their sensitivity to noise was also experimentally evaluated; the influence of common mode noise on their performance and the effect of common ground connection between the EMAT and the specimen on the noise reduction were tested. Their lift-off limitations were validated; the thickness of the glass envelope, under which is the stainless steel pipe of the absorber tubes, will attenuate the signal generated by the EMAT and will perform as a fixed lift-off between the EMAT and the stainless steel pipe. Hence, the maximum lift-off EMATs can reach when they are used for GWT needs to be known. Their power requirements were also investigated, since only Lorentz force is generated in stainless steel, making the EMATs less efficient and more power demanding.

The experimental setup used is shown in Figure 6. The specimen used is a 316Ti stainless steel square plate of 1.25 m length and 3 mm thickness. The EMAT coil is made of lacquered copper wires of 0.315 mm diameter and the maximum operating temperature of the EMATs is 250 °C. Ritec RAM 5000 SNAP pulser/receiver (RITEC Inc., Warwick, RI, USA) was used for driving the EMAT transmitter with a six-cycle, Hanning-windowed pulse of 256 kHz frequency. Ritec was also used for amplifying the signal received with a gain of 80 dB and filtering it within the bandwidth of 10 kHz and 20 MHz. The signal is finally collected, averaged, and recorded in a 2-channel Agilent oscilloscope (Keysight Technologies Inc., Santa Rosa, CA, USA).

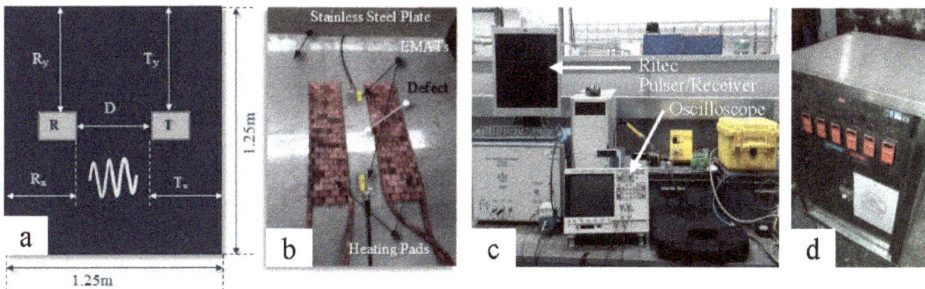

Figure 6. (**a**) Schematic of experimental setup; (**b**) experimental setup (EMATs, specimen, heating pads); (**c**) pulser/receiver—oscilloscope; (**d**) heating unit.

During the room-temperature experiments both defect-free and defective areas were tested. The distance between the transmitter and the receiver was equal to 30 cm. Five defects were created with different length and mass loss each and located 10 cm away one from the other. The defect tested was 20 mm long and with 66.6% mass loss; the transmitter was 15 cm away from one edge of the defect and the receiver was 15 cm away from the other edge of the defect. The effect of the voltage difference between the EMAT and the specimen on the quality of the signal received has been also investigated; in fact, an additional thin, stainless steel cover was also placed all around the transducers, touching both the EMATs and the specimen, for establishing a common ground connection. The influence of lift-off on EMAT response was investigated from zero to 1 mm lift-off with a step of 0.1 mm. A study regarding the power supply requirements of these EMATs was also accomplished by gradually decreasing the power output of Ritec with a step of 5% starting from its maximum power level (5000 W) and stopping at 20% of its maximum power, where no useful information could be retrieved anymore from the signal received. For the high-temperature experiments, the distance between the transmitter and the receiver was equal to 30 cm with the defect located 15 cm away from the transmitter and 15 cm away from the receiver as well (similar to the room temperature setup). The temperature was increased from ambient to 180 °C with a step of 10 °C, using a three-phase heating unit. During the high-temperature experiments the EMATs were continuously in contact with the specimen.

6. Experimental Results and Discussion

6.1. Room Temperature Experiments

During this set of experiments, the EMATs were tested for their sensitivity to common mode noise. Four case studies were investigated; the EMATs were employed in both a defect-free and a defective area, with and without common ground connection with the specimen (shielding). More particularly, the connection/interaction between the EMAT and the specimen is differential, since the specimen induces to the EMAT coil an alternating, differential mode current. Parasitic capacitance exists between the specimen and the EMAT as a result of their physical spacing and the presence of dielectric between them [35]. Tranformers perform in a similar way and they also suffer from common mode noise; in real transformers, a small capacitance links the primary to the secondary winding and also serves as a path for the common mode current across the transformer. As a result, in both cases the common current flows to the ground via the parasitic capacitance and thus no current flows to the EMAT coil/specimen or secondary winding. Nevertheless, an autotransformer acts as a high-value parallel impedance that does not attenuate the differential current significantly but presents zero impedance to the common mode signals by shorting them to ground potential [42]. Autotransformers have smaller resistance and leakage reactance compared to conventional two winding transformers; therefore, the former is more efficient than the latter [43]. Hence, if we presume that the interaction between the EMAT and the specimen is equivalent to a transformer and we connect them so as to perform as an autotransformer, then the common mode noise should be cancelled and the EMAT receiver should work more efficiently. In this case, the alternating, differential mode current will be induced to the EMAT coil and the signal received will have an enhanced SNR and valid information would be retrieved from it. The autotransformer connection is also more robust, reasulting in an increase of the amplitude of the "wanted" signal. If an extra layer of stainless steel is attached to EMATs, touching both the EMAT housing and the specimen, then a common ground connection (autotransformer) is established. Figure 7 shows the equivalent electrical circuit of the EMAT/specimen connection.

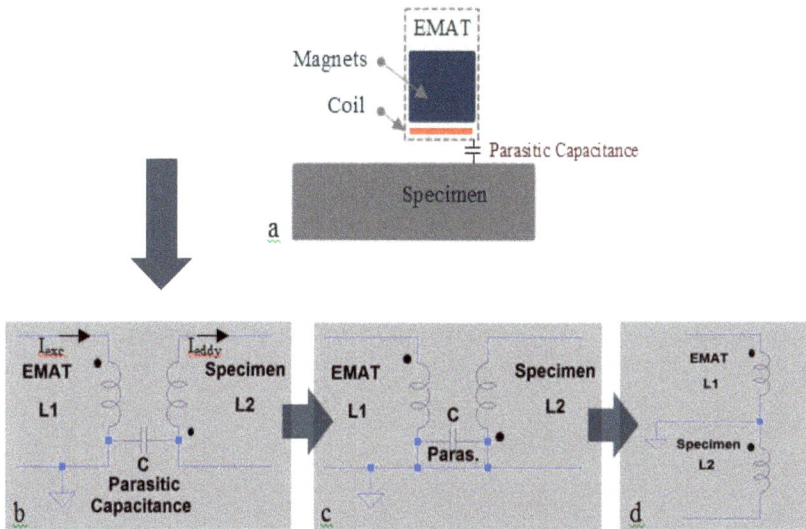

Figure 7. (a) Schematic of EMAT/specimen connection; (b) equivalent electrical circuit of EMAT/specimen without common ground connection; (c) common ground connection; (d) EMAT/specimen—autotransformer schematic.

Figure 8a shows the signal received from a defect-free area when the EMAT and the specimen do not have any common ground connection. In this figure the first reflection is the signal transmitted from the transmitter to the receiver and the other three are coming from the edges of the plate. In this case the noise level is high and the amplitude of all the reflections is low, leading to a low SNR. Figure 8b illustrates the signal received when the defective area was tested without the common ground connection. Similar to Figure 8a, the signal transmitted and the three reflections from the edges of the plate are clearly obvious in the signal received. However, no reflections from the defect are obvious. Figure 9a demonstrates the signal received when the EMATs test the defect-free area while a common ground connection between the EMATs and the specimen has been established. In this case, the amplitude of the reflections increased, the noise level decreased, and as a result the SNR increased by five times. Figure 9b shows the signal received from the defective area when the EMATs and the specimen had a common ground connection. Similar to Figure 9a, the SNR of the signal increased four times more compared to the signal shown in Figure 8b. Actually, in Figure 9b both reflections from the crack are clearly obvious.

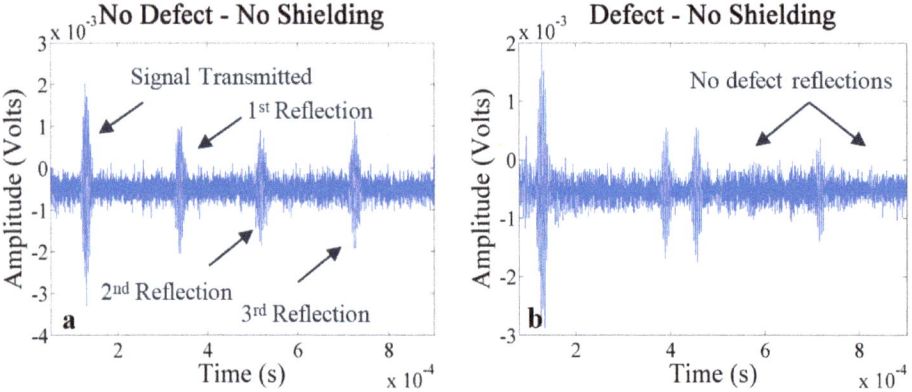

Figure 8. (**a**) Signal received from the defect-free area; (**b**) signal received from the defect.

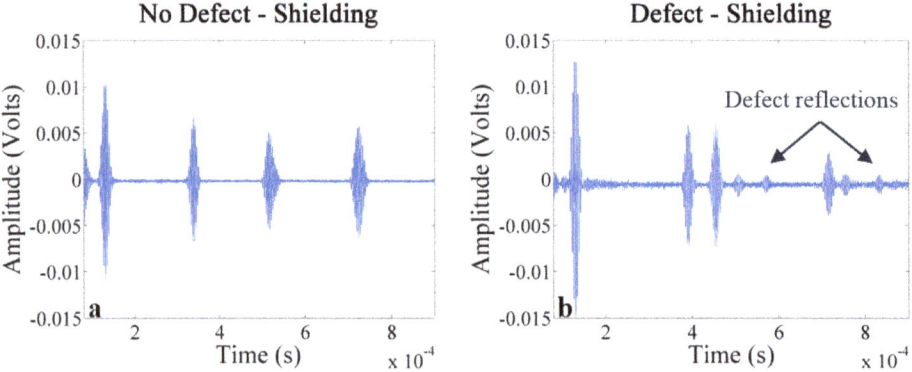

Figure 9. (**a**) Signal received from the defect-free area with shielding; (**b**) signal received from the defect with shielding.

Hence, the shielding has resulted in an enhanced SNR due to the noise cancellation. The electromagnetic coupling between the specimen and the EMAT receiver is weak when they are connected as a two-winding transformer (no shielding); in this case, the electromagnetic losses between the EMAT and the specimen are greater than in a conventional two-winding transformer, since no ferrite connects the EMAT and the specimen. The air between them increases the noise level and attenuates the electromagnetic coupling/"wanted" signal (there is no ferrite to drive the electromagnetic wave; on the contrary, the electromagnetic wave scatters when air is between the EMAT and the specimen). When the specimen/EMAT connection performs as an autotransformer, the noise level decreases and the amplitude of the "wanted" signal (ultrasonic response of the specimen) increases; no noise interferes with the EMAT receiver and thus more current is induced to the coil. The autotransformer is more efficient than a two-winding transformer and as a result more current is induced in the coil when the autotransformer connection is established between the EMAT and the specimen. The unshielded EMAT has greater losses than a conventional transformer would have due to the weak connection between the EMAT and the specimen, while an autotransformer has smaller losses compared to both the unshielded EMAT and a conventional transformer. Thus, a significant increase in the amplitude of the "wanted" signal is observed when the EMAT is shielded. Consequently, the voltage difference between the EMAT and the specimen significantly affects the quality of the signal received and when there is no voltage difference and both components are connected to the ground, the probability of defect detection increases as well. Also, the frequency selected based on the size defect and the dispersion curves is proven to be correct for the detection of the 20 mm long defect.

The attenuation of the signal and the ToF of the main four reflections from the edges of the plate remain features that enable us to distinguish the defective from the defect-free areas when there was no common ground connection. Although the velocity of SH_0 was not to change during the experiments, for no temperature rise occurred, the ToF of both the first and the second reflection from the edges of the plate change from one case study to the other. It is likely that the defect causes this time delay by trapping a portion of the energy/spectrum of the wave propagating inside the plate.

From all the above figures and more especially from Figure 9a,b, it is obvious that only the SH_0 wave mode had been received from the EMAT; the time of arrival of the reflections matches with the SH_0 velocity, as it was calculated from the dispersion curves and the electromagnetic model. Hence, the experimental results match with the theoretical. However, further experimental investigation should be conducted regarding the wave purity characteristics of this EMAT design. In this set of experiments the EMAT receiver was mainly evaluated regarding its ultrasonic potentials; laser interferometry tests, during which the actual displacement generated from the EMAT transmitter can be observed, may also be required for further experimental evaluation of the EMAT transmitter regarding its wave purity characteristics.

Figure 10a illustrates how the amplitude of the signal transmitted changes (%) with respect to the lift-off. According to the literature, EMATs are sensitive to lift-off [28] and therefore their efficiency decreases with the increase of lift-off. The amplitude of the signal transmitted decreases almost linearly with the lift-off increase; when the lift-off is equal to 1 mm only the signal transmitted can be clearly observed. This confirms the high sensitivity of EMATs to lift-off; more particularly, the lift-off limitations of EMATs differ depending on the application the EMAT is designed for. EMATs for thickness measurements can still operate efficiently when the lift-off exceeds 1 mm; however, EMATs for guided waves are more sensitive to lift-off. A parameter that influences the performance of EMAT regarding lift-off is the impedance of the coil. The impedance changes with lift-off as well as with the material properties of the specimen. The inductance due to the magnetization of the specimen is smaller when the EMAT is employed on paramagnetic materials compared to ferromagnetic materials. Consequently, an alternative coil design should be used so that EMATs can be efficiently used in the inspection of absorber tubes, since the thickness of the glass envelope is 7 mm; stacked coils may be more efficient.

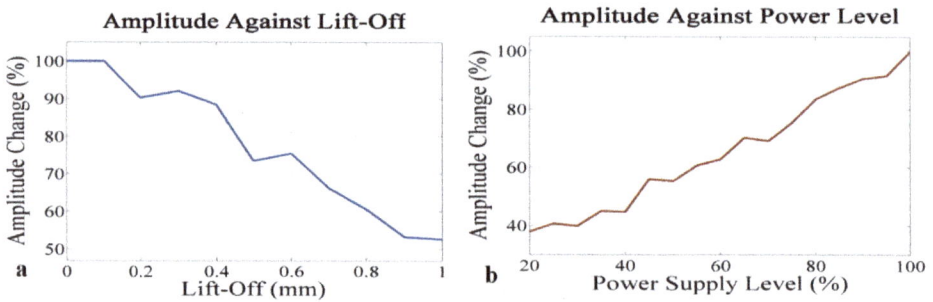

Figure 10. Amplitude change of signal transmitted against (**a**) lift-off increase; (**b**) power increase.

An experimental evaluation of these EMATs regarding their power requirements was also conducted. The power level decreased gradually from 100% to 20% with a step of 5%; we stopped there as no useful information could be retrieved from the signal received when the power level was smaller than 20%. Figure 10b shows how the amplitude of the signal transmitted increases with power supply increase; it can be observed that the amplitude increases almost linearly with the power increase. Similar to lift-off, the impedance of the coil affects the power requirements of EMATs.

If the pulser/receiver unit drives the EMAT transmitter through an output resistor, the magnitude of the impedance of the coil should be equal to the output resistor of the pulser unit, so that the voltage drop in the coil will be minimized. Therefore, impedance matching is always required, which means that either a unit with zero output impedance should be chosen or an impedance matching network should be added between the pulser and the EMAT transmitter. Impedance matching should be used for the coil to be driven with the maximum power possible so that strong signals will be obtained. Nevertheless, the power supply level of EMATs remains high, leading to the conclusion that EMATs are considered to be more efficient as receivers rather than transmitters.

6.2. High-Temperature Experiments

As was mentioned in the theoretical section, high-temperature EMATs have been designed so far only for thickness measurements and thus an EMAT that can withstand high temperatures and excite guided waves is still required. A first approach for that would be the selection of the suitable high Curie magnets and high-temperature coil or the design of a cooling system so that the EMAT would be as efficient as possible at high temperatures. However, all of the above may result in a more complicated design. Hence, a further study about room temperature, guided wave EMATs, and their performance at high temperatures should be conducted prior to the design of a new EMAT.

Hence, the EMATs were tested from ambient temperature to 180 °C with a step of 10 °C; the maximum operating temperature of these EMATs is 250 °C and therefore they were tested up to 180 °C only, so that any serious and irreversible damage will be avoided. The EMATs were continuously exposed to the heat source with zero lift-off during the rise in temperature, while the overall time they were exposed to the heat was equal to 15 min. This set of experiments was conducted three times. Figure 11a–d show the signal received at room temperature, 60 °C, 100 °C, and 180 °C, respectively. Firstly, in the signal obtained at room temperature, both the reflections from the plate edges and the first two reflections from the defect are clear. However, it is obvious that the amplitude of the second reflection from the plate diminishes greatly after 60 °C, while at 100 °C and 180 °C it can hardly be noticed. Similarly, the amplitude of the first reflection from the defect and the forth reflection from the plate decreases with the increase in temperature.

Figure 12 shows how the amplitude of the signal transmitted decreases with temperature rise; it is clear that the amplitude dwindles almost linearly with the rise in temperature. However, the amplitude of the signal transmitted in 30 °C and 40 °C was slightly larger than the amplitude at room temperature in the areas marked in red in Figure 12. Also, the amplitude error alters with temperature rise. A reason for that may be the ground connection between the EMAT and the specimen. The thermal conductivity of stainless steel is low and therefore the specimen was not heated up uniformly; as a result, the plate bended and the mechanical connection between the EMAT ground and the specimen altered with temperature rise due to the gradient of the bend. This mechanical/electrical connection significantly influences the amplitude of the signal transmitted and thus may be the reason for the amplitude increase at 30 °C and 40 °C.

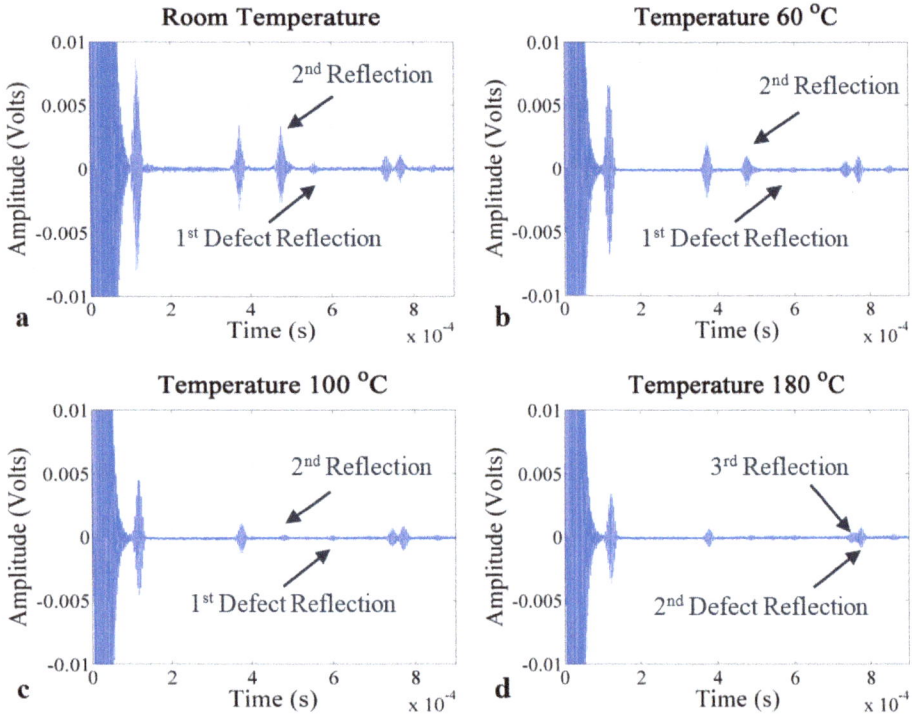

Figure 11. Signal received at (**a**) room temperature; (**b**) 60 °C; (**c**) 100 °C; (**d**) 180 °C.

Time shifting is also observed, as was expected due to the change in the wave velocity at high temperatures, presented in the theoretical study. The third reflection from the plate shifts in time and starts coming closer to the second reflection from the crack, leading to an increase of the magnitude of the latter. Consequently, the experimental results match the theoretical. Temperature compensation should take place and both the ToF of the reflections as well as the drop of their amplitude may be two features that can be further processed and used for the identification of the temperature of the structure being tested. Room-temperature EMATs cannot be used for the monitoring of high-temperature structures (>100 °C); however, a high-temperature EMAT specifically designed for the monitoring of high-temperature structures (>200 °C) should be compared with the performance of this EMAT up to 100 °C.

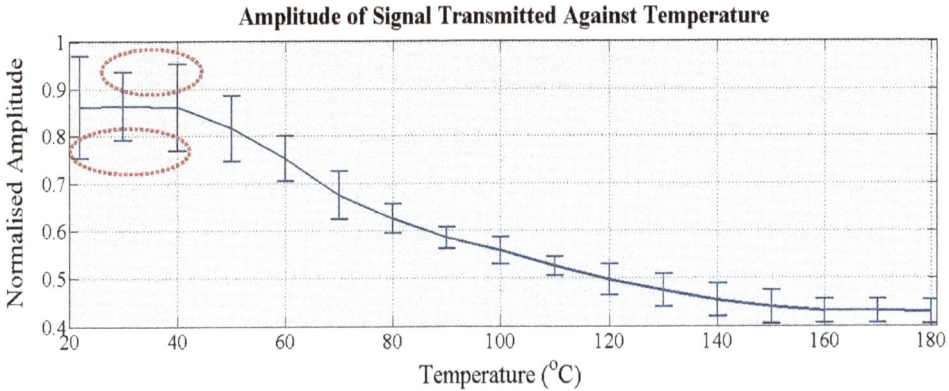

Figure 12. Amplitude of the signal transmitted against temperature.

7. Conclusions and Future Work

The absorber tube is an essential part of Parabolic Trough CSPs and is very likely to get damaged due to its hostile operating conditions. Hence, NDT techniques are required for their monitoring and/or inspection. A promising technique for this application is the use of high-temperature EMAT transducers for the excitation of guided waves and more particularly of the T(0,1) wave mode. A theoretical study about the GWT of absorber tubes, their dispersion curves, and the wave mode that should be applied for their inspection were calculated. The ultrasonic response of a PPM EMAT and its wave purity characteristics were also presented. A pair of PPM EMATs was experimentally evaluated regarding its wave purity, its sensitivity to noise, its lift-off limitations, its power requirements, and its performance at high temperatures while it was exciting/receiving guided waves. It was found that a PPM EMAT receiver can mainly detect SH_0 and thus it can be used for the GWT of the absorber tubes in terms of wave purity characteristics, as the theoretical study also showed. However, laser interferometry tests are also required for the transmitter to be validated. Also, a common ground connection between the EMAT and the specimen can significantly enhance the SNR of the signal received. The current EMAT design is not efficient enough to inspect a stainless steel pipe with a lift-off larger than 1 mm and therefore it cannot be employed for the inspection of absorber tubes. A room-temperature EMAT cannot be applied at high temperatures (<200 °C) and thus a high-temperature EMAT is still required for the guided wave monitoring of high-temperature structures.

The design and manufacturing of a high-temperature PPM EMAT with racetrack coil operating efficiently at temperatures higher than 300 °C for long-term inspection or monitoring is our next step. Several thermal, electromagnetic, and mechanical simulations have already been carried out and have given encouraging results. Moreover, a further experimental investigation regarding the impedance of the EMAT and its relationship with the lift-off and the material properties of the specimen being tested is also another part of our future research.

Acknowledgments: The authors are indebted to the European Commission for the provision of funding through the INTERSOLAR FP7 project. The INTERSOLAR project is coordinated and managed by Computerized Information Technology Limited and is funded by the European Commission through the FP7 Research for the benefit of SMEs program under Grant Agreement Number: GA-SME-2013-1-605028. The INTERSOLAR project is a collaborative research project between the following organizations: Computerized Information Technology Limited, PSP S.A., Technology Assistance BCNA 2010 S.L., Applied Inspection Limited, INGETEAM Service S.A., Brunel University, Universidad De Castilla—La Mancha (UCLM), and ENGITEC Limited.

Author Contributions: M.K. performed the literature review and the finite element simulations, conducted all the experiments and wrote the paper; A.M., V.K. and C.S. reviewed the manuscript; T.-H.G., W.B. and L.C. supervised the research.

Conflicts of Interest: The authors declare no conflict of interest.

References

1. Poullikkas, A. Economic analysis of power generation from parabolic trough solar thermal plants for the Mediterranean region—A case study for the island of Cyprus. *Renew. Sustain. Energy Rev.* **2009**, *13*, 2474–2484.
2. Staff, C.W. Industry Experts Trumpet Untold Solar Potential. 2013. Available online: http://www.constructionweekonline.com/article-23701-industry-experts-trumpet-untold-solar-potential/ (accessed on 12 March 2015).
3. Taggart, S. Parabolic troughs: Concentrating Solar Power (CSP)'s Quiet Achiever. 2008. Available online: http://www.renewableenergyfocus.com/view/3390/parabolic-troughs-concentrating-solar-power-csp-s-quiet-achiever (accessed on 12 March 2015).
4. Steinfeld, A. Topics for Master/Bachelor Thesis and Semester Projects: Spectral Optical Properties of Glass Envelopes of Parabolic trough Concentrators. Available online: http://www.pre.ethz.ch/teaching/topics/?id=87 (accessed on 18 August 2015).
5. Guillot, S.; Faika, A.; Rakhmatullina, A.; Lambert, V.; Verona, E.; Echegut, P.; Bessadaa, C.; Calvet, N.; Py, X. Corrosion effects between molten salts and thermal storage material for concentrated solar power plants. *Appl. Energy* **2012**, *94*, 174–181.
6. Herrmann, U.; Kearney, D.W. Survey of Thermal Energy Storage for Parabolic Trough Power Plants. *J. Sol. Energy Eng.* **2002**, *124*, 145–152.

7.   Papaelias, M.; Cheng, L.; Kogia, M.; Mohimi, A.; Kappatos, V.; Selcuk, C.; Constantinou, L.; Muñoz, C.Q.G.; Marquez, F.P.G.; Gan, T.H. Inspection and structural health monitoring techniques for concentrated solar power plants. *Renew. Energy* **2015**, *85*, 1178–1191.

8.   Lee, J.L.; Hu, L.W.; Saha, P.; Kazimi, M.S. Numerical analysis of thermal striping induced high cycle thermal fatigue in a mixing tee. *Nucl. Eng. Des.* **2009**, *239*, 833–839.

9.   Noguchi, Y.; Okada, H.; Semba, H.; Yoshizawa, M. Isothermal, thermo-mechanical and bithermal fatigue life of Ni base alloy HR6W for piping in 700 °C USC power plants. *Procedia Eng.* **2011**, *11*, 1127–1132.

10. Jinu, G.R.; Sathiya, P.; Ravichandran, G.; Rathinam, A. Comparison of thermal fatigue behaviour of ASTM A 213 grade T-92 base and weld tubes. *J. Mech. Sci. Technol.* **2010**, *24*, 1067–1076.

11. Sunny, S.; Patil, R.; Singh, K. Assessment of thermal fatigue failure for BS 3059 boiler tube experimental procedure using smithy furnace. *Int. J. Emerg. Technol. Adv. Eng.* **2012**, *2*, 391–398.

12. Posteraro, K. Thwart Corrosion under Industrial Insulation. *Chem. Eng. Prog.* **1999**, *95*, 43–47.

13. Halliday, M. Preventing corrosion under insulation-new generation solutions for an age old problem. *J. Prot. Coat. Linings* **2007**, *24*, 24–36.

14. De Vogelaere, F. Corrosion under insulation. *Process. Saf. Prog.* **2009**, *28*, 30–35.

15. Kumar, M.S.; Sujata, M.; Venkataswamy, M.A.; Bhaumik, S.K. Failure analysis of a stainless steel pipeline. *Eng. Fail. Anal.* **2008**, *15*, 497–504.

16. Stine, W.B.; Harrigan, R.W. *Solar Energy Systems Design*; John Wiley and Sons, Inc.: New York, NY, USA, 1986.

17. Hutchins, D.A.; Saleh, C.; Moles, M.; Farahbahkhsh, B. Ultrasonic NDE Using a Concentric Laser/EMAT System. *J. Nondestruct. Eval.* **1990**, *9*, 247–261.

18. Kirk, K.J.; Lee, C.K.; Cochran, S. Ultrasonic thin film transducers for high-temperature NDT. *Insight* **2005**, *47*, 85–87.

19. Momona, S.; Moevus, M.; Godina, N.; R'Mili, M.; Reynauda, P.; Fayolle, G. Acoustic emission and lifetime prediction during static fatigue tests onceramic-matrix-composite at high temperature under air. *Compos. A Appl. Sci. Manuf.* **2010**, *41*, 913–918.

20. Shen, G.; Li, T. Infrared thermography for high temperature pipe. *Insight* **2007**, *49*, 151–153.

21. Beattie, G. Acoustic emission principles and instrumentation. *J. Acoust. Emiss.* **1983**, *2*, 95–128.

22. Hagemaier, D.J. *Fundamentals of Eddy Current Testing*; American Society for Nondestructive, Testing: Columbus, OH, USA, 1990.

23. Malmo, J.T.; Jøkberg, O.J.; Slettemoen, G.A. Interferometric testing at very high temperatures by TV holography (ESPI). *Exp. Mech.* **1998**, *28*, 315–321.

24. Silk, M.G.; Bainton, K.F. The propagation in metal tubing of ultrasonic wave modes equivalent to Lamb waves. *Ultrasonics* **1979**, *17*, 11–19.

25. Pfander, M.; Lupfert, E.; Pistor, P. Infrared temperature measurements on solar trough absorber tubes. *Sol. Energy* **2007**, *81*, 629–635.

26. Cawley, P.; Alleyne, D. The use of Lamb waves for the long range inspection of large structures. *Ultrasonics* **1996**, *34*, 287–290.

27. Hirao, M.; Ogi, H. A SH-wave EMAT technique for gas pipeline inspection. *NDT E Int.* **1999**, *32*, 127–132.

28. Bottger, W.; Schneider, H.; Weingarten, W. Prototype EMAT system for tube inspection with guided ultrasonic waves. *Nucl. Eng. Des.* **1986**, *102*, 369–376.

29. Wilcox, P.; Lowe, M.; Cawley, P. Omnidirectional guided wave inspection of large metallic plate structures using an EMAT array. *IEEE Trans. Ultrason. Ferroelectr. Freq. Control* **2005**, *52*, 653–665.

30. Andruschak, N.; Saletes, I.; Filleter, T.; Sinclair, A. An NDT guided wave technique for the identification of corrosion defects at support locations. *NDT E Int.* **2015**, *75*, 72–79.

31. Burrows, S.E.; Fan, Y.; Dixon, S. High temperature thickness measurements of stainless steel and low carbon steel using electromagnetic acoustic transducers. *NDT E Int.* **2014**, *68*, 73–77.

32. Hernandez, V.F. Pulsed-Electromagnet EMAT for High Temperature Applications. Ph.D. Thesis, University of Warwick, Warwick, UK, 2011.

33. Idris, A.; Edwards, C.; Palmer, S.B. Acoustic wave measurement at elevated temperature using a pulsed laser generator and an Electromagnetic Acoustic Transducer detector. *Nondestruct. Test. Eval.* **1994**, *11*, 195–213.

34. Dixon, S.; Edwards, C.; Palmer, S.B. High-accuracy non-contact ultrasonic thickness gauging of aluminium sheet using electromagnetic acoustic transducer. *Ultrasonics* **2001**, *39*, 445–453.

35. Jian, X.; Dixon, S.; Edwards, R.S.; Morrison, J. Coupling mechanism of an EMAT. *Ultrasonics* **2006**, *44*, 653–656.

36. Ribichini, R.; Cegla, F.; Nagy, P.B.; Cawley, P. Study and Comparison of Different EMAT Configurations for SH Wave Inspection. *IEEE Trans. Ultrason. Ferroelectr. Freq. Control* **2011**, *58*, 2571–2581.

37. Gaultier, J.; Mustafa, V.; Chahbaz, A. EMAT Generation of Polarized Shear Waves for Pipe Inspection. In Proceeding of the 4th PACNDT, Toronto, ON, Canada, 14–18 September 1998.

38. Rose, J.L.; Lee, C.M.; Hay, T.R.; Cho, Y.; Park, I.K. Rail Inspection with Guided Waves. In Processing of the 12th A-PCNDT—Asia-Pacific Conference on NDT, Auckland, New Zealand, 5–10 November 2006.

39. Jackel, P.; Niese, F. EMAT Application: Corrosion Detection with Guided Waves in Rod, Pipes and Plates. In Proceedings of the 11th European Conference on Non-Destructive Testing (ECNDT 2014), Prague, Czech, 6–10 October 2014.

40. Hübschen, G. Generation of Horizontally Polarized Shear Waves with EMAT Transducers. *e-J. Nondestruct. Test.* **1998**, *3*, 1–7.

41. Kogia, M.; Mohimi, A.; Liang, C.; Kappatos, V.; Selcuk, C.; Gan, T.H. High temperature Electromagnetic Acoustic Transducer for the inspections of jointed solar thermal tubes. In Proceedings of the First Young Professionals International Conference, Budapest, Hungary, 17–20 September 2014.

42. Pulse Electronics, Understanding Common Mode Noise. Available online: www.pulseelectronics.com/download/3124/g204/pdf (accessed on 10 November 2015).

43. Winders, J. Autotransformers and Three-Winding Transformers. In *Power Transformers: Principles and Applications*, 1st ed.; Marcel Dekker: New York, NY, USA, 2002.

Acoustic Emission Activity for Characterizing Fracture of Marble under Bending

Eleni Tsangouri, Dimitrios G. Aggelis, Theodore E. Matikas and
Anastasios C. Mpalaskas

Abstract: The present paper occupies with the acoustic emission (AE) monitoring of fracture of marble. The specimens belong to two different material types and were tested in three-point bending after being ultrasonically interrogated. Consequently, they were repaired by means of suitable epoxy agent and mechanically re-loaded. Apart from the well-known correlation of pulse velocity to strength, which holds for the materials of this study as well, AE provides some unique insight in the fracture of the media. Parameters like the frequency content of the waveforms, and their duration among others show a transition in relation to the load. According to their strength class, the specimens exhibit distinct AE characteristics even at low load, enabling to judge their final strength class after having sustained just a small percentage of their ultimate capacity. More importantly, the AE activity during reloading indicates the quality of repair; specimens with good restoration of strength, exhibited similar AE activity to the intact specimens, while specimens with lower repaired capacity exhibited random behavior. This work discusses the passive monitoring of marble fracture and shows that AE parameters that have been used to successfully characterize cementitious materials, provide good results in monolithic materials like marble as well. It is suggested that AE monitoring during a proof loading can provide good information on the potential strength class, which is especially useful for repaired specimens, where the pulse velocity cannot be easily used.

Reprinted from *Appl. Sci.* Cite as: Tsangouri, E.; Aggelis, D.G.; Matikas, T.E.; Mpalaskas, A.C. Acoustic Emission Activity for Characterizing Fracture of Marble under Bending. *Appl. Sci.* **2016**, *6*, 6.

1. Introduction

Acoustic emission (AE) is a technique used for many decades in structural health monitoring (SHM) applications. It uses sensors to record the elastic motion that follows crack nucleation and propagation events within the materials and transform them into electric waveforms. Their intensity (energy), number and other parameters reveal information on the crack density and propagation rate, the location and the mode of fracture [1–4]. The use of the technique in a typical laboratory experiment is shown in Figure 1. In typical rock materials like marble and granite, acoustic wave techniques have been used for quite some time. The cumulative AE activity was studied in relation to the fracture process of pre-cracked marble under compression. It was shown that specimens with "zigzag" cracks failed in a more unpredictable way since most of the activity was registered at the moments of macroscopic failure [5]. The evolution of AE activity was very indicative of the strain at creep experiments of granite [6]. The AE activity and energy release during compression was used to indicate the damage level showing that in marble, main damage comes with the peak stress, while earlier only micro-cracking is formed compared to granite and rock salt [7]. The frequency content of AE has shown to increase with the load, being essentially a precursor of failure in confining pressure experiments, while close to the peak load a strong drop was noticed [8]. The Kaiser effect (absence of detectable AE before the previous maximum load is overpassed) in Brazilian and bending tests was investigated to reveal the previous tensile stress imposed in marble [9]. Energy rate, hit counts, amplitude and location of AE events were used for characterization of crack propagation during the indentation test [10] and compression test in rock materials [11].

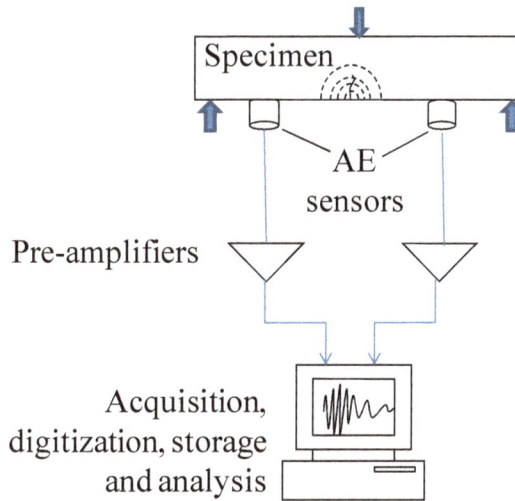

Figure 1. Typical acoustic emission (AE) setup.

AE monitoring of rock material under compression showed mainly two time zones of high "avalanche-like" activity, interrupted by a long period of acoustic silence [12]. The effect of temperature on brittleness of marble was studied with AE corresponding very well to the "rupture evolution law" [13]. Complementary to AE, another elastic wave method, ultrasonic pulse velocity (UPV) has also been used for characterization of marble condition and mechanical performance. Specifically

UPV has been used to derive elastic constants of marble [14] and has been positively correlated to the mean grain size of marble [15], while the anisotropy has also been highlighted due to the noticeable difference between the velocities at the three different axes [16]. The influence of water saturation in granite has also been examined by ultrasonic measurements. The effect of microstructure is revealed by the systematic increase of velocity for higher ultrasonic frequencies (150 kHz compared to 54 kHz) [17].

The present study is a continuation of preliminary tests on granite and marble. Previously, different fracture modes were targeted showing that AE is quite sensitive to the loading conditions [18,19]. Specimens loaded in pure bending emitted higher frequencies and signals of shorter duration relatively to shear-loaded ones. After encouraging results were found, in this case, a large population of specimens of two marble types was tested in bending with concurrent AE recoding to increase the statistical adequacy in an effort to check possible correlations between passively monitoring parameters and mechanical properties. The AE behavior proved very consistent exhibiting three distinct stages with different AE rate. Specific parameters can be used from the early loading to characterize the capacity of the material, like average frequency and RA value, which have been adopted from concrete practice. The fractured specimens were consequently repaired and reloaded to failure. The AE activity of repaired specimens showed capability to characterize the repair efficiency since the specimens with high reloading capacity exhibited similar AE stages with the intact material, while the ones with low strength restoration exhibited a more random behavior driven by the weak bond between the epoxy and the marble. To the authors knowledge, this is the first effort to relate low load AE (approximately one-third of the ultimate load) to the final load bearing capacity of marble both in the intact and the repaired state.

2. Experimental Section

2.1. Materials and Mechanical Details

Two types of marble were used, one named California Honey from Morocco with an average measured density of 2280 kg/m^3 and the other named Crema Mocha with an average measured density of 2508 kg/m^3 from Spain [20]. For convenience the first marble type will be addressed as MA and the next as MB. The specimens were prismatic of size $40 \times 30 \times 160$ mm. Thirty-three specimens of the first type of marble and forty specimens of the second type of marble were subjected to three-point bending according to EN 13892-2:2002, see Figure 2a,b. The load was applied at a constant rate of 50 N/s and the loading was automatically terminated at the moment of instant load drop. At the three-point bending test, the crack starts from the central point of the bottom side due to the tensile stresses (Figure 2b).

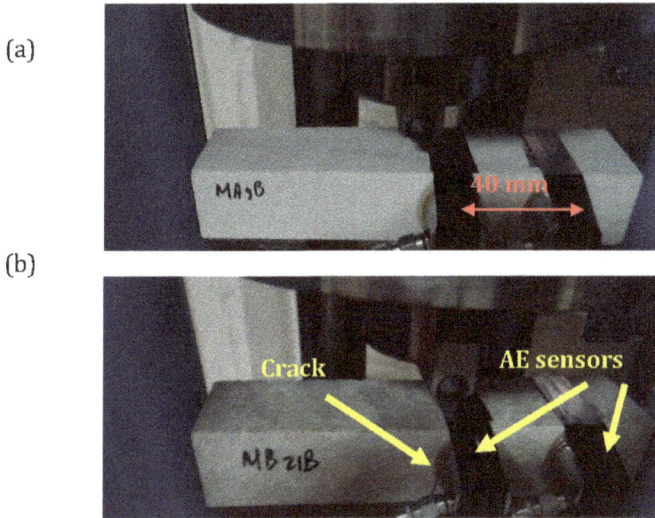

Figure 2. (**a**) Three point bending of (**a**) MA and (**b**) MB marble with concurrent AE monitoring by two sensors.

After bending fracture has occurred all the specimens were repaired in the crack surface with a two-component solvent-free bonding system based on epoxy resin and hardener named "Epoxol" in its commercial product name. It is a specially formulated, high viscosity, polyester marble adhesive-putty, used to bond and fill for repair purposes with hardening time of approximately 5–6 h [21]. The bonding was carefully established so that the original geometry of the specimen was not altered. Figure 3a,b show the populations of healthy and repaired specimens for MA marble, respectively.

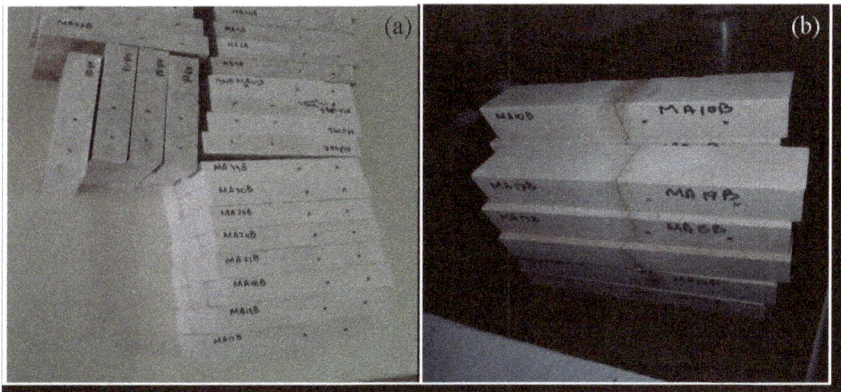

Figure 3. The population of (**a**) healthy and (**b**) repaired specimens for MA marble.

2.2. Ultrasonic Inspection

The whole population of specimens was examined by ultrasound before the fracture test. Two AE transducers (pulser and receiver) were placed on the longitudinal axis of the specimens as seen in Figure 4, having a wave path of 160 mm. A layer of grease between the transducers and the specimen enhanced acoustic coupling. The transducers were of 150 kHz resonance (R15, Mistras Group). The excitation came by a pulse generator with an electric cycle of 150 kHz directly fed to the pulser as well as the acquisition board to act as reference. The pulse velocity (UPV) was calculated by the wave path over the time delay to the first detectable disturbance of the received waveform. The signal was clear and there was no need for extra processing like stacking or filtering.

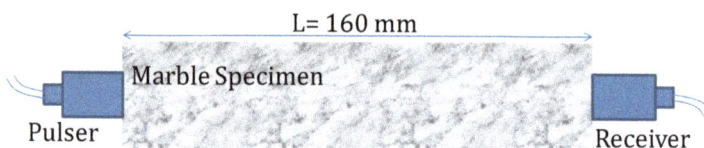

Figure 4. Schematic representation of ultrasound testing.

2.3. Acoustic Emission

AE monitoring took place by means of two piezoelectric sensors, (R15, Mistras). Their resonance comes at 150 kHz and their positions on the specimens are shown in Figure 2. The center of one sensor was placed 15 mm away from the point of the expected crack, which was secured by a notch. The second sensor was placed 40 mm away. Both sensors were positioned at the same side of the crack in order to be able to characterize the distortion of the signals as they propagate for the additional distance of 40 mm. Nevertheless, the effect of distortion will not be discussed herein and will be treated separately in another study. Acoustic coupling was improved by silicon grease between the sensors face and the specimens' surface. AE activity was captured by a two-channel PCI-2 Mistras board with sampling rate of 5 MHz. The threshold was 40 dB, as well as the pre-amplification. A schematic representation of a waveform is seen in Figure 5. Some of the main features are the maximum amplitude, AMP (usually in dB), and the duration, DUR (period between the first and the last threshold crossing). The "rise time" (RT) (which is the time between the first threshold crossing and the point of peak amplitude in μs) is related to the fracture mode of the crack and so is the inverse of the slope of the initial part of the signal (RA value, RT/A in μs/V). The energy (ENE) is defined as the area under the rectified signal envelope. Frequency content can be measured by AF (average frequency), which is the total number of threshold crossings divided by the duration, while there are other indices based on the spectrum of the FFT.

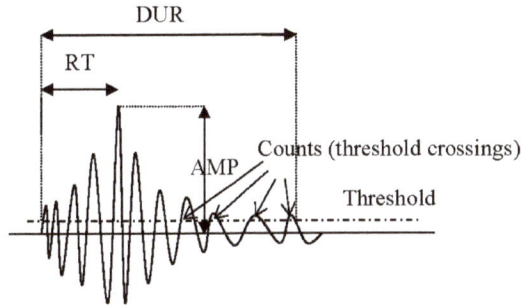

Figure 5. Typical AE waveform and its basic characteristics.

3. Results and Discussion

The mechanical results will be discussed along the monitoring ones in the corresponding ultrasonic and AE paragraphs.

3.1. Ultrasonics

The ultrasonic velocity results are summarized in Figure 6 in correlation to the maximum load of the specimens. Clearly, the two different types of marble formed two clusters. Type MB, which exhibited higher maximum load, also showed much higher pulse velocity. The differences are quite strong, as there is no overlap between the clusters, while there is also a weak (but not negligible) correlation within each cluster. In any case, the sensitivity of pulse velocity to the stiffness and indirectly to the strength of the engineering materials is well known [22,23].

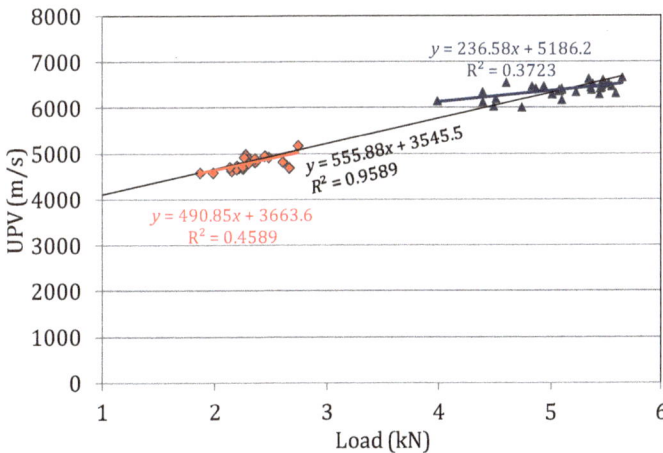

Figure 6. Correlation between pulse velocity and maximum load for the marble specimens.

3.2. Acoustic Emission

Figure 7a,b show typical examples of cumulative AE activity as received by both sensors for marble type MA and MB, respectively. In all of the cases, three distinct stages of activity can be defined. At the start of the experiment a quite high rate of events is recorded (Stage A) followed by a long time window of moderate activity (Stage B). At the end and quite close to the macroscopic fracture of the material, the rate increases again (Stage C). Similar AE activity patterns have been seen recently in similar specimens during compression. The initial saturation was attributed to the depletion of a "weakpoint reservoir in the deforming material", while the increase at the final stage was attributed to clustering of the individual defects due to decreased distance among the sources [12]. It seems that after application of load, formation of micro-cracks is quite fast. However, coalescence of them into larger ones that lead to final fracture takes more time and accumulated energy. This energy is released at the end with another "avalanche" like activity. Although the number of AE hits of Stage C are just a small fraction of the total activity, their energy is much higher.

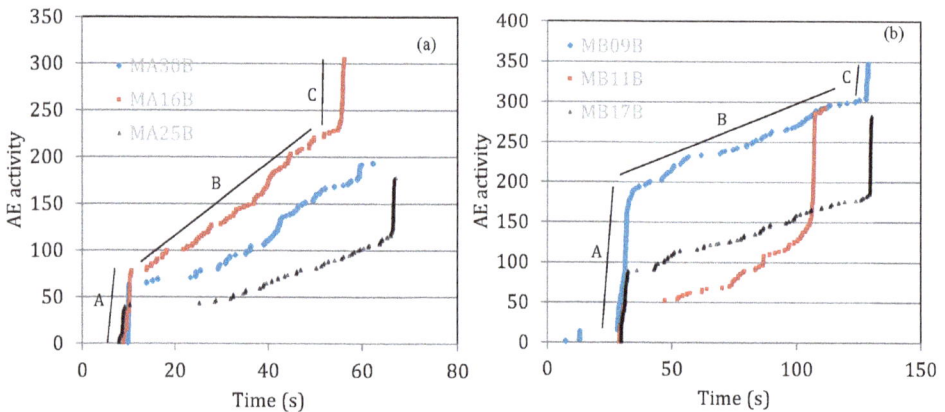

Figure 7. Cumulative AE activity history for marble type (**a**) MA and (**b**) MB.

The behavior of the different marble types was quite repeatable. Apart from the three indicative cases included in the figures, nearly all specimens exhibited the three stages of high, moderate and high activity again.

However, in addition to the total number of hits, high importance is carried by the AE waveform parameters. To give an idea of the transient changes of AE characteristics during loading, Figure 8 shows the RA values for two representative specimens of type MA marble (Figure 8a) and other two from type MB marble (Figure 8b). The data corresponding to the last Stage C are included in a transparent pattern since it is intended to focus on the behavior prior the macroscopic failure. It is obvious that the RA initial level is at much higher levels for type MA than type MB, since values up to 50 ms/V are quite frequent, while for MB specimens values are almost restricted below 30 ms/V. As loading continues high values become scarce resulting in a decreasing trend for RA for both types. At the end of Stage B, and just before the macroscopic failure, the RA values of both types of specimens seem to deepen below 10 ms/V. Concerning the whole population of specimens (more than 30), the average values are shown in Table 1.

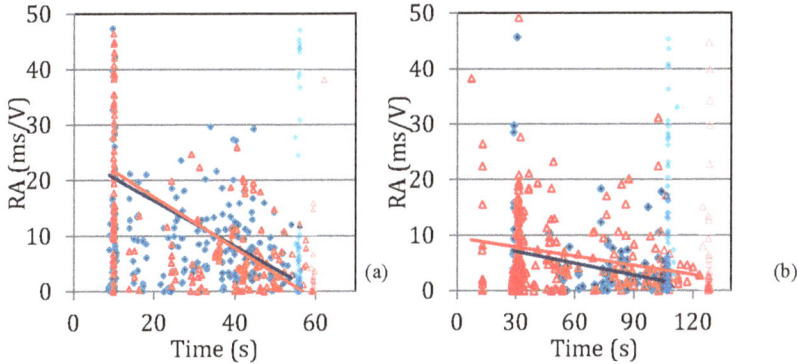

Figure 8. RA values for two specimens of marble type (**a**) MA and (**b**) MB.

Table 1. Results of the bending tests for two types of marble.

Type	Maximum Load (kN)	UPV (m/s)	Fracture Stage	RT (μs)	ENE (-)	DUR (μs)	AMP (dB)	AF (kHz)	RA (μs/V)
			A	455	10.9	1321.0	47.0	25.8	20305
MA	2.30	4798	B	167	7.9	820.8	46.8	30.6	8222
			C	380	8.7	1158.7	46.8	31.0	15829
			A	231	8.6	832.6	48.3	35.8	8452
MB	5.08	6383	B	95	5.8	508.3	47.9	41.0	4344
			C	964	29.4	2904.5	50.7	38.2	22355

From the above table, a general conclusion is that the AE features differ for the successive stages. As an example, the rise time and duration of the second stages (Stage B) is clearly lower than the first (Stage A), while the final stage (Stage C) produces again higher values. Inversely, the frequency content as measured by AF increases for Stage B relative to Stage A. Duration of AE signals is related to the intensity of the source and it shows again the same transition as the total AE activity. It would be premature to interpret these fluctuations of AE parameters in terms of shift between shear and tensile phases, as it should be combined with the knowledge of the microstructure. However, this lowering of RA in the intermediate stage, follows the rate of damage accumulation, which is slower in Stage B, as revealed

135

by the low AE activity rate denoting the creation of micro-cracks, the coalescence of which leads to the final fracture at Stage C.

Apart from the differences between the successive stages, there are considerable differences in between the different marble types. Type MB exhibits for the first two fracturing stages, AF higher by as much as 10 kHz than marble MA. This difference is quite high considering that it was recorded by resonant sensors. Correspondingly, it exhibits lower duration, RT and RA values. In addition, while MA exhibits lower RA values for the last stage, type MB exhibits a strong increase. These different trends are attributed to the different fracturing behavior of the two media since the specimen geometry and the whole experimental lay-out was constant. As aforementioned changes of RA are due to the fracture mode. Although the macroscopic stress field is similar, the microstructure of the different marbles may induce different motion of the crack tips strongly influencing the AE parameters received.

It has been seen that high RA values correspond to material closer to fracture while the inverse is measured for frequencies. The frequencies drop just before the final failure. It comes as no surprise therefore, that the material type with higher frequency and lower RA exhibited higher density and load bearing capacity. What should be stressed is that these differences are obvious from the initial fracture stage; MB exhibits 39% higher frequency and 58% lower RA allowing predictions as to the final capacity of the beams.

By separating the activity at the different stages we can study the AE activity at low, moderate and high load. Figure 9a shows the graph of AF *vs.* RA for the first stage of fracturing (Stage A) as explained in the previous section, while the corresponding graph of the intermediate Stage B is given in Figure 9b. The whole group of points extends from frequencies of approximately 20 kHz to 45 kHz and from RA values of 3 to 30 ms/V. However, it is important to notice that the type MB occupies most of the spots of higher frequency and low RA, while the two populations (MA and MB) could even be separated by a single line (see dashed diagonal line) having only three overlapping points. For the next Stage B of moderate activity, the whole population shifts to higher frequencies and lower RA values (see Figure 9b) but there is still a substantial difference of the centers of the two populations. Marble MB again exhibits higher frequencies and lower RA values and the separation line attains a much higher slope. The separation lines are certainly not unique and successful clustering could achieved by means of pattern recognition algorithms. However, this is left for the next stage, when larger populations will be examined.

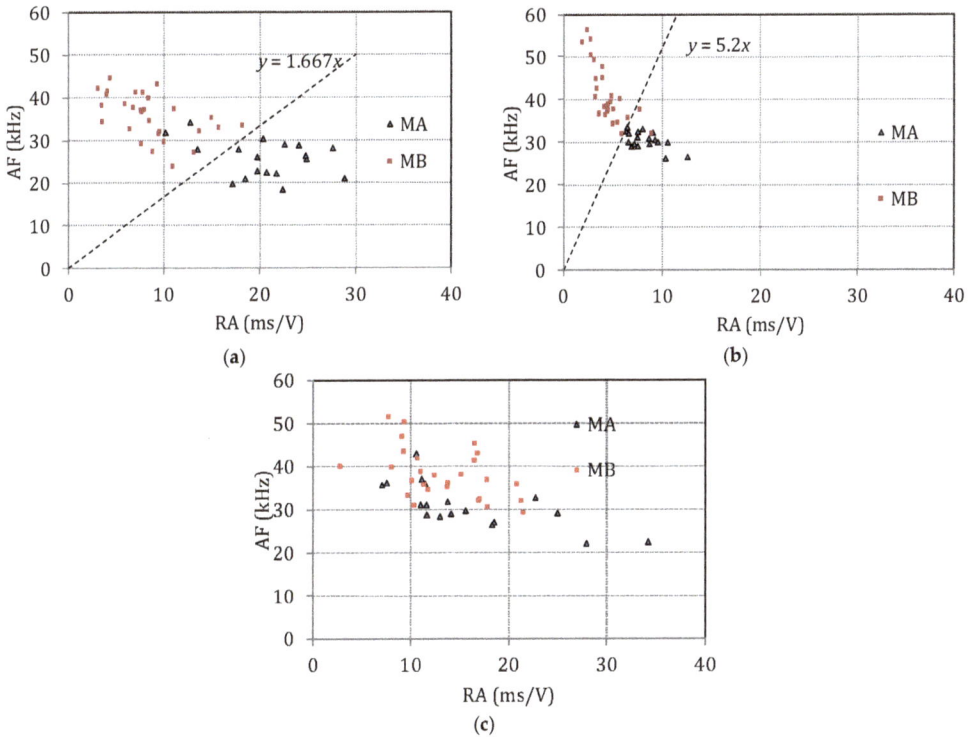

Figure 9. Correlation plot between AF and RA for (**a**) the initial loading Stage A; (**b**) the intermediate loading Stage B; and (**c**) the final Stage C.

Based on the AE activity of the preliminary loading stage it can be concluded that if the ratio of the average AF over RA is more than 1.67, the ultimate load will be higher than 4 kN (denoting type MB) while if it is lower the load capacity will be less than 3 kN (type MA). The importance of such a correlation lies in the fact that the strength capacity of the material can be assessed by a quite reliable way based on AE data resulted from loading the specimen, without however, inflicting any serious damage: the initial activity (Stage A) is recorded while the load is less than 10% of the ultimate.

Results are quite similar for the intermediate stage where the load is between 20% and 90%. The analogy between AF and RA now changes to 5.2 but again the separation with a straight line is similarly successful.

For the last stage of macroscopic fracture, the results are more mixed, showing that in macro-fracture the intensity of the phenomena "covers" delicate changes that are imposed by the microstructure (Figure 9c).

3.3. Behavior of Repaired Specimens

As aforementioned, the fractured specimens were repaired by means of adhesive epoxy, suitable for cultural heritage projects. In order to check the strength restoration they were tested again with the same bending setup. Their AE activity was once again monitored. Table 2 shows the average value of maximum load for both repaired marble types. The maximum load is approximately 1.63 kN and 2.46 kN for types MA and MB, respectively, corresponding to 70% and 48% of their original strengths. The marble with higher original strength (MB type) exhibited the higher capacity even after repair, although it suffered larger loss as a percentage of the original strength.

Table 2. Results of the bending tests for two types of repaired marble (reloading).

Marble Type	Maximum Load (kN)	UPV * (m/s)	Fracture Stage	RT (μs)	ENE (-)	DUR (μs)	AMP (dB)	AF (kHz)	RA (ms/V)
MA	1.63	4798	A	146	5.6	629	47.6	38	7.33
			B	61	4.6	416	48.1	43.9	2.93
			C	372	15.2	1153	49.5	36.7	9.45
MB	2.46	6383	A	179	7.2	740	48.8	37.2	5.60
			B	71	4.5	454	47.6	39.9	3.07
			C	344	16.4	1022	50.9	43.9	9.08

* UPV was measured before fracture.

Compared to the nearly uniform behavior of the sound specimens, the AE behavior of the repaired MA type specimens showed mixed trends. Most of the specimens (20 out of 33) followed a similar behavior to the sound (first loading) with the same stages (high initial rate-plateau-high final rate) but the rest of the specimens exhibited a more random behavior where the specific stages could not be defined. Examples are seen in Figure 10a,b for the aforementioned cases. The absence of fracture stages can be attributed to the inefficient interfacial bonding between the marble and the epoxy.

Table 3 shows the average values of ultimate load and reload for the specimens of marble MA that exhibited the three stages at the reloading (20 specimens), the specimens with random behavior (13 specimens) as well as the average of the whole population (33 specimens). It also includes the repair factor (n) as the ratio of ultimate load of the repaired to the initial load of the intact specimens.

It is worth noticing that the specimens that during reloading followed a behavior similar to the initial loading exhibit substantially higher repaired load. Specifically this group of specimens exhibited an average maximum load of 1.75 kN substantially higher than the group of the random behavior (1.46 kN). This shows that the bonding was of better quality and therefore the structural capacity, the fracture process and as a result the AE behavior was closer to the behavior of the intact material achieving a restoration factor of 76% which is quite high for similar cases [24].

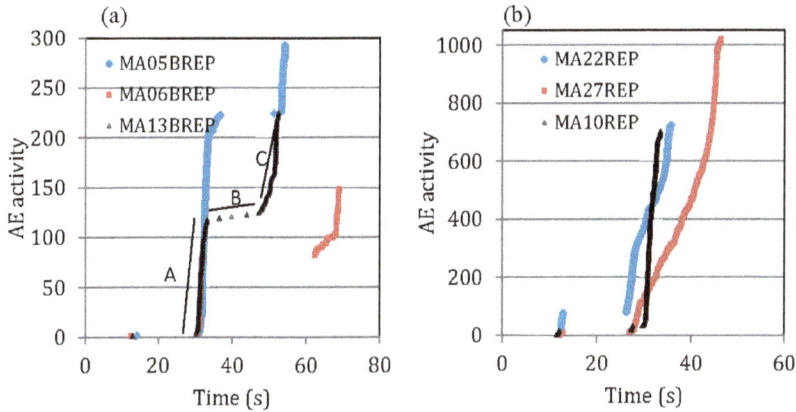

Figure 10. (a) Cumulative AE history for repaired MA specimens following three stages. (b) Cumulative AE history for repaired MA specimens without similar three stages.

Table 3. Results of the bending tests for MA type of repaired marble.

MA	Maximum Load (kN)	UPV * (m/s)	Maximum Load(REP) (kN)	n = Load(REP)/Load (-)
no stages in AE	2.24	4713	1.46	0.66
3 stages in AE	2.33	4846	1.75	0.76
total	2.30	4798	1.63	0.72

* UPV was measured before fracture.

Concerning type MB, nearly all specimens (38 out of 40) behaved similarly to the sound specimens. This has to be related to the bond between the epoxy and the matrix. For the case of MB, the epoxy and its bond to the marble works well and results in the same behavior like the initial monolithic material. This is confirmed by the maximum load after repair.

Additionally, a general observation is that the plateaus (silent periods) are longer for the repaired specimens, while stage B is characterized by lower AE rate than the initial loading. Specifically Stage B exhibits an activity of 3.3 hits/s for marble MA and 2.2. hits/s for marble MB. For the repaired version of the same specimens, the corresponding rates was negligible with only sparse AE hits for Stage B (check again Figure 10a). This lower activity (and the longer plateaus in the beginning of the test) can be attributed to the lower elasticity of epoxy. The stiffness of this layer is much lower than marble, allowing it to deform without cracking for longer period than marble. Therefore, it withstands the imposed deflection with limited AE. Additionally it has been seen that the type of epoxy may influence this behavior. In recent experiments in concrete repaired by flexible healing agent, the AE

139

development was smooth since the bonding took place in small and frequent events. On the other hand, rigid epoxy follows a slightly different trend by fewer but larger debonding events [24].

Excluding the specimens that exhibited random behavior (13 of type MA and two of type MB) one can again separate the AE behavior in the successive stages. Results are included in Table 2 above. It is evident that the parameter values are much closer between the different marble types for each stage. For the first stage received at low load, average frequencies are almost identical and RA values are very close. It should be recalled that in the first loading, marble MB exhibited 39% higher frequency, while the RA value of B was 58% lower than MA (see Table 1). Now these differences are nearly eliminated; frequencies are essentially the same and the average RA of MB is only 23% lower than MA. Differences in rise time and duration are similarly compressed. The AE parameters show indeed that in the repaired specimens, the importance lies primarily in the bonding between epoxy matrix and less in the marble itself. The difference of 23% in RA although may seem substantial, does not allow successful separation of the two classes, due to the overlap. Indeed in AE, the experimental scatter of the parameter values is high and a difference between averages of the order of 20% of even 50% is not enough to separate different populations [25]. However, it can still be seen that type MA with higher RA value in the preliminary stage of loading exhibited lower maximum load. Therefore, there is certainly potential in the study of AE in the repaired specimens as to predict the final behavior. It should be kept in mind that it is possible to apply pattern recognition combining not only two but many features which offer the highest characterization capacity or even apply principal component analysis.

4. Discussion

The reason why AE inspection acquires even more significance is that ultrasound cannot reliably assess the condition of the repaired specimen. As shown in Figure 4, the nominal wave path is 160 mm. In the repaired material, the layer of epoxy resin is approximately 1 mm corresponding to less than 1% of the total. Therefore, the stiffest marble (in this case type MB) would always exhibit much higher velocities and the sensitivity to the epoxy layer and its adhesion to the matrix would be negligible. Despite the fact that ultrasound shows very good correlation to the material properties and the final strength, it would not be possible to characterize by the same test the capacity of the repaired specimens.

It should also be mentioned that although AE gets signals from fracturing of the material it is not necessarily "close to failure", as micro-cracking activity is induced very early in the loading. There are several other examples in literature where AE starts in a small fraction of the ultimate load [7,11,12,25,26]. Therefore, AE can be used in the concept of "proof loading". This kind of test is a usual application in different types of structures (e.g., pressure vessels). The structure is loaded up to a safety point and the AE is monitored. According to the received AE the structure can be categorized either as safe for continuous use, as safe for a period of time before another monitoring, or unfit and should be replaced. AE, due to the sensitivity of the sensors records the very early fracturing incidents in the micro-scale before the structural capacity of the member/specimen is compromised. In the specific case, the loading rate was constant. Taking into account the accumulated AE activity for marble MA, the initial AE stage is completed well before 15 or 20 s in a total of 50 to 60 s until fracture (check Figure 7a). This means that the load until that moment was approximately one-third of the maximum load. For marble MB, the initial activity (first stage) was completed at about 30 s while the duration of the loading until fracture was more than 100 s (Figure 7b). Again this means that the loading for the first AE stage is well below one-third of the maximum. Therefore, it can be supported that the AE data used for correlation with the final load capacity, as presented in Figure 9a correspond to a stage when the material is still macroscopically intact since the load was about 30% of the ultimate. For the repaired specimen, the initial stage of AE represents approximately half the duration of the experiment, so we can say that the "early" AE data were recorded until 50% of the capacity (Figure 10a). After the initial stage, the specimen continued to receive increasing load within a period of acoustic silence before it was fractured 20 to 40 s later. Therefore, it can again be argued that the early loading did not effectively compromise the load bearing capacity of the specimens.

Concerning the localization of the AE sources, the present setting with two sensors enables only linear location. More sensors would not add much to the accuracy of the AE events. This is because the accuracy is compromised by different factors. The specific type of sensors is widely applied in practice due to their high sensitivity but on the other hand they are far from being considered point sensors, as their diameter is approximately 16 mm. This is comparable to the propagation distance between the crack (middle of specimen) and the 1st sensor and also comparable to the specimen dimensions (thickness of 40 mm). Therefore, AE location would enable to separate events that occurred centimeters away but within a single damage zone it would be troublesome to make more accurate statements. As an example, in our study the thickness of the epoxy in the repaired specimens is approximately 1 mm. AE location results would not be reliable enough to check if an event occurred within this mm (crack in epoxy) or 0.5 mm to the right or left (marble crack), as the typical error is certainly higher. In addition to the above mentioned geometry reasons, the heterogeneity of the material should be kept in mind that creates a variation in the pulse velocity and further compromises the accuracy. Therefore, in the specific case, AE location could not be used to clarify the exact point of the fracture and specifically discriminate between fracture in the epoxy or in the marble matrix.

5. Conclusions

The present paper occupies mainly with the AE behavior of large populations of two marble types under bending and the repair efficiency as tested by a second loading after application of adhesive epoxy. The main conclusions of the work are summarized below:

 (i) The fracture of the material follows three successive stages of high, low and high AE activity.
 (ii) The AE parameters at each stage vary showing that the fracture pattern changes throughout the damage process even though the macroscopic failure of both marble types is derived from the same three-point bending setup.
 (iii) Different marble types exhibit fairly different behavior, even from the initial stage of loading. Based on AE parameters from the early stage (AF and RA), an almost complete separation of the two populations (high and low strength) is possible, allowing the prediction of the strength group that each specimen belongs.
 (iv) Repaired specimen with adhesive epoxy exhibited a restoration of flexural strength between 50% and 75% depending on the bonding efficiency.
 (v) The repaired specimens, which followed AE behavior similar to the intact ones (same three distinct fracture stages), exhibited much higher strength restoration (20% higher ultimate load at the test after repair), something attributed to the good adhesion that dictates the specimens to fracture similar to the intact ones.

Study should continue with different fracture modes apart from bending, as well as with other natural stones like granite (different morphology, greater flexural strength and density than marble). In addition, pattern recognition approaches should be applied in order to achieve even better characterization or classification of the different AE populations.

Author Contributions: Author Contributions: All authors contributed to the work presented in this paper, and to the writing of the final manuscript. E.T. contributed to the data analysis. D.G.A. was responsible for the ultrasonic experiment and the analysis of the data. T.E.M. provided discussion on the testing procedure and the results. A.M. organized the study and was responsible for the mechanical testing and AE monitoring.

Conflicts of Interest: Conflicts of Interest: The authors declare no conflict of interest.

References

1.   Grosse, C.U.; Ohtsu, M. *Acoustic Emission Testing*; Springer: Berlin, Germay, 2008.

2. Zitto, M.E.; Piotrkowski, R.; Gallego, A.; Sagasta, F.; Benavent-Climent, A. Damage assessed by wavelet scale bands and *b*-value in dynamical tests of a reinforced concrete slab monitored with acoustic emission. *Mech. Syst. Signal Process.* **2015**, *60*, 75–89.

3. Shiotani, T.; Oshima, Y.; Goto, M.; Momoki, S. Temporal and spatial evaluation of grout failure process with PC cable breakage by means of acoustic emission. *Constr. Build. Mater.* **2013**, *48*, 1286–1292.

4. Carpinteri, A.; Lacidogna, G.; Accornero, F.; Mpalaskas, A.C.; Matikas, T.E.; Aggelis, D.G. Influence of damage in the acoustic emission parameters. *Cem. Concr. Compos.* **2013**, *44*, 9–16.

5. Li, Y.; Chen, L.; Wang, Y. Experimental research on pre-cracked marble under compression. *Int. J. Solids Struct.* **2005**, *42*, 2505–2516.

6. Chen, L.; Wang, C.P.; Liu, J.F.; Liu, Y.M.; Liu, J.; Su, R.; Wang, J. A damage-mechanism-based creep model considering temperature effect in granite. *Mech. Res. Commun.* **2014**, *56*, 76–82.

7. Zhang, Z.; Zhang, R.; Xie, H.; Liu, J.; Were, P. Differences in the acoustic emission characteristics of rock salt compared with granite and marble during the damage evolution process. *Environ. Earth Sci.* **2015**, *73*, 6987–6999.

8. Zhang, L.; Ren, M.; Ma, S.; Wang, Z.; Wang, J. Acoustic emission and fractal characteristics of marble during unloading failure process. *Chin. J. Rock Mech. Eng.* **2015**, *34*, 2862–2867.

9. Fu, X.; Xie, Q.; Liang, L. Comparison of the Kaiser effect in marble under tensile stresses between the Brazilian and bending tests. *Bull. Eng. Geol. Environ.* **2015**, *74*, 535–543.

10. Yin, L.J.; Gong, Q.M.; Ma, H.S.; Zhao, J.; Zhao, X.B. Use of indentation tests to study the influence of confining stress on rock fragmentation by a TBM cutter. *Int. J. Rock Mech. Min. Sci.* **2014**, *72*, 261–276.

11. Carpinteri, A.; Corrado, M.; Lacidogna, G. Heterogeneous materials in compression: Correlations between absorbed, released and acoustic emission energies. *Eng. Fail. Anal.* **2013**, *33*, 236–250.

12. Chmel, A.; Shcherbakov, I. A comparative acoustic emission study of compression and impact fracture in granite. *Int. J. Rock Mech. Min. Sci.* **2013**, *64*, 56–59.

13. Chen, G.-F.; Yang, S.-Q. Study on failure mechanical behavior of marble after high temperature. *Eng. Mech.* **2014**, *31*, 189–196.

14. Prassianakis, I.N.; Prassianakis, N.I. Ultrasonic testing of non-metallic materials: concrete and marble. *Theor. Appl. Fract. Mech.* **2004**, *42*, 191–198.

15. Sarpun, H.; Kilickaya, M.S.; Tuncel, S. Mean grain size determination in marbles by ultrasonic velocity techniques. *NDT E Int.* **2005**, *38*, 21–25.

16. Sáez-Pérez, M.P.; Rodríguez-Gordillo, J. Structural and compositional anisotropy in Macael marble (Spain) by ultrasonic, XRD and optical microscopy methods. *Constr. Build. Mater.* **2009**, *23*, 2121–2126.

17. Vasconcelos, G.; Lourenco, P.B.; Alves, C.A.S.; Pamplona, J. Ultrasonic evaluation of the physical and mechanical properties of granites. *Ultrasonics* **2008**, *48*, 453–466.

143

18. Aggelis, D.G.; Mpalaskas, A.C.; Matikas, T.E. Acoustic signature of different fracture modes in marble and cementitious materials under flexural load. *Mech. Res. Commun.* **2013**, *47*, 39–43.

19. Mpalaskas, A.C.; Matikas, T.E.; van Hemelrijck, D.; Iliopoulos, S.; Papakitsos, G.S.; Aggelis, D.G. Acoustic signatures of different damage modes in plain and repaired granite specimens. In Proceedings of the SPIE—The International Society for Optical Engineering, San Diego, CA, USA, 9–11 March 2015.

20. Skandalis marbles. Available online: http://www.skandalis.gr/el/products/marbles.html (accessed on 19 December 2015).

21. Neotex, S.A. Technical Data Sheet. Available online: http://www.neotex.gr/frontoffice/portal.asp?cpage= NODE&cnode=122&clang=1 (accessed on 15 December 2015).

22. Naik, T.R.; Malhotra, V.M.; Popovics, J.S. The Ultrasonic Pulse Velocity Method. In *CRC Handbook of Nondestructive Testing of Concrete*; Malhotra, V.M., Carino, N.J., Eds.; CRC: Boca Raton, FL, USA, 2004.

23. Kourkoulis, S.K.; Prassianakis, I.; Agioutantis, Z.; Exadaktylos, G.E. Reliability assessment of the NDT results for the internal damage of marble specimens. *Int. J. Mater. Prod. Technol.* **2006**, *26*, 35–56.

24. Tsangouri, E. Experimental Assessment of Fracture and Autonomous Healing of Concrete and Polymer Systems. Ph.D. Thesis, MEMC-Vrije Universiteit Brussel, Brussel, Belgium, 2015.

25. Aggelis, D.G.; Verbruggen, S.; Tsangouri, E.; Tysmans, T.; van Hemelrijck, D. Characterization of mechanical performance of concrete beams with external reinforcement by acoustic emission and digital image correlation. *Constr. Build. Mater.* **2013**, *47*, 1037–1045.

26. Lockner, D. The role of acoustic emission in the study of rock fracture. *Int. J. Rock Mech. Min. Sci. Geomech. Abstr.* **1993**, *30*, 883–889.

Lamb Wave Interaction with Adhesively Bonded Stiffeners and Disbonds Using 3D Vibrometry

Ryan Marks, Alastair Clarke, Carol Featherston, Christophe Paget and
Rhys Pullin

Abstract: There are many advantages to adhesively bonding stiffeners onto aircraft structures rather than using traditional mechanical fastening methods. However there is a lack of confidence of the structural integrity of adhesively bonded joints over time. Acousto-ultrasonic Lamb waves have shown great potential in structural health monitoring applications in both metallic and composite structures. This paper presents an experimental investigation of the use of acousto-ultrasonic Lamb waves for the monitoring of adhesively bonded joints in metallic structures using 3D scanning laser vibrometry. Two stiffened panels were manufactured, one with an intentional disbonded region. Lamb wave interaction with the healthy and disbonded stiffeners was investigated at three excitation frequencies. A windowed root-mean-squared technique was applied to quantify where Lamb wave energy was reflected, attenuated and transmitted across the structure enabling the size and shape of the defect to be visualised which was verified by traditional ultrasonic inspection techniques.

Reprinted from *Appl. Sci.* Cite as: Marks, R.; Clarke, A.; Featherston, C.; Paget, C.; Pullin, R. Lamb Wave Interaction with Adhesively Bonded Stiffeners and Disbonds Using 3D Vibrometry. *Appl. Sci.* **2016**, *6*, 12.

1. Introduction

Within the aerospace industry there is a constant drive to produce more environmentally friendly aircraft. Although at present the aerospace sector is not a major contributor to greenhouse emissions, historical trends show that global air travel is doubling every fifteen years [1]. One way of reducing the environmental impact of such an increase in air travel is to produce aircraft that are lighter in mass through the use of advanced composite materials and construction techniques which provide improved strength and stiffness to weight ratios when compared to traditional materials and methods.

Aircraft structures are typically constructed from thin load bearing skins with stiffeners attached to provide the structure with the required amount of rigidity [2]. There are two predominant methods of attaching the stiffeners to the skin; mechanical fasteners such as rivets, or bonding with an adhesive. Traditionally, mechanical fasteners have been used in primary structure of the aircraft (*i.e.*, the structure in which failure would be catastrophic) because of their ability to transfer loads between components with predictable performance. However mechanical fasteners create high stress concentrations which can lead to crack initiation [3,4].

Adhesively bonded metallic joints have been used in the secondary structure of aircraft for the past fifty years [5]. Bonding stiffeners to the aircraft's skin is advantageous when compared to riveting as it can reduce the mass of the joints, distribute the stresses more evenly and be more resistant to environmental effects (such as temperature, humidity and vibration) which an aircraft experiences [6]. Adhesively bonding stiffeners is also less labour intensive and more cost effective than using mechanical fasteners as it does not require the structure to go through repetitive drilling operations [7]. Adhesive bonding is particularly suited to composite structures where the use of mechanical fasteners presents additional difficulties. Due to the hardness and high tensile strengths of some composites, drilling can cause rapid deterioration of the drill bits [8] resulting in high machining costs and thus skilled operatives are required to carry out the operation. Drilling can also cause damage to the composite such as interlaminar crack propagation, micro cracking, fibre breakage, fibre pull-out, matrix cracking, thermal damage and delaminations. The presence of the hole also forms a reduction in the strength of the material due to the fibres no longer being continuous.

Though adhesives bonds have been used in the secondary structure without arrestment fasteners (*i.e.*, fasteners that would still bear the load should the adhesive fail) they are rarely used in the primary structure without the use of arrestment fasteners due to the lack of confidence in the reliability of the joint [9]. From the experience of the Royal Australian Air Force (RAAF) 53% of defects detected in aircraft structures such as the F-111 were found to be bond failures [10]. Many cases of adhesive bond failure have been found to be the result of improper application of the adhesive or insufficient surface preparation [9] *i.e.*, human error in the manufacture of the joint.

Though it has been proven that bonded joints are more resistant to fatigue loading as well as being able to withstand higher peak loading [10], the degradation of the adhesive layer over time is not yet fully understood [11]. This uncertainty in the condition of the bond throughout the aircraft's in-service life requires monitoring using non-destructive techniques to ensure airworthiness. Traditional non-destructive (NDT) techniques can be used to inspect the structural integrity of bonds on aircraft structures. However with the increasing size of commercial aircraft, the influence of human error and pressures from aircraft operators to reduce maintenance times [12], alternative solutions to traditional NDT techniques have to be found to enable the application of airworthy adhesively bonded stiffeners. If a sensor network could be installed to monitor damage and degradation of the bonded joint an "as required" inspection program could be used. The most straightforward method of implementing such a structural health (SHM) network onto an aircraft is during the design stage [13]. It has been suggested that by implementing structural health monitoring during design, weight savings of up to 15% could be achieved at component level [14].

A long-established technique for detecting damage in structures is by using acousto-ultrasonic induced Lamb waves. The principle involves exciting a piezoelectric transducer mounted to the structure's surface which induces a Lamb wave that is then detected by another transducer mounted at a different location on the structure. If damage occurs within the field between the two sensors, the signal propagation is altered resulting in a quantifiable difference in the signal received. This technique can be extended to networks of multiple sensors to improve detection capability.

To enable complex adhesive bond defects such as kissing bonds (where bond fails but still remains in close contact with the bond face hence appearing to still be bonded. These are notoriously difficult to detect [15]) to be detected using acousto-ultrasonic techniques, sensor networks need a high probability of detection. An understanding of Lamb wave interaction with stiffeners and their defects will enable better sensor network design whilst also driving towards minimizing the weight penalty the sensors add to the overall structure.

Following a review of the use of Lamb waves in SHM and their study using laser vibrometry, this paper presents the results of an experimental investigation of the interaction of Lamb waves with adhesively bonded stiffeners (with and without disbonds) using 3D scanning laser vibrometry. Results demonstrating Lamb wave interaction are shown, together with post-processing of the vibrometry data with a windowed root-mean-square technique. The results are further compared with the results of ultrasonic inspection of the disbond and its associated geometry is shown to correlate very closely with disturbances of the Lamb waves as they interact with the stiffener and the damage, demonstrating the power of Lamb wave-based SHM systems to detect the presence of adhesive failure.

2. Lamb Waves

Lamb waves are traction free surface waves that are induced in thin plates. There are two main types of modes of Lamb wave, the symmetrical (S) modes which can be represented by cosine functions and the asymmetric (A) modes which can be represented by sine functions. A diagrammatic presentation of these two wave modes are presented in Figure 1. The full numerical solutions are presented in [16].

Theoretically there are an infinite number of number of wave modes [17]. Though high order modes have shown potential for monitoring structures, particularly for small defects in their microstructure, their low amplitudes [18] and higher frequency content mean that they experience greater attenuation limiting their use in SHM systems [19]. It is for this reason that for monitoring larger structures the S_0 and A_0 modes are only typically excited in an SHM system to reduce the number of sensors required and limit the processing costs.

Wave direction

Asymmetric i.e. transverse in centreline

Symmetric i.e. longitudinal in centreline

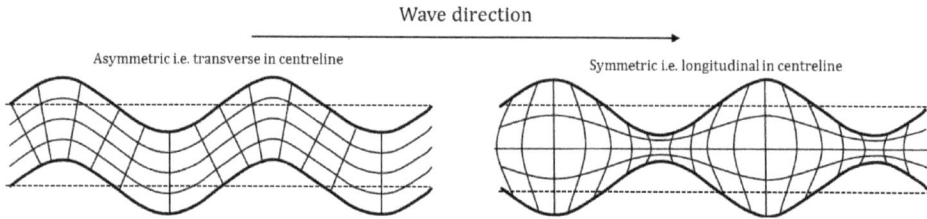

Figure 1. A diagrammatic representation of symmetrical and antisymmetric wave modes.

Lamb waves exhibit a phenomenon known as dispersion where the velocity of the wave through a plate is a function of its frequency [20]. Typically, the symmetrical mode has a higher wave velocity than the antisymmetric mode of the same order. Due to velocity and wavelength being a function of the frequency of a Lamb wave, excitation frequency is an important consideration for pulse-receive type SHM systems.

There have been extensive studies carried out on the behaviour of Lamb waves in plate like structures [21,22]. A comprehensive experimental study supported by mathematical prediction of Lamb wave behaviour in metallic plate-like structures with adhesively bonded joints was performed by Rokhlin [23]. This work investigated the Lamb wave interaction and transmission through lap-shear joints. Mode conversion was studied as the primary phenomena to determine the condition of the joint. This work goes some distance towards determining appropriate modes and frequencies to be used for inspecting the condition of adhesively bonded lap shear joints. However, it only considered higher order wave modes for detecting the presence of disbonds [23].

The mechanics and theory behind the reflection and transmission of plate waves by vertical stiffeners was presented in [24]. This work highlights the complexity of Lamb wave interaction with vertical stiffeners. As Lamb waves interact with a stiffener some of the wave energy is transmitted through the stiffener while a percentage of the wave energy is reflected by the stiffener back towards the source and a small amount is attenuated by the stiffener. As a Lamb wave interacts with an adhesively bonded stiffener, it is possible for mode conversion to also occur due to changes in thickness and hence boundary conditions. This was numerically investigated using a boundary element method and further experimentally investigated by Cho [25]. Cho demonstrated that changes in thickness, particularly through a joint or a change in thickness due to a defect can result in different guided wave behaviours including reflection, transmission and mode conversion. Lemistre and Balageas [26] also experimentally investigated the mode conversion phenomenon of Lamb waves interacting with a delamination in a composite panel based on a working hypothesis derived from the work of Han *et al.* [27]. The work demonstrated that the diffraction effect exhibited by interaction with the defect did indeed induce mode conversion. It is also possible for mode conversion to occur at other changes in acoustic impedance [28] such as a change in material (*i.e.*, interaction with adhesive layer). Changes in wave mode can cause issues with the detection of Lamb waves, in particular for systems that primarily detect one wave component.

Ramadas *et al.* [29] numerically investigated the interaction of the fundamental antisymmetric mode with a bonded stiffened section made from glass fibre reinforce plastic. Transmitters and receivers were positioned on different sections of the structure to investigate the turning modes as the A_0 interacted with the structural discontinuity. The numerical simulations were validated with experimental data and it was found that the interaction resulted in a fundamental symmetrical mode being generated. It is worth noting that this work only considered the stiffened region of the plate and did not consider the full transmission of the Lamb wave through the stiffened section.

Generally, Lamb waves exhibit different and usually more complex behaviour in composites compared to metallic materials due to weave and fibre orientation. This makes the problem of implementing an SHM system more difficult. Traditionally, in metallic aircraft structures stiffeners are not bonded but are attached using mechanical fastening methods whereas composite stiffeners are either bonded or are integral to the structure. To reduce the complexity in both the Lamb wave behaviour and specimen design in this investigation, a metallic stiffener bonded to a metallic plate is considered with a view to applying the techniques presented to more complex composite components in future work.

3. Lamb Wave Measurement Using 3D Laser Vibrometry

One of the earliest set of studies to use scanning laser vibrometry to sense Lamb waves and their interaction with damage was presented in [30–32]. In these studies, 1D scanning laser vibrometry was used to measure the out-of-plane velocities of A_0 Lamb wave propagation through aluminium plates with various defects. The Vibrometry data recorded was validated using the responses of piezoelectric transducers bonded to the surface. Numerical simulations were also found using the local interaction simulation analysis (LISA). This set of studies demonstrated the potential of using scanning laser vibrometry for investigating Lamb wave propagation and interaction with damage.

149

Out-of-plane scanning laser Doppler vibrometry has been previously used to successfully detect damage in a stiffened panel [33]. Experimental work was carried out to detect the voids left by the absence of rivets used to attach a t-shaped aluminium stiffener to a flat aluminium plate. The plate was excited by a lead zirconate titanate (PZT) transducer with a 5-cycle sine wave multiplied by a Hann-window at frequencies of 5 kHz, 35 kHz, 100 kHz and random noise. By calculating the route mean square of the signal received at each measurement point it was possible to generate a visualisation that clearly showed the absence of the rivets. The best visualisations were found to be obtained from results taken at higher frequencies.

Out of plane measurements were taken by Sohn *et al.* [34] using a 1D scanning laser vibrometer to measure Lamb wave interaction with delaminations in a composite plate after impact as well as Lamb wave interaction with a disbond of a composite spar. In this work, frequency-wavenumber domain and Laplacian image filters were placed over the wave field images to enhance imaging of the defect. This successfully located the presence of the damage and highlighted the potential for image-based techniques for use in SHM applications.

3D scanning laser vibrometry has served as a successful method for validating results from finite element analysis models by Olson *et al.* [35]. This study demonstrated the capability of 3D scanning laser vibrometry to measure accurately in-plane and out-of-plane Lamb modes in thin plates for SHM applications. The advantages of using 3D scanning laser vibrometer over a 1D system were demonstrated by using the data of the in-plane modes to qualitatively and quantitatively more accurately validate finite element models.

4. Experimental Study

4.1. Panel Manufacture and Geometry

A 3 mm thick 6082-T6 aluminium plate was bonded to a 6082-T6 aluminium unequal angle stiffener to construct a stiffened panel with the dimensions shown in Figure 2. The dimensions of the panel were chosen to reduce the effects of edge reflections. The stiffener was bonded to the plate using commercially available Araldite® 420 (Huntsman, Woodlands, TX, USA) adhesive. The film thickness of the adhesive was regulated using 0.1 mm copper wire gauges to achieve the best shear strength [36].

A geometrically similar second panel was also manufactured with an induced disbonded region of 25.4 mm length across the width of the stiffener. The disbonded region was induced by installing PTFE tape prior to applying the adhesive. This was then removed once the adhesive had cured.

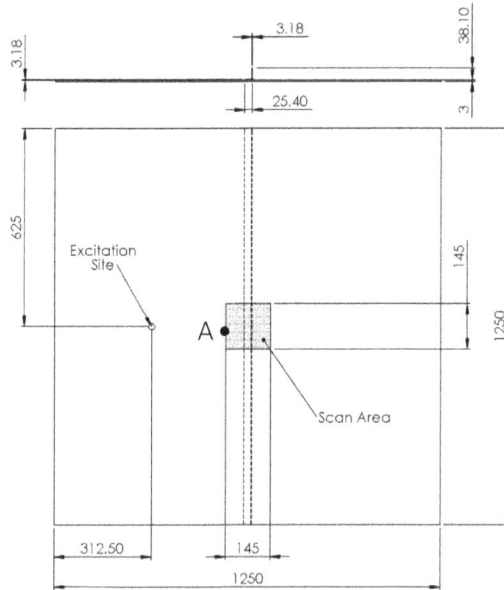

Figure 2. Dimensions of the stiffened panel (all dimensions in mm).

A commercially available PANCOM Pico-Z (200 kHz–500 kHz) transducer (PANCOM, Huntingdon, UK) was acoustically coupled to the panel using Loctite® (Loctite, Düsseldorf, Germany) Ethyl-2-Cyanoacrylate adhesive. This transducer was selected because of its flat broadband frequency response in the frequency range under investigation and its relatively small face (5 mm) as previous studies have shown that sources that are representative of a point source produce good results [37]. The transducer was located at the mid-point between the panel edge and the stiffener centreline to reduce the effects of edge reflections on the transmitted wave. This also was sufficient distance from the scan area to ensure that that the Lamb waves had fully formed.

4.2. Experimental Setup

A 10-cycle sine wave was generated by the Mistras Group Limited (MGL) WaveGen function generator software (MGL, Cambridge, UK) which was connected to the MGL µdisp/NB-8 hardware (MGL). This allowed the transducer to input sufficient energy into the structure. The peak-to-peak amplitude of the excitation signal was 160 V. Three frequencies were selected for this experiment; 100 kHz, 250 kHz and 300 kHz. The frequencies investigated primarily excited the two fundamental modes as shown in the calculated dispersion curves for a 3 mm aluminium plate in Figure 3.

151

100 kHz was chosen to investigate how a longer wavelength interacted with the stiffener and the disbonded region. The wavelengths for the A_0 and S_0 were calculated to be 15 mm and 55 mm respectively. Though this frequency fell outside the operating resonance window of the chosen transducer, the calibration certificate for the transducer showed that it would function at this frequency but at a reduced amplitude. Therefore, frequencies of 250 kHz and 300 kHz were also selected to investigate the interaction of smaller wavelengths. The wavelengths for the A_0 and S_0 were calculated to be 9 mm and 22 mm respectively for the 250 kHz excitation and 8 mm and 18 mm respectively for the 300 kHz excitation. Frequencies above 300 kHz were not considered due to the constraint of the sampling frequency of the acquisition card in the laser vibrometer used. A 10 V peak-to-peak wave was also generated and used as a reference signal for triggering the acquisition of the vibrometer. A repetitive trigger rate of 20 Hz was used as this gave sufficient time for the induced wave energy to fully dissipate before the next measurement was taken.

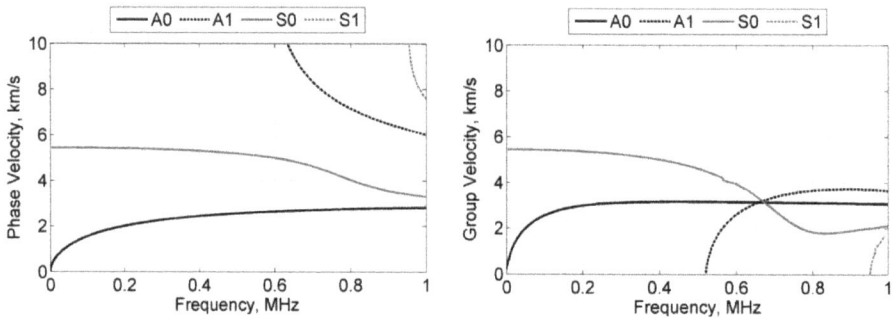

Figure 3. Calculated dispersion curves for a 3 mm aluminium plate, **Left:** Phase Velocity; **Right:** Group Velocity.

The vibrometer used for this experiment was the Polytec PSV-500-3D-M (Polytec, Waldbronn, Germany). This vibrometer uses three laser heads to measure each scan point. The measurements recorded by each laser head can then be used to calculate both the in-plane and out-of-plane modes using trigonometry. Since measurements are taken from three heads it is less important for the lasers to be perpendicular to the structure under test with a 3D scanning system than it is with a 1D system.

The sampling frequency was chosen as 2.56 MHz, which gave sufficient resolution for reconstructing the wave. A drawback of the vibrometry system used however is that at this sampling frequency, the lowest sensitivity range is 200 mm/s. This is not ideal as the amplitude of the measured velocities was of the order of 1 mm/s and thus the measured signal had a low signal to noise ratio. This was overcome by taking 200 measurements at each point and averaging the signal.

152

4096 samples were taken giving an overall sample time of 1.6 ms and a resolution of 390.625 ns. This sample length was considerably longer than required by this study resulting in the waveforms recorded capturing the Lamb wave interaction with the stiffener followed by the edge reflections with a period of no activity in between. Therefore, the sample time was reduced to capture only the first interaction of the Lamb wave with the stiffener in post-processing.

A scan area of 145 mm × 145 mm was measured which comprised of 5329 measurement points. This area was coated with retro-reflective glass beads which have a diameter of the order of 10 μm. The glass beads improved the back-scatter of the laser light and hence improved the quality of the signal. The beads were applied to the surface using a spray adhesive and a thin coverage was ensured by removing excess beads.

5. Results and Discussion

The velocity measurements were integrated using Polytec PSV software to obtain the displacement values. This was carried out to be representative of the signal that a sensor bonded to the structure would measure. The magnitude of each of the displacement components for each excitation frequency are presented in turn. For each set of a results a 0 μs datum point was taken immediately before the S_0 mode reached the measurement field. In each plot presented the presence of the stiffener is denoted by the dotted lines.

5.1. 100 kHz Results

The results from the healthy panel and the panel with the disbonded region are shown in Figure 4. The results from the healthy panel show that the higher wave velocity S_0 Lamb mode is relatively unaffected by the presence of the stiffener and is mostly transmitted straight through. The slower wave velocity A_0 mode however is greatly affected. At 105 μs in Figure 4 the A_0 mode interacts with the stiffener resulting in a reduction in amplitude on the other side demonstrating that the majority of the wave energy is being either attenuated or reflected by the stiffeners rather than transmitted.

The results from the panel with the disbond show that in this case the S_0 Lamb mode interacts with the disbonded region of the stiffener. This is seen in Figure 4 at 30 μs as an area of higher amplitude in the disbonded region. Similarly, a higher level of wave amplitude is observed as the A_0 mode also interacts with the disbonded region shown at 75 μs. As the A_0 mode continues to interact with the disbonded region, transmitted (right of the stiffener) and reflected (left of the stiffener) conical diffraction fringes are observed, clearly seen from 105 μs onwards. The behaviour of the Lamb wave outside of the disbonded region is similar to that of the healthy panel.

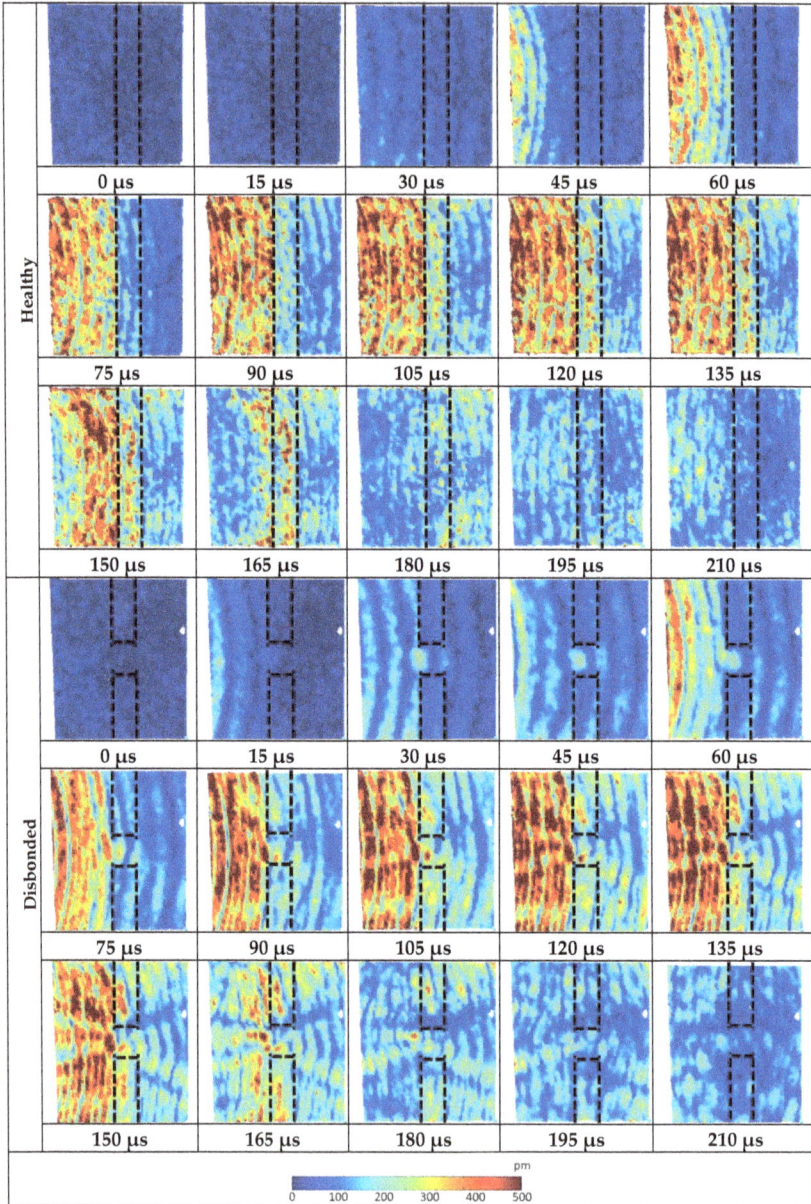

Figure 4. 100 kHz three-component displacement magnitude showing Lamb wave interaction with the healthy and disbonded stiffened panel.

5.2. 100 kHz Windowed Cross-Correlation Analysis

Example waveforms of the out-of-plane component from the corresponding measurement points in the area of investigation are plotted in Figure 5 for the healthy

154

and disbonded panel respectively. The measurements presented were at the left hand boundary on the centreline denoted by point A in Figure 2. This measurement point lay equidistant between the excitation site and the stiffener.

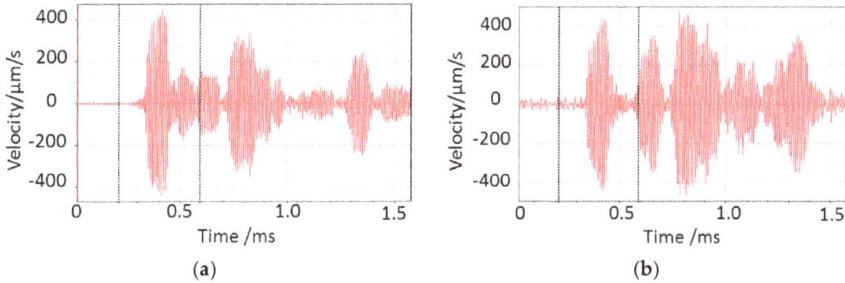

Figure 5. Waveforms from the left hand boundary of the area of investigation in-line with the disbonded region for each panel—(**a**) Healthy (**b**) Disbonded. The dashed lines denote the section of waveforms that was used for time windowing.

From visual comparison, it is evident that there is a significant difference between the two waveforms. The signal measured from the disbonded panel had a higher level of noise than the signal measured from the healthy panel. This is mostly likely due to a difference in the level of backscattered light received. Even by simple visual comparison, the incoming wave is similar. However, by 0.45 ms there is a significant change in the waveforms. The healthy panel shows the reflected wave-packet from the stiffener whereas this is not present in the waveform from the disbonded panel as the wave is transmitted through the disbond.

To quantify this difference in the measured waves, a windowed cross-correlation analysis was conducted. Starting at 0.21 ms, both waveforms were windowed into respective 100 μs time windows and the cross-correlation coefficient was calculated. The windows were transposed by 20 μs forward and the cross-correlation coefficient was once again calculated. The cross-correlation coefficients from this analysis are plotted in Figure 6.

The first two time windows show a low value for the cross-correlation coefficient. This is due to random noise mostly being present in these windows. As the wave arrives in the 0.25–0.35 ms time window, the value of the cross-correlation coefficient significantly increases to 0.93 and continues to increase until it peaks in the 0.31–0.41 ms time window at 0.97. It is worth noting that in an installed acousto-ultrasonic system, the cross-correlation coefficient may be higher than this. However, due to two similar panels being used in this experiment and small differences in experimental setup this indicates a good correlation of the two wave forms.

155

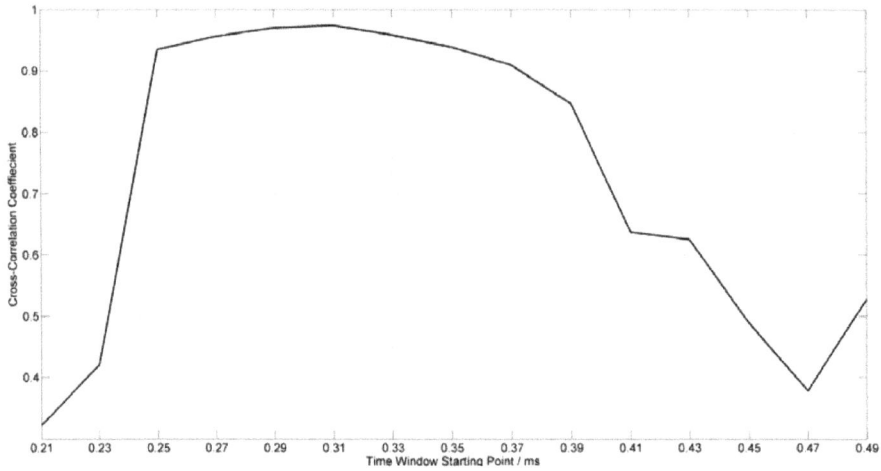

Figure 6. Moving window cross-correlation coefficient plot.

As the waveforms start to differ due to the reflections, the cross-correlation coefficient starts to reduce significantly. In the 0.47–0.57 ms time window the cross-correlation coefficient was found to be 0.38 indicating a large difference in the waveforms and hence, the presence of the disbonded region.

This analysis demonstrates that wave transmitted from the excitation site was comparable on both panels during this experiment. For an installed SHM system this technique demonstrates that reflected waves from the stiffener can be used to identify the presence of disbonds.

5.3. 250 kHz Results

The results from the healthy panel and the panel with the disbonded region are shown in Figure 7. As with the 100 kHz results, there is a significant reduction in wave amplitude as the A_0 mode interacts with stiffener. The presence of the stiffener is clearly seen in Figure 7 at 60 μs. As with the results from the 100 kHz excitation, the S_0 mode interacts with the disbonded region illustrated at 30 μs in Figure 7 by the increase in wave amplitude. The S_0 mode wave energy transmitted through the disbonded region is also shown by a fringe of increased amplitude to the right of the disbonded region.

The A_0 mode is transmitted through the disbonded region with a lower reduction in amplitude than the bonded regions of the stiffener. This has led to a fringe of high amplitude to the right of the stiffener.

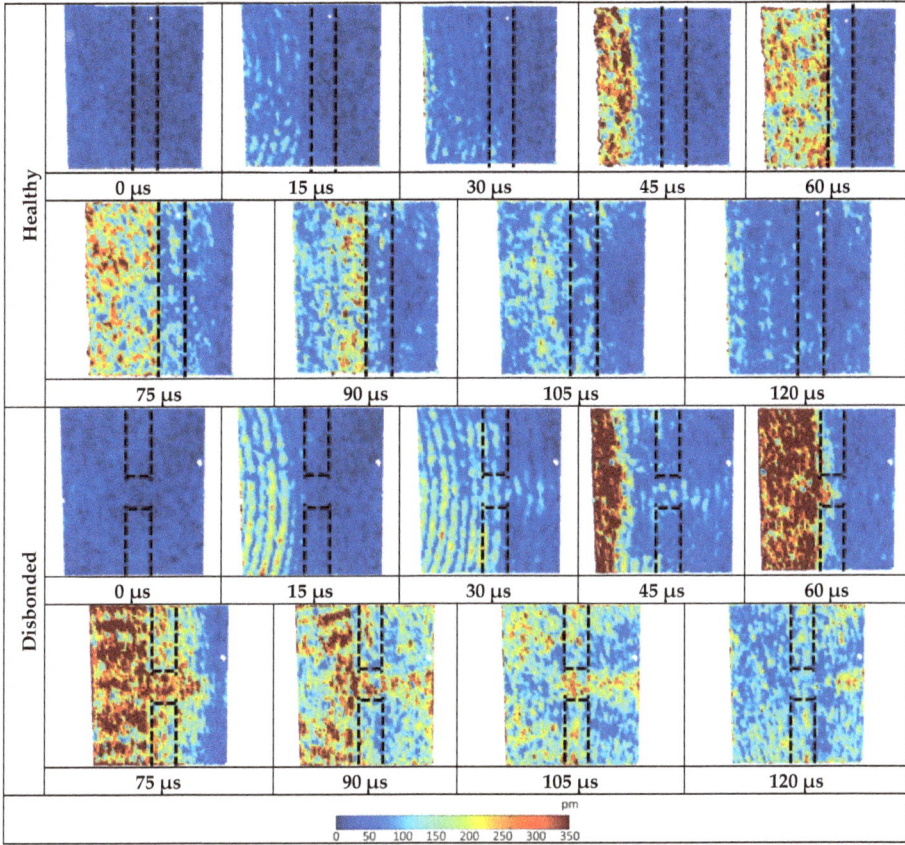

Figure 7. 250 kHz three-component displacement magnitude showing Lamb wave interaction with the healthy disbonded stiffened panel.

5.4. 300 kHz Results

The results from the healthy panel and the panel with the disbonded region when excited at 300 kHz are shown in Figure 8. As with the other two excitation frequencies, the presence of the stiffener significantly reduces the amplitude of the A_0 mode as it interacts with it as shown at 60 μs in Figure 8.

Again, as with the other two excitation frequencies, the S_0 wave transmits through the disbonded region more effectively than through the regions that are adhesively bonded as shown at 30 μs in Figure 8.

At 60 μs the A_0 mode also interacts with the disbonded region. This interaction results in a transmitted conical fringe pattern to the right hand side of the stiffener as the wave energy transmits through the disbonded region.

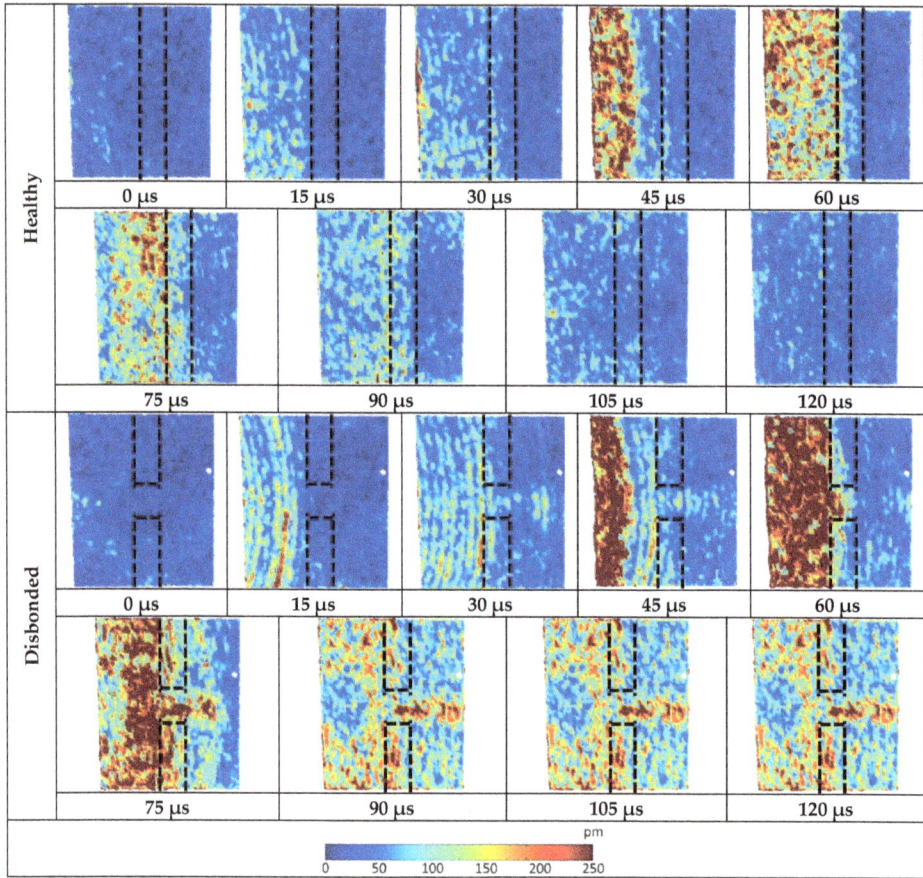

Figure 8. 300 kHz three-component displacement magnitude showing Lamb wave interaction with the healthy and disbonded stiffened panel.

5.5. Comparative Discussion

Lamb waves exhibit a phenomena known as dispersion where the speed of the wave varies with the excitation frequency. When comparing these results, particularly the 100 kHz and 250 kHz results, the higher frequency waves can be seen to travel further in a shorter space of time.

It is due to the dispersion phenomena that the amplitude of the Lamb wave is seen to reduce as it moves across the area of investigation in the higher frequency excitations; therefore demonstrating that the higher frequencies attenuate more.

Comparing the results from the two different panels, the S_0 mode is seen to have a greater amplitude in the disbonded panel results. This increase in wave amplitude can be attributed to a difference in sensor coupling. Despite using the same method to couple the sensor to each panel, it is very difficult to achieve a consistent and repeatable sensor coupling.

One significant difference between the three frequencies is the transmitted conical fringe patterns produced by the Lamb wave interaction with the disbonded region. The results from the 100 kHz excitation show a reduction in wave amplitude to the right of the disbonded region. This differs greatly to what is observed in the results from the 250 kHz and 300 kHz excitations where an increased amplitude is seen to the right of the disbonded region. It has been suggested that this may be a function of the difference in the resultant wavelengths of the different excitation frequencies [38]. This is due to the shorter wavelengths of the higher frequencies being more sensitive to interacting with damage. In terms of the design of a damage detection system however there is a trade-off between the minimum damage size detectable and the global coverage required due to the higher attenuation of the higher frequency Lamb waves.

It may also be noted that the shape of the reflected conical fringe pattern to the left of the stiffener in the 100 kHz excitation results appears to diffract from the entire width of the disbonded region whereas in the other two sets of results the fringe pattern appears to diffract from a much smaller area. It is worth noting that this problem cannot be simply modelled in a similar manner to that of a fluid mechanics problem due to the transmission of the wave through the stiffener and the added complexity of differing wave modes.

The quality of the results from the 250 kHz and 300 kHz excitations are not as clear as the results from the 100 kHz excitation. This could be attributed to the spatial resolution of the scan points. The number of scan points was kept consistent throughout the study in order obtain results that were comparable and to allow the study to be completed in a timely fashion. By increasing the number of scan points more detailed 3D vibrometry fringe plots could be achieved allowing a better comparison to be made between the experimental results.

6. Windowed RMS Analysis

An understanding of the effects of wave interaction is fundamental for use in designing sensor networks to monitor stiffener disbonds on aerospace structures. It therefore is advantageous to obtain an understanding of the amplitude of Lamb wave modes as they interact with defects.

It has been demonstrated by Lee and Staszewski that windowing the time signal and calculating a quantitative measure of the wave amplitude can provide a useful insight for sensor positioning [39]. This is particularly important for developing SHM systems with a high probability of detection.

A root mean-squared (RMS) value for a 10 μs time window of the magnitude of the three displacement signals was calculated for each measurement point. For each data set three plots of the time windows have been produced which show three stages of interaction: the S_0 mode interacting prior to the A_0 mode interacting with the stiffener, the first interaction of the A_0 mode and the subsequent transmission and reflection of the A_0 mode. Due to dispersion, the higher frequency Lamb waves have higher wave velocities meaning that they interact with the stiffener earlier. Therefore, for each frequency, the three stages of Lamb wave interaction occur at different times, dependant on the wave velocity. Ten microsecond time windows between each plot have not been plotted for simplicity.

6.1. 100 kHz RMS Plots and Discussion

The RMS plots for the healthy and disbonded panels with a 100 kHz excitation are presented in Figure 9. The time window of 10 μs–20 μs shows the wave packet starting to interact with the stiffeners. In this time window, the only mode present is the S_0 mode. The healthy panel shows clearly the presence of the stiffener as very little energy is transmitted through the stiffener. The panel with the disbonded region however shows an indication of a defect but not as clear or definitive as the later time windows. There is evidence of wave energy "bleeding" into the disbonded region which is initiating a low energy transmitted conical pattern. This is most likely a result of the S_0 mode interacting with the defect.

As the wave continues to interact with the stiffener it is apparent from both the healthy panel and the panel with the disbonded region that some of the energy is reflected back from the stiffener towards the source. This is shown in Figure 9 by the "tiger-stripe" interference fringes on the left of the stiffener. In the results from the panel with the disbonded region, this interference pattern is disrupted. This is clearly shown in the 50 μs–60 μs window by the circular shape of the reflection pattern to the left of the disbonded region. This reflected conical pattern to the left of the disbonded region has higher RMS values than the neighbouring reflected signal and the reflected signal on the healthy panel. Conversely, as with the plots presented in Figure 4 there is an area of low energy to the right of the disbonded region. This area of low energy is shown by a transmitted conical region of low RMS values as the A_0 mode interacts with the stiffener. This transmitted conical fringe pattern suggest that at this excitation frequency, a greater amount of the Lamb wave energy is reflected back towards the source than is transmitted through the disbonded region. The shape of the defect is also more prominent as the shorter wavelength of the A_0 mode interacts.

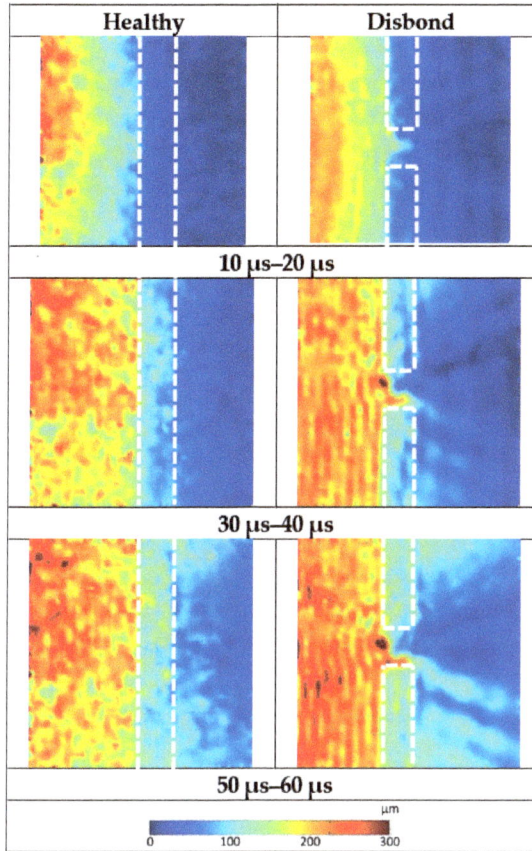

Figure 9. Windowed root mean-squared (RMS) plots of the three-component displacement magnitude for 100 kHz excitation.

6.2. 250 kHz RMS Plots and Discussion

The RMS plots for the healthy and disbonded panel with a 250 kHz excitation are presented in Figure 10. The 0–10 μs time window shows an indication that the lower amplitude S_0 mode interacts with the disbonded region. This is shown as two areas of higher RMS above and below the disbonded region.

As with the 100 kHz excitation, as time continues the presence of the stiffener in the healthy panel reduces Lamb wave transmission. This demonstrates that significantly more wave energy is reflected back towards the source and attenuated by the stiffener than is transmitted through it. The 20–30 μs window clearly shows the presence of the disbonded region as the short wavelength of the A_0 mode interacts however, the high RMS values observed in the disbonded region do not coincide with the supposed shape of the disbonded region. Instead only a small percentage of the

161

25.4 mm length of the disbonded region has a high RMS across the whole width of the stiffener. This is seen in the 40–50 μs window to be the source of the transmitted conical fringe pattern. In this window a transmitted conical fringe pattern of higher Lamb wave energy is seen to be "leaking" through this part of the disbonded region. As a result of this there is a significant decrease in RMS values to the left of the disbonded region demonstrating that at this part of the stiffener more wave energy is being transmitted through the stiffener.

The difference in amplitudes of the two datasets, as with results presented in the time domain, can be attributed to a difference in sensor coupling. Despite of this, the results can still be compared qualitatively.

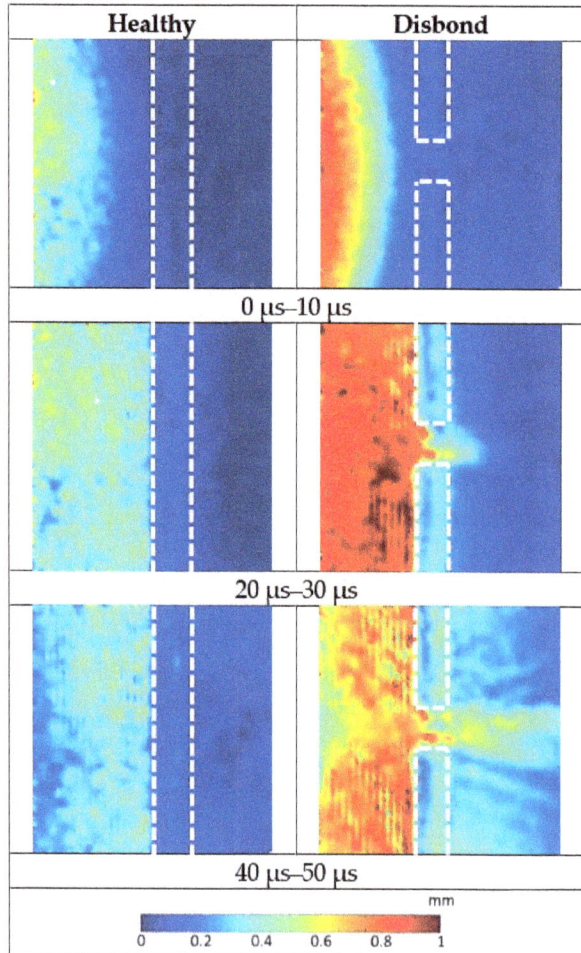

Figure 10. Windowed RMS plots of the three-component displacement magnitude for 250 kHz excitation.

6.3. 300 kHz RMS Plots and Discussion

The RMS plots for the healthy and disbonded panel with a 300 kHz excitation are presented in Figure 11. As with the 250 kHz a small amount of S_0 Lamb wave energy is seen to be interacting with the disbonded region in the 0 µs–10 µs window. This is shown by a small disruption in the low level RMS values in the disbonded region which are not present in the healthy plot. By adjusting the colour scales it may be possible to improve the clarity of this low energy interaction. However, this low level energy lies in the noise floor of the signal and thus the clarity of the rest of the image may be compromised.

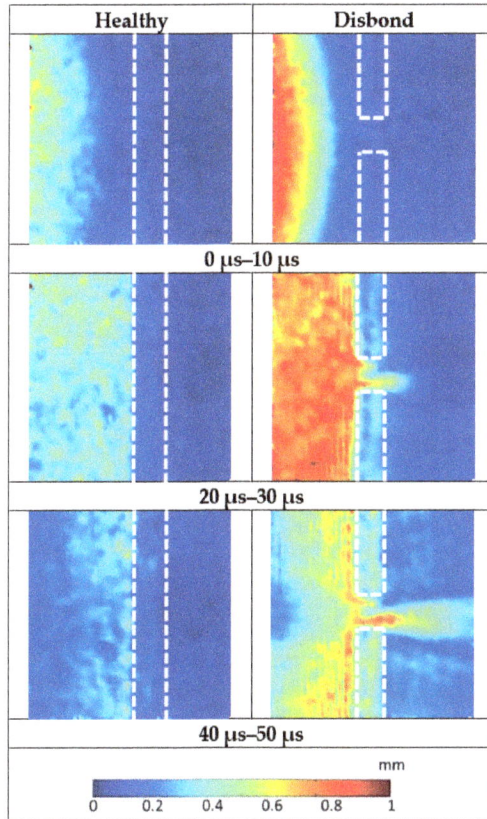

Figure 11. Windowed RMS of the three-component displacement magnitude plots for 300 kHz excitation.

As time progresses again the influence of the stiffener on the Lamb wave transmission is apparent. In the 20–30 µs time window, a clear division of energy is again observed as the A_0 mode interacts. The plot of the disbonded panel of the

RMS values in this time window again shows the Lamb wave clearly interacting with the induced defect in the disbonded region. As with the 250 kHz excitation, the indication does not appear to be as expected as the values of high RMS are not formed in a square. There is again a line of high RMS values which appear to transmit across the entire width of the stiffener to the bottom of the disbonded region. There is also a smaller region of the high RMS values at the top of the disbonded region.

Studying the 40 μs–50 μs time window, the transmitted conical fringe pattern appears to initiate from the line of high RMS values. This is similar to what was observed with the 250 kHz excitation. A secondary fringe also appears to initiate from the top of the disbonded region. This is can be observed when studying the front of the conical fringes where there are two distinct cones of high RMS values.

As with the 250 kHz excitation, there is a reflected conical fringe of low RMS values to the left of the disbonded region. This indicates that again more Lamb wave energy is being transmitted through the stiffener at this point than is being reflected and attenuated.

6.4. Comparative Discussion

There are distinct differences in the windowed-RMS results from the three excitation frequencies. The longer wavelength of the 100 kHz excitation interacts differently with the defect when compared to the shorter wavelengths of the 250 kHz and 300 kHz excitations. The primary difference is that the 100 kHz Lamb wave does not transmit through the disbonded region like the higher frequency excitations. This is demonstrated by the low RMS values to the right of the disbonded region and the higher values to the left indicating that the wave is reflected instead of being transmitted through the disbonded region.

All three sets of RMS plots indicate that the defect in the disbonded region is not a perfect square as was intended. Results at all three excitation frequencies suggest that there is a discrepancy in the defect allowing the Lamb wave to transmit better through the lower part of the disbonded region. This is seen in the RMS plots for the 100 kHz excitation however it is clearer on the higher frequency plots which also suggest that there are other inconsistencies present in the disbonded region.

There is a difference in the amplitudes between the 100 kHz vibrometry data and the higher frequencies shown by the differing values of the colour bars. This can be attributed to the higher frequencies attenuating more, thus the Lamb wave energy decreases significantly in the higher frequency plots.

7. Ultrasonic Inspection of Disbonded Region

As discussed, on reviewing the vibrometry results it was noted that the conical diffraction fringes were not as expected. From the displacement results, particularly the higher frequency excitation, the transmitted conical fringe appeared to originate from a smaller area at the bottom of the disbonded region. To investigate the condition of the induced disbonded region an ultrasonic C-scan (Midas NDT Systems Ltd, Ross-on-Wye, UK). was completed. This method uses a water-coupled ultrasonic probe which pulses out an ultrasonic wave through the structure which reflects off a glass plate on the other side of the structure. The amplitude of the reflected signal is recorded as the probe transverses the structure. Differences in received amplitude allow a plot of the inspected area to be produced [40].

7.1. Experimental Setup

The panel with the induced disbonded was cut down to dimensions of 465 mm × 626 mm to enable it to be positioned in to the C-scan apparatus while ensuring that the cutting operation would not affect the disbonded region.

A MIDAS NDT water coupled C-scan unit was used with Zeus software (Midas NDT Systems Ltd, Ross-on-Wye, UK) to record the results.

7.2. Results and Discussion

The results from the C-scan are plotted in Figure 12. It is apparent that there was something present in the disbonded region which had had an effect on the interaction of the Lamb wave, shown in Figure 12 as a light grey area in the disbond region.

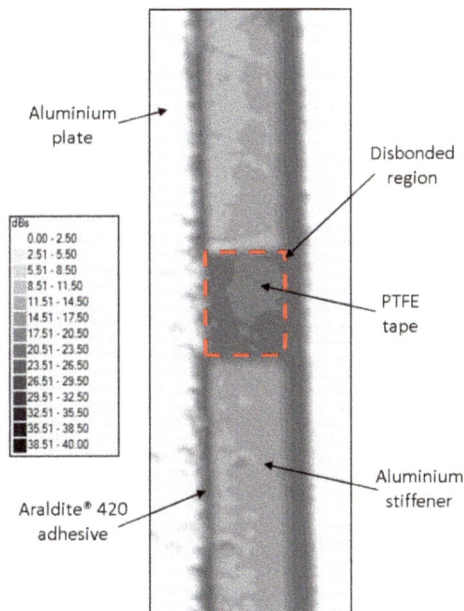

Figure 12. C-Scan results of the panel with the disbonded region. Note the region with the induced disbond containing what appears to be remnants of PTFE tape.

165

It was suspected that this defect occurred during the manufacturing process when the intentional disbond was being created. The disbonded region was induced by using PTFE). When the adhesive was cured the PTFE tape was removed however it was possible that some tape was left inside the bond due to the difficulty of removing it. Therefore it is believed that the conical fringe patterns were caused by the Lamb wave interaction with this PTFE tape.

With that stated, it is highly unlikely that a disbond that occurs on an aircraft during its in-service life or its manufacture would be perfectly square and is more likely to be irregular in shape. Therefore, it is fair to state that the disbonded region that has been produced in this study is potentially more representative of what is likely to be found on a real aircraft structure.

The C-scan results also show that there is a small discontinuity in the adhesive film thickness of the healthy region of the stiffener. This is shown by a difference in the amplitude of the frequency response. This highlights the issue of the inconsistency of adhesive bonds.

8. Swept Sine Wave

The RMS results presented in Figures 7–9 demonstrated that the excitation frequency of a Lamb wave has a significant influence on interaction with defects due to the difference in wavelength. An additional experiment was undertaken using a swept sine wave excitation to further investigate the Lamb wave interaction.

8.1. Experimental Setup

A conical transducer developed by the National Physical Laboratory was acoustically coupled in line with the disbonded region at a distance of 312.5 mm from the midpoint. This excitation source was chosen for its flat broadband response across a range of frequencies [41]. The same excitation hardware was used for this investigation as previously. A swept sine wave ranging from 100 kHz to 500 kHz over a 200 μs duration was used for the excitation signal. This time window allowed the entire frequency sweep to be excited while reducing the effect of reflections.

The same vibrometry setup was used for this investigation although the area of investigation was changed to 250 mm × 90 mm with 5805 velocity measurement points. This not only improved the spatial resolution required for the higher frequency waves but also allowed the focus to be solely on the wave interaction with the disbonded region.

A sample of duration of 800 μs was used for this investigation as it allowed the whole transmitted wave to be captured at all measurement points while minimising the recording of edge reflections.

8.2. Results and Discussion

The RMS value of each of the 5805 velocity measurement points was calculated and plotted as shown in Figure 13. Though measurements were taken for all three planes, the out-of-plane measurements most clearly show the interaction (most likely due to the source primarily exciting out-of-plane modes).

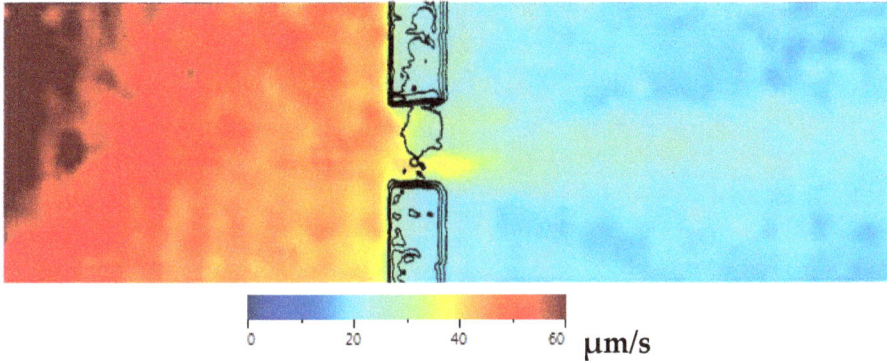

Figure 13. 100–500 kHz swept sine wave excitation out-of-plane RMS plot with an outline of the C-Scan over laid. Note the reduction in wave energy surrounding the PTFE tape region causing a reduction in high amplitude waves transmitting through the disbonded region.

It is apparent that the presence of the stiffener has a significant effect on the amplitude of the Lamb wave. On the left side of the stiffener, RMS amplitudes in excess of 60 μm/s were observed. As the Lamb wave is transmitted through the stiffener the RMS amplitude reduces to less than 15 μm/s; an approximate decrease of 75%.

As observed with the time domain results, a conical transmitted diffraction fringe is present to the right of the disbonded region. This is seen in Figure 13 as a region of higher amplitude (light blue).

As with the time domain results, particularly at 90 μs in Figure 8 the transmitted conical fringe pattern originates from the lower part of the disbonded region. This is particularly highlighted in Figure 13 where in the disbonded region the areas of highest amplitude (yellows) are observed in the lower part of the disbonded region.

When the results from the C-scan are superimposed onto this RMS plot it is evident that the region suspected of being remnants of PTFE tape has a significant influence on the propagation of the wave as the amplitude significantly reduces in this region.

As with the 300 kHz plots in Figure 11, areas of high RMS values are observed at the top and the bottom of the disbonded region. When comparing this to the results of the C-Scan it is evident that no or very little PTFE tape (or other obstruction) is present across the width of the stiffener in these areas.

9. Discussion

It is apparent that the induced Lamb waves all interacted with the disbonded region of the stiffener. There was a quantifiable difference in the Lamb wave interaction when compared with the results for the healthy panel. This demonstrates that by using a statistical measure of the difference of the received Lamb wave such as the cross-correlation technique [42] it would be possible to use an acousto-ultrasonic system to routinely inspect the condition of bonded stiffeners. By combining this technique with a statistical measure of the received energy, such as calculating the RMS value as demonstrated in this study, the location and characterisation of the damage could also be potentially assessed.

The results presented also demonstrate that the excitation frequency of an ultrasonic system is an important consideration for inspecting adhesively bonded stiffeners for disbonds. The 100 kHz excitation did not transmit through the disbonded region whereas the higher frequencies did. For damage detection systems, particularly those which are triggered on an amplitude threshold, this is particularly significant to fully ensure a high probability of detection.

This study only considered a bond which had areas where the disbond spanned the whole width of the stiffener. In the case of real aircraft structure, this may not be the case. It would be beneficial to also be able to detect this type of disbond. The results from the 100 kHz excitation suggest that this could be possible by considering the response of the reflected signal. Therefore an acousto-ultrasonic system capable of pulsing and receiving at different frequencies would be beneficial.

All three excitation frequencies used in this study demonstrated that a low amount of the induced Lamb wave energy was transmitted through the stiffener. This is clearly a consideration for acousto-ultrasonic SHM systems but should also be considered important for passive SHM systems such as acoustic emission (AE). AE induced Lamb wave propagation will be affected by the presence of adhesively bonded stiffeners on a structure. AE events occurring on one side of a stiffener may not be detected by sensors on the other side due to the decrease in Lamb wave amplitude. In addition to this, signals reflected from the stiffener may be interpreted incorrectly leading to added complication when locating and characterising damage using passive techniques. Therefore, the geometry of a structure is an important consideration for the design of a passive AE sensor network.

The results from the RMS plots demonstrate that laser vibrometry is not only a powerful tool in increasing understanding of Lamb wave interaction but can also be used to indicate the size and shape of a defect. The RMS plots serve as a useful aid in considering optimal sensor locations for an active sensor network by highlighting areas of high Lamb wave energy.

The transmission of the S_0 mode through the disbonded region seems mostly unaffected by the presence of the PTFE tape regardless of excitation frequency. The S_0 mode is mostly constructed of in-plane components which gives reason to suggest that in-plane modes maybe less influenced by the presence of kissing bonds. Further investigation using in-plane excitation would be beneficial to further investigate in-plane Lamb wave interaction with stiffeners.

10. Conclusions

This paper has demonstrated how 3D scanning laser vibrometry can be used to conduct a thorough investigation of Lamb wave interaction with adhesively bonded stiffeners and adhesive disbonds. The work presented has demonstrated that the way in which Lamb waves interact with disbonds is dependent on the excitation frequencies.

Analysis performed using a windowed RMS technique has revealed areas of high and low Lamb wave energy. This has revealed that, dependent on excitation frequency, conical fringes with high levels of Lamb wave energy can be found to be either reflected or transmitted through an adhesive disbond.

RMS plots are also beneficial in revealing optimal structural areas for locating sensors to ensure a high probability of damage detection using Lamb waves. These plots demonstrate the complexity of sensor network design and show that considerations such as amplitude threshold are important for use in active sensing systems.

The C-scan inspection revealed the suspected presence of PTFE tape within the disbonded region which had a significant influence on the interaction of the induced Lamb wave. By inducing a frequency swept Lamb wave and plotting the RMS of the velocity measurements taken using the 3D scanning vibrometer it was possible to clearly observe the Lamb wave interaction. This revealed the shape of the PTFE tape which correlated well with the results from the C-scan.

This study has only considered the wave interaction when the actuator is in line with disbond. In reality, this may not be the case. Therefore, it would be beneficial to investigate the effect of the actuator being at different angles and distances relative to the disbond. Other considerations could also be taken into account including defect size in order to work towards achieving an optimisation platform for sensor network design. By using the 3D scanning laser vibrometer as a validation tool, it is possible to validate computational models of a Lamb wave interaction. Therefore it would be of significant benefit to be able to model several damage scenarios which would then allow a platform for sensor location optimisation to be produced.

The work presented has shown that acousto-ultrasonically induced Lamb waves are a suitable method for the in-service monitoring of adhesively bonded structures for SHM applications. Careful consideration must be taken however when planning sensor location due to the complexity of the interaction with adhesive disbonds, in particular disbonds with irregular shapes.

Acknowledgments: The authors would like to acknowledge the support of Engineering and Physical Sciences Research Council (EPSRC) Doctoral Training Grant (DTG) together with Airbus UK Ltd. for the funding of this project as well as Mark Eaton, Mathew Pearson and Stephen Grigg for their advice and assistance during this project. In addition, the authors wish to acknowledge the staff at Polytec Ltd. for their guidance of best practice for using the laser vibrometer.

Author Contributions: Ryan Marks, Carol Featherston, Alastair Clarke and Rhys Pullin conceived the experiment; Christophe Paget provided significant, industrially relevant design contribution to the project; Ryan Marks was the principle researcher on this project, conducted the experiments, analysis and wrote the paper; Rhys Pullin, Carol Featherston and Alastair Clarke quality checked the written manuscript.

Conflicts of Interest: Conflicts of Interest: The authors declare no conflict of interest.

References

1. Airbus. *Global Market Forecast—Futures Journeys 2013–2032*; Airbus: Blagnac, France, 2013.
2. Megson, T.H.G. *Aircraft Structures for Engineering Students*, 3rd ed.; John Wiley & Sons: London, UK, 1999.
3. Madenci, E.; Shkarayev, S.; Sergeev, B.; Oplinger, D.; Shyprykevich, P. Analysis of composite laminates with multiple fasteners. *Int. J. Solids Struct.* **1998**, *35*, 1793–1811.
4. Krishnamoorthy, A.; Boopathy, S.R.; Palanikumar, K. Delamination analysis in drilling of CFRP composites using response surface methodology. *J. Compos. Mater.* **2009**, *43*, 2885–2902.
5. Higgins, A. Adhesive bonding of aircraft structures. *Int. J.Adhes. Adhes.* **2000**, *20*, 367–376.
6. Arenas, J.M.; Alía, C.; Narbón, J.J.; Ocaña, R.; González, C. Considerations for the industrial application of structural adhesive joints in the aluminium-composite material bonding. *Compos. B Eng.* **2013**, *44*, 417–423.
7. Petrie, E.M. Adhesives for the assembly of aircraft structures and components: Decades of performance improvement, with the new applications of the horizon. *Met. Finish.* **2008**, *106*, 26–31.
8. Wang, X.; Kwon, P.Y.; Sturtevant, C.; Kim, D.D.W.; Lantrip, J. Tool wear of coated drills in drilling CFRP. *J. Manuf. Process.* **2013**, *15*, 87–95.
9. Bossi, R.; Piehl, M.J. Bonding primary aircraft structure: The issues. *Manuf. Eng.* **2011**, *146*, 101–102,104–109.
10. Davis, M.; Bond, D. Principles and practices of adhesive bonded structural joints and repairs. *Int. J. Adhes. Adhes.* **1999**, *19*, 91–105.

11. Quaegebeur, N.; Micheau, P.; Masson, P.; Castaings, M. Methodology for optimal configuration in structural health monitoring of composite bonded joints. *Smart Mater. Struct.* **2012**, *21*.

12. Wu, H.; Liu, Y.; Ding, Y.; Liu, J. Methods to reduce direct maintenance costs for commercial aircraft. *Aircr. Eng. Aerosp. Technol.* **2004**, *79*, 15–18.

13. Assler, H.; Telgkamp, J. Design of aircraft structures under consideration of NDT. In *WCNDT-World Conference of Non-Destructive Testing*; ECNDT: Montreal, QC, Canada, 2004.

14. Speckmann, H.; Roesner, H. Structural health monitoring: A contribution to the intelligent aircraft structure. In *ECNDT 2006*; European Conference on Non-Destructive Testing: Berlin, Germany, 2006.

15. Brotherhood, C.J.; Drinkwater, B.W.; Dixon, S. The detectability of kissing bonds in adhesive joints using ultrasonic techniques. *Ultrasonics* **2003**, *41*, 521–529.

16. Rose, J.L. *Ultrasonic Waves in Solid Media*; Cambridge University Press: Cambridge, UK, 1999.

17. Diligent, O. *Interaction between Fundamental Lamb Modes and Defects in Plates*; Imperial College London: London, UK, 2003.

18. Rauter, N.; Lammering, R. Numerical simulation of elastic wave propagation in isotropic media considering material and geometrical nonlinearities. *Smart Mater. Struct.* **2015**, *24*.

19. Eaton, M. *Acoustic Emission (AE) Monitoring of Buckling and Failure in Carbon Fibre Composite Structures*; Cardiff University: Cardiff, UK, 2007.

20. Rindorf, H.J. Acoustic emission source location in theory and in practice. In *Bruel and Kjaer Technical Review*; Bruel and Kjaer: Naerum, Denmark, 1981.

21. Viktorov, I.A. *Rayleigh and Lamb Waves—Physical Theory and Applications*; Plenum: New York, NY, USA, 1967.

22. Lamb, H. On waves in an elastic plate. *Proc. R. Soc. Lond. Ser. A Contain. Pap. Math. Phys. Character* **1917**, *93*, 114–128.

23. Rokhlin, S.I. Lamb wave interaction with lap-shear adhesive joints: Theory and experiment. *J. Acoust. Soc. Am.* **1991**, *89*, 2758–2765.

24. Reusser, R.S.; Holland, S.D.; Chimenti, D.E.; Roberts, R.A. Reflection and transmission of guided ultrasonic plate waves by vertical stiffeners. *J. Acoust. Soc. Am.* **2014**, *136*, 170–182.

25. Cho, Y. Estimation of ultrasonic guided wave mode conversion in a plate with thickness variation. *IEEE Trans. Ultrason. Ferroelectr. Freq. Control* **2000**, *47*, 591–603.

26. Lemistre, M.; Balageas, D. Structural health monitoring system based on diffracted lamb wave analysis by multiresolution processing. *Smart Mater. Struct.* **2001**, *10*, 504–511.

27. Han, J.-B.; Cheng, J.-C.; Wang, T.-H.; Berthelot, Y. Mode analyses of laser-generated transient ultrasonic lamb waveforms in a composite plate by wavelet transform. *Mater. Eval.* **1999**, *57*, 837–840.

28. Alleyne, D.N.; Cawley, P. The interaction of lamb waves with defects. *IEEE Trans. Ultrason. Ferroelectr. Freq. Control* **1992**, *39*, 381–396.

29. Ramadas, C.; Balasubramaniam, K.; Joshi, M.; Krishnamurthy, C.V. Interaction of lamb mode (A_o) with structural discontinuity and generation of "turning modes" in a T-joint. *Ultrasonics* **2011**, *51*, 586–595.

30. Staszewski, W.J.; Lee, B.C.; Mallet, L.; Scarpa, F. Structural health monitoring using scanning laser vibrometry: I. Lamb wave sensing. *Smart Mater. Struct.* **2004**, *13*, 251–260.

31. Mallet, L.; Lee, B.C.; Staszewski, W.J.; Scarpa, F. Structural health monitoring using scanning laser vibrometry: II. Lamb waves for damage detection. *Smart Mater. Struct.* **2004**, *13*, 261–269.

32. Leong, W.H.; Staszewski, W.J.; Lee, B.C.; Scarpa, F. Structural health monitoring using scanning laser vibrometry: III. Lamb waves for fatigue crack detection. *Smart Mater. Struct.* **2005**, *14*, 1387–1395.

33. Radzienski, M.; Dolinski, L.; Krawczuk, M.; Palacz, M. Damage localisation in a stiffened plate structure using a propagating wave. *Mech. Syst. Signal Process.* **2013**, *39*, 388–395.

34. Sohn, H.; Dutta, D.; Yang, J.Y.; Desimio, M.; Olson, S.; Swenson, E. Automated detection of delamination and disbond from wavefield images obtained using a scanning laser vibrometer. *Smart Mater. Struct.* **2011**, *20*.

35. Olson, S.; DeSimio, M.; Davies, M.; Swenson, E.; Sohn, H. Computational Lamb Wave Model Validation Using 1D and 3D Laser Vibrometer Measurements, Proceedings of the SPIE, the International Society of Optical Engineering, Redondo Beach, CA, USA, 25 March 2010; SPIE: Redondo Beach, CA, USA, 2010.

36. Huntsman. Aerospace Adhesives: Araldite 420 a/b Two Component Epoxy Adhesive. Availabel online: https://www.silmid.com/MetaFiles/Silmid/0f/0fe65000-9138-4ebb-8e64-4829b5a52e13.pdf (accessesed on 5 January 2016).

37. Yan, T.; Theobald, P.; Jones, B.E. A conical piezoelectric transducer with integral sensor as a self-calibrating acoustic emission energy source. *Ultrasonics* **2004**, *42*, 491–498.

38. National Aeronautics and Space Administration (NASA). *Ultrasonic Testing of Aerospace Materials*; NASA: Huntsville, AL, USA, 1998.

39. Lee, B.C.; Staszewski, W.J. Sensor location studies for damage detection with lamb waves. *Smart Mater. Struct.* **2007**, *16*, 399–408.

40. Adams, R.D.; Drinkwater, B.W. Nondestructive testing of adhesively-bonded joints. *NDT E Int.* **1997**, *30*, 93–98.

41. Theobald, P.; Thompson, A. *Towards a Calibrated Reference Source for in-situ Calibration of Acoustic Emission Measurement Systems*; National Physics Laboratory: Teddington, UK, 2005.

42. Pullin, R.; Eaton, M.J.; Pearson, M.R.; Featherston, C.; Lees, J.; Naylon, J.; Kural, A.; Simpson, D.J.; Holford, K. On the development of a damage detection system using macro-fibre composite sensors. *J. Phys. Conf. Ser.* **2012**, *382*.

Proof of Concept of Crack Localization Using Negative Pressure Waves in Closed Tubes for Later Application in Effective SHM System for Additive Manufactured Components

Michaël F. Hinderdael, Dieter De Baere and Patrick Guillaume

Abstract: Additive manufactured components have a different metallurgic structure and are more prone to fatigue cracks than conventionally produced metals. In earlier papers, an effective Structural Health Monitoring solution was presented to detect fatigue cracks in additive manufactured components. Small subsurface capillaries are embedded in the structure and pressurized (vacuum or overpressure). A crack that initiated at the component's surface will propagate towards the capillary and finally breach it. One capillary suffices to inspect a large area of the component, which makes it interesting to locate the crack on the basis of the pressure measurements. Negative pressure waves (NPW) arise from the abrupt encounter of high pressure fluid with low pressure fluid and can serve as a basis to locate the crack. A test set-up with a controllable leak valve was built to investigate the feasibility of using NPW to localize a leak in closed tubes with small lengths. Reflections are expected to occur at the ends of the tube, possibly limiting the localization accuracy. In this paper, the results of the tests on the test set-up are reported. It will be shown that the crack could be localized with high accuracy (millimeter accuracy) which proves the concept of crack localization on basis of NPW in a closed tube of small length.

Reprinted from *Appl. Sci.* Cite as: Hinderdael, M.F.; De Baere, D.; Guillaume, P. Proof of Concept of Crack Localization Using Negative Pressure Waves in Closed Tubes for Later Application in Effective SHM System for Additive Manufactured Components. *Appl. Sci.* **2016**, *6*, 33.

1. Introduction

1.1. Additive Manufacturing

Additive manufacturing (AM), or 3D printing, is a new group of manufacturing technologies whereby physical objects are built by sequential addition of material on basis of a virtual 3D model.

These additive manufacturing techniques allow for completely different designs than the subtractive manufacturing techniques such as milling, grinding, drilling, *etc.*, which require that a tool can reach the spot that has to be shaped (which is not always possible). With 3D printing, the cost is merely related to the amount of printed material, not to the complexity of the design. As the tendency exists to produce components at the lowest possible cost, there exists a stimulus to reduce the weight of 3D printed components, which is a clear benefit in several sectors such as aeronautics. Three-dimensional printed components also possess the property of embedding smart technologies during the printing process, generating overall smart solutions.

As additive manufactured components are constructed layer by layer, their metallurgic structure is different from conventionally produced materials. Fatigue and crack propagation properties of additive manufactured materials are not very well known yet [1], because a lack of process understanding and a lack of *in situ* process monitoring and control, especially in metal AM systems, currently results in unknown porosity levels and distributions [2]. Multiple sources report a reduced fatigue lifetime. In the work of Chan *et al.* [3], fatigue lifetime of titanium Ti6Al-4V alloys fabricated by means of additive manufacturing were compared to conventionally produced Ti6Al-4V alloys (rolled or cast). Conventionally rolled Ti6Al-4V alloys showed approximately twice the fatigue lifetime of the best additive manufactured Ti6Al-4V alloys. Gong *et al.* [4] stated that as-built surfaces of selective laser melted (SLM) Ti-6Al-4V samples become crack initiation sites, resulting in a reduced fatigue lifetime when the stress level exceeds 500 MPa. Similar results were also recorded in the paper of Strantza *et al.* [5], where the SLM Ti6Al-4V samples indicated a stress level at failure under fatigue loading of 590 MPa and 570 MPa.

Despite the numerous benefits, the reduced fatigue behavior is currently jeopardizing the introduction of additive manufactured primary components in industries where safety and product availability are very important. Inspection intervals must then be shortened, increasing inspection costs and product unavailability, which inevitably induce large economic losses. Including an effective structural health monitoring system that monitors the fatigue behavior of 3D printed components can largely eliminate this major drawback.

1.2. Effective Structural Health Monitoring System

Structural health monitoring (SHM) is the integration of sensing and possibly also actuation devices to allow the damaging conditions of a structure to be recorded, analyzed, localized and predicted in a way that non-destructive testing becomes an integral part of the structure [6].

Many structural health monitoring principles have been developed in laboratory conditions but suffered problems when introduced in real applications. Grease, environmental changes, electrical noise, vibrations and many other external influences have shown to have a severe impact on the performance of these structural health monitoring systems. Despite the clear benefits of an SHM system, but due to this lack of robustness, only a few systems have made the transition from the laboratory to a real application [7,8]. A successful structural health monitoring system must be robust in the first place: very low false alarm rates, high detectability of the cracks and insensitive to external influences.

Bearing this in mind, a new effective structural health monitoring philosophy (eSHM) was recently developed [7]. Small subsurface capillaries are embedded in the structure by means of additive manufacturing techniques, close to the region where fatigue cracks are expected to grow. The capillaries are then pressurized (vacuum or overpressure) and the capillary pressure is monitored continuously during the operation of the component. A crack, that initiated at the surface of the component, propagates through the material until it finally breaches the capillary. The resulting leak changes the internal capillary pressure. The presence of a fatigue crack is then derived from the moment that the capillary pressure bypasses a certain pre-set limit. In Figure 1, a possible application of the new effective SHM system is depicted. The embedded capillaries are indicated in red.

The capillaries of the eSHM system must be embedded in the regions where fatigue cracks are expected to grow. In order not to reduce the strength of the component, one must take into account the presence of the capillaries during the design phase. It was concluded in the paper of Strantza *et al.* [9] that—for the considered structure in the paper—the presence of the capillary did not influence the fatigue life negatively, although the cracks initiated from the capillary when positioned close to the specimen edge. Another paper of Strantza *et al.* [10] reported on four point bending test results of test samples with a different layout of the embedded capillaries. It was concluded that the capillary had no crucial influence on the fatigue initiation location for the test specimen and test conditions under investigation. Future studies are being focused on robustness improvements of the eSHM to delay crack initiation at the capillary location.

Figure 1. Embedded capillaries (in red) of the effective Structural Health Monitoring System (eSHM) [7].

Embedding the capillary in a way that it connects all locations prone to fatigue allows monitoring the critical areas without the need for multiple sensing devices. In the example of the cogwheel of Figure 1, only one pressure measurement is required to inspect all teeth. This largely facilitates the installation, data storage and post-processing of the data of the eSHM system. After detection of a fatigue crack, further analysis has to be performed on the critical area. It is therefore interesting to know, on basis of the pressure measurement, where the cracked region is located (e.g., which tooth suffers from a crack).

1.3. Negative Pressure Waves

Negative pressure waves (NPW) serve as a basis for the inspection of pipelines. As explained in the paper of Wan *et al.* [11], when a leak happens along the pipeline, the fluid density of the leak point declines immediately due to the fluid medium losses and the pressure drops. Then the pressure wave source spreads out from the leak point to both ends of the pipeline. The analysis of the time-of-arrival of these wave fronts at known locations at both sides of the leak then allows for locating the leak with high accuracy. In the paper of Silva *et al.* [12], NPW are analyzed to locate leaks in pipelines under operation (with flow). The paper reports that the leak in the pipeline could be located with an accuracy comparable to the theoretical uncertainty related to the finite sampling rate that was used.

In this paper, the authors will analyze if NPW can also be used to locate a leak in a closed volume (without flow), as will be the case in the embedded capillaries of the eSHM system. The static pressure inside the volume will continuously change as a consequence of the leak flow, which is not the case in the well known pipeline applications. First, simulations are discussed to understand the nature and amplitude of the negative pressure waves. Secondly, measurements on a test set-up are discussed and the crack localization capability on the basis of negative pressure waves in a closed volume is analyzed.

By installing a pressure sensor at each side of the closed tube, one can measure the difference in time-of-arrival of the two waves. The time needed for the NPW to travel to the two sensors is easily deduced from Figure 2. For the latter part of this paper, the sensor with name 'S1' will always be referring to the sensor which is closest to the leak (it receives the signal first), while sensor 'S2' will be the sensor located far from the leak (receiving the signal later).

$$\text{Sensor S1} \quad t_1 \;=\; \frac{x_1}{c} \tag{1}$$

$$\text{Sensor S2} \quad t_2 \;=\; \frac{L - x_1}{c} \tag{2}$$

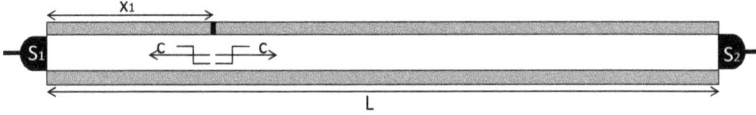

Figure 2. The difference in time-of-arrival of the negative pressure waves at the two pressure sensors mounted at both ends of the tube allows for locating the crack.

Because t_1 and t_2 are only known relatively, only combination of Equations (1) and (2) leads to a localization formula:

$$x_1 = \frac{L - c(t_2 - t_1)}{2}.$$ (3)

The theoretical uncertainty on the crack location (Δx) is related to the sampling rate ($f_s = \frac{1}{\Delta T_s}$) of the data acquisition unit through

$$\Delta x = \frac{c\Delta T_s}{2} = \frac{c}{2f_s}.$$ (4)

Because of the high sonic speeds (for air at room temperature: $c = \sqrt{kRT} = 343\frac{m}{s}$), it will be required to use a high sampling rate to reduce the uncertainty on the estimated location and to deduce the localization feasibility. A sampling rate of 204.8 kHz, for instance, results in a theoretical uncertainty on the location of the leak of 0.84 mm.

For the ease of practical implementation, the capillaries are represented by 3/4" tubes (diameter of 16.9mm) and with a length of 462 mm. The circular leak had a diameter of 1mm. The initial pressure level is set at approximately 20 kPa (absolute) while the outside environment is at ambient conditions (101.3 kPa). The temperatures are expected to be all in equilibrium at room temperature (293.15 K). The performed simulations and measurements discussed in the following parts of the paper will always be referring to this test set-up.

We can estimate the amplitude of the NPW on basis of the formula derived by Rocha [13] (and summarized in the paper of Loth *et al.* [14]):

$$\Delta P_{cap} = 0.3 P_{atm} \left(\frac{D_{leak}}{D_{tube}}\right)^2.$$ (5)

Given the diameters of the tube ($D_{tube} = 16.9mm$) and leak ($D_{leak} = 1mm$), and the ambient conditions around the tube ($P_{atm} = 101300Pa$), we expect the amplitude of the NPW to be:

$$\Delta P_{cap} = 0.3 \times 101300 \text{ Pa} \times \left(\frac{1 \text{ mm}}{16.9 \text{ mm}}\right)^2 = 106.5 \text{ Pa}.$$ (6)

177

It must be noted here that one calls the acoustic waves "NPW" as they were first analyzed and used in pipeline applications. These pipelines are mostly put under overpressure to distribute liquids over large distances. A leak lowers the internal pressure, which is the reason why the occurring waves are referred to as "NPW". In a vacuum application, which we will be using in the eSHM system, the amplitude of the wave will be positive (increasing pressure), so that the name "Positive Pressure Waves" is more opportune. However, this name has never been used in the literature before. We will therefore stick to the name NPW, although the sign of the pressure change depends on the pressurization case.

2. Simulations

Simulations in "COMSOL Multiphysics" were performed in order to understand the nature of the NPW in more detail. As we are interested in the development and propagation of the NPW, a time-dependent analysis was chosen. Because of a small leak, laminar flow was assumed. The tube with the crack, as described in Section 1.3, is represented by a large cylinder (the tube) connected to another smaller cylinder (the crack). Both cylinders contain air that is assumed to be initially at rest. The air is at an absolute pressure of 20 kPa (vacuum). When the crack is opened, air will enter through the small cylinder (the crack). Therefore, a pressure condition (equal to the ambient pressure) is set to this inlet surface. The no slip condition is applied to all other walls. In order to interpret the development and propagation of the NPW, the plots only contain the pressure information in the range of 20,000–20,200 Pa (as is expected from Equation (6)).

Figure 3 shows the pressure results of the computational fluid dynamics analysis (CFD) obtained in a section in the longitudinal direction of the tube. The abrupt opening of the crack results in a discontinuity of the pressure at the crack location. Assuming that the capillary pressure is below the ambient pressure, the low pressure on the inside of the capillary meets the relative higher pressure of the atmosphere. That discontinuity results in a local increase of density and pressure at the crack location. The pressure wave further propagates through the crack and expands as a spherical wave front in the tube. Once in the tube, the spherical wave front forms two plane waves that travel in opposite directions at the speed of sound (c) through the capillary.

The last time step in Figure 3 clearly indicates that when the NPW arrive at the end of the tube (acoustically seen as a hard wall), the amplitude of the NPW doubles because the incident wave is in phase with the reflected wave. The reflected wave then travels back to the other side of the tube where it will be detected by the other sensor. The amplitude of the arriving NPW is 100 Pa (doubled at the wall).

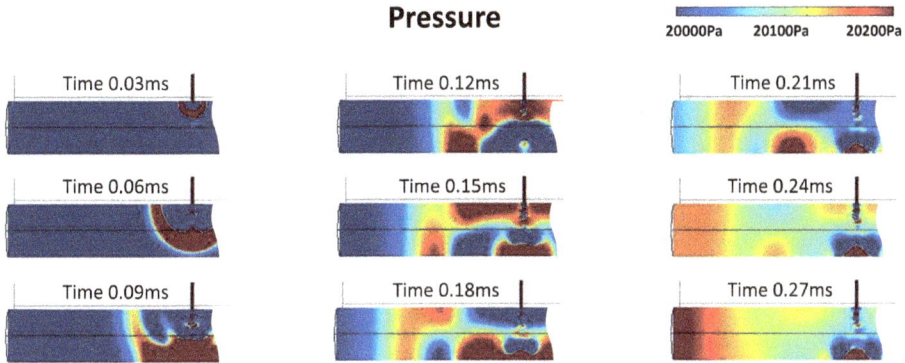

Figure 3. Computational fluid dynamics analysis of negative pressure waves.

3. Experimental Set-Up

A test set-up was built to investigate the crack localization capability on the basis of the NPW. For the ease of practical implementation, the capillaries are represented by 3/4" tubes (internal diameter of 16.9mm) and with a length of 462mm. Along the longitudinal direction of the tube, multiple connections are foreseen to simulate different leak positions and to pull vacuum. A detailed drawing of the tube can be seen in Figure 4 and a graphical presentation of the test-setup is given in Figure 5. The circular leak had a diameter of 1 mm. The initial pressure level is set at approximately 20 kPa (absolute) while the outside environment is at ambient conditions (101.3 kPa). The temperatures are expected to be all in equilibrium at room temperature (293.15 K).

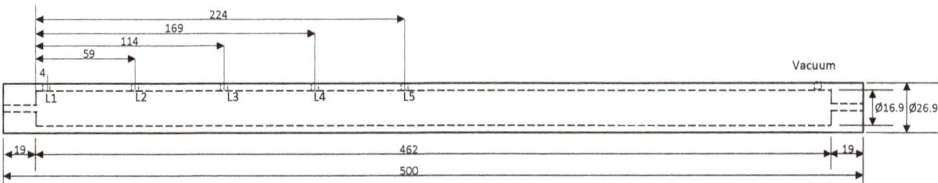

Figure 4. Drawing of the 3/4" tube of the test set-up (dimensions in mm).

Along the length of the tube, multiple holes were foreseen to simulate different leak positions (called L1 to L5 from left to right in Figure 4). During a test, all but one were closed while the remaining hole was connected to a solenoid on–off valve to simulate the opening of the crack with a known diameter of 1mm. The last hole, on the right along the length direction of the tube, was used to pressurize at the required initial pressure level before the measurement.

179

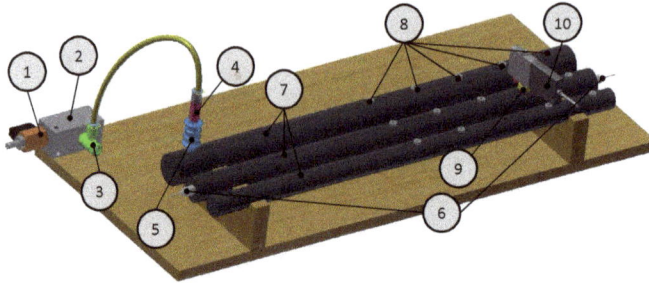

①	Shut-off valve
②	Venturi
③	Flow controller
④	Check valve
⑤	Shut-off valve
⑥	Pressure sensors
⑦	Tubes
⑧	Leak positions
⑨	Leak size
⑩	Solenoid valve

Figure 5. Graphical presentation of the test set-up.

The tube was closed at both ends and each side was equipped with a pressure sensor (Endevco MEGGIT 8540-200). The sensor was selected such that it could measure static pressure changes with a sufficiently high sensitivity (noise level about 10–20Pa, much below the expected amplitude of the NPW (see Equation (6))). The sensors were connected to a LMS Scadas III mobile data acquisition system with a maximum sampling rate of 204.8 kHz. Such a high sampling rate is needed to lower the theoretical uncertainty on the location of the leak to a value of 0.84 mm (see Equation (4)).

4. Results and Discussion

In the following sections, the measurements on the test set-up will be analyzed to derive the crack localization feasibility on basis of the NPW. First, the measurement at leak location L2 will be explained in detail and the location of the leak will be estimated on basis of the pressure measurements. ($x_1 = 59$ mm as defined in Figure 2). Secondly, the measurements at different leak locations will be compared and the localization results on these measurements will be summarized.

4.1. Detailed Analysis of Leak Position L2

Figure 6 presents the pressure measurements for leak location L2 obtained from sensors S1 (full line) and S2 (dashed line). The plot is focused at the moment of time-of-arrival of the NPW at the two sensors and is zeroed at the initial pressure level. Vertical line 0 corresponds to the theoretical moment of opening the leak. Vertical lines 1 and 2 then correspond to the moment of time-of-arrival at sensors S1 and S2 of which the times are given below the figure.

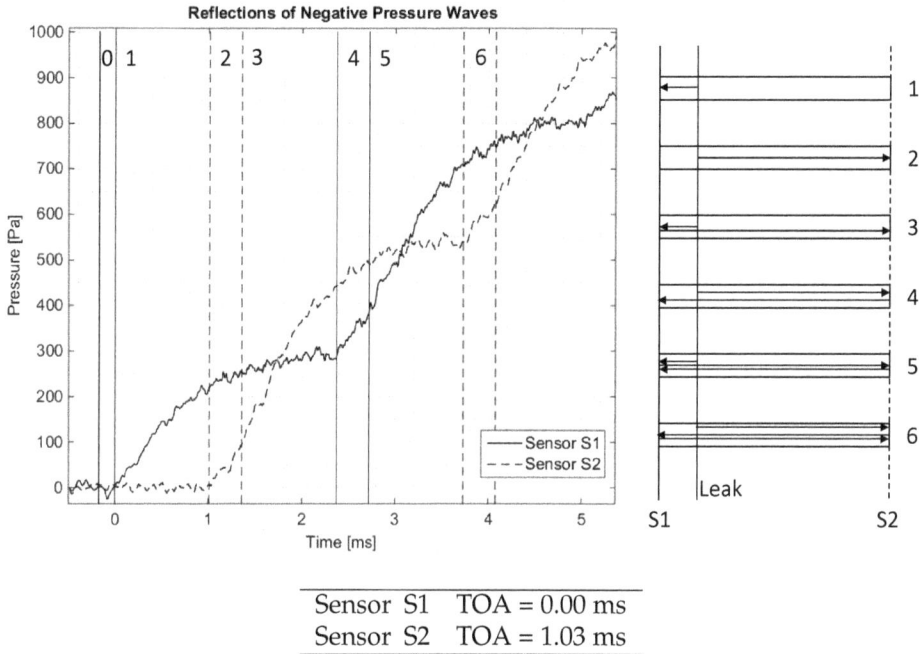

Figure 6. Time-of-arrival of the (reflections of) negative pressure waves.

Sensor S1	TOA = 0.00 ms
Sensor S2	TOA = 1.03 ms

It must be noted that Figure 6 shows the presence of pressure steps, rather than a continuous increase of the pressure. This is due to the reflections of the NPW at the hard walls of the tube (where the sensors are located). Vertical line 4, for instance, corresponds to the moment of arrival at sensor S1 of the NPW that first went to the side of sensor S2 (and was measured there at the moment corresponding to vertical line 2) and then reflected back to the side of sensor S1.

The static pressure change can thus be seen as a succession of exponential pressure steps. The measurement of sensor S1 clearly shows a first exponential pressure rise between vertical lines 1 and 4. A second, less clear exponential pressure step starts at vertical line 4 and ends at vertical line 5 where a third exponential pressure step starts. That exponential behavior is the reason why the breakpoint at vertical line 4 is more clear than at vertical line 5: the more the exponential has flattened, the more the slope of the two successive exponentials differ and the clearer the breakpoint becomes. The recognition of a breakpoint in the pressure steps also allows location of the leak, just by using one pressure sensor [15].

To evaluate the amplitude of the NPW and to compare with the theoretical prediction (Equation (6)), one must must take into account that the sensor measures twice the amplitude of the NPW because of the hard wall at which the wave reflects and the pressure adds up (see section 2). Together with this effect, it is important

181

to note that, due to the contribution of a direct wave and the contribution of a reflected wave that arrives a bit later, multiple NPW might contribute to the pressure rise we are investigating. The pressure rise on sensor S2 between vertical line 2 and 6 flattens out when approaching vertical line 6. We can thus assume that the amplitude of that pressure step can be approximated to the value at vertical line 6, i.e., 530 Pa. Breakpoint 3, however, corresponds to the moment that a second, reflected, wave arrives at the location of sensor S2. The pressure between vertical lines 2 and 6 thus corresponds to the addition of two NPW with amplitude 530 Pa/2/2 $= 132.5$ Pa. This is in rather good correspondence with the theoretical prediction 106Pa (Equation (6)) and the simulations 100 Pa (see Section 2).

Given the moments of time-of-arrival of the direct waves below the figure, it is possible to estimate the crack location on basis of Equation (3).

$$x_1 = \frac{0.462 - 343 \times 0.00103}{2} \text{ m} = 54.4 \pm 0.84 \text{ mm.} \tag{7}$$

This estimate has to be compared to the real location of the leak, which is 59 mm. We can conclude that the estimation of the leak position is rather good but still larger than the theoretical minimum (related to the finite sampling rate (204.8 kHz)). The remaining error is due to noise present on the measurements, which makes it difficult to select the exact moment of arrival of the NPW.

4.1.1. Sensitivity of Leak Localization Technique

It must be highlighted that the measurements are sensitive to the selection of the point of time-of-arrival which is not always easy to select due to noise. Therefore, a sensitivity analysis of the leak localization technique on basis of the NPW was performed. We therefore assume that the moment of time-of-arrival on Sensor S1 is correct, and we show the theoretical interval of detection of the time-of-arrival on Sensor S2 to have a localization of the leak with an error lower than ± 1 cm. The following figure, Figure 7, presents the same measurement as Figure 6 but is more focused on the moments of time-of-arrival at the sensors. The region corresponding to an inaccuracy of ± 1 cm on the localization of the leak was added around the theoretical time-of-arrival at sensor S2.

It must be clear from this measurement that the presence of noise negatively affects the localization accuracy. It is feasible to locate the leak with relative high accuracy, but we must allow an uncertainty in the approximation of a centimeter, which is an order of magnitude larger than the theoretical limit related to the finite sampling rate. Improvement of the test set-up to reduce the noise (e.g. by use of lower range pressure sensors with a higher sensitivity and resolution, *etc.*) or better processing techniques can solve this issue and increase the localization accuracy. In the following section, one example of a better processing technique is given.

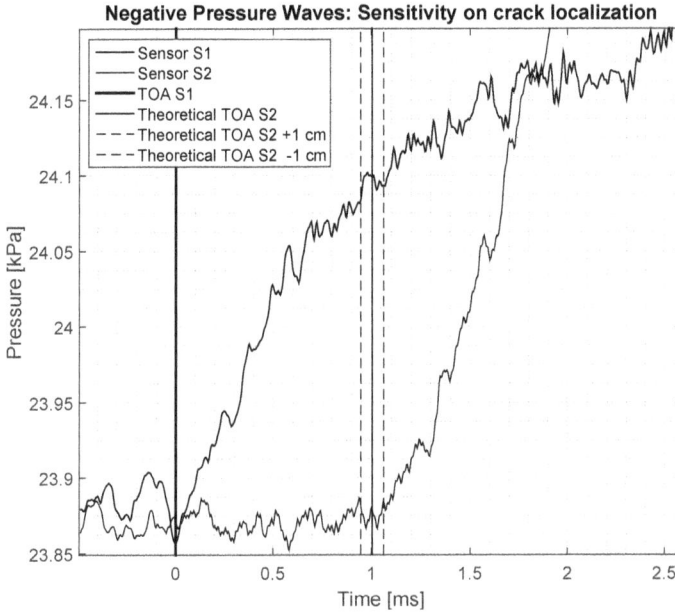

Figure 7. Sensitivity analysis for time-of-arrival point selection for crack localization.

4.1.2. Improvement of the Leak Localization Technique Due to Least Squares Method

Up to now, the difference in time-of-arrival of the NPW was defined as the difference in time between the points where the pressure starts increasing. That principle is very prone to the effects of noise as the selection of these points can be different on basis of the noise that is present on the signal. We therefore aim for a different method to estimate the difference in time-of-arrival of the NPW. The principle is based on a least squared method on the first rising part of the NPW. The signal of sensor S2 is shifted in time and compared with the measurement of sensor S1. The optimum shift of the signal is found when the least squares analysis reports the smallest error between the two signals. The plot in the top right corner of Figure 8 indicates the part of the signal of Figure 6 that will be used for the analysis. The bottom right corner shows that the optimum was found when the signal of sensor S2 was shifted 207 datapoints to the left. The resulting match is plotted on the left of the same figure.

Based on this analysis, we can now estimate the location of the leak. We therefore again use Equation (3):

$$x_1 = \frac{0.462 - 343 \times \frac{207}{204800}}{2} \text{ m} = 57.7 \pm 0.84 \text{ mm.} \tag{8}$$

183

This location should again be compared to the exact distance, 59mm. This method clearly reduces the effect of noise on the localization of the leak. The error now falls within the theoretical uncertainty related to the finite sampling rate. Improving the localization accuracy (e.g., by using better processing techniques) enables the reduction of the sampling rate of the data acquisition while keeping the error on the localization constant, which is an important cost reduction when considering a later application in the eSHM system. Much research has been performed in the past to improve the leak localization accuracy for pipeline applications such as that presented in the papers of Ostapkowicz [16] and Srirangarajan *et al.* [17].

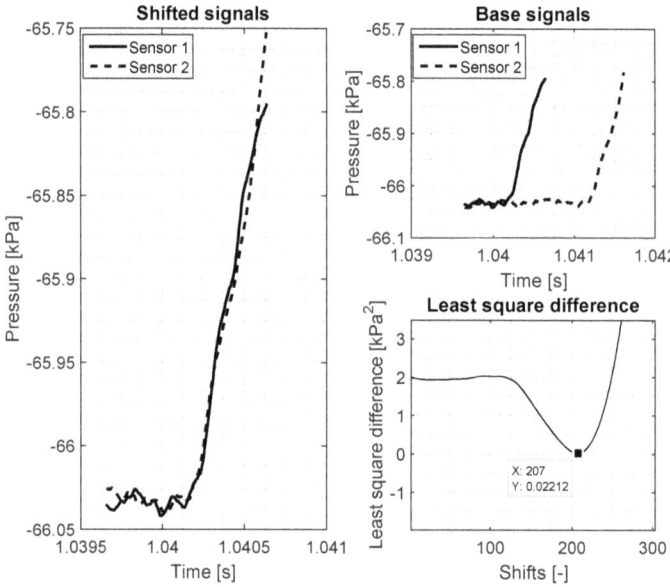

Figure 8. Least square analysis on time-of-arrival of negative pressure waves originating from leak position L2.

4.2. Different Leak Positions

As stated in Section 3 and indicated in Figure 4, multiple leak locations (L1–L5) were foreseen on the tube. As to evaluate the leak localization feasibility, it must be possible to distinguish the different leak locations (L1 to L5) from the obtained difference in time-of-arrival of the NPW. Figure 9 contains five different measurements for the five different leak locations (L1–L5). The thick vertical line corresponds for all five measurements to the moment of time-of-arrival of the NPW at sensor S1. The figure only shows the data measured at sensor S2. The difference in time-of-arrival between sensor S1 (reference) and S2 (increasing pressure) is compared for the different leak locations.

The more the leak is positioned near the middle of the tube (L5), the smaller the difference in time-of-arrival at the sensor locations because the NPW travelled approximately equal times to both ends of the tube. The NPW that originated from L1, the most extreme leak position, clearly arrived with a large time delay at sensor S2.

The theoretical moments of time-of-arrival corresponding to the different leak locations are highlighted through the thin vertical lines. It is clear from the analysis of Figure 9 that the different leak locations can be separated on basis of the interpretation of the NPW.

The least squares method allows to estimate the leak position with high accuracy. Table 1 summarizes the localization estimations and errors of all five measurements (L1–L5).

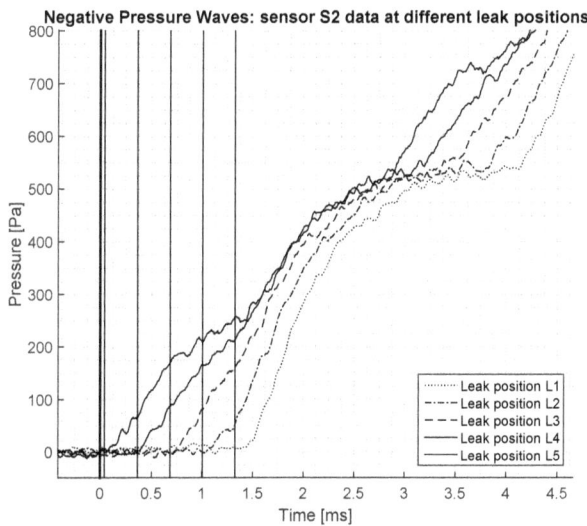

Figure 9. Comparison of the difference in time-of-arrival for different leak positions.

The average error on the location of the leak is found to be

$$\text{Error} = \frac{0.528}{100} \times 0.462 \text{ m} = 2.4 \text{ mm}. \tag{9}$$

This error is not much more than the theoretical limit due to the finite sampling rate, which allows us to conclude that the NPW analysis can be used to locate the crack with high accuracy.

185

Table 1. Localization feasibility on basis of time-of-arrival of the negative pressure waves using the least squares analysis. (L_{tube} = 462 mm).

N°	Location $x_{1,real}$ (mm)	Localization Estimate $x_{1,est}$ (mm)	Error $\frac{x_{1real} - x_{1,est}}{L_{tube}}$ %
L1	4	7.4	0.736
L2	59	57.6	0.303
L3	114	109.6	0.952
L4	169	167.4	0.346
L5	224	222.6	0.303
Average			0.528

5. Conclusions

Recently, a new effective structural health monitoring methodology was presented whereby capillaries are embedded in a 3D printed structure in order to monitor the presence of fatigue cracks. As multiple critical regions of the components can be inspected by means of just one capillary, it is of interest to investigate the crack localization possibilities on the basis of the pressure signals measured in the capillaries. Therefore, a feasibility study was performed on a test set-up with a controllable leak valve using closed tubes that imitate the capillaries of the presented system. The static pressure change in the closed tube is found to be a succession of pressure steps, corresponding to the arrivals of (the reflections of) NPW. As with pipelines, the difference in time-of-arrival of NPW is investigated in order to locate the leak in closed tubes. It was found that the difference in time-of-arrival allows to locate the leak, but that the selection of the exact moment of time-of-arrival is prone to noise fluctuations which significantly reduced the accuracy of the localization. Better processing techniques, such as a leasts squares analysis, allowed for the reduction of the localization uncertainty to approximately the theoretical limit related to the finite sampling rate (millimeters). We can therefore conclude that we validated the proof-of-concept of crack localization on the basis of NPW in closed tubes.

Acknowledgments: The research work was financed by the Agency for Innovation by Science and Technology in Flanders (IWT) and the Strategic Basic Research project SBO 110070 eSHM with AM.

Author Contributions: The authors contributed equally to the presented work through discussions, improvements and analysis of the obtained measurements.

Conflicts of Interest: The authors declare no conflict of interest.

References

1. Wycisk, E.; Solbach, A.; Siddique, S.; Herzog, D.; Walther, F.; Emmelmann, C. Effects of Defects in Laser Additive Manufactured Ti-6Al-4V on Fatigue Properties. *Phys. Proced.* **2014**, *45*, 371–378.

2. Slotwinski, J.A.; Garboczi, E.J.; Hebenstreit, K.M. Porosity Measurements and Analysis for Metal Additive Manufacturing Process Control. *J. Res. Natl. Inst. Stand. Technol.* **2014**, *119*, 494–528.

3. Chan, K.S.; Koike, M.; Mason, R.L.; Okabe, T. Fatigue Life of Titanium Alloys Fabricated by Additive Layer Manufacturing Techniques for Dental Implants. *Metall. Mater. Trans. A* **2013**, *44*, 1010–1022.

4. Gong, H.; Rafi, K.; Starr, T.; Stucker, B. Effect of defects on fatigue tests of as-built Ti-6Al-4V parts fabricated by selective laser melting. In Proceedings of the 23rd Annual International Solid Freeform Fabrication Symposium, Austin, Texas, USA, 6–8 August 2012, 499–506.

5. Strantza, M.; De Baere, D.; Rombouts, M.; Clijsters, S.; Vandendael, I.; Terryn, H.; Guillaume, P.; Van Hemelrijck, D. 3D Printing for Intelligent Metallic Structures. In Proceedings of the EWSHM 7th European Workshop on Structural Health Monitoring, Nantes, France, 8–11 July 2014.

6. Boller, C.; Meyendorf, N. State-of-the-art in Structural Health Monitoring of Aerospace Structures. In Proceedings of the International Symposium on NDT in Aerospace, Fürth, Bavaria, Germany, 3–5 December 2008.

7. De Baere, D.; Strantza, M.; Hinderdael, M.; Devesse, W.; Guillaume, P. Effective Structural Health Monitoring with Additive Manufacturing. In Proceedings of the EWSHM 7th European Workshop on Structural Health Monitoring, Nantes, France, 8–11 July 2014.

8. Farrar, C.R.; Worden, K. An introduction to Structural Health Monitoring. *Philos. Trans. R. Soc. A: Math. Phys. Eng. Sci.* **2007** *365*, 303–315.

9. Strantza, M.; Vafadari, R.; De Baere, D.; Rombouts, M.; Vandendael, I.; Terryn, H.; Hinderdael, M.; Rezaei, A.; van Paepegem, W.; Guillaume, P.; *et al.* Evaluation of Different Topologies of Integrated Capillaries in Effective Structural Health Monitoring System Produced by 3D Printing, In Proceedings of the 10th International Workshop on Structural Health Monitoring, Stanford University, Stanford, CA, USA, 1–3 September 2015.

10. Strantza, M.; De Baere, D.; Rombouts, M.; Maes, G.; Guillaume, P.; van Hemelrijck, D. Feasibility study on integrated structural health monitoring system produced by metal three-dimensional printing. *Struct. Health Monit.* **2015**, doi:10.1177/1475921715604389.

11. Wan, J.; Yu, Y.; Wu, Y.; Feng, R.; Yu, N. Hierarchical Leak Detection and Localization Method in Natural Gas Pipeline Monitoring Sensor Networks *Sensors* **2012**, *12*, 189–214.

12. Silva, R.A.; Buiatti, C.M.; Cruz, S.L.; Pereira, J.A.F.R. Pressure wave behaviour and leak detection in pipelines. *Comput. Chem. Eng.* **1996**, *20*, S491–S496.

13. Rocha, M.S. Acoustic Monitoring of Pipeline Leaks. In Proceedings of the Instrument Society of America (ISA) Symposium, Calgary, Alberta, Canada, 3–5 April 1989.

14. Loth, J.L.; Morris, G.J.; Palmer, G.M.; Guiler, R.; Mehra, D. Technology Assessment of On-Line Acoustic Monitoring for Leaks/Infringements in Underground Natural Gas Transmission Lines. In *Task 3 for DOE Contract DE-FC26-02NT413424*; Daniel Driscoll: NETL Morgantown, WV, USA, 2003.

15. Misiunas, D.; Vitkovsky, J.; Olsson, G.; Simpson, A.; Lambert, M. Pipeline Break Detection Using Pressure Transient Monitoring. *J. Water Resour. Plan. Manag.* **2005**, *131*, 316–325.

16. Ostapkowicz, P. Leakage detection from liquid transmission pipelines using improved pressure wave technique. *Eksploat. Niezawodn.* **2014**, *16*, 9–16.

17. Srirangarajan, S.; Allen, M.; Preis, A.; Iqbal, M.; Lim, H.B.; Whittle, A.J. Wavelet-Based Burst Event Detection and Localization in Water Distribution Systems. *J. Signal Process. Syst.* **2012**, *72*, 1–16.

Identification of a Critical Time with Acoustic Emission Monitoring during Static Fatigue Tests on Ceramic Matrix Composites: Towards Lifetime Prediction

Nathalie Godin, Pascal Reynaud, Mohamed R'Mili and Gilbert Fantozzi

Abstract: Non-oxide fiber-reinforced ceramic-matrix composites are promising candidates for some aeronautic applications that require good thermomechanical behavior over long periods of time. This study focuses on the behavior of a SiC_f/[Si-B-C] composite with a self-healing matrix at intermediate temperature under air. Static fatigue experiments were performed below 600 °C and a lifetime diagram is presented. Damage is monitored both by strain measurement and acoustic emission during the static fatigue experiments. Two methods of real-time analysis of associated energy release have been developed. They allow for the identification of a characteristic time that was found to be close to 55% of the measured rupture time. This critical time reflects a critical local energy release assessed by the applicability of the Benioff law. This critical aspect is linked to a damage phase where slow crack growth in fibers is prevailing leading to ultimate fracture of the composite.

Reprinted from *Appl. Sci.* Cite as: Godin, N.; Reynaud, P.; R'Mili, M.; Fantozzi, G. Identification of a Critical Time with Acoustic Emission Monitoring during Static Fatigue Tests on Ceramic Matrix Composites: Towards Lifetime Prediction. *Appl. Sci.* **2016**, *6*, 43.

1. Introduction

Ceramic Matrix Composites (CMCs), and more particularly SiC/SiC composites are very attractive candidates for many high-temperature structural applications, because of their excellent creep resistance, high-temperature strength and light weight. Damage tolerance is achieved through the use of low shear strength fiber coatings that deflect cracks along the interfaces [1–7]. The multi-layered matrix was introduced in the new generations of SiC_f/[Si-B-C] composites in order to improve the lifetime under medium and high temperatures thanks to the formation of sealant glasses [8,9].

Future engine applications in civil aircrafts are foreseen for such composites. These applications require that the material can resist both severe conditions (temperatures up to 1200 °C under air) and very long static loadings at intermediate temperatures (400–700 °C during tens of thousands of hours). Various authors have studied the mechanical behavior of SiC_f/SiC composites and the damage

189

mechanisms occurring at high temperatures [1–7]. Now more information is needed at intermediate temperatures: the oxidation kinetics of the different constituents are complex, and the effect of matrix sealing on the lifetime of the composite has to be examined. Expected lifetimes in use conditions are several thousands of hours, which can hardly be reached by laboratory tests for practical reasons. Therefore, a real-time prediction of the remaining lifetime during tests is necessary. It requires the monitoring of damage evolution for which AE measurement is a suitable technique. In fact, the AE technique consists in the recording and analysis of elastic waves created during material damage. It provides real-time data on initiation and evolution of damage in terms of location and mode.

Common analysis of AE is based on diagrams of cumulative hits or counts or histograms of amplitudes. Nevertheless, in the case of composite material, this approach is inadequate in order to identify the acoustic emission signature of several damage mechanisms. It was improved by grouping signals of similar shapes into clusters using classifier parameters [10]. Many works [11–16] have shown that AE techniques and multivariable classification techniques are the basis of pattern recognition tools. Kostopoulos [12], Godin [13,14], Moevus [15,16] have identified different classes of AE signals which were attributed to damage modes in oxide/oxide, glass/polyester and SiCf/[Si-B-C] composites. Moevus [16] analyzed the AE data using an unsupervised multi-variable clustering technique. This analysis showed that several types of signals may be distinguished, and that the obtained clusters were consistent with the expected damage modes in 3D SiC/SiC composites. In this way, a careful analysis of acoustic emission signals can lead to the discrimination of the different damage mechanisms occurring in a composite material. It is a possible solution for identification of damage during service with a view to component lifetime control.

A second analysis is based on a global AE analysis, on the investigation of liberated energy, with a view to identify a critical point. Many researchers investigated the elastic energy release during the failure process of materials [17–24]. They all observed that the energy release accelerated following a power law. These models are representative of an avalanche behavior very similar to that observed in seismicity.

This study focuses on static fatigue experiments realized at intermediate temperature under air on a SiC_f/[Si-B-C] composite with a self-healing matrix. A lifetime diagram is presented, and the evolution of several damage indicators is discussed. The objective of this approach is to propose a method based on acoustic energy in order to evaluate the remaining lifetime during long-term mechanical tests. This approach is based on the determination of energy released and identification of a critical point, in energy release during mechanical test. Therefore, two criteria have been defined which allow predicting the end of a static fatigue test knowing the AE

190

activity emitted during the first half of the test. Therefore, beyond this characteristic point the criticality can be described by a power-law in order to evaluate time to failure. Moreover, a supervised classification method was used to differentiate the signals generated during fatigue tests performed on composites at intermediate temperature and build a specific library [16]. This library was used to identify the damage modes generated during fatigue tests performed at various temperatures, in order to establish a link between this critical time and the damage mechanisms.

2. Experimental Procedure

2.1. Material

The material, manufactured by SAFRAN-HERAKLES Group (Bordeaux, France), had a multilayered [Si-B-C] reinforced matrix obtained by chemical vapor infiltration on SiC fibers and a PyC interface layer (Figure 1). This 2.5D woven composite is composed of Hi-Nicalon fibers, a pyro-carbon interphase layer and a self-healing matrix. This matrix has been processed by means of several chemical vapor infiltrations with various compositions, based on the ternary Si-B-C system. The external surface is protected by a seal-coat. The composite contains a volumic fraction of fibers equal to 35 vol.% and 20 vol.% of porosity. Dog-bone shaped specimens were used with the following dimensions: thickness 4 mm, width 16 mm, working length 40 mm with a constant cross section of 64 mm^2.

2.2. Static Fatigue Tests

Static fatigue tests were performed under uniaxial tensile loading with one direction of fibers parallel to the loading axis. Tests were carried out at 450 °C, 500 °C and 560 °C under air, on a pneumatic testing machine. Before loading, specimens were heated up to the testing temperature at a rate of 50 °C/min and held during 30 min to get a uniform temperature distribution in the gauge section. Specimens were loaded at a constant rate of 600 N/min up to a constant stress chosen in the range of 44% to 95% of the rupture stress. Specimen elongation was measured using a high temperature extensometer.

Seldom unload-reload cycles were applied every twelve hours in order to measure the secant elastic modulus changes during static fatigue and to evaluate the damage parameter D (Figure 2). E_0 is the elastic modulus in the undamaged state.

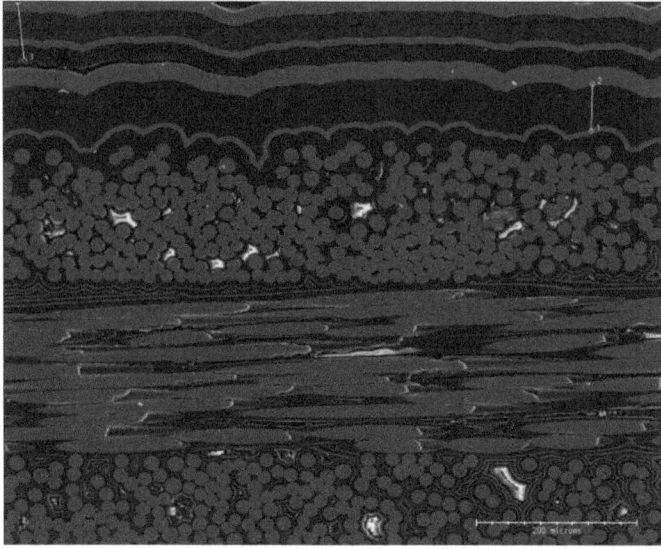

(a)

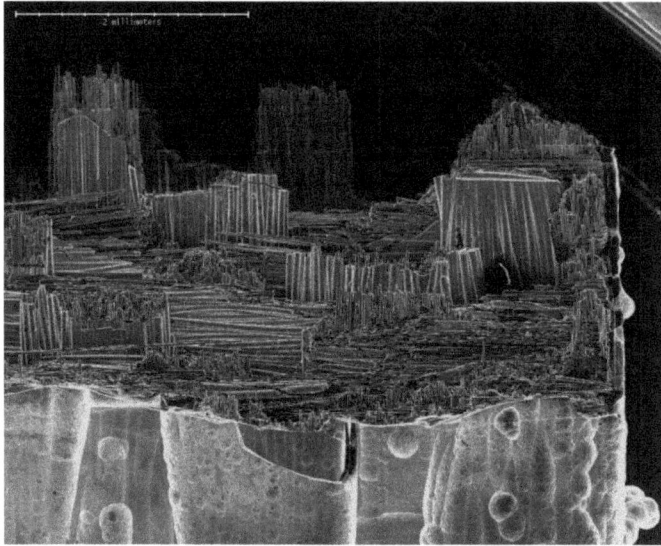

(b)

Figure 1. (a) SEM micrograph of the cross-section of $SiC_f/[Si-B-C]$ composites; (b) Typical fracture surface of $SiC_f/[Si-B-C]$ composite.

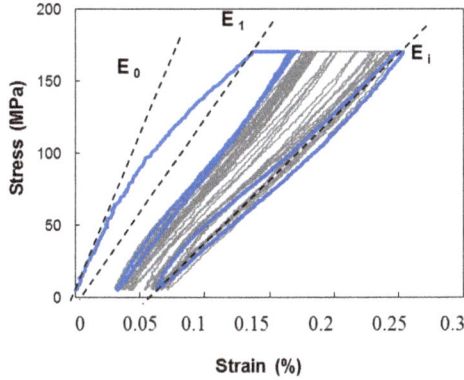

Figure 2. Schematic hysteresis loop and measure of the hysteresis loops modulus, E_0 is the initial Young's modulus, E_i is the secant elastic modulus for the unloading-reloading cycle i.

2.3. Acoustic Emission Monitoring

To detect acoustic emission, two MICRO-80 sensors (MISTRAS Group, Sucy en Brie, France) were put on the specimen inside the grips with vacuum grease as a coupling agent. Acquisition parameters were set as follows: preamplification 40 dB, threshold 32 dB, peak definition time 50 μs, hit definition time 100 μs, hit lockout time 1000 μs. AE signal descriptors such as amplitude, energy, duration, counts, average frequency, rise time and location as well as time, load and strain were recorded in real time by a MISTRAS 2001 data acquisition system.

For AE sources localization, AE wave velocity Ce_0 was determined before tests using a pencil lead break procedure on as received composites. This velocity was measured equal to 10,000 m/s. Since the elastic modulus decreases as damage occurs in the material, it is important to take into account the evolution of Ce during the mechanical test in order to better evaluate the location of the AE sources. As proposed by Morsher [25], the elastic secant modulus during unloading $E_i(\varepsilon)$ was measured during a cycled tensile test, where hysteresis loops were obtained at different strains. The velocity $Ce(\varepsilon)$ was then determined by using the Equation (1):

$$\frac{Ce(\varepsilon)}{Ce_0} = \sqrt{\frac{E_i(\varepsilon)}{E_0}} \tag{1}$$

where Ce_0 and E_0 are respectively the velocity and the elastic modulus in the undamaged state and $C_e(\varepsilon)$, $E_i(\varepsilon)$ respectively the velocity and the elastic modulus under a maximum strain ε. At the end of the tensile test, the velocity on SiC/SiC composite was found to be equal to 6480 m/s, instead of 10,000 m/s on the undamaged state. This decrease of wave velocity is thus not negligible. The location of sources has been calculated using the difference in times of arrival on each sensor. Only the signals coming from the working length of the specimens are then analyzed.

3. Acoustic Emission Analysis

3.1. Definition of the Acoustic Energy

Energy of an AE signal includes the energy released by the source at crack initiation and is affected by various parameters: wave propagation distance, damage attenuation, sensor/material surface coupling and sensor frequency response. Wave theory states that the energy of an acoustic wave decreases exponentially with the increase of propagation distance. Therefore, the following equation was proposed to describe the energy of recorded AE signals (for instance, at sensor 1) received from the source n [26,27]:

$$E_1(n) = E_s(n) \cdot A_1 \cdot e^{-B(L+x(n))} \tag{2}$$

$E_s(n)$ is the energy released at source n in the form of elastic waves. Due to differences in coupling between sensors and material surface or in sensors frequency responses, for a source at equal distance, the sensors may record significantly different amounts of energy. Thus, A_1 is the proportion of source energy that is recorded by sensor 1. It is a constant characteristic of sensor. $L + x(n)$ is the propagation distance from source n to the sensor 1 (2L is the distance between two sensors). The attenuation coefficient B is linked to the propagation medium, which may change with damage evolution. Similarly, AE signal energy received at sensor 2 is expressed as:

$$E_2(n) = E_s(n) \cdot A_2 \cdot e^{-B(L-x(n))} \tag{3}$$

The source energy is then defined as the square root of the product of the amounts of energy received at both sensors for each source:

$$E(n) = \sqrt{E_1(n) \cdot E_2(n)} \tag{4}$$

3.2. Attenuation Coefficient B

In order to evaluate energy attenuation [28], the ratio of AE signal energies recorded at both sensors is calculated for each source n. For an easier identification of attenuation coefficient B, $x(n)$ is defined as the natural logarithm of this ratio. From Equations (2) and (3), it comes:

$$X(n) = \log\frac{E_1(n)}{E_2(n)} = \log\frac{A_1}{A_2} - 2 \cdot B \cdot x(n) \tag{5}$$

The second term corresponds to the effect of propagation, related to source location $x(n)$ and to attenuation coefficient B. Both parameters can be estimated from $x(n)$ for various sources (1,2,..., n,...) since $x(n)$ is a linear function of x. The coefficient of attenuation B can be determined from the slope of the linear fit.

Because attenuation varies with damage, parameters estimation is done at successive time intervals. For a given time interval, values of $x(n)$ are analyzed in many space intervals in order to consider uncertainties in localization of AE sources. Estimation of attenuation constants was performed as follows. Attenuation is evaluated for each time interval with the median values of $x(n)$ in all space interval (width: 10 mm, overlapping: 5 mm). Each time interval is defined by 2000 AE successive sources. Each median value of $x(n)$ corresponds to a few hundred AE sources located in the same space-time interval. The linear approximation is repeated in consecutive time intervals. At every increment of 500 sources, a new time interval is considered. The overlapping is set in order to accurately monitor evolution of both attenuation parameters.

3.3. Coefficient of Emission R_{AE}

The coefficient of emission R_{AE} is defined [29], as the increment of energy ΔE recorded during a time increment Δt, divided by the total energy emitted during the initial loading of the sample:

$$R_{AE}(t) = \frac{1}{E_{loading}} \frac{\Delta E}{\Delta t} \tag{6}$$

where $E_{loading}$ is the cumulative AE energy for all the signals recorded during the initial loading up to the nominal load of the test, ΔE is the cumulative AE energy for all signals recorded during the interval $[t; t + \Delta t]$.

3.4. Power Law

Benioff law [30], suggested initially for precursory phenomena of earthquakes, has been applied originally on composites [29]. Based on Equation (7), the increase of AE during fatigue is analyzed.

$$\Omega(t) = \sum_{i=1}^{N(t)} \sqrt{E_i} = \Omega_R + B (t_R - t)^{1-\gamma} \tag{7}$$

where E_i is the energy of the ith AE signal detected and $N(t)$ is the number of AE signals recorded and located along the gauge length until time t.

Ω_R is the value of $\Omega(t) = \sum_{i=1}^{N(t)} \sqrt{E_i}$ when $t = t_R$, t_R is the failure time. $B = -\dfrac{\phi}{1-\gamma}$ is negative, $1 - \gamma$ is an exponent and φ is a constant.

The optimum circle method (OCM) [31], also developed in seismology, is used here to assess the relevance of the Benioff law. In addition, it determines which AE sources should be considered in order to achieve the best approximation. The OCM method was used to evaluate the time t_{start} when the AE energy is well simulated by the Benioff law. Two approximations are carried out on the energy release resulting from each time interval $[t_{start}; t_r]$: a power-law approximation using the Benioff law and a linear approximation $(\Omega(t) = \alpha t + \beta)$ used as reference. The c-value is defined by the ratio of the root mean square error of the approximation by the Benioff law over that of the linear fit. When the c-value is lower than 1, there is a positive contribution of the Benioff law since the approximation error is lower than that of a linear fit. It is a relative validation of the relevance of the approximation by the Benioff law. Therefore, to ensure quality of the approximation, only c-values lower than 0.5 are considered to be relevant.

3.5. Identification of Damage Mechanisms with Supervised Clustering

In such CMCs, matrix cracking occurs in several modes of cracking [16]. First, cracks initiate in the external seal-coat and in macropores inside the composite, then propagate through the inter-yarn matrix. Afterwards cracks propagate inside the transverse yarns through fiber/matrix interfaces. Multiple matrix cracking finally occurs inside the axial yarns. These cracks are deviated by the fiber/matrix interphase layer, yielding to fiber debonding and overloading. Some fiber breaks occur under high stresses. They rapidly lead to unstable fracture of entire yarns and of the composite. Hence, the sources of AE are the various matrix cracks, fiber/matrix interfacial debonding, individual fiber fractures and yarn ruptures.

The supervised classification technique requires a data base of signals that have been labelled: the training set. This training set was created by merging data collected during tensile test [16]. As described in the previous paper [16], the analysis of AE signals, observation of microstructures and analysis of the mechanical behavior of the composite led to the identification of 4 types of AE signals and to the following labelling of classes:

Class A: Signals of cluster A are assigned to two damage mechanisms which are chronologically separated: seal coat cracking and tow breaks.

Class B: cluster B contains also signals from two damage mechanisms which are chronologically well separated: longitudinal matrix cracking and individual fiber breaks in the fracture zone just before failure.

Class C: Cluster C contains signals with relatively short duration, short rise time and low amplitude when compared to the others: transversal matrix cracking.

Class D: this cluster is the last one to be activated and it becomes more active as strain increases and the D-type signals have a longer rise time when compared to other signals: sliding at fiber/matrix interfaces, fiber/matrix debonding.

In order to establish the training set of labelled signals for the supervised analysis, the same amount of signals (500 signals) in each class (A, B, C and D) was used. This training set included all the damage modes that may operate in this composite.

The supervised method consists in comparing each detected signal to those of the library and to assign it to the class of the K nearest neighbors [32] in the library. First, it is necessary to determine the number K of nearest neighbors required to obtain the best classification. For this purpose, a self-validating procedure was used which consists in comparing each signal of the training set to all the others. The classification error rate was estimated for all the values of K using the "leave-one-out" method [33]. The optimal value of K corresponded to the lowest value of error, and in the present study $K = 11$, and error = 2.5%. Then, for each AE signal of a given test, the Euclidean distance to all signals of the training set was calculated and the signal was labelled according to the class the most frequent class among its 11 nearest neighbors. The supervised clustering procedure was carried out on the different fatigue static tests.

4. Results and Discussion

4.1. Mechanical Analysis

The lifetime diagram, obtained at several temperatures and under various stresses, of the composite for static fatigue at intermediate temperature is plotted in Figure 3a. All the points are aligned in the logarithmic representation and follow the power-type law:

$$t \cdot \sigma^n = A \tag{8}$$

where t is the lifetime, σ is the applied stress, A and n are constants depending on the material, the temperature and the environment. The stress exponent n is estimated at 3.2 ± 0.3. A comparison between our results on composite and on the fibers in literature [34] under static fatigue at 500 °C (Figure 3b) shows that at intermediate and high stresses the lifetime of the composite is longer than the lifetime of the fibers. This is characterized by two different values of n, n is equal to 3.2 for the composite and to 8.45 for the fibers [34].

For this range of applied stress, slow crack growth of the fibers needs the diffusion of oxygen but is controlled by the chemical reaction at the crack tip. As shown by Gauthier and Lamon [34], the activation energy 129 kJ/mol corresponds to the reactivity of carbon (present in the fibers) to oxygen. In the composite the diffusion of oxygen toward the fiber surfaces is slackened by the oxygen consumption due to reaction with self-healing matrices and with interphases. Under lower stresses, the lifetimes of the composite and of the fiber bundles are similar. For this low level of stresses reaction kinetics with fiber surfaces are slower than diffusion kinetics of oxygen into the matrix crack. Hence the subcritical crack growth kinetics in dry bundles and in the fibers of the composites are similar.

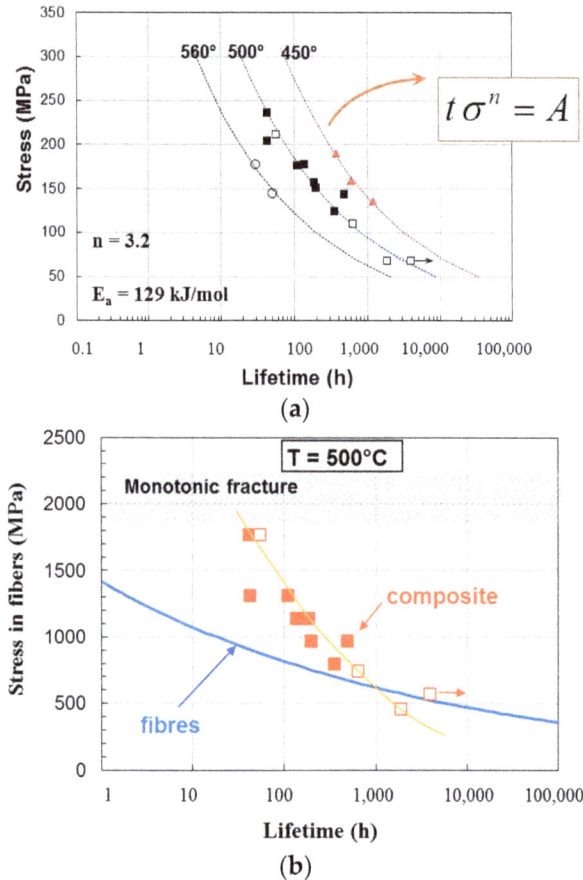

(a)

(b)

Figure 3. (a) Lifetimes obtained during static fatigue at 450 °C, 500 °C and 560 °C under air on SiC$_f$/[Si-B-C] composites for several applied stresses; (b) Lifetimes on the fibers bundles [34] and on SiC$_f$/[Si-B-C] composites in static fatigue at 500 °C under air.

Stress as a function of strain is shown in Figure 4. The grey curve represents the monotonic tensile curve whereas the other curves correspond to the static fatigue curves for various ratios of stress and with some loading and unloading sequences. One may notice that strain increases significantly during the test, and that hysteresis secant loops evolve as well: the mean elastic modulus decreases whereas the loops width increases. The increase in the loops area is linked to a change in the interfacial shear stress, suggesting that some debonding occur during static fatigue. The evolution of the Young's modulus was evaluated during the tests by measuring the hysteresis loops modulus E_i. Then the damage parameter $D = 1 - E_i/E_0$ was calculated, E_0 being the initial Young's modulus of the composite. The evolution of D is plotted *versus* time in Figure 5, for several σ/σ_R ratios. The biggest increase of D occurs during the loading step and the first 24 h of static fatigue. Then D is observed to rise monotonically up to the failure of the specimen. The general trend is that the highest the σ/σ_R ratio, the highest the D parameter for similar durations of test. The final D values obtained at failure are not exactly the same: they vary in the range 0.6–0.8, and it seems that this value decreases when the σ/σ_R ratio decreases. However it seems difficult to define a failure criterion based on the damage parameter.

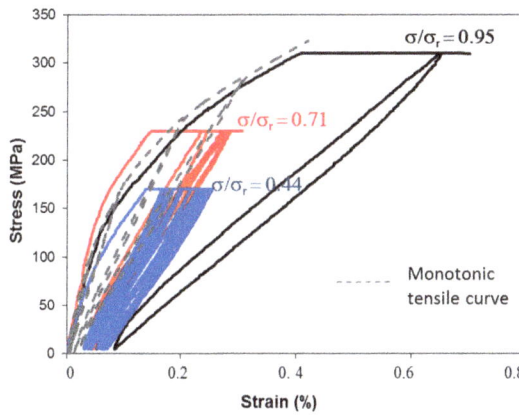

Figure 4. Stress *versus* strain during static fatigue experiments for different σ/σ_R ratios: 0.44, 0.71 and 0.95 at 500 °C on $SiC_f/[Si\text{-}B\text{-}C]$ composites. The grey curve represents the monotonic tensile curve.

The evolution of strain and acoustic emission events is shown in Figure 6 for a specimen loaded at $\sigma/\sigma_R = 0.44$. Under the steady loading the strain rate is important at the beginning then decreases before reaching a constant value. At this time, strain rises monotonically up to the final failure for the majority of the specimens (Figure 6b). The AE activity recorded during the constant load hold is also plotted in Figure 6b. The acoustic emission activity evolves with the same trend. Figure 6c shows the evolution of the cumulated acoustic energy. The main activity occurred at the beginning of the test and then reached a plateau. The renewal of AE activity (Figure 6c) just before failure seems to be an interesting way to anticipate the fracture but was not always noticeable on the AE energy *vs.* time plot.

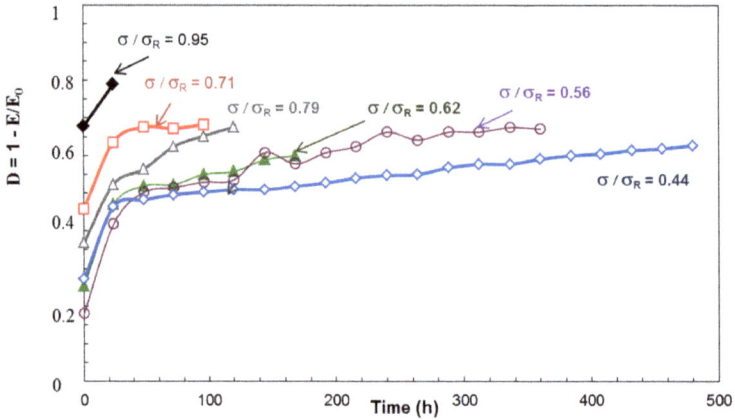

Figure 5. Evolution of the damage parameter D during static fatigue experiments for various σ/σ_R ratios 0.44, 0.56, 0.62, 0.71, 0.79 and 0.95 at 500 °C on $SiC_f/[Si\text{-}B\text{-}C]$ composites.

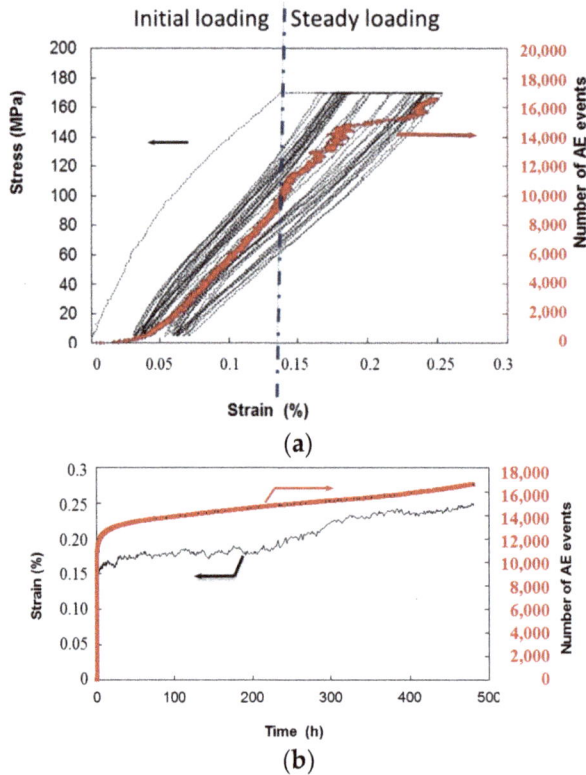

(a)

(b)

Figure 6. *Cont.*

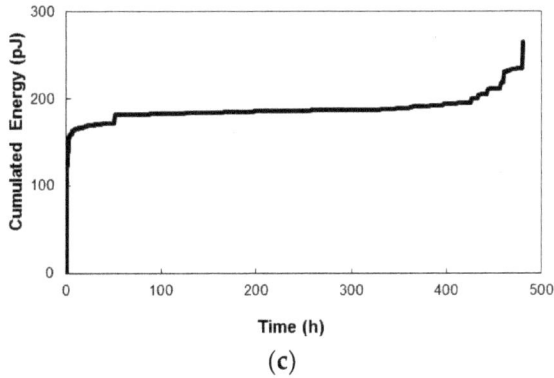

(c)

Figure 6. (a) Stress-strain curve and cumulated number of acoustic emission (AE) events recorded during initial loading up to the nominal load, and during the static loading at a constant load ($\sigma/\sigma_R = 0.44$ and $T = 500\,°C$); (b) Strain and cumulated number of AE events *versus* time; (c) Cumulated acoustic energy *versus* time.

The main activity occurs at the beginning of the static fatigue. Then the AE rate decreases without stopping. During the initial loading up to the nominal load, the number of events increases with the applied stress, which is consistent with the increasing number of cracks in the matrix (the main source of acoustic emission) as the applied stress increases. Then, during the static loading, the number of events before failure is lower under high stress than under low stress (Figure 7): damage needs to progress more under low stress to cause the final failure of the composite, than under high stress where the matrix already contains a lot of cracks.

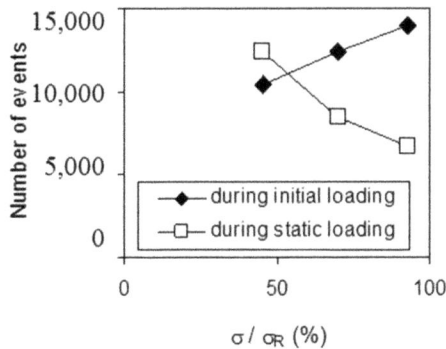

Figure 7. Number of AE events recorded during initial loading up to the nominal load and during the static loading at a constant load.

4.2. Identification of Critical Time

The evolution of R_{AE} is plotted *versus* time in Figure 8 for three experiments. This coefficient is observed to decrease first, then to reach a minimum value for $t = t_m$, and then to increase up to the failure of the composite at time t_R. This ultimate increase in AE activity was not always visible on the curves of AE energy *versus* time, but it was revealed by this representation. The ratio t_m/t_R appears to be quite reproducible: t_m/t_R =0.57 $\pm$ 0.06 (Figure 9). Therefore, the R_{AE} ratio may be used as a criterion for predicting the remaining lifetime of a specimen under static loading. During static fatigue test, attenuation coefficient B growths significantly during the first part of tests up to a plateau value near 50% of the rupture time (Figure 10). For each test, the slight increase of B observed during initial loading is very low compared to the one occurring during static fatigue. Therefore, the increase of attenuation coefficient B may be related to matrix crack opening and to the recession of interfaces.

The detection of the plateau during the typical evolution of attenuation coefficient B may be considered as an indicator for lifetime prediction. As the minimum of coefficient R_{AE}, the value plateau of B indicates the beginning of the critical damage phase and provides an estimation of the remaining lifetime.

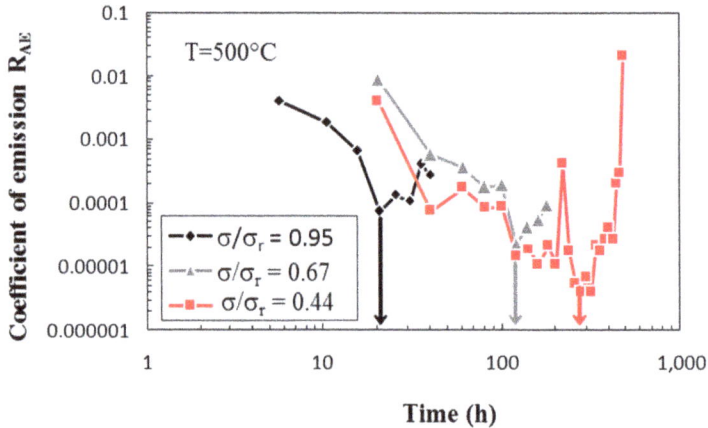

Figure 8. Evolution of the R_{AE} coefficient during the static load hold on SiC$_f$/[Si-B-C] composites.

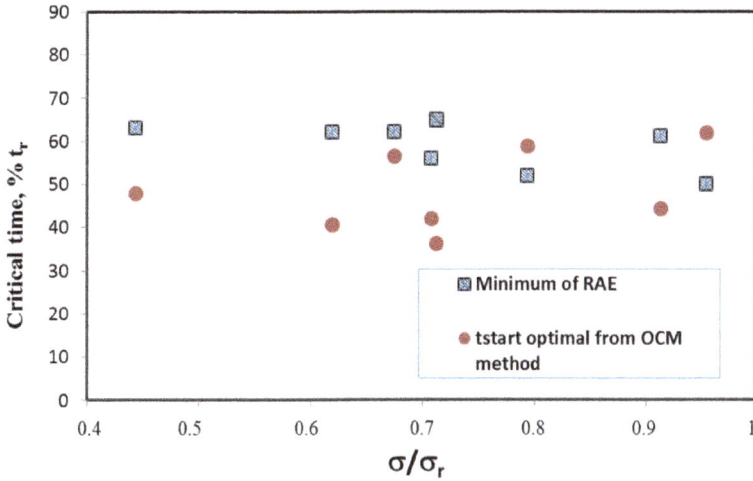

Figure 9. Critical times obtained with the minimum value of R_{AE} and the optimum t_{sart} for the Benioff law obtained with the OCM method.

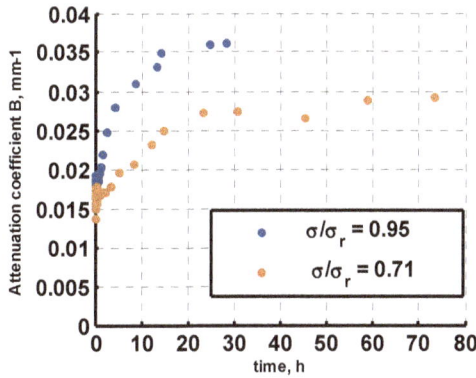

Figure 10. Evolution of the attenuation coefficient B during the static load hold ($\sigma/\sigma_R = 0.95$, $\sigma/\sigma_R = 0.71$ and $T = 500\,^{\circ}C$).

The critical damage phase prior to final rupture may be attributed to the avalanche fibers breaks, controlled by fibers oxidation and by recession of interfaces. In order to confirm this hypothesis, the coefficient R_{AE} is calculated for several damage mechanisms identified with clustering analysis of AE data. The Figure 11 show the activities of several damage mechanisms identified with the supervised analysis. During static fatigue, clusters B and D are more active. The evolution of the coefficients B obtained for the two classes A and B go through a minimum, contrary to those of classes C and D (Figure 12). For the B class (mainly fiber failure at the end of the tests), the minimum value of the coefficient B is observed around 65% of the lifetime. For the A class (mainly yarn fractures or collective fiber breaks at the end of the tests), the minimum is also observed at 65%. One may noticed that the minimum is observed only for clusters A and B corresponding to fibers breaks during the second part of the test. If the growth of the attenuation coefficient B is linked to matrix crack opening, coefficient B also allows considering the plateau observed on the evolution of attenuation coefficient B shows that matrix crack opening leads to an equilibrium state near 40%–50% of the rupture time. The significant increase of matrix crack opening pointed out before 50% of the rupture time is linked to carbon oxidation in the interphases provoking an increase in length of the debonded region of fibers in the vicinity of matrix cracks. Beyond 50% of the rupture time, the oxygen flux, determined by the degree of matrix crack opening, controls the rate of fibers break by slow crack growth. This critical time corresponds to the beginning of a second damage phase where slow crack growth in fibers is prevailing, leading to the ultimate fracture of the composite.

The minimum of coefficient RAE and the value plateau of B appear to be a promising tool to perform "short" static fatigue experiments which could be stopped once the minimum value of R_{AE} is reached, rather than waiting the final failure of the specimen. In this way the test duration could be divided by 1.5 to 2.

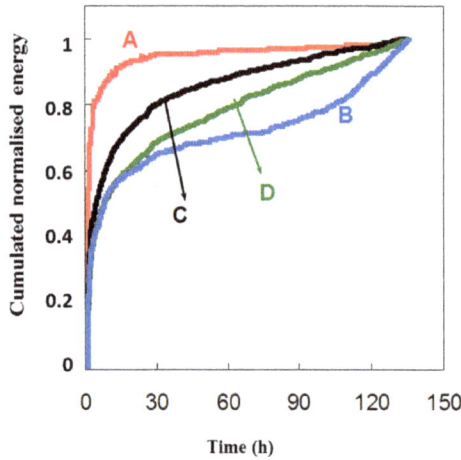

Figure 11. Activities of several clusters during static fatigue test at 500 °C and $\sigma/\sigma_R = 0.79$.

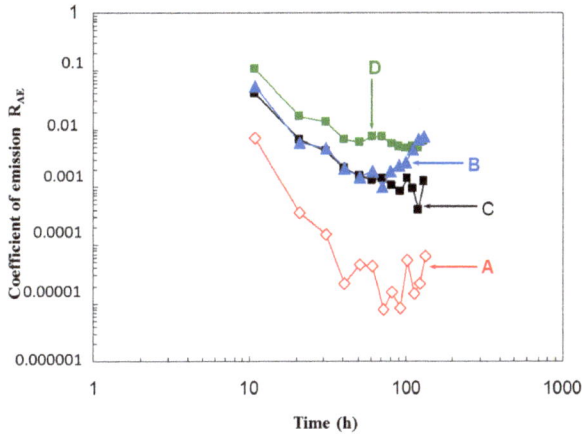

Figure 12. Evolution of the coefficient of emission for several clusters at 500 °C and $\sigma/\sigma_R = 0.79$.

4.3. Toward Lifetime Prediction

The Benioff law was applied to AE data recorded during the static fatigue tests in order to evaluate the lifetime of the studied material. This law is based on a power law description of the cumulative energy (Ω) released by the composite (Equation (2)) and characterized by 4 parameters (Ω_r, γ, φ and t_r).

Momon [11] studied the applicability on AE signals recorded for C/SiC and SiC/SiC composites. Data beyond the minimum of R_{AE} were in better agreement with the Benioff law than those before the minimum. This point was confirmed with the estimation of the c-value.

The Optimum Circle Method was applied to the fatigue tests. Optimum values (leading to the minimum c-value) of t_{start} are reported on the Figure 9, which means that the approximation by the Benioff law was relevant for the data collected behind this time. A value of t_{start} was taken every hour in time interval of 10% to 90% of rupture time t_r. The minimum c-value appears clearly around on average 40%–50% of rupture time. The energy release prior to rupture under static fatigue exhibits a critical evolution at 50% of rupture time regardless of the applied stress level. Thus, the Benioff law, initially used to study the activation of earthquakes, may also be applied to the damage of composites. However, this procedure requires preliminary tests until rupture to determine γ and φ for the studied material.

5. Conclusions

Static fatigue experiments were performed at intermediate temperatures under air to assess lifetimes of a $SiC_f/[Si-B-C]$ composite. The lifetime as a function of applied stress follows a power-type law, which can be used to predict lifetimes.

Additional information is obtained from strain measurement and AE monitoring during the tests. Two criterion based on the AE cumulative energy has been defined, which can be used to predict the final failure of a specimen if the AE activity during the first half of the test is known. This could be a way to shorten such static fatigue experiments, and to divide their duration by 1.5. The coefficient R_{AE} and the attenuation coefficient B confirm the existence of two distinct phases during damage of CMCs in static fatigue at intermediate temperatures. The first phase being mainly attributed to interfacial changes and the second one to the predominance of subcritical crack growth in fibers. Beyond this characteristic point, energy release may be modelled with the Benioff law in order to extrapolate AE activity and hence evaluate time to failure. The same analysis is in progress for the behavior during cyclic fatigue tests. Future works will focus on the use of the Benioff law as a predictive model.

Acknowledgments: The authors gratefully acknowledge Snecma Propulsion Solide, CNRS and DGA for supporting this work in the frame of the research program: "Modélisation, extrapolation, validation de la durée de vie des composites à matrice céramique".

Author Contributions: All the authors have contributed to perform mechanical tests, to analyze data and to write this paper.

Conflicts of Interest: The authors declare no conflict of interest.

References

1. Fantozzi, G.; Reynaud, P.; Rouby, D. Thermomechanical behaviour of long fibres ceramic-ceramic composites. *Silic. Ind.* **2002**, *66*, 109–119.
2. Reynaud, P.; Rouby, D.; Fantozzi, G. Effects of temperature and of oxidation on the interfacial shear stress between fibres and matrix in ceramic-matrix composites. *Acta Mater.* **1998**, *46*, 2461–2469.
3. Naslain, R.; Christin, F. Sic-matrix composite materials for advanced jet engines. *MRS Bull.* **2003**, *28*, 654–658.
4. Lamon, J. Micromechanics-based approach to the mechanical behavior of brittle matrix composites. *Compos. Sci. Technol.* **2002**, *58*, 2259–2272.
5. Ohnabe, H.; Masaki, S.; Onozuka, M.; Miyahara, K.; Sasa, T. Potential applications of ceramic matrix composites to aero-engine components. *Compos. A* **1999**, *30*, 489–496.
6. Lamon, J. CVI SiC/SiC composites. In *Handbook of Ceramics Composites*; Bansal, N.P., Ed.; Springer: Berlin, Germany, 2005; pp. 55–76.
7. Fantozzi, G.; Reynaud, P. Mechanical behaviour of SiC fiber reinforced ceramic matrix composites. In *Comprehensive Hard Materials*; Sarin, V.K., Ed.; Elsevier: Philadelphia, PA, USA, 2014; pp. 345–366.
8. Quemard, L.; Rebillat, F.; Guette, A.; Tawil, H.; Louchet-Pouillerie, C. Self-healing mechanisms of a SiC fiber reinforced multi-layered ceramic matrix composite in high pressure steam environments. *J. Eur. Ceram. Soc.* **2007**, *27*, 2085–2094.

9. Forio, P.; Lamon, J. Fatigue behavior at high temperatures in air of a 2D SiC/SiBC composite with a self-sealing multilayered matrix. In *Ceram Transactions*; Bansal, N.P., Singh, J.P., Lin, H.T., Eds.; American Ceramic Society: Westerville, OH, USA, 2001; Volume 128, pp. 127–141.

10. Ono, K.; Huang, Q. Pattern Recognition Analysis of Acoustic Emission Signals. *NDT and E Int.* **1997**, *30*, 109.

11. Yan, T.; Holford, K.; Carter, D.; Brandon, J. Classification of acoustic emission signatures using a self-organization neural network. *J. Acoust. Emiss.* **1999**, *17*, 49–59.

12. Kostopoulos, V.; Loutas, T.H.; Kontsos, A.; Sotiriadis, G.; Pappas, Y.Z. On the identification of the failure mechanisms in oxide/oxide composites using acoustic emission. *NDT E Int.* **2003**, *36*, 571–580.

13. Pappas, Y.Z.; Kontsos, A.; Loutas, T.H.; Kostopoulos, V. Failure mechanisms analysis of 2D carbon/carbon using acoustic emission monitoring. *NDT E Int.* **1998**, *31*, 571–580.

14. Godin, N.; Huguet, S.; Gaertner, R. Integration of the Kohonen's self-organising map and *k*-means algorithm for the segmentation of the AE data collected during tensile tests on cross-ply composites. *NDT E Int.* **2005**, *38*, 299–309.

15. Moevus, M.; Rouby, D.; Godin, N.; R'Mili, M.; Reynaud, P.; Fantozzi, G.; Fayolle, G. Analyse of damage mechanisms and associated acoustic emission in two SiC/[Si-B-C] composites exhibiting different tensile curves. Part I: Damage patterns and acoustic emission activity. *Compos. Sci. Technol.* **2008**, *68*, 1250–1257.

16. Moevus, M.; Godin, N.; Rouby, D.; R'Mili, M.; Reynaud, P.; Fantozzi, G.; Farizy, G. Analyse of damage mechanisms and associated acoustic emission in two SiC/[Si-B-C] composites exhibiting different tensile curves. Part II: Unsupervised acoustic emission data clustering. *Compos. Sci. Technol.* **2008**, *68*, 1258–1265.

17. Smith, R.L.; Phoenix, S.L. Asymptotic distributions for the failure of fibrous materials under series-parallel structure and equal load-sharing. *J. Appl. Mech.* **1981**, *48*, 75–82.

18. Curtin, W.A. Theory of mechanical properties of ceramic-matrix composites. *J. Am. Ceram. Soc.* **1991**, *74*, 2837–2845.

19. Newman, W.I.; Phoenix, S.L. Time dependent fiber-bundles with local load sharing. *Phys. Rev. E Stat. Nonlin. Soft Matter Phys.* **2001**, *63*, 021507.

20. Turcotte, D.; Shcherbakov, R. Can damage mechanics explain temporal scaling laws in brittle fracture and seismicity? *Pure Appl. Geophys.* **2006**, *163*, 1031–1045.

21. Guarino, A.; Garcimartin, A.; Ciliberto, S. An experimental test of the critical behaviour of fracture precursors. *Eur. Phys. J. B* **1998**, *6*, 13–24.

22. Nechad, H.; Helmstetter, A.; El Guerjouma, R.; Sornette, D. Andrade and critical time-to-failure laws in fiber-matrix composites: Experiments and model. *J. Mech. Phys. Solids* **2005**, *53*, 1099–1027.

23. Deschanel, S.; Vanel, L.; Vigier, G.; Godin, N.; Ciliberto, S. Experimental study of crackling noise: Conditions on power law scaling correlated to fracture precursors. *J. Stat. Mech. Theory Exp.* **2009**.

24. Shaira, M.; Godin, N.; Guy, P.; Vanel, L.; Courbon, J. Evaluation of the strain-induced martensitic transformation by acoustic emission monitoring in 304L austenitic stainless steel: Identification of the AE signature of the martensitic transformation and power-law statistics. *Mater. Sci. Eng. A* **2008**, *492*, 392–399.

25. Morscher, G.N. The velocity and attenuation of acoustic emission waves in SiC/SiC composites loaded in tension. *Compos. Sci. Technol.* **2002**, *62*, 1171–1180.

26. Maillet, E.; Godin, N.; R'Mili, M.; Reynaud, P.; Lamon, J.; Fantozzi, G. Analysis of Acoustic Emission energy release during static fatigue tests at intermediate temperatures on Ceramic Matrix Composites: Towards rupture time prediction. *Compos. Sci. Technol.* **2012**, *72*, 1001–1007.

27. Maillet, E.; Godin, N.; R'Mili, M.; Reynaud, P.; Lamon, J.; Fantozzi, G. Determination of acoustic emission sources energy and application towards lifetime prediction of ceramic matrix composites. In *Mechanical Properties and Performance of Engineering Ceramics and Composites VII*; Singh, D., Salem, J., Halbig, M., Mathur, S., Eds.; Wiley: Hoboken, NJ, USA, 2012; Volume 33, pp. 15–25.

28. Maillet, E.; Godin, N.; R'Mili, M.; Reynaud, P.; Fantozzi, G.; Lamon, J. Real-time evaluation of energy attenuation: A novel approach to acoustic emission analysis for damage monitoring of ceramic matrix composites. *J. Eur. Ceram. Soc.* **2014**, *34*, 1673–1679.

29. Momon, S.; Moevus, M.; Godin, N.; R'Mili, M.; Reynaud, P.; Fantozzi, G.; Fayolle, G. Acoustic emission and lifetime prediction during static fatigue tests on ceramic-matrix-composite at high temperature under air. *Compos. A Appl. Sci. Manuf.* **2010**, *41*, 913–918.

30. Bufe, C.G.; Varnes, D.G. Predictive modeling of the seismic cycle of the Greater San Francisco Bay Region. *J. Geophys. Res.* **1993**, *98*, 9871–9883.

31. Bowman, D.D.; Ouillon, G.; Sammis, C.G.; Sornette, A.; Sornette, D. An observational test of the critical earthquake concept. *J. Geophys. Res.* **1998**, *103*, 24359–24372.

32. Likas, A.; Vlassis, N.; Verbeek, J. The global *k*-means clustering algorithm. *Pattern Recognit.* **2003**, *36*, 451–461.

33. Lee, M.S.; Keerthi, S.; Jin Ong, C.; Decoste, D. An efficient method for computing leave one out error in support vector machines with gaussian kernels. *IEEE Trans. Neural Netw.* **2004**, *15*, 750–757.

34. Gauthier, W.; Lamon, J. Delayed failure of Hi-Nicalon and Hi-Nicalon S multifilament tows and single filaments at intermediate temperatures. *J. Am. Ceram. Soc.* **2009**, *92*, 702–709.

Wavelet Packet Decomposition to Characterize Injection Molding Tool Damage

Tomaž Kek, Dragan Kusić and Janez Grum

Abstract: This paper presents measurements of acoustic emission (AE) signals during the injection molding of polypropylene with new and damaged mold. The damaged injection mold has cracks induced by laser surface heat treatment. Standard test specimens were injection molded, commonly used for examining the shrinkage behavior of various thermoplastic materials. The measured AE burst signals during injection molding cycle are presented. For injection molding tool integrity prediction, different AE burst signals' descriptors are defined. To lower computational complexity and increase performance, the feature selection method was implemented to define a feature subset in an appropriate multidimensional space to characterize the integrity of the injection molding tool and the injection molding process steps. The feature subset was used for neural network pattern recognition of AE signals during the full time of the injection molding cycle. The results confirm that acoustic emission measurement during injection molding of polymer materials is a promising technique for characterizing the integrity of molds with respect to damage, even with resonant sensors.

Reprinted from *Appl. Sci.* Cite as: Kek, T.; Kusić, D.; Grum, J. Wavelet Packet Decomposition to Characterize Injection Molding Tool Damage. *Appl. Sci.* **2016**, *6*, 45.

1. Introduction

Injection molding is still regarded as one of the most important and very popular manufacturing process because of its simple operation steps. A typical production cycle begins with the mold closing, followed by the injection of melt into the mold cavity. Once the cavity is filled, holding (packing) pressure is applied to compensate for the material shrinkage. In the next step, the screw turns, feeding the material for the next shot to the screw tip. This causes the screw to retract as the next cycle is almost ready. Once the molded part is sufficiently cooled, the mold opens and the part is finally ejected.

Chen and Turng [1] defined three categories of variables that characterize the hierarchy and dependencies of an injection-molding control system. The first-level variables are defined as machine variables and include barrel temperatures in several zones, pressure, sequence and motion, *etc.* The second-level variables are process-dependent variables that include melt temperature in the nozzle, runner and mold cavity, melt pressure in the nozzle and cavity, melt-front advancement, maximum shear stress, and rate of heat dissipation and cooling. The third level incorporates quality measures such as part weight and thickness, shrinkage and warpage, *etc.* Current research mostly deals with the control tasks at the first two levels. For quality control purposes, this is often achieved indirectly through online dynamic process control or offline statistical process control (SPC). The majority of commercially available in-mold sensors for monitoring second level variables are pressure and temperature sensors.

Zhang *et al.* [2] applied a mechanical wireless data transmission technique using ultrasonic waves as the information carrier for online injection mold cavity pressure measurements. Ultrasonic transmitters with specific frequency characteristics were designed, modeled, simulated, and prototyped for pressure data retrieval from an enclosed machine environment. Lijuan Zhao *et al.* [3] developed an ultrasonic monitoring technique using a high-temperature ultrasonic transducer for real-time diagnostics in polymer processing and tracking of morphologic changes in injection molding processing.

The location and propagation of a crack in the metal can be detected using the acoustic emission technique, as already reported by many researchers. The major AE signal sources detected by NDT are crack growth and plastic deformation of the steel [4]. Acoustic emission (AE) as transient elastic waves has drawn a great deal of attention because of its applicability to online surveillance and capability to acquire data with high sensitivity [5–7]. Cao *et al.* [8] investigated acoustic emission (AE) signals during four-point bending fatigue crack propagation in the base metal and weld of -16 Mn steel. In the first stage of fatigue damage process micro-crack initiation was a dominant AE source. During stable crack extension (intermediate stage), stacking and slipping of dislocations ahead of the crack tip are sources of AE signals. During unstable crack extension (final stage) shearing of ligaments and connectivity between dimples are AE sources. Input parameters of their neural network model are based on burst energy, peak amplitude, duration, and counts. Mukhopadhyay, *et al.* [9] investigated AE signals during fracture toughness tests of steel CT specimens. The AE generated during fracture toughness tests has been attributed to the plastic zone formation at the crack tip and initiation and/or extension of crack. They measured burst peak amplitude and *RA* value and used the differentiation of the AE burst signals' RMS voltage, count, and energy with respect to time by taking the average of the slopes of adjacent data points to evaluate fracture toughness. Drummond *et al.* [10] investigated enhancement of proof and fatigue testing procedures for wire ropes by incorporating AE signals. They suggested that to differentiate between signals from wire breaks and other sources, the most effective AE signal discriminators are peak amplitude and (cumulative) burst energy. Also, the rope's condition can be ascertained without the necessity of continuous monitoring and the AE emission is indicative of the level of damage. Ki-Bok Kim *et al.* [11] analyzed fatigue crack growth in standard CT specimens with a hydraulic loading machine. The objective of their research was the development of a neural network-based model for the prediction of stress intensity factor as a function of five basic AE parameters (ring down count, rise time, energy, event duration, peak amplitude). Their research proved that AE energy increases gradually as the number of fatigue cycles increases. Change in the AE energy with the fatigue cycle number could be one of the effective parameters to estimate the activity of the crack propagation and the stress intensity factor.

The AE method provides a huge potential for extensive *in situ* failure analysis of materials, since it provides information on sub-macroscopic failure phenomena as well as on the overall damage accumulation [12]. Generated AE signal contains useful information on the damage mechanism. One of the main issues of AE is to discriminate the different damage mechanisms from the detected AE signals. Each signal can be associated to a pattern composed of multiple relevant descriptors [13–16]. Then the patterns can be divided into clusters representative of damage mechanisms according to their similarity by the use of multivariable data analyses based on pattern recognition algorithms [17].

Piezoelectric sensors can measure ultrasound with a sensitivity of about 1 V/nm as displacement sensors and a few V/(mm/s) as velocity sensors. The dynamic amplitude range of piezoelectric sensors is around 120 dB. Another useful property when measuring ultrasound is that they measure only dynamic events and so automatically compensate for low-frequency motion of measuring objects that are caused by environmental vibrations. This intrinsic low-frequency cutoff is a consequence of the leakage of the accumulated charge. This acts as a high-pass filter that determines the low-frequency cutoff (compensational bandwidth) through a time constant given by the capacitance and resistance of the device [18]. Additionally, piezoelectric sensors are also insensitive to electromagnetic fields and radiation, enabling measurements under harsh conditions.

Correlation of basic AE burst signal parameters (peak amplitude, energy, *RMS* value, *RA* value, burst count, *etc.*) and crack propagation in steel are well described in the literature. The detection of damage in molds during the injection molding process is a topic that has not gained sufficient coverage in the literature, although injection molding is one of the most widely used production process. In this paper we present the possibility of damage detection and process steps based on time domain, frequency domain variables, and wavelet packet transform descriptors of AE bursts.

The first steps towards characterizing injection molding tool damage and process phases are the determination and calculation of AE descriptors for acquired AE bursts (Section 3.1). These descriptors are stacked to form a feature (measurement) vector Z. The next step is unsupervised fuzzy C means (FCM) clustering to separate the bursts characteristic for damage (cluster 5) of injection molding tool from the AE bursts characteristic for injection of the melt and holding stage of the injection molding process (Section 3.2). AE bursts of cluster 5 can be acquired only during injection molding with a damaged tool. Next we introduced a feature selection method to select the 10-dimensional feature subset from the 73-dimensional feature vector to improve suitability for classification (Section 3.3). Next, the feature subset was used for neural network pattern recognition of AE signals during the full time of the injection molding cycle (Section 3.4). At the end of the paper we present a classification of all input vectors and an evaluation of the proposed method to characterize injection molding tool damage and process phases.

2. Experimental Section

Acoustic emission signals were captured during the production cycle of standard test specimens, designed for shrinkage evaluation of various plastic materials. We used an injection molding tool (TECOS, Celje, Slovenia) with one cavity for production of standard D2 ISO test specimens with square dimensions of $60 \times 60 \times 2$ mm^3, which can be used for a variety of tests according to ISO 294-3. The injection molding tool is made of OCR12VN steel. A polypropylene material isofil H40 C2 F NAT manufactured by Sirmax (Cittadella, Italy), mainly used in the automotive industry, was employed. A Vallen-Systeme GmbH acoustic emission measurement system AMSY-5 (Vallen-Systeme GmbH, Icking, Germany) was used to capture and analyze the AE signals at 5 MHz sampling frequency. Two piezoelectric AE sensors VS150-M with a measuring range between 100 and 450 kHz and resonance at 150 kHz were mounted on both sides of the tool via waveguides, as shown in Figure 1. These sensors are widely used in AE testing and monitoring and are low cost. The sensor also covers frequencies characteristic for AE signals generated during plastic deformation and fatigue [19]. We used an 18-kHz high pass filter (Vallen-Systeme GmbH, Icking, Germany). AE waveguides protect the sensors against the high surface temperature of injection molds during the production cycle, possible fumes exhaled out of the mold, and damage in case operator intervention is needed to help remove the specimen. The length of the waveguide is determined based on experience to protect sensors against the aforementioned hazards. PZT sensors were connected to the AMSY-5 measurement system via two AEP4 preamplifiers (Vallen-Systeme GmbH, Icking, Germany) with a fixed gain of 40 dB. The amplitude threshold was set at 40 dB. Experiments were conducted on a hydraulic injection molding machine KM 80 CX-SP 380 (KraussMaffei Group, Munich, Germany), for a broad spectrum of injection molding process parameters.

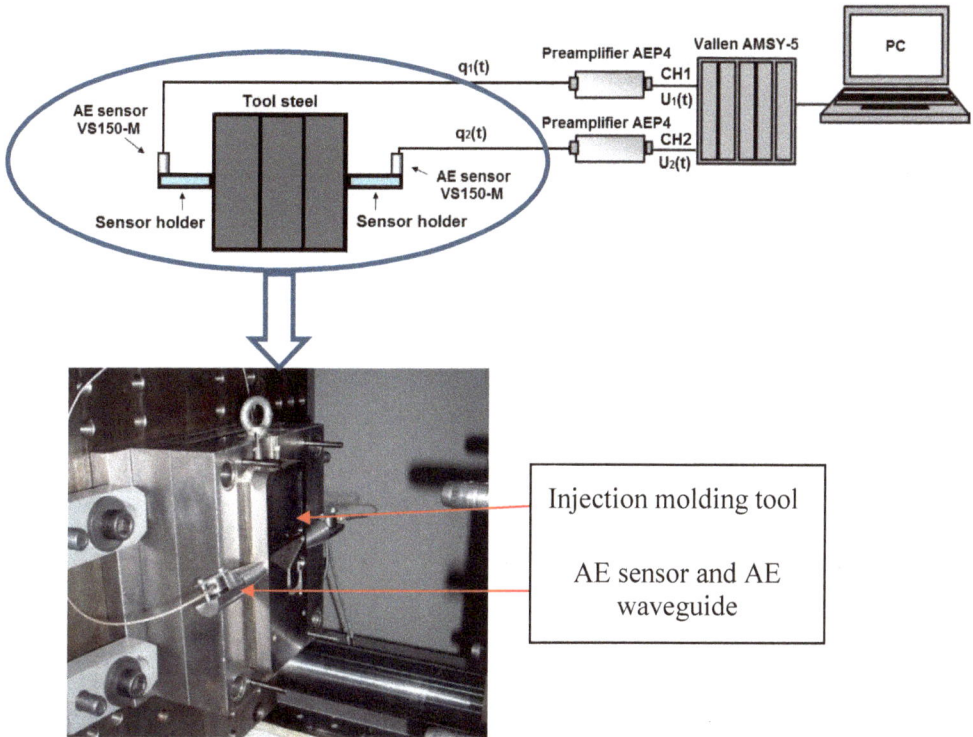

Figure 1. Experimental setup for AE monitoring during the injection molding process.

3. Results and Discussion

AE monitoring is a very promising method for the detection of injection mold defects. Namely, the rapid increase of mold pressure during the filling and holding stages of the injection molding process can trigger defects (acoustic emission sources) in the tool steel. Acoustic emission testing can detect dynamic processes associated with the degradation of structural integrity.

Experiments were conducted using a new injection mold tool and a damaged tool, respectively. Surface cracks were generated by successive local laser surface hardening of the tool steel. Both tools were tested using magnetic particle testing with fluorescent magnetic suspension. The cracks detected on the damaged tool are presented in Figure 2. After 24 h, the standard injection-molded polypropylene test specimens were scanned with an optical ATOS II SO 3D-digitizer. The ATOS II SO 3D digitizer is based on the triangulation principle whereby a sensor unit first projects different fringe patterns onto the scanned object's surface, which is then recorded by two cameras. Each measurement generates up to 4 million data points. In order to digitize a complete object, several individual measurements from different angles

213

are required. The cracks on the damaged tool are also reflected on the surface of injection molded specimens. Shape deviations (Figure 2) in two dimensions from the ideal shape are stated in mm.

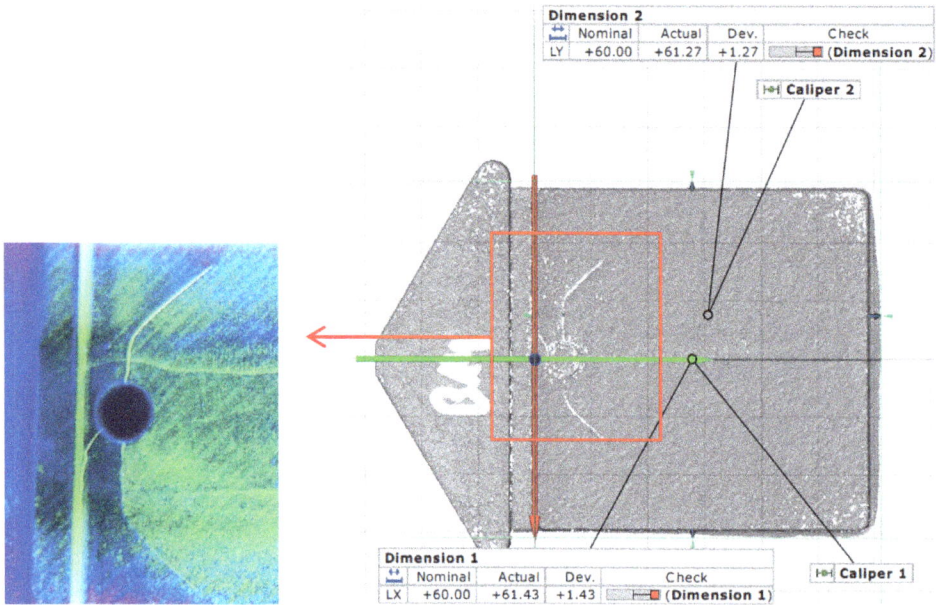

Dimension 2			
↔ Nominal	Actual	Dev.	Check
LY +60.00	+61.27	+1.27	(Dimension 2)

⊢⊣ Caliper 2

Dimension 1			
↔ Nominal	Actual	Dev.	Check
LX +60.00	+61.43	+1.43	(Dimension 1)

⊢⊣ Caliper 1

Figure 2. Cracks in the injection molding tool and accompanying PP specimen.

Figure 3 shows peak amplitudes and energy of AE bursts during the manufacturing of test specimens produced in the new and damaged mold, respectively. Absence of damage in the new tool offers us the opportunity of classifying measured AE bursts as process-orientated signals. On the other hand, the presence of a damaged area in the injection molding tool causes a considerable increase in the number of AE bursts detected during the filling and holding phases, above all in an amplitude range between 40 and 60 dB. Consequently, we can ascribe the additional AE bursts measured during the filling and holding phase to active AE sources from injection mold defects. A rapid increase in mold pressure during the filling phase could stimulate plastic zone formation, stacking, and slipping of dislocations ahead of the crack tip and crack extension [8,9] in the tool steel. Plastic deformations in the vicinity of the crack in the steel, mostly related to sliding dislocations, and crack growth cause distinctive AE bursts [20]. We connected AE signals with propagation of damage in the insert but this incorporates, based on the research of other authors, several physical phenomena that we described in the article as possible sources. The injection molding cycle ends after 16 s when the part is ejected out of the mold.

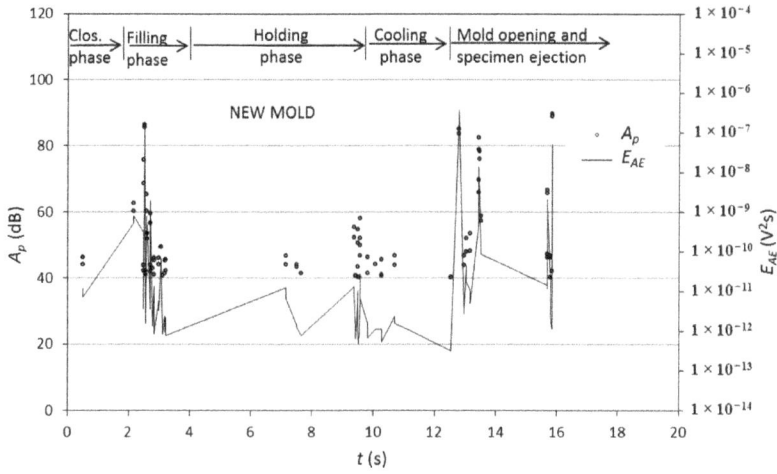

(a)

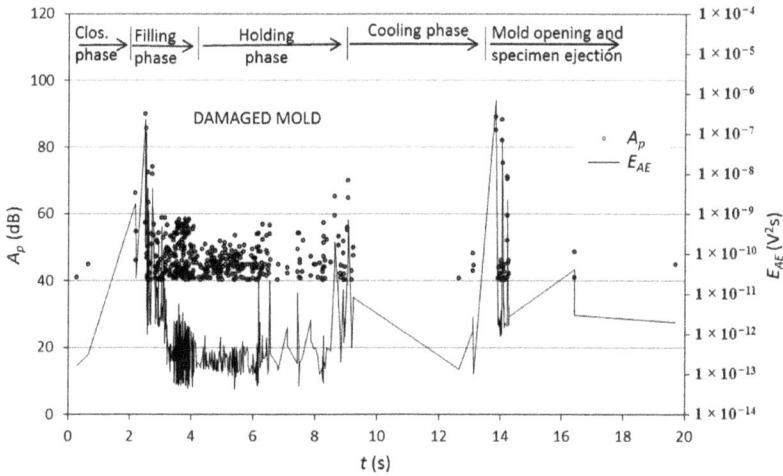

(b)

Figure 3. AE bursts' amplitude and energy during injection molding with (**a**) a new mold and a (**b**) damaged mold.

3.1. AE Burst Descriptors

Measured AE bursts are used for derivation of AE descriptors. These descriptors are stacked to form a feature (measurement) vector $Z = (z_1, z_2, \ldots, z_{73})$, representing acquired AE bursts during different phases of injection molding process. The union of 704 feature vectors, in this paper, composes feature space. AE burst is extracted from acquired AE waveform based on defined rearm time; in our research this was

215

0.46 ms. An AE burst is over when the rearm time elapses due to the absence of threshold crossings.

Calculated time domain descriptors based on acquired AE bursts are: peak amplitude value A_p, duration d, rise time Rt, and RA value. RA value is the ratio of rise time and amplitude of AE burst [15]. We calculated frequency domain descriptors. The P_L parameter characterizes the lower part of the power spectrum of each AE burst (a = 90 kHz, b = 190 kHz). Parameter P_H characterizes the higher part of the power spectrum (a = 250 kHz, b = 350 kHz), and parameter P_S covers all power spectrum frequencies (a = 50 kHz, b = 550 kHz). The parameters P_L, P_H, and P_S carry information about the energies at various frequencies in a sensor's frequency spectrum:

$$g_i = \frac{\overline{yyi}}{m} \tag{1}$$

$$m = \max\left(\mathbf{Y} \cdot \overline{\mathbf{Y}}\right) \tag{2}$$

$$P_L = \sum_{a=90}^{b=190} g_i \tag{3}$$

$\mathbf{Y} = (y_1, y_2, \dots, y_n)$ represents the vector of the discrete Fourier transform using FFT.

We introduced the kurtosis and skewness parameters describing the amplitude distribution of recorded AE burst. Kurtosis is a measure of the "peakedness" of the probability distribution for a real-valued random variable. In our case we calculated the kurtosis of measured AE burst amplitude values. Kurtosis was used in our research to characterize infrequent extreme deviations of amplitude values in AE burst as:

$$K = \frac{E(a_i - \overline{a})^4}{s^4} \tag{4}$$

where s is a sample standard deviation, a_i is an amplitude value.

Skewness S is defined as a third central moment of amplitude values in AE burst about its mean value. Skewness is in our research was used as a measure of asymmetry about the mean amplitude value of AE burst and is defined as:

$$S = \frac{E(a_i - \overline{a})^3}{s^3} \tag{5}$$

Wavelet Packet Analysis Descriptors

Wavelet analysis overcomes the pitfalls of Fourier methods and is suitable for processing AE signals during injection molding. The wavelet packet analysis is a subtler multiresolution analysis, which decomposes signal into multi-levels in the whole frequency band [21]. The detail of the signal is decomposed as well, so the frequency resolution is improved [22]. The wavelet packets include information on signals in different time windows at different resolutions. Each packet corresponds to some frequency band. Some packets contain important information while others are relatively unimportant.

Wavelet packet transform of energy limited signal $X(t)$ can be computed using the algorithm below:

$$P_0^1(t) = X(t) \tag{6}$$

$$P_j^{2i-1}(t) = HP_{j-1}^i(t) \tag{7}$$

$$P_j^{2i}(t) = GP_{j-1}^i(t) \tag{8}$$

where $P_j^i(t)$ is the ith packet on j^{th} resolution, with time parameter $t = 1, 2, \ldots, 2^{J-j}, i = 1, 2, \ldots, 2^j, j = 1, 2, \ldots, J, J = \log_2 N$. Operators H and G are convolution sum defined as:

$$H\{\cdot\} = \sum_k h(k - 2t) \tag{9}$$

$$G\{\cdot\} = \sum_k g(k - 2t) \tag{10}$$

where $h(t)$ and $g(t)$ are a pair of quadrature mirror filters and a time parameter t is taken as a series of integers k ($t \rightarrow k = 1, 2, \ldots$).

The energy of each wavelet packet is defined as:

$$E_j^i = \sum_{t=1}^{2^{J-j}} \left(P_j^i(t)\right)^2 \tag{11}$$

The energy of wavelet packets on selected resolution j is processed by normalization, and normalization value of each packet is

$$R_j^i = E_j^i / \sum_{i=1}^{2^j} E_j^i \tag{12}$$

Based on described AE signal descriptors, we have set the feature vector as $Z = (P_H, P_L/P_H, P_L/P_S, K, S, A_p, Rt, RA, d, R_5^1, R_5^2, \ldots, R_5^{32}, E_5^1, E_5^2, \ldots, E_5^{32})$.

3.2. Unsupervised Fuzzy C Means Clustering

We used unsupervised fuzzy C means clustering (FCM) for classification of feature vectors Z_j ($j = 1, n$) into $D = 2$ clusters. FCM clustering was found to be an effective algorithm for separating AE patterns composed of multiple features extracted from the AE waveforms [17,23]. Cluster 2 represents AE bursts during injection of the melt, cluster 3 represents process-orientated AE bursts during the holding stage, and cluster 5 bursts indicate damage during the injection and holding stages of the injection molding process. We used AE bursts measured during the injection molding process with a damaged tool, since we can also acquire AE signals representing cluster 5. We used fuzzy partitioning to connect feature vector Z_j ($j = 1, n$) to cluster ω_i ($i = 1, D$) with different membership grades $u_i(Z_j)$ between 0 and 1.

With FCM we find cluster centers $V = [K_1 \mid K_2]$ that minimize the function J:

$$J(U, V) = \sum_{j=1}^{n} \sum_{i=1}^{D} [u_i(Z_j)]^f d_e^2(Z_j, K_i) \tag{13}$$

U is the membership matrix with D lines and n columns with elements $u_i(Z_j)$ representing the membership value of the jth feature vector Z_j to the cluster ω_i with a condition:

$$\sum_{1=1}^{D} u_i(Z_j) = 1 \tag{14}$$

f is a scalar representing the fuzzy degree. The square of Euclidian distance d_e between the feature vector and the cluster is:

$$d_e^2(Z_j, K_i) = (Z_j - K_i)'(Z_j - K_i) \tag{15}$$

The algorithm of FCM has the following steps [17]:

(1) the membership matrix U is randomly initialized with values between 0 and 1 that represent the membership values $u_i(Z_j)$ of the n feature vectors Z_j ($j = 1, n$) to each cluster ω_i ($i = 1, D$).

(2) The new cluster centers K_i are calculated according to:

$$K_i = \frac{\sum_{j=1}^{n} [u_i(x_j)]^f x_j}{\sum_{j=1}^{n} [u_i(x_j)]^f} \tag{16}$$

(3) The membership matrix U is updated such that:

$$u_i(Z_j) = \left[\sum_{k=1}^{D} \left(\frac{d(Z_j, K_i)}{d(Z_j, K_k)} \right)^{2/(f-1)} \right]^{-1} \tag{17}$$

Steps 2 and 3 are iterated until the improvement over the previous iteration is below a threshold θ where β are the iteration steps:

$$\| U^\beta - U^{\beta-1} \| < \theta, \, 0 < \theta < 1 \tag{18}$$

Figure 4 shows examples of signals designated as process-orientated AE burst (cluster 3) and AE burst indicating damage (cluster 5). Figure 5 shows wavelet packet coefficients associated with all 32 nodes on 5th resolution of the wavelet packet tree for signals represented on Figure 4. With a red color, coefficients associated with nodes 8, 16, and 24 on 5th resolution are emphasized.

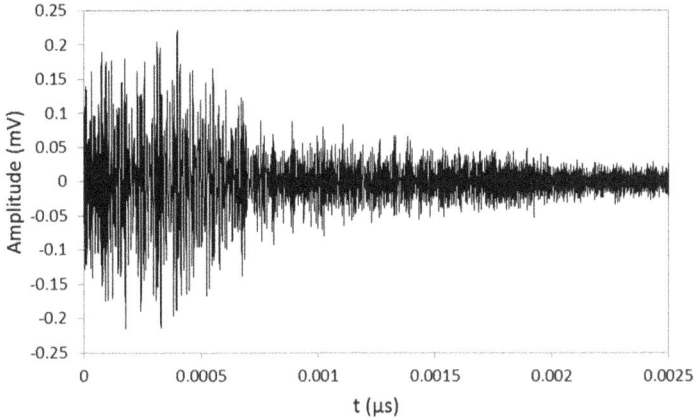

(a)

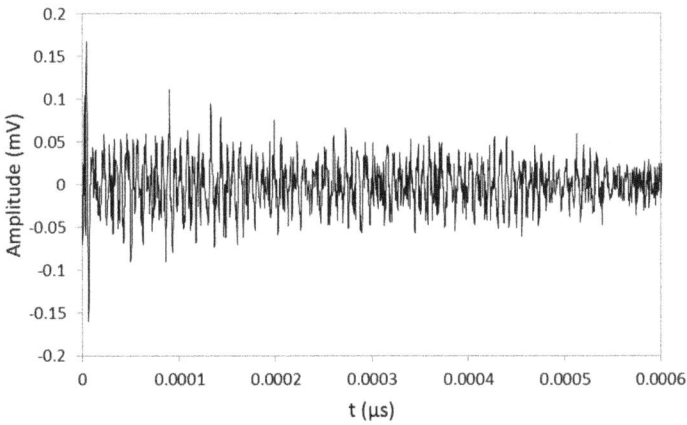

(b)

Figure 4. Signals designated as process-orientated AE burst (**a**) and AE burst indicating damage (**b**).

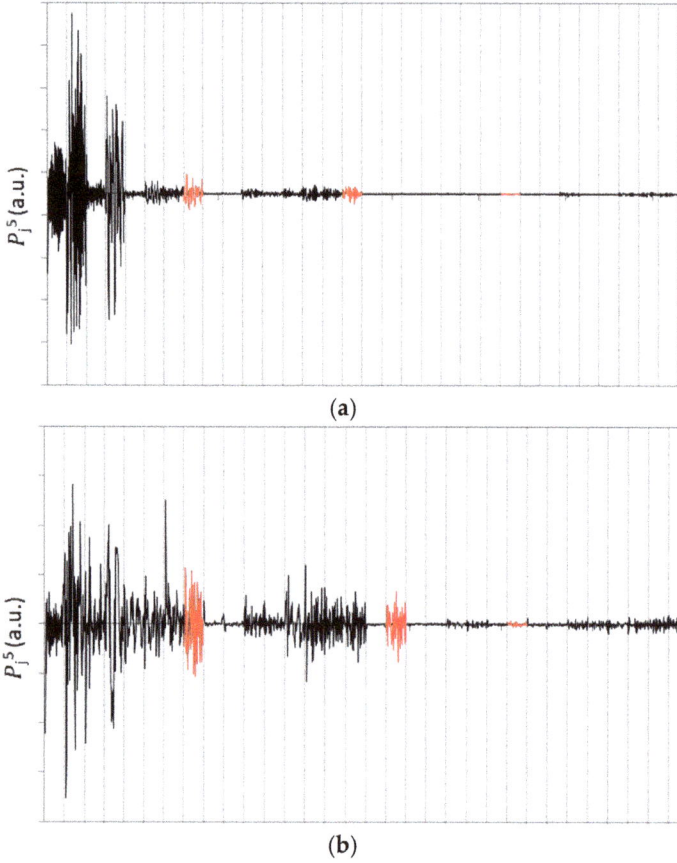

Figure 5. Wavelet packet coefficients associated with all the nodes on 5th resolution of the wavelet packet tree for above process-orientated AE burst (**a**) and AE burst indicating damage (**b**).

Figure 6 show feature vectors representing process-oriented AE bursts (cluster 3) and feature vectors indicating damage (cluster 5) during the holding stage of the injection molding process. Axes of 3D plot are normalization values of wavelet packets energy at 5th resolution. A black circle indicates the position of the feature vector of the AE burst from Figure 4a and a black square the position of the feature vector of the AE burst from Figure 4b. Cluster 3 feature vectors are concentrated around the origin of the coordinate system while Cluster 5 feature vectors are spread away from the origin.

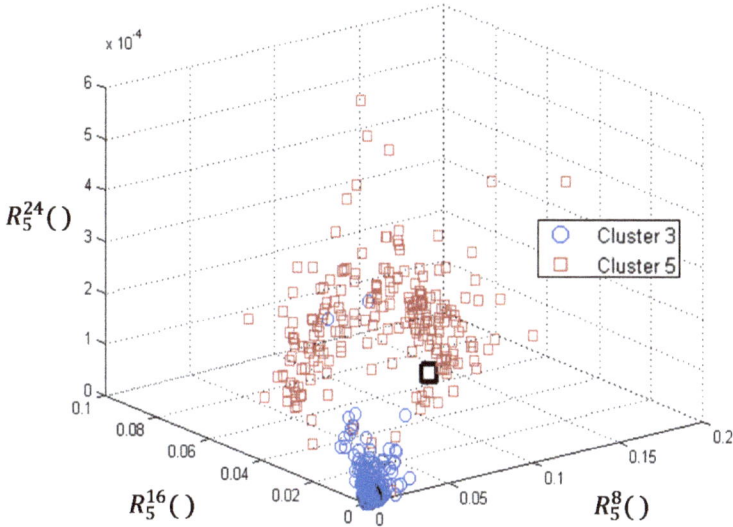

Figure 6. Visualization of feature vectors designated as cluster 3 and 5.

3.3. SFS Feature Selection

The dimensions of the feature vector cannot be arbitrary large. One reason is the computational complexity, and the second reason is that the dimension ultimately causes a decrease of performance [14,24]. Our goal is to reduce the dimensions of the data by finding a small set of important features that can give good classification performance. With feature selection we find a subset $F_j(W) = \{z_w \mid w=1, ..., W\}$ that outperforms all other subsets with dimension W as:

$$J(F_i(W)) \geqslant J(F_j(W)) \, for \; all \;\; j \epsilon \{1, \ldots, q(W)\} \tag{19}$$

where $q(W)$ is the number of evaluations of performance measure $J(F_j(W))$.

We used sequential forward selection (SFS). The method starts at the bottom (the root of the tree) with an empty set and proceeds its way to the top without backtracking. At each level of the tree, SFS adds one feature to the current feature set by selecting from the remaining available measurements the element that yields a maximal increase of the performance. A structure of feature selection is shown in Figure 7, where z_n is the designation of a feature.

We used a k-Nearest Neighbor (k-NN) classification algorithm as a criterion. With k-NN the input consists of the k closest training examples in the feature space. In our research $k = 3$. The output of the algorithm is a class membership. An object is classified by a majority vote of its neighbors, with the object being assigned to the class most common among its k nearest neighbors.

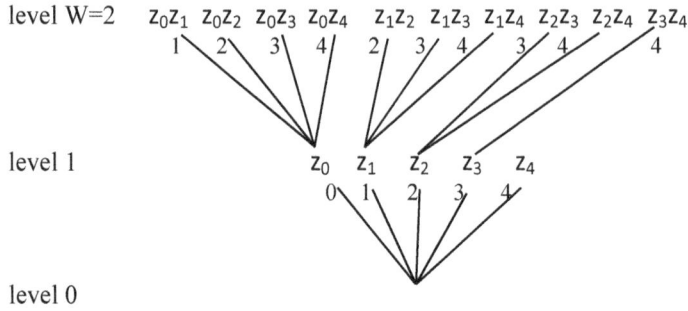

Figure 7. A bottom-up tree structure for feature selection [24].

The major focus of our research was detection of damage in an injection molding tool during the injection molding process. This was the reason for using SFS with k-NN on feature vectors of AE burst signals acquired during the holding phase of the injection molding cycle on intact and damaged molds. We selected feature subset Z_s with a size of 10. Feature subset was used for neural network pattern recognition of AE signals during the full time of the injection molding cycle. The selected features based on SFS with k-NN are: $Z_s = (R_5^8, R_5^{16}, R_5^{24}, R_5^{21}, R_5^{18}, R_5^{20}, d, R_5^6, R_5^{32}, A_p)$.

3.4. Classification of All Input Vectors Based on AE Signals

For the purposes of presenting all feature vectors Z, we introduced principal component analysis (PCA) as an unsupervised feature reduction method. PCA offers us transformation of high-dimensional feature vector Z to a much lower dimensional vector *i.e.*, P. PCA is mathematically defined [25] as an orthogonal linear transformation that transforms the data into a new coordinate system such that the greatest variance by some projection of the data comes to lie on the first coordinate (called the first principal component), the second greatest variance on the second coordinate, and so on. Figure 8 shows the distribution of feature vectors Z in a 3D coordinate system of the first three principal components. Feature vectors Z are divided into five classes; Class 1—the closing of the mold, Class 2—the injection of the melt with maximum pressure, Class 3—holding (packing) stage, Class 4—opening of the tool, Class 5—AE signals characterizing damage of the tool.

Based on the measured AE bursts, we set 10-dimensional feature subset vector Z_s with instances that are real-valued descriptors (variables). Vector Z_s offers us defining points in appropriate multidimensional space to characterize the process. We used a classification to identify the categories or sub-populations based on a training dataset, containing instances whose category membership is known. The training data consist of a set of 704 input vectors Z_s that have been designated based on the time moment of their acquisition during the injection molding cycle and unsupervised FCM for vectors of the damaged tool. For neural network pattern recognition we used a feed-forward network with the F Tan-Sigmoid transfer function in both the hidden and the output layer. We used 10 neurons in the hidden layer and five output neurons, because there are five target categories associated with each input vector. The network is trained with scaled conjugate gradient back-propagation.

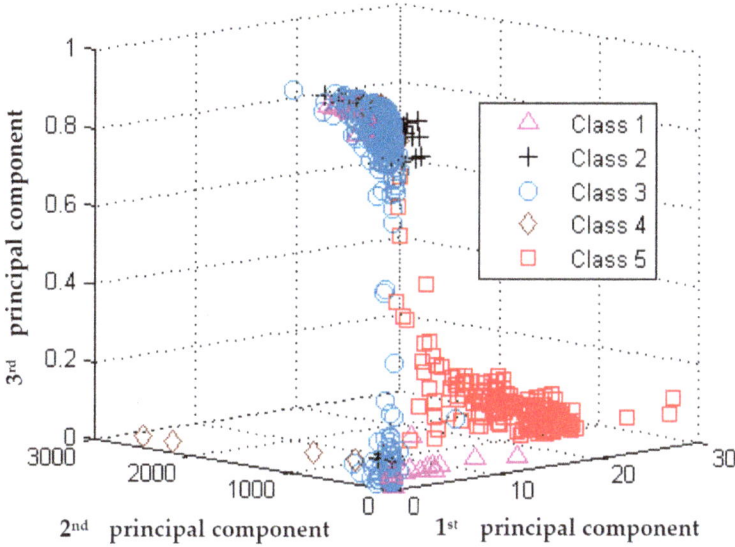

Figure 8. PCA visualization of all feature vectors Z during injection molding cycles.

Table 1. Classification of all input vectors in all confusion matrix.

		\multicolumn{6}{c}{**All Confusion Matrix**}					
	1	8 1.1%	0	1 0.1%	0	3 0.4%	33.3%
	2	0	27 3.8%	0	4 0.6%	0	12.9%
Output Class	3	32 4.5%	0	378 53.7%	8 1.1%	4 0.6%	10.4%
	4	0	3 0.4%	1 0.1%	26 3.7%	0	13.3%
	5	2 0.3%	0	1 0.1%	0	206 29.3%	1.4%
	E	81.0%	10.0%	0.8%	31.6%	3.3%	8.4%
	o.c./t.c.	1	2	3	4	5	E
		\multicolumn{6}{c}{**Target Class**}					

E—Means error.

For training of our network, data were randomly distributed into three subsets. A training subset (70% of the data) is used for computing the gradient and updating the network weights W and biases B. In the validation subset (15% of the data) the error is monitored during the training process. The network weights and biases are saved at the minimum of the validation subset error. The testing subset (15% of the data) has no effect on training but provides an independent measure of network performance during and after training. The results of classification of all input data are shown in Table 1. The table has five output classes (o.c.) and five target classes (t.c.) with a number of vectors classified into appropriate classes.

The presented data show that we have on average 91.6% correctly classified subset feature vectors Z_s. The lowest error $E = 0.8\%$ is for Class 3 vectors, representing process-orientated AE signals during the holding (packing) stage. Detection of damage (Class 5) in the injection molding tool is 96.7% correct. Misclassified vectors representing damage are almost evenly distributed between Class 1 and 3. The highest error, 81%, is for detection of tool closing, *i.e.*, Class 1 vectors Z_s. Also, the class 4 vectors representing opening of the tool show relatively high error $E = 31.6\%$.

4. Conclusions

In this work we presented an AE monitoring system with a resonant sensor that is able to detect the damaged area in an injection molding tool. We made a close comparison of AE signals captured from a new mold and a damaged mold under the same injection molding process conditions.

The most useful information about the injection molding tool integrity based on AE bursts is gained during the filling and holding stage of the injection molding process, when the mold pressure is high enough.

We introduced time domain, frequency domain, and wavelet packet transform descriptors to classify feature vectors of AE bursts during the injection molding cycle. A 10-dimensional subset feature vector with real-valued explanatory variables is proposed for classification of AE burst signals.

Based on SFS, the feature subset consists mainly of normalization values of the energy of wavelet packets, which turned out to be the most informative for prediction of damaged areas in the injection molding tool.

Neural network pattern recognition results show very good prediction probability of tool damage and very good prediction of process-orientated signals during the filling and holding phases of the injection molding cycle.

Acknowledgments: The research work was partly financed by the European Union, European Social Fund, and by the Slovenian Ministry for Higher Education, Science and Technology.

Author Contributions: All the authors conceived and designed the experiments; Tomaž Kek performed the experiments, analyzed the data and wrote the paper; Dragan Kusić performed the experiments and contributed materials and injection molding machine; Janez Grum led the research work and reviewed the work.

Conflicts of Interest: The authors declare no conflict of interest.

References

1. Chen, Z.; Turng, L.S. A Review of current developments in process and quality control for injection molding. *Adv. Polym. Technol.* **2005**, *24*, 165–182.

2. Zhang, L.; Theurer, C.B.; Gao, R.X.; Kazmer, D.O. Design of ultrasonic transmitters with defined frequency characteristics for wireless pressure sensing in injection molding. *IEEE Trans. Ultrason. Ferroelectr. Freq. Control* **2005**, *52*, 1360–1371.

3. Lijuan, Z.; Yu, L.; Chen, P.; Cheng-Kui, J.; Kuo-Ding, W. Real-time diagnosing polymer processing in injection molding using ultrasound. *J. Appl. Polym. Sci.* **2012**, *126*, 2059–2066.

4. Pollock, A.A. Fundamentals of Acoustic Emission Testing. In *Nondestructive Testing Handbook—Acoustic Emission Testing*, 3rd ed.; Moore, P.O., Miller, R.K., Hill, E.V.K., Eds.; American Society for Nondestructive Testing: Columbus, OH, USA, 2009; pp. 31–108.

5. Davoodi, S.; Mostafapour, A. Modeling acoustic emission signals caused by leakage in pressurized gas pipe. *J. Nondestruct. Eval.* **2013**, *32*, 67–80.

6. Harri, K.; Guillaume, P.; Vanlanduit, S. On-line damage detection on a wing panel using transmission of multisine ultrasonic waves. *NDT E Int.* **2008**, *41*, 312–317.

7. Mazal, P.; Pazdera, L.; Kolar, L. Basic acoustic emission signal treatment in the area of mechanical cyclic loading. *Int. J. Microstruct. Mater. Prop.* **2006**, *1*, 341–352.

8. Cao, J.; Luo, H.; Han, Z. Acoustic emission source mechanism analysis and crack length prediction during fatigue crack propagation in 16Mn steel and welds. *Procedia Eng.* **2012**, *27*, 1524–1537.

9. Mukhopadhyay, C.K.; Sasikala, G.; Jayakumar, T.; Raj, B. Acoustic emission during fracture toughness tests of SA333 Gr.6 steel. *Eng. Fract. Mech.* **2012**, *96*, 294–306.

10. Drummond, G.; Watson, J.F.; Acarnley, P.P. Acoustic emission from wire ropes during proof load and fatigue testing. *NDT E Int.* **2007**, *40*, 94–101.

11. Kim, K.-B.; Yoon, D.-J.; Jeong, J.-C.; Lee, S.-S. Determining the stress intensity factor of a material with an artificial neural network from acoustic emission measurements. *NDT E Int.* **2004**, *37*, 423–429.

12. Baensch, F. Damage Evolution in Wood and Layered Wood Composites Monitored *in Situ* by Acoustic Emission, Digital Image Correlation and Synchrotron Based Tomographic Microscopy. PhD Thesis, ETH Zürich, Zürich, Switzerland, 17 December 2014.

13. Zaki, A.; Chai, H.K.; Aggelis, D.G.; Alver, N. Non-Destructive evaluation for corrosion monitoring in concrete: A review and capability of acoustic emission technique. *Sensors* **2015**, *15*, 19069–19101.

14. Momon, S.; Godin, N.; Reynaud, P.; Mili, M.R.; Fantozzi, G. Unsupervised and supervised classification of AE data collected during fatigue test on CMC at high temperature. *Compos. Part A* **2012**, *43*, 254–260.

15. Aggelis, D.G. Classification of cracking mode in concrete by acoustic emission parameters. *Mech. Res. Commun.* **2011**, *38*, 153–157.

16. Svečko, R.; Kusić, D.; Kek, T.; Sarjaš, A.; Hančič, A.; Grum, J. Acoustic emission detection of macro-cracks on engraving tool steel inserts during the injection molding cycle using PZT sensors. *Sensors* **2013**, *13*, 6365–6379.

17. Marec, A.; Thomas, J.H.; El Guerjouma, R. Damage characterization of polymer-based composite materials: Multivariable analysis and wavelet transform for clustering acoustic emission data. *Mech. Syst. Signal Process.* **2008**, *22*, 1441–1464.
18. Požar, T.; Možina, J. Detection of Subnanometer Ultrasonic Displacements. In *Fundamentals of Picoscience*; Sattler, K.D., Ed.; CRC Press: Boca Raton, FL, USA, 2014; pp. 553–578.
19. Kumar, J.; Punnose, S.; Mukhopadhyay, C.K.; Jayakumar, T.; Kumar, V. Acoustic emission during tensile deformation of smooth and notched specimens of near alpha titanium alloy. *Res. Nondestruct. Evaluation* **2012**, *23*, 17–31.
20. Hase, A.; Mishina, M.; Wada, M. Correlation between features of acoustic emission signals and mechanical wear mechanisms. *Wear* **2012**, *292–293*, 144–150.
21. Kunpeng, Z.; San, W.Y.; Soon, H.G. Wavelet analysis of sensor signals for tool condition monitoring: A review and some new results. *Int. J. Mach. Tools Manuf.* **2009**, *49*, 537–553.
22. Wang, X.H.; Zhu, C.M.; Mao, H.L.; Huang, Z.F. Wavelet packet analysis for the propagation of acoustic emission signals across turbine runners. *NDT E Int.* **2009**, *42*, 42–46.
23. Oskouei, A.R.; Heidary, H.; Ahmadi, M.; Farajpur, M. Unsupervised acoustic emission data clustering for the analysis of damage mechanisms in glass/polyester composites. *Mater. Des.* **2012**, *37*, 416–422.
24. Van der Heijden, F.; Duin, R.P.W.; de Ridder, D.; Tax, D.M.J. *Classification, Parameter Estimation and State Estimation*; John Wiley & Sons Ltd: West Sussex, UK, 2004.
25. Jolliffe, I.T. *Principal Component Analysis*, 2nd ed.; Springer: New York, NY, USA, 2002; pp. 10–28.

Dynamic Characterization of Cohesive Material Based on Wave Velocity Measurements

Wojciech Sas, Katarzyna Gabryś, Emil Soból and Alojzy Szymański

Abstract: The paper presents a description of the dynamic properties of cohesive material, namely silty clays, obtained by using one of the applied seismology methods, the bender elements technique. The authors' aim was to present the dynamics of a porous medium, in particular an extremely important passage of seismic waves that travel through the bulk of a medium. Nowadays, the application of the bender element (BE) technique to measure, e.g., small strain shear stiffness of soils in the laboratory is well recognized, since it allows for reliable and relatively economical shear wave velocity measurements during various laboratory experiments. However, the accurate estimation of arrival time during BE tests is in many cases unclear. Two different interpretation procedures (from the time domain) of BE tests in order to measure travel times of waves were examined. Those values were then used to calculate shear and compression wave velocities and elastic moduli. Results showed that the dynamic parameters obtained by the start-to-start method were always slightly larger (up to about 20%) than those obtained using the peak-to-peak one. It was found that the peak-to-peak method led to more scattered results in comparison to the start-to-start method. Moreover, the influence of the excitation frequency, the mean effective stress and the unloading process on the dynamic properties of the tested material was studied. In addition, the obtained results highlighted the importance of initial signal frequency and the necessity to choose an appropriate range of frequencies to measure the shear wave velocity in clayey soils.

Reprinted from *Appl. Sci.* Cite as: Sas, W.; Gabryś, K.; Soból, E.; Szymański, A. Dynamic Characterization of Cohesive Material Based on Wave Velocity Measurements. *Appl. Sci.* **2016**, *6*, 49.

1. Introduction

The determination of seismic velocities, as well as the elasticity modulus and structural properties of porous materials plays an extremely important role in the development of various engineering projects [1]. The small strain and elastic moduli, such as the shear modulus (G_{max}) and Young's modulus (E_{max}), are key parameters during, for example, the site response analysis of an earthquake, the design of machine foundations and soil dynamics problems [2]. Burland *et al.* [3] even emphasized the significance of G_{max} in static deformation analysis of geotechnical problems. Usually, the above-mentioned elastic modulus of soil deposits is measured *in situ* by means of so-called seismic exploration [4]. In order to evaluate the shear modulus (G) or Young's modulus (E) values in the laboratory, triaxial or torsional shear testing machines, which employ what is generally known as the static loading methods [5], are applied. Other kinds of evaluation methods, commonly referred to as vibration test methods, are those relying on applying wave motions to the test specimens and then observing their behavior during the resonance, including free oscillation time. A great example is a resonant column apparatus. Apart from this, there are also pulse transmission techniques, which include the ultrasonic pulse test, bender element (BE), *etc.* They allow calculating G_{max} at small strains on the basis of the wave velocity [6].

Some of the laboratory tests and field studies, such as cross-hole seismic [7,8], down-hole seismic [9], suspension logging [10], seismic cone [11], seismic flat dilatometer [12] and spectral analysis of seismic waves [13], are indirect tests [14]. Compared to direct tests, the indirect tests enable measuring quantities different from the more desired ones and relate them to each other through mathematical relationships. Seismic techniques are classified as indirect testing methods for small strain stiffness [7]. They yield profiles of wave propagation velocity. When we assume that the behavior of the material is linear-elastic, this means that elastic stiffness relates to wave propagation velocity following the equations below:

$$\upsilon_P = \sqrt{\frac{\lambda + 2G}{\rho}} \tag{1}$$

$$\upsilon_S = \sqrt{\frac{G}{\rho}} \tag{2}$$

where υ_P is the propagation velocity of pressure, which might as well be called a primary (P-) wave velocity, υ_S is the propagation velocity of shear or the secondary (S-) wave and λ and G are Lame's constants (another term for G is the shear modulus). For engineers, it is more convenient to express Lame's constant as a function of Young's modulus and Poisson's ratio in accordance with these equations:

$$G = \frac{E}{2(1 + \nu)} \tag{3}$$

$$\lambda = \frac{\nu E}{(1 + \nu)(1 - 2\nu)} \tag{4}$$

Figure 1 illustrates the particle motion in P- and S-waves.

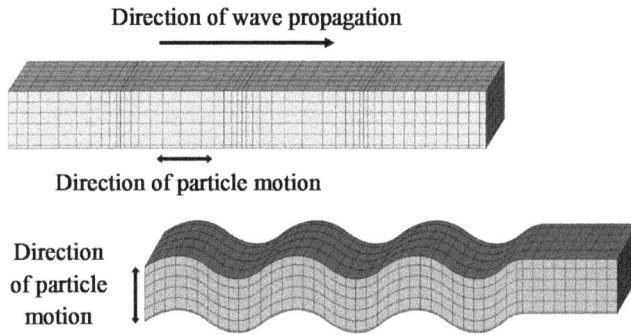

Figure 1. P-wave (top) and S-wave (bottom) particle motion. The particle motion in P-waves is longitudinal, while in S-waves, it is transverse. The particle motion vector in the plane perpendicular to the direction of propagation is referred to as polarization (only in the case of S-waves) [8].

Over the years, bender elements have become increasingly popular in order to determine the S-wave velocity [9,10] through experiments. Nowadays, they are commonly available [11,12]. Bender elements consist of two sheets produced from piezoceramic plates, which are rigidly bonded, to a center brass shim or stainless steel plate. Because of their piezoelectric properties, they are capable of converting mechanical excitation into electrical output and *vice versa*. When excited by an input voltage, the bender element changes its shape, which is accompanied by mechanical excitation. Because of this fact, it acts as a signal transmitter. Being subjected to mechanical excitation causes the emission of an electrical output by another bender element, which is, in this way, acting as a signal receiver [9]. The first person to use piezoceramic elements bonded to shear plates was probably Lawrence [13,14]. His idea was to measure the velocity of the shear wave in materials, such as clay and sand. In 1978, Shirley [15] proposed the introduction of piezoceramic bender elements to determine S-wave velocity in the laboratory. Other scientists, like Dyvik and Madshus [16], described the incorporation of BE for geotechnical laboratory testing in the 1980s.

The appeal of the BE technique lies in its apparent simplicity: one of the transducers is excited at one end of a specimen by a single pulse excitation, and the second, located at the other end of a specimen, is receiving it. The time required for this process can be simply read off from an oscilloscope. On the basis of this value, shear wave velocity can be obtained [17]. Therefore, BEs are an inexpensive and versatile solution, which can be used for laboratory seismic measurement. Furthermore, their capabilities of monitoring the process of stiffening with effective stresses, namely cementation, load stabilization, curing or consolidation, are particularly appealing [18]. However, there is a critical drawback connected with a BE test that contributes to many errors, namely the determination of the travel time. A number of researchers, including Viggiani and Atkinson (1995), attempted to use various methods in order to find the arrival time in order to reduce the degree of subjectivity [19]. In 2005, Lee and Santamarina [20], as well as Leong *et al.* [21] reported a method of travel time determination, by performing a bender element test. Nevertheless, there are still some differences between the first arrival of the S-wave and the preferred input wave. Therefore, the BE technique still needs to be verified with respect to how reliable it is, when it comes to calculation of the arrival time [2,22].

The intention of the authors of this paper was to recognize the dynamic characteristics of the cohesive medium through the performance of the BE tests on this material, *i.e.*, clayey soil, obtained from the area of Warsaw, the capital of Poland, at various effective confining pressures under saturated conditions. Two methods, namely peak-to-peak and start-to-start, were used to determine the travel time using the BE technique. The υ_S was calculated on this basis. Subsequently, shear modulus (G) was obtained using the value of shear wave velocity. Additionally, the authors used piezoceramic elements to determine compression wave velocity (υ_P). Furthermore, they tried to evaluate the values of dynamic Poisson's ratio (ν_d). The dynamic Poisson's ratio of soil deposits is a matter that has attracted rather little attention so far [23]. Nevertheless, it significantly affects stresses, strains, wave propagation characteristics in a mass of soils and other important types of soil deposit behavior during dynamic excitation [4]. In this paper, ν_d was calculated from the results of measurements of longitudinal and shear wave velocities in seismic exploration and then used to establish Young's modulus (E). The authors also studied the possible impact of unloading on the received values of elastic modulus.

2. Methods and Materials

2.1. Test Equipment

For the experiments, the authors employed a Stokoe fixed-free type of resonant column, manufactured in 2009 by the British company GDS Instruments Ltd, with its office in Hook, Hampshire, U.K. A detailed description of this device is contained in the authors' other publications [24,25] and will not be discussed in this paper. The BEs, produced as well by GDS Instruments Ltd in 2009, were installed in the above-mentioned resonant column apparatus. The GDS bender elements are made from piezoelectric ceramic bimorphs. Two sheets are bonded together with a metal shim in between them. Excitation voltage is used to produce a displacement in the source transducer, resulting in a wave being sent through the sample. This wave generates a displacement in the receiver, which induces a voltage that can be measured.

The GDS bender elements come in the form of an insert that can be mounted in a top cap or a pedestal (Figure 2). This makes them easy to replace and able to be used with a wide range of triaxial cells. The insert for the base pedestal is made of stainless steel, while the one for the top cap is made of titanium. This reduces the weight by half and minimizes the axial load caused by the top cap, which is being applied to the sample. The inserts are embedded in a modified Perspex top cap and stainless steel base pedestal. At the same time, the top cap and the base pedestal are mounted on the specimen in the same way, as during a conventional triaxial test [26].

Figure 2. Bender element inserts mounted in a top cap and base pedestal.

The S-wave source comprises two piezoceramic strips both polarized in the same direction. The application of an excitation voltage induces the extension of one strip and the contraction of the other, causing the strip to bend. Turning the excitation voltage in the opposite direction contributes to strips bending reversely. The two strips in the receiver are polarized in opposite directions. The P-wave source consists of two piezoceramic strips polarized conversely. When an excitation voltage is applied, strips extend simultaneously. Reversion of the excitation voltage causes the strips to contract [26].

In Figure 3, the BE arrangement adopted for the presented experiments is shown. Elements are manufactured to allow both S- and P-wave testing to be performed (in opposite propagation directions) on the same sample. The software switches input gain levels (of the received signal), sets the level of the output signal voltage and controls switching between P- and S-wave modes for our combined wave type elements [26]. In the current research, bender elements were used to obtain v_{VH} velocities, propagating in the vertical plane (index V), polarized in the horizontal (index H). When performing a bender element test, one of its most important aspects is the phase orientation of the elements. The authors always checked the relationship between the received signals with respect to the source signal. The desirable orientation was "in-phase". If the orientation was correct, the source and received traces were exactly "in-line". An example of non-desirable output is presented in Figure 4. The source signal and the received one are not simultaneously upward.

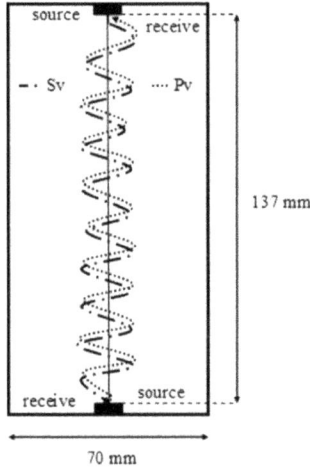

Figure 3. Frontal sample view with bender element (BE) arrangement and an example of the waves' path.

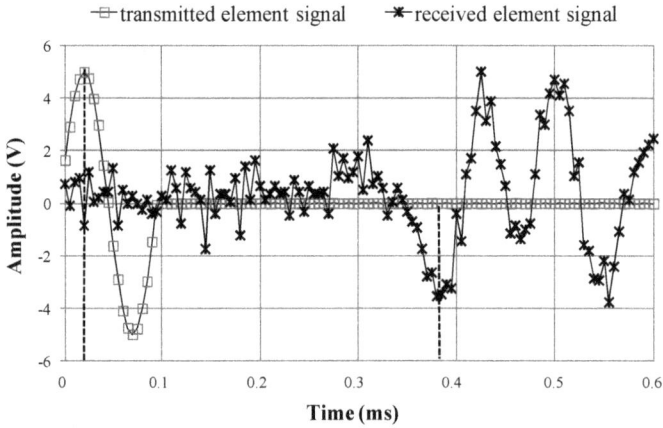

Figure 4. Non-desirable orientation of the top cap and base pedestal element.

2.2. Methods of Interpretation of BE Results

The BE technique has many advantages, which have been mentioned so far, but the interpretation of BE results can often be questionable [18,22,27]. A number of methods for the analysis of bender element results are commonly used in the time or frequency domain [17,28]. The time domain methods are classified as direct measurements, which use plots of electrical signals *versus* time [19,29]. These methods employ a time-based axis in order to identify the propagation time, hence [6]. In frequency domain approaches, on the other hand, the spectral breakdown of the signals is analyzed, and the shift in the phase angle between the trigger and response signal is calculated [19,30]. The time domain techniques are generally simpler and more straightforward, as the travel time can be directly defined by studying the time interval between characteristic points in the transmitted and received wave traces. However, the frequency domain methods appear to be more detailed, since they employ the support of signal processing and spectrum analysis tools. Additionally, they enable automated data acquisition and processing [18]. It is important to note that no method of interpreting the BE test results is yet proven superior to the others [6,19,31]. Reliable determination of travel time is very significant, due to BEs being installed in laboratory equipment and their application for studying relatively small specimens, which means that the travel distance is quite small. Below, the authors briefly present the methods of determining the arrival time of the shear wave that were employed in the paper. These are further illustrated in Figure 5.

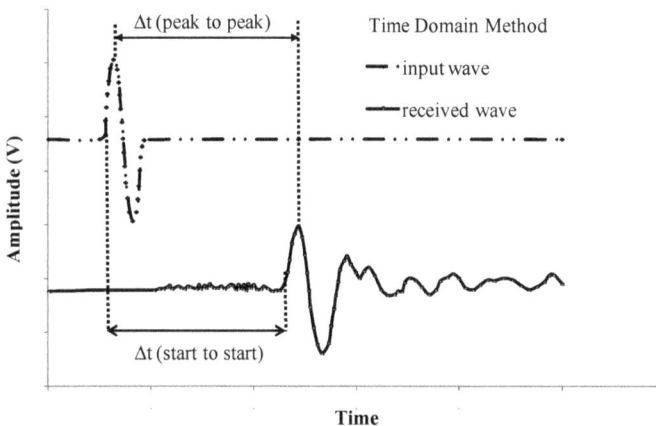

Figure 5. Typical identification techniques of travel time in the time domain method.

The peak-to-peak method [19,28] is based on the assumption that the received signal bears a high resemblance to the transmitted one. The travel time may be treated as the time elapsed between any two corresponding characteristic points in the signals. By doing this, the near field problems should be reduced. The characteristic points that are most commonly used to identify the wave travel time are the first positive peak in the input of S-wave signals and in the output [32]. In some circumstances, however, properly defining the first major peak becomes quite difficult, as the received signal has several consecutive peaks, which differ

233

slightly in amplitude. The reliability of choosing the right first peak as the first major one can be influenced by sample geometry and size or, for example, by the energy-absorbing nature of the soil. The arriving signal is then distorted to various extents due to the increase of damping with distance. The quality of received signals significantly affects this technique [28].

The principle of the start-to-start method, also known as the visual picking method, is the acceptance of the moment, when the first major deflection of the received signal occurs, as the shear wave arrival time [19,33]. The presented method is characterized by simplicity, which is the main reason behind its popularity. Depending on the installation and polarity of the bender elements, the first important deviation from zero amplitude may be positive or negative [28]. In many cases, it is simply a single sine pulse. This represents the start of energy transfer from the source to the soil. However, in case of the output wave signal, it is the moment when the investigated receiver begins exhibiting motion. Said moment represents the instant of the energy transfer from the soil to the receiving BE [32].

During the bender element test, shear wave velocity is calculated from the simple measurement of propagation distance (Δs) and propagation time (Δt). Based on a number of previous works, it is generally accepted that the travel distance is the distance between the tips of two BEs [11]. There are various waveforms, for example sine and square waves, with various frequencies, that have been recommended as an excitation signal [2].

2.3. Characterization of Test Material

In order to investigate the elastic modulus of cohesive material, associated with the passage of seismic waves (an S-wave and a P-wave), silty clay (siCl) samples, from the test site located in the center of Warsaw (the capital of Poland), recognized by particle size analysis using the sieves and hydrometer methods (European Standard Eurocode 7, [34]), were used. The description of the research area can be found in [25]. The index properties of the examined soils are summarized in Table 1. In order to compute the fundamental indices (w_P, w_L), standard test methods were employed, namely plastic limit test of soil, Casagrande liquid limit test and fall cone liquid limit test. The grain size distribution of the soil specimens is shown in Figure 6. The proportional content of each fraction is as follows: Gr = 0%, Sa = 13%, Si = 66%, Cl = 21%. The soil used during the tests, which is of Quaternary origin, was sampled in an undisturbed state using a standard Shelby tube. All of the samples were acquired at a depth of around 6.0 m and selected cautiously with the consideration of the soil structure uniformity, their physical properties and their double phase. Tubes were pressed carefully and gently into the pre-drilled holes. Subsequently, the samples were sealed and stored in a humidity room until needed [35].

Table 1. Basic properties of the tested soils.

Parameter	Value
w (%)	17.52
w_P (%)	17.14
w_L (%)	33.00
I_P (%)	15.86
I_L (%)	0.02
I_C (%)	0.98
ρ (kg/m^3)	2140

w is the water content; w_P is the plastic limit; w_L is the liquid limit; I_P is the plastic index; I_L is the liquidity index; I_C is the consistency index; and ρ is the mass density.

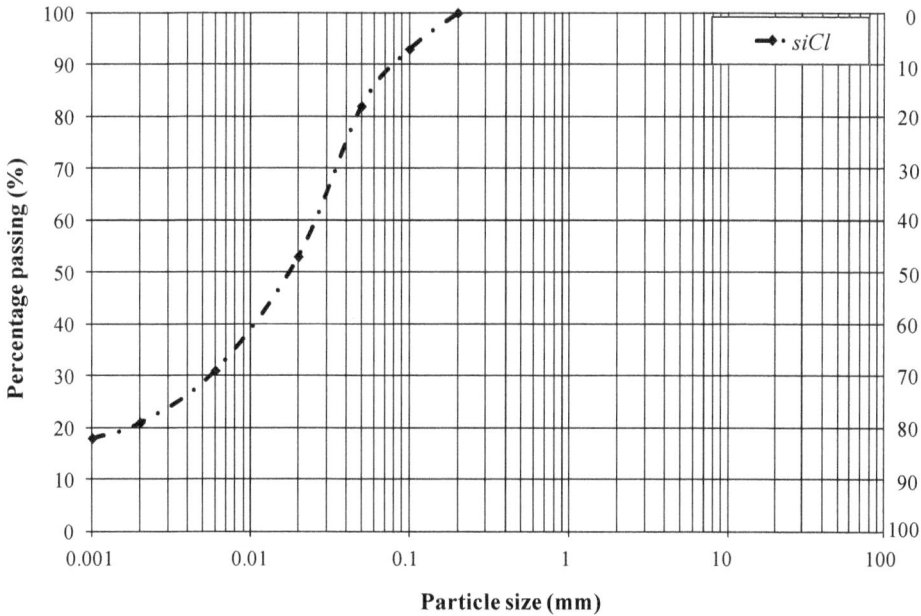

Figure 6. Grain size distribution of the tested soils.

2.4. Test Setup and Procedure

The test procedure followed that of a drained isotropic resonant column test. An upright cylindrical specimen of soil, with an aspect ratio of 2:1 (length:diameter), was employed. The proportions between length and diameter were as follows: 140 mm × 70 mm. Undisturbed material was set up in the resonant column cell, then saturated by the back pressure method in order to achieve the level of full saturation and subsequently consolidated to predetermined isotropic stress levels of 30, 120, 180, 240, 360 and 410 kPa. Back pressure was increased slowly, in order to ensure the proper saturation of the sample, until Skempton's B value reached 0.84. This represents, with respect to Head [36], the saturation level being equal to approximately 97%. During the consolidation stage, the axial deformation and volume change of the sample were measured. The back pressure was kept at a constant level of 290 kPa. At the end of each consolidation stage, seismic wave velocities were checked, in the undrained conditions, as well as using the bender elements located at the top and bottom of the soil specimen. The BE transmitter was excited with a sine pulse, whose magnitude equaled 15 V. A change of voltage applied to the transmitter caused bending and transmission of a shear wave through the sample. The receiver, located at the other end of the specimen, registered the arrival of the shear wave as a change in voltage [37–39]. For each isotropic confining pressure level, a range of input frequencies between 100 and 1 kHz was tested, in order to identify the value of greater amplification in the received signals. This value should represent the clearest output signals. In the case of some input frequencies, particularly the higher ones, compression wave velocity was measured additionally. Due to the limitations of the software and hardware (S- and P-waves were transmitted by the same piezoelectric element) provided by the GDS company, the authors were unable to conduct the measurements in the same period (T) of S- and P-waves. Hence, there is an apparent limit in the predetermined period or frequency of propagating various waves through the soil sample.

In Figure 7, the oscilloscope data obtained during the discussed studies are shown. The input signals at different frequencies were sent under the same stress conditions. During the analysis of this figure, it appeared to the authors, that at higher frequencies, the determination of the arrival time was more precise. Changes in the input signal frequency did not produce any change of polarization in the shear waves. For all applied frequencies, the input and output signals indicated the same amount of polarization.

To interpret the BE results, the authors applied two commonly-known techniques used in time domain methods, which they described in the earlier subsection. For the start-to-start method, the noise level is very significant. Leong *et al.* [40] verified the criteria for the improvement of the interpretation of the BE test, among which there is the signal-to-noise ratio (SNR) criterion. According to their research, the SNR should be of at least 4 dB for the receiver signal. In the authors' research, the average SNR for the start-to-start method was equal to 10.44 dB. This level of the SNR allowed picking a very reliable and comfortable wave onset. Values of the wave velocity were calculated on the basis of the tip-to-tip distance between the transmitter and the receiver bender element [41]. The shear wave velocity was evaluated from the relationship presented below [11,42]:

$$v_S = \frac{h}{t} \qquad (5)$$

where h is the distance between the transmitter and the receiver and t is the travel time.

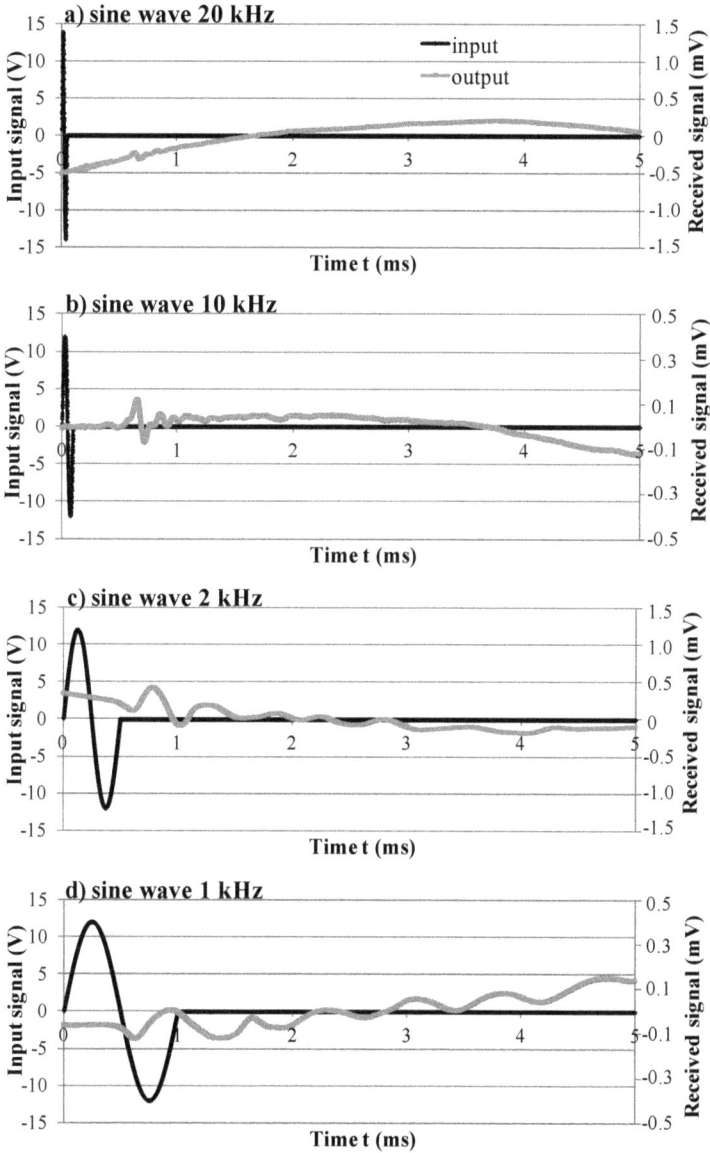

Figure 7. The oscilloscope signals at various frequencies of input signals: (**a**) 20 kHz, (**b**) 10 kHz, (**c**) 2 kHz, (**d**) 1 kHz and at the same effective stress value.

Finally, the unloading process was carried out. After each unloading stage, BE tests were also performed, followed by the measurement of travel times, which were determined using peak-to-peak and start-to-start interpretation procedures.

BE tests were conducted to provide information on the shear (G) and Young's (E) moduli. From the S-wave velocity (v_S), the small strain shear modulus (G_{max}), was determined, using the elastic wave velocity according to the equation [43,44]:

$$G_{max} = \rho \cdot v_S{}^2 \tag{6}$$

where ρ is the soil mass density.

By transforming Equation (3) and combining it with Equation (5), the authors estimated the value of the small strain Young's modulus (E_{max}) as follows [45]:

$$E_{max} = 2 \cdot \rho \cdot v_S{}^2 \cdot (1 - v) \tag{7}$$

where v is Poisson's ratio.

According to a well-known relation [4], Poisson's ratio is equal to:

$$v = \frac{\left(\dfrac{v_P}{v_S}\right)^2 - 2}{2\left[\left(\dfrac{v_P}{v_S}\right)^2 - 1\right]} \tag{8}$$

3. Results and Discussion

3.1. Frequency Dependency

In Figures 8 and 9 the variation of the measured shear wave velocity (v_S) with a predetermined period of propagating wave at different specified stress levels is shown. In Figure 10, however, the equivalent plot for measured compression wave velocity (v_P) is illustrated. The presented results are related to both procedures adopted for determining the travel time.

Inspection of Figures 8 and 9 indicates that, regardless of the method used to measure the travel time, the shear wave velocities decrease with the predetermined period (T), which transfers to frequency (f). Oscillation frequency is in fact the inverse of period. The linear approximation fits accurately most of the presented data obtained using the peak-to-peak and start-to-start methods. The coefficient of determination (R^2) reaches high values, from 70% upwards, which shows that for the studied cohesive material, the linear regression functions explain the majority of v_S variation. In general, the differences in the shear wave velocities with different frequencies are not large, from 18 to 27 m·s^{-1} and from 32 to 38 m·s^{-1}, depending on the technique used to measure the travel time.

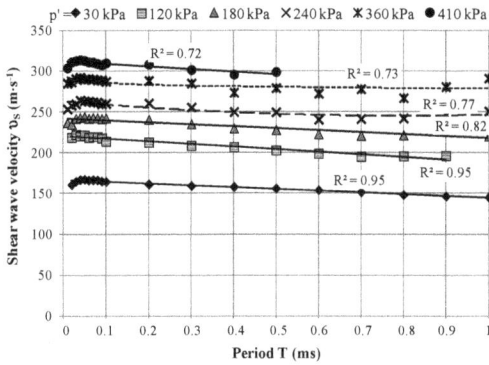
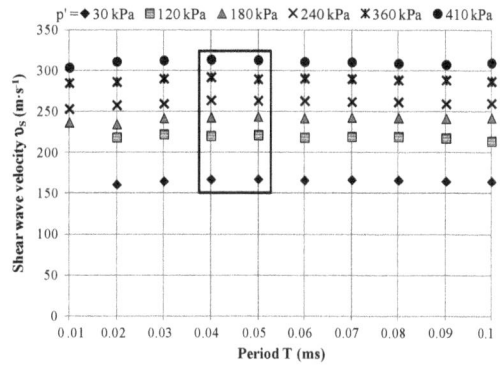

(a)   (b)

Figure 8. Variation of measured υ_S with period: peak-to-peak method. (**a**) Period's range from 0 to 1 ms; (**b**) Period's range from 0.01 to 0.1 ms.

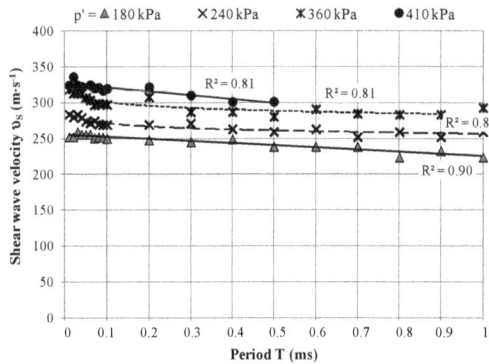
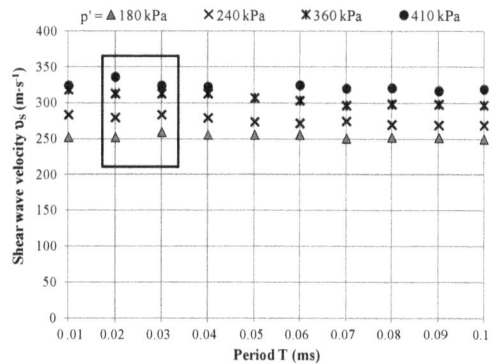

(a)   (b)

Figure 9. Variation of measured υ_S with period: start-to-start method. (**a**) Period's range from 0 to 1 ms; (**b**) Period's range from 0.01 to 0.1 ms.

In said graphs, the influence of stress level (p') is visible, as well. The average effective stress level significantly affects υ_S. For the highest pre-set pressure, *i.e.*, 410 kPa, the authors received the highest values of shear wave velocity, with average values around 307 m·s^{-1} (peak-to-peak method) and 319 m·s^{-1} (start-to-start method). The lowest values of ν_S were recorded for the smallest given stress (30 or 180 kPa), their average being around 160 m·s^{-1} (peak-to-peak method) and 246 m·s^{-1} (start-to-start method). The difference in υ_S decreases with the increasing pressure value.

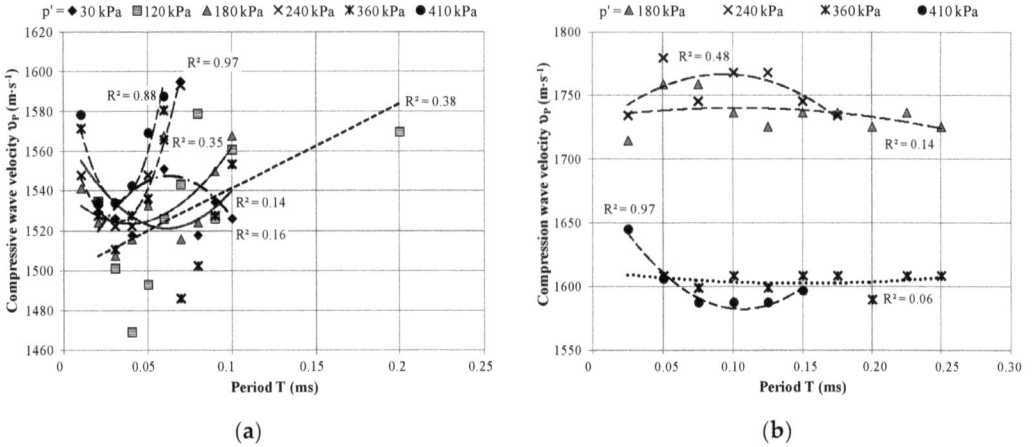

Figure 10. Variation of measured υ_P with period. (**a**) Peak-to-peak method; (**b**) start-to-start method.

The results of the current study show also that for the period of propagation lasting up to 0.1 ms (*i.e.*, f = 10 kHz), both selected procedures used in the time domain methods consistently produce the values of υ_S, which are less dependent on frequency than in the case of a larger period of wave (Figures 8 and 9 on the right). The frame in these figures indicates the highest values of shear wave velocity for each mean effective stress level. The authors suggest in this way certain frequencies at which the bender elements tests should be performed. Hence, the recommended frequency values, for which the highest shear wave velocities can be obtained, are in the range of 20 to 50 kHz. Certainly, it should be remembered that the optimum frequency of excitation depends on many factors, for example on the pressure applied in the cell during tests [46]. There is no rule at all for choosing the best wave [47]. In the literature, some authors have reported their experience with choosing a specific frequency. Brignoli *et al.* in 1996 [38] noted that when bender elements are measuring the samples, whose height is between 100 and 140 mm, the most interpretable output waves occur in the range from 3 to 10 kHz. Jovočič *et al.* [33], however, observed that with a higher frequency, the near field effect is less discernible. For stiffer materials, the required frequency must be higher than the reported 10 kHz [33], although the problem of overshooting becomes more important. In addition to this, a high frequency wave attenuates faster than the low frequency one, provided the medium is the same. The fact that the use of a high frequency gives the clearest output signals remains undisputed.

In the case of compression waves (Figure 10), the authors carried out research in the frequency scope of up to 10 kHz. A very large dispersion of the results and a lack of dependence of υ_P on the frequency ratio, as well as on mean effective stress can be clearly noticed. The authors could not find a proper relationship of v_P *vs.* frequency and mean effective stress. Empirical equations expressing the variation of P-wave velocity with frequency are characterized, however, by the low coefficients of determination (R^2), listed in Figure 10. In general, the differences in compression wave velocities with the different frequencies reach high values, from 54 to 110 m·s^{-1} and from 19 to 57 m·s^{-1}, depending on the technique used to measure the travel time.

In Figures 11 and 12 the effect of excitation frequency on shear modulus (Figure 11) and Young's modulus (Figure 12) is presented. The excitation frequencies varied from 1 to 100 kHz. The specimens were subjected to various isotropic effective confining pressures. It was observed (black frame in the figures) that the best signal, and consequently the greatest values of shear modulus (G), were received at the excitation frequency, whose values were equal 20 kHz for the peak-to-peak method (Figure 11a) and 50 kHz for the start-to-start method (Figure 11b). The excitation frequency being equal to (Figure 12b) or smaller than 50 kHz (Figure 12a) gave the greatest values of Young's modulus (E), as well. It is worth paying attention to a decrease in the modulus values along with the decreasing frequency, although G_{max} and E_{max} values were not obtained for the highest frequency, as already mentioned above. The average G_{max} value obtained from all applied values of pressure ranged from 139 MPa to 194 MPa, depending on the measurement of the travel time, while average E_{max} was around 405 MPa and 557 MPa. In the case of the minimal dynamic properties of the studied soils, using the lowest frequencies resulted in the smallest values of G and E, with the average minimal G ranging from 116 MPa to 152 MPa and the minimal E ranging from 379 MPa to 509 MPa.

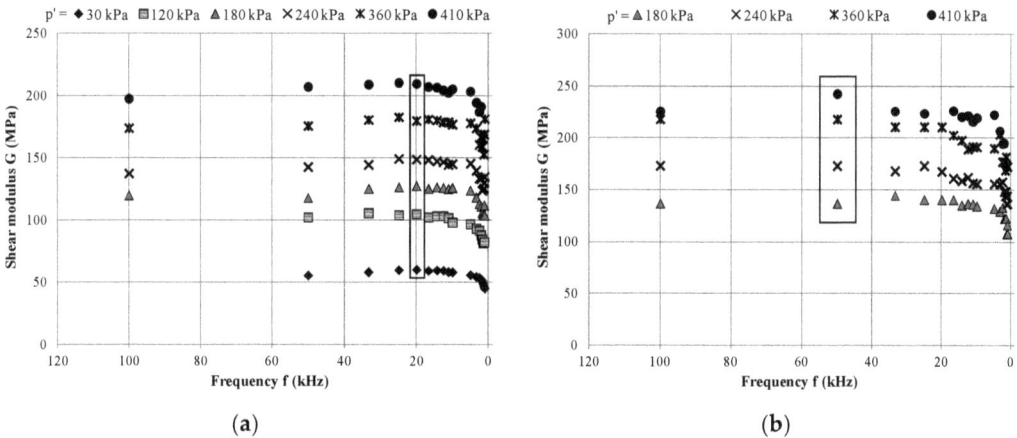

Figure 11. Variation of dynamic stiffness with frequency. (**a**) Peak-to-peak method; (**b**) start-to-start method.

It should be also pointed out that the values of E are arranged in some trend with increasing frequency, whereas the values of v_P do not show any dependence on frequency. This may seem incorrect, because Young's modulus is relevant to the compression wave velocity, due to the motion of the particles. However, it must be remembered that the authors used the results of v_P only indirectly, in order to determine the proper value of v.

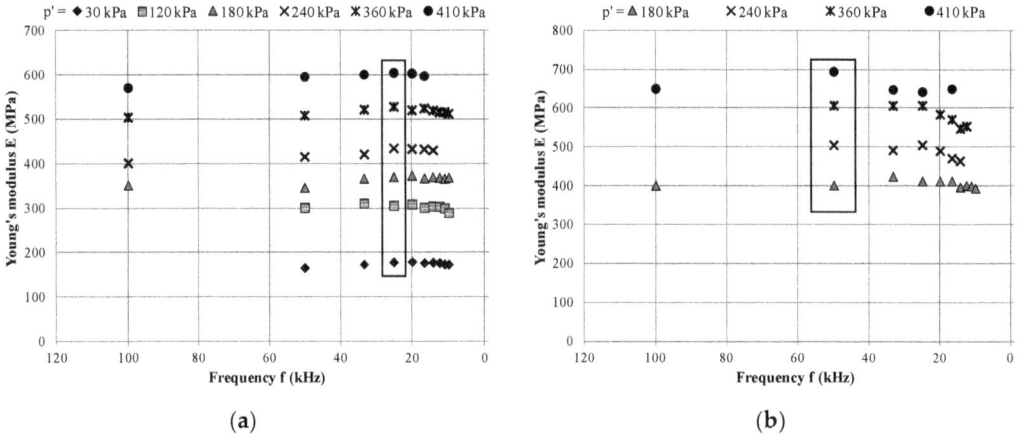

Figure 12. Variation of Young's modulus with frequency. (**a**) Peak-to-peak method; (**b**) start-to-start method.

3.2. The Effect of Stress Level

In Table 2, the authors compiled the range of received S-wave and P-wave velocities after each consolidation stage. On the basis of Table 2, it can be concluded that the values of shear wave velocity obviously increase with applied pressure. At the same time, it is visible that the differences between the shear wave velocity values and each p' amount to 20 to 30 m·s^{-1}. The values of compression wave velocity confirmed the earlier observation about the lack of dependence between v_P and stress level. The typical P-wave velocity of the tested material reached the value of 1400 to 1500 m·s^{-1} for the peak-to-peak method and 1600 to 1700 m·s^{-1} for the start-to-start method.

In Table 2, additionally, the average ratio list of seismic wave's velocities is enclosed. It is evident that the values of v_P are several times higher, when compared with v_S. The v_P/v_S velocity ratio has the variation interval ranging from 5.0 to 9.3. The authors obtained larger values of P-wave velocity than is observed in certain positions in the literature. Das and Ramana [48], for example, reported the characteristic compression wave velocity as ranging from 1220 m·s^{-1} to 1370 m·s^{-1} for saturated clay, although the ratio v_P/v_S was in the range of 8.0 to 9.0, thus similar to those obtained by the authors. Gary Mavko [49], on the other hand, in his study on Parameters That Influence Seismic Velocity mentions P-wave velocity for saturated clays as being in the range of 1100 to 2500 m·s^{-1}. Campanella and Steward [50] detailed typical speeds of longitudinal and transversal waves to a depth of 40 m. According to their research, compressive wave velocity for saturated soils may have values between 1500 and 1900 m·s^{-1}.

Table 2. Seismic waves velocities for various mean effective stress levels. (a) Peak-to-peak method; (b) start-to-start method.

(a)

Mean Effective Stress	S-Wave Velocity	P-Wave Velocity	Average Ratio of Seismic Waves Velocities
p' (kPa)	υ_S (m·s^{-1})	υ_P (m·s^{-1})	υ_P/υ_S (-)
30	145–168	1518–1595	9.3
120	195–222	1469–1579	7.0
180	221–224	1508–1568	6.4
240	240–264	1522–1593	5.9
360	267–292	1486–1581	5.3
410	296–314	1435–1588	5.0

(b)

Mean Effective Stress	S-Wave Velocity	P-Wave Velocity	Average Ratio of Seismic Waves Velocities
p' (kPa)	υ_S (m·s^{-1})	υ_P (m·s^{-1})	υ_P/υ_S (-)
180	224–260	1715–1759	6.8
240	252–284	1734–1779	6.3
360	281–319	1590–1609	4.7
410	301–336	1588–1645	4.9

In Table 3, the results of the average dynamic Poisson's ratio computed for each applied mean effective stress is shown. ν_d was determined on the basis of the relationship between S-wave and P-wave velocities (Equation (8)). The values of ν_d seem to be consistent with the literature [4,51,52]. The authors described the values of this parameter in another publication [23], in which the results obtained for saturated clay were at the level of 0.40 to 0.50, which is similar to the value obtained during the current studies. It is interesting to note that there is no significant change in ν_d regarding the pressure. Variation in Poisson's ratio amounts to approximately 4% and approaches 0.5. Moreover, a number of different techniques for interpreting the BE results causes no definite difference in the values of Poisson's ratio. Based on the experiments conducted, a general trend can be observed: as far as tested clayey material is concerned, Poisson's ratios represent little appreciable change with regard to shear modulus (Figure 13). The values of ν_d were used afterwards in order to determine the dynamic modulus of elasticity (E).

Table 3. Dynamic Poisson's ratio for various mean effective stress levels. (a) Peak-to-peak method; (b) start-to-start method.

(a)

Mean Effective Stress	Average Dynamic Poisson's Ratio
p' (kPa)	ν_d (-)
30	0.48
120	0.47
180	0.46
240	0.46
360	0.44
410	0.44

(b)

Mean Effective Stress	Average Dynamic Poisson's Ratio
p' (kPa)	ν_d (-)
180	0.47
240	0.46
360	0.44
410	0.44

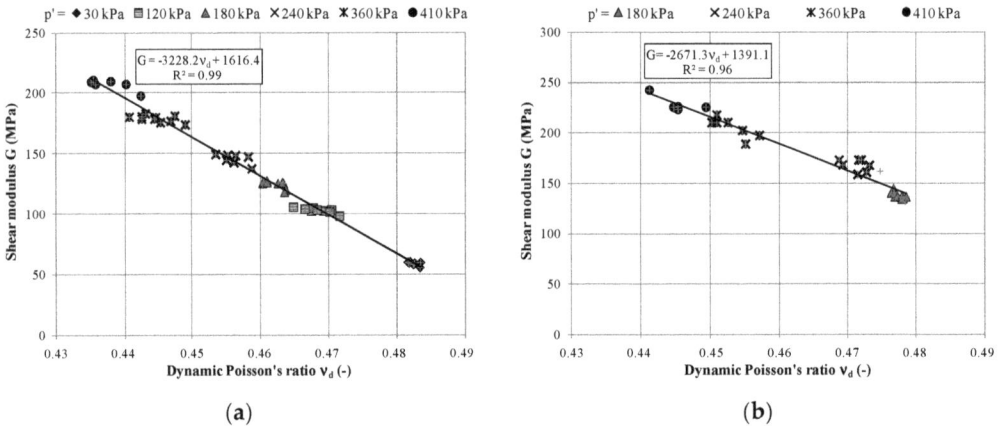

Figure 13. Shear modulus *versus* dynamic Poisson's ratio. (a) Peak-to-peak method; (b) start-to-start method.

The effect of mean effective stress on the small strain shear modulus and on the small strain modulus of elasticity of the examined silty clays is displayed in Figures 14 and 15. The presented data were obtained for one excitation frequency, selected on the basis of its greatest values of dynamic properties regarding the analyzed soil specimens, as described in the previous section. BE measurements show that G_{max} and E_{max} values increase with confining pressure as a linear function. Shear modulus at small strains, which was calculated for the tested material using Equation (6), does not exceed the interval of 60 to 242 MPa, regarding the travel time measurement. The modulus of elasticity at small strains, on the other hand, varies from 178 MPa to 694 MPa.

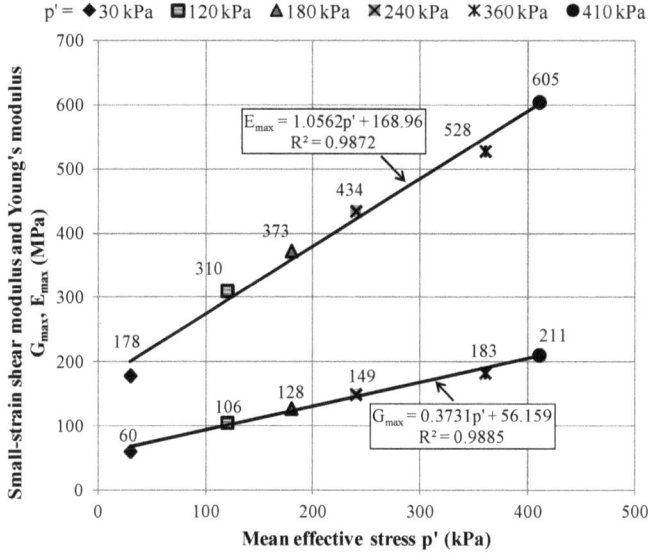

Figure 14. Elastic moduli from the peak-to-peak method under different stress levels.

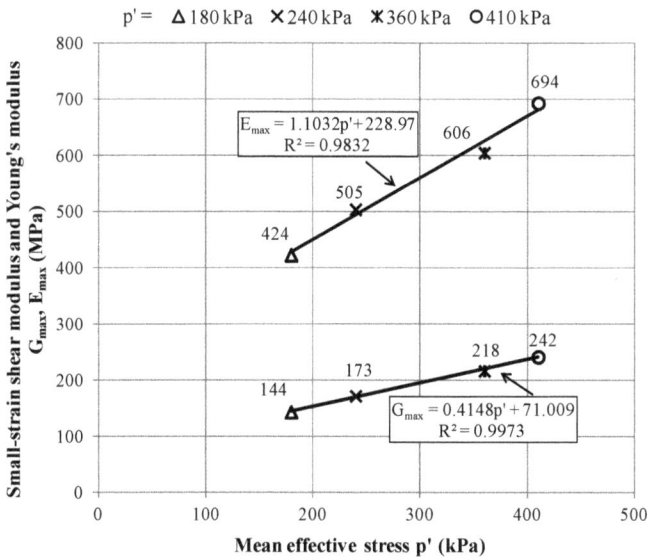

Figure 15. Elastic moduli from the start-to-start method under different stress levels.

3.3. Comparison of Results from Different Methods

The input signals employed in all of the tests were single sinusoidal pulses with different frequencies, obtained using peak-to-peak and start-to-start methods. The amplitude of input signals in both cases was fixed at ± 15 V. The tip-to-tip distances between the transducers were used to calculate the shear wave velocity of the samples (Equation (5)). All of the test traces were examined in the time domain, in order to obtain the travel time of the wave through the specimens.

In Figure 16, S-wave velocities determined by means of the two analyzed time domain techniques are portrayed. It is clearly visible that travel time derived from the start-to-start method must be longer due to the presence of the greatest values of the shear wave velocities. The difference in the obtained values of v_S between these two methods remains at the average level of 21.25 m$\cdot$s^{-1}, in favor of the start-to-start method (Figure 17a). Naturally, these spotted differences vary, depending on the mean effective stress.

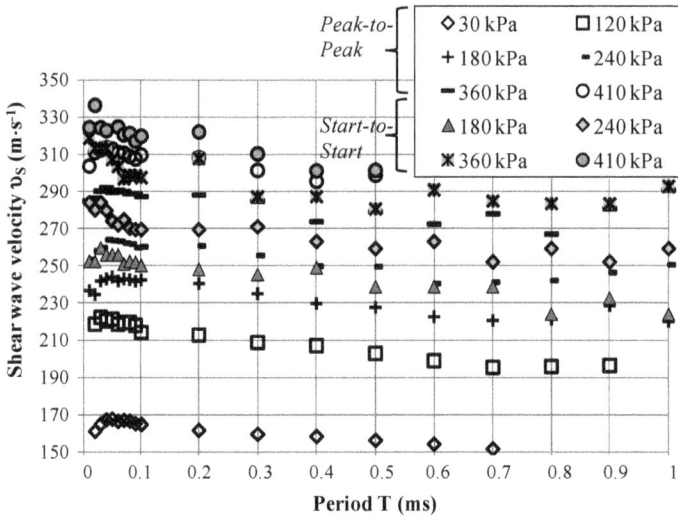

Figure 16. Comparison of v_S obtained using various methods of travel time determination.

In Figure 17b, the evolution of the shear stiffness modulus based on travel time deduced by means of the peak-to-peak and start-to-start techniques is summarized. In this comparison, the second of the aforementioned methods comes off as the better one. The average difference in stiffness of the examined cohesive material falls in the range of about 33.44 MPa. For the E_{max} (Figure 17c), the discrepancy between the results of two different interpretation methods achieves a maximum percentage of about 16% (89.2 MPa).

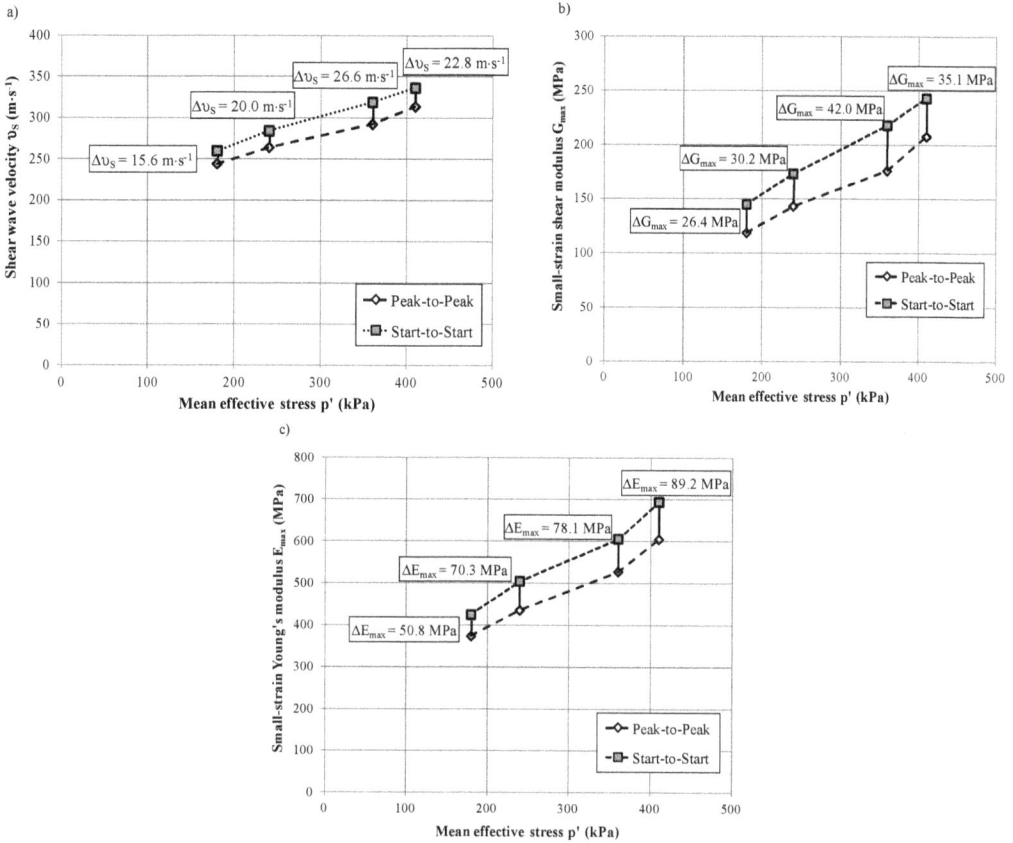

Figure 17. Comparison of results obtained by means of the peak-to-peak and start-to-start methods. (**a**) Shear wave velocity; (**b**) small strain shear modulus; (**c**) small strain Young's modulus.

In conclusion, it is worth noting that in the case of the start-to-start method, the findings of v_S and G_{max} lie on a straight line, but of a different gradient. This observation is not reflected in the outcomes regarding E_{max}. The results obtained with the use of the peak-to-peak method appear to be more scattered. On the basis of Figure 17, it can be noted that the S-S method (start-to-start method) of identification results in a relatively smaller variation, when compared to the P-P technique (peak-to-peak method). The latter method yields slightly smaller values of the dynamic parameters.

3.4. The Effect of Unloading

The aim of the current study, besides finding S-wave velocity and the small strain shear modulus when loading soil samples, was also to find v_S and G_{max} while unloading. The unloading process of the tested material was carried out to the pressure level of 120 kPa. In Figures 18 and 19 the values of computed parameters from the loading and the subsequent unloading stages are presented.

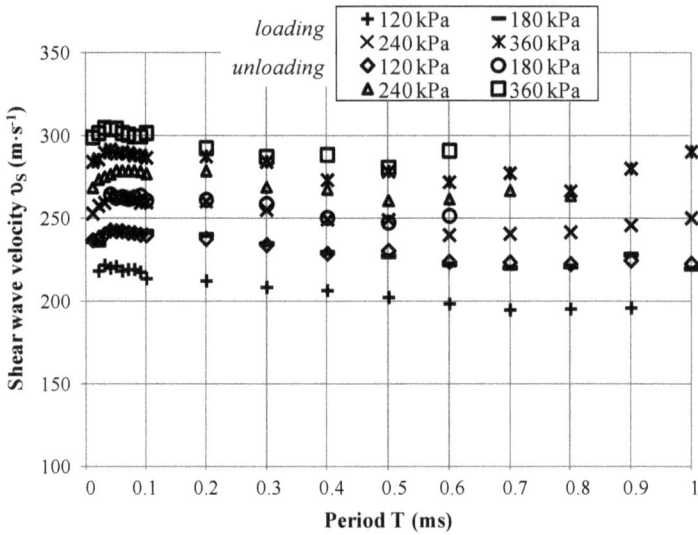

Figure 18. Shear wave velocity *versus* period at the loading and unloading stage.

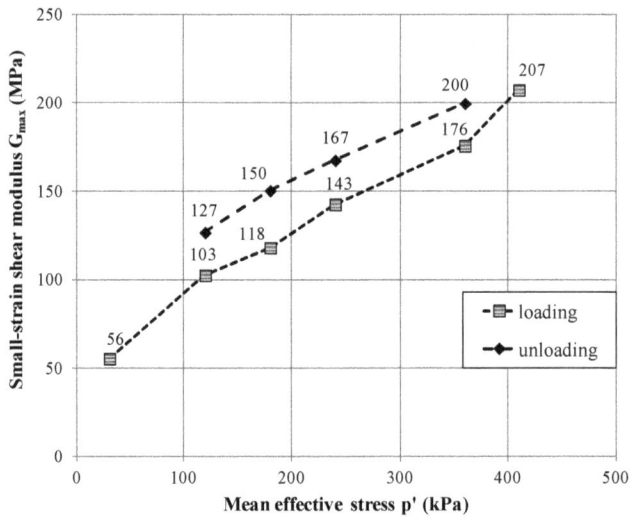

Figure 19. Small strain shear modulus *versus* mean effective stress at the loading and unloading stage.

During the inspection of Figures 18 and 19 the authors noticed that $\upsilon_{S,unloading}$ and consequently $G_{max, unloading}$ carry greater values than $\upsilon_{S, loading}$ and $G_{max, loading}$. The variance between silty clay dynamic stiffness during loading and unloading stages falls within the range between 16.8 and 22.8 MPa. The biggest difference was obtained for the mean effective stress equal to 180 kPa, the smallest one for $p' = 360$ kPa. The unloading process causes the increase of the results from 8% to 16%. Thus, it should be highlighted that the unloading of soil considerably improves its stiffness.

4. Conclusions

In this study, the characterization of the dynamic properties of cohesive material, namely silty clays, was inspected. The conducted experiments included the dynamic tests in the GDS Resonant Column Apparatus equipped with bender elements. The authors aimed to verify the dynamic behavior of clayey soil in connection with wave propagation produced by piezoelectric transducers. Shear wave velocity is related to the shear modulus of soil. Consequently, the measurements of the S-wave velocity provided a convenient method for determining soil stiffness. Additionally, the calculation of the P-wave velocity and, subsequently, the dynamic Poisson's ratio allowed for the examination of Young's modulus with regard to the tested cohesive material.

The findings of this research, in line with the literature in relation to the methods of interpretation of bender element test results, indicate that, for tests with piezoelectric transducers installed in the resonant column apparatus on silty clays, both analyzed techniques used as a part of the time domain methods, *i.e.,* peak-to-peak and start-to-start, yield similar results. They are also considered as the most comprehensive and appropriate methods to identify the travel time of the shear waves propagating through the sample. The authors, however, did not inspect any other methods, like the cross-correlation or π-point phase, but the velocities established by those methods are comparable to the findings from the resonant column measurements [25]. The dynamic parameters obtained by the start-to-start method were always slightly larger (up to about 20%) than those obtained using the peak-to-peak method. According to the authors, this is a consequence of the broadening of the excitation pulses after they are propagated through the sample. The lowest wavelength in the presented experiments was about 1.0 mm, which was at least 10-times bigger than the largest grain size of soil sample (that was less than or equal to 100 μm). In addition, no inhomogeneous pockets were found in the sample constitution after a destructive analysis of the material. In these conditions the macroscopic sample was expected to behave as a conducting media, and no enhanced scattering was expected at the grain boundaries. Since the studied sample was homogeneous, in terms of the wavelength, the difference between the results from the two discussed techniques can be explained by taking into account different wave velocities and different attenuations of the frequency components that constitute the excitation peaks. Although each excitation pulse used in the experiments was considered to have a well-defined frequency, in reality, the pulses are composed of a continuity of different frequencies with different amplitudes. This happens due to the finite size of the excitation pulse in the time domain, which adds additional frequencies to it. In the start-to-start method, the results were strongly influenced by the fastest travelling wave, while in the peak-to-peak method, the frequency component with the highest amplitude was the one considered. From the Figures 9 and 10 it can be clearly noticed that the wave velocity from the S-S method had an almost flat dependency as a function of frequency; however, in the P-P method, it has a change of about 20% in the frequency range experimented. This suggests that, in the S-S method, the frequency component responsible for the excitation of the detector is independent of the frequency chosen for the excitation pulse (due to the unintentional frequencies present in the excitation pulse). In certain experimental circumstances, such as during the measurements of G, the peak-to-peak method led to more scattered results, which were probably connected to the interpretation of the findings made by the authors. It must be always kept in mind, that even when using the same identification method, reading points differed with the laboratory [22]. The scatter in the S-S method was definitely smaller in comparison to another method. Presented results show and confirm observations made by previous researchers [6,22], acknowledging the status of the start-to-start method as the most prevalent and consistent method for the determination of travel time.

With reference to the dependence of the dynamic behavior of clayey material on various factors, the following conclusions can be drawn:

- The data show that the input signal frequency, in particular frequencies, which are below the level of $T = 0.1$ ms, does not significantly affect the stiffness values obtained by both adopted identification methods. The decrease in frequency results in greater dependence on frequency.
- The best measurements of travel time for the tested soils were received at the frequencies of the input signal in the range of 50 to 20 kHz. In most of the considered cases, the greatest values of the dynamic parameters were obtained when the frequencies from this range were applied.
- Mean effective stress has a visible influence on the seismic wave velocities and on the elastic modulus of the examined soils.
- There are linear relationships between the mean effective stress and the small strain shear modulus and the small strain modulus of elasticity.
- The compression waves propagate through the analyzed clayey material with far larger velocities than in the case of shear waves; at certain pressure levels, these velocities can be even up to 10-times larger.

- The dynamic Poisson's ratio for the tested cohesive soils is almost constant, being approximately equal to 0.46.
- The dynamic Poisson's ratio linearly increases with the decrease of the shear modulus.
- Concerning the unloading process, it profitably affects the stiffness of silty clays. The dynamic properties rise in their values, from 8% to 16%, owing to unloading.

Based on the results as described above, the standard procedure for the BE tests was proposed here in order to obtain appropriate test results in the present conditions. The authors discussed the important issue regarding the applied sciences, precisely the seismic research and dynamic response of porous material due to dynamic loads, such as earthquakes, traffic loads or machine vibrations. They attempted to characterize the dynamic properties of a fine-grained material, whose average porosity is approximately 0.44. This differs from the commonly-encountered in the literature dynamic characteristics of granular materials, non-cohesive soils, with a larger porosity value, described for example in [53,54]. The authors used generally-applied laboratory tests in order to investigate a still new material, as well as to expand the knowledge of the interpretation of this new material's results.

The acquired knowledge of the behavior of the tested material should be verified in the process of modeling, using an appropriate model, such as the one proposed in [55].

Acknowledgments: The discussion about the results obtained was carried out with the valuable support of Bruno Cury Camargo (LNCMI—Laboratoire National des Champs Magnetiques Intenses, Toulouse, France).

Author Contributions: Wojciech Sas was the inventor of the research presented in this paper. Katarzyna Gabryś and Emil Soból conducted the described research and also performed geotechnical laboratory studies. Wojciech Sas and Katarzyna Gabryś prepared the manuscript. Katarzyna Gabryś and Emil Soból performed the analysis of the tests results. Alojzy Szymański consulted the research program and proofread the manuscript.

Conflicts of Interest: The authors declare no conflict of interest.

References

1. Salacak, M.; Pinar, S.B.S. Soil parameters which can be determined with seismic velocities. *Jeofizik* **2012**, *16*, 17–29.
2. Youn, J.-U.; Choo, Y.-W.; Kim, D.-S. Measurement of small-strain shear modulus G_{max} of dry and saturated sands by bender element, resonant column, and torsional shear tests. *Can. Geotech. J.* **2008**, *45*, 1426–1438.
3. Burland, J.B.; Longworth, T.I.; Moore, J.F.A. Study of ground and progressive failure caused by deep excavation in Oxford clay. *Géotechnique* **1977**, *27*, 557–591.
4. Oshaki, Y.; Iwasaki, R. On dynamic shear moduli and Poisson's ratio of soil deposits. *Soils Found.* **1973**, *13*, 61–73.

5. Sas, W.; Głuchowski, A.; Radziemska, M.; Dzięcioł, J.; Szymański, A. Environmental and Geotechnical Assessment of the Steel Slags as a Material for Road Structure. *Materials* **2015**, *8*, 4857–4875.

6. Yamashita, S.; Kawaguchi, T.; Nakata, Y.; Mikami, T.; Fujiwara, T.; Shibuya, S. Interpretation of international parallel test on the measurement of G_{max} using bender elements. *Soils Found.* **2009**, *49*, 631–650.

7. Benz, T. *Small-Strain Stiffness of Soils and Its Numerical Consequences*; Universität Stuttgart: Stuttgart, Germany, 2007.

8. Menzies, B. Near-surface site characterization by ground stiffness profiling using surface wave geophysics. In *H. C. Verma Commemorative Volume*; Indian Geotechnical Society: New Delhi, Indian, 2000; pp. 1–14.

9. Leong, E.C.; Cahyadi, J.; Rahardjo, H. Measuring shear and compression wave velocities of soil using bender-extender elements. *Can. Geotech. J.* **2009**, *46*, 792–812.

10. Ha Giang, P.H.; van Impe, P.; van Impe, W.F.; Menge, P.; Haegeman, W. Effects of grain size distribution on the initial small strain shear modulus of calcareous sand. In Proceedings of the 16th ECSMGE Geotechnical Engineering for Infrastructure and Development, Edinburg, UK, 13–17 September 2015; pp. 3177–3182.

11. Sas, W.; Gabryś, K.; Szymański, A. Laboratoryjne badanie sztywności gruntu według Eurokodu 7 [In Polish] Laboratory tests of soil stiffness by Eurocode 7. *Acta Sci. Pol. Ser. Archit.* **2013**, *12*, 39–50.

12. Sas, W.; Gabryś, K.; Szymański, A. Comparison of resonant column and Bender elements tests on selected cohesive soil from Warsaw. *Electron. J. Pol. Agric. Univ. EJPAU* **2014**, *17*, #07. Available online: http://www.ejpau.media.pl/volume17/issue3/art-07.html (accessed on 23 July 2014).

13. Lawrence, F.V. *Propagation of Ultrasonic Waves through Sand*; Research Report R63-08; Massachusetts Institute of Technology: Cambridge, MA, USA, 1963.

14. Lawrence, F.V. *Ultrasonic Shear Wave Velocity in Sand and Clay*; Research Report R65-05; Massachusetts Institute of Technology: Cambridge, MA, USA, 1965.

15. Shirley, D.J. An improved shear wave transducer. *J. Acoust. Soc. Am.* **1978**, *63*, 1643–1645.

16. Dyvik, R.; Madshus, C. *Lab Measurement of G_{max} Using Bender Elements*; Norwegian Geotechnical Institute: Oslo, Norway, 1985; pp. 186–196.

17. Styler, M.A.; Howie, J.A. Continuous Monitoring of Bender Element Shear Wave Velocities During Triaxial Testing. *Geotech. Test. J.* **2014**, *37*, 218–230.

18. Ferreira, C.; Martins, J.P.; Correia, A.G. Determination of the small-strain stiffness of hard soils by means of bender elements and accelerometers. In *Proceedings of the 15th European Conference on Soil Mechanics and Geotechnical Engineering*; Athens, Greece, 12–15 September 2013, Andreas, A., Michael, P., Christos, T., Eds.; IOS Press: Amsterdam, The Netherlands, 2011; Volume 1, pp. 179–184.

19. Viggiani, G.; Atkinson, J.H. Interpretation of bender element tests. *Géotechnique* **1995**, *45*, 149–154.

20. Lee, J.S.; Santamarina, J.C. Bender elements: Performance and signal interpretation. *J. Geotech. Geoenviron. Eng. ASCE* **2005**, *131*, 1063–1070.

21. Leong, E.C.; Yeo, S.H.; Rahardjo, H. Measuring shear wave velocity using bender elements. *Geotech. Test. J.* **2005**, *28*, 488–498.

22. Camacho-Tauta, J.F.; Cascante, G.; da Fonseca, A.V.; Santos, J.A. Time and frequency domain evaluation of bender element systems. *Géotechnique* **2015**, *65*, 548–562.

23. Sas, W.; Gabryś, K.; Szymański, A. Determination of Poisson's ratio by means of resonant column tests. *Electron. J. Pol. Agricu. Univ. EJPAU* **2013**, *16*, #03. Available online: http://www.ejpau.media.pl/volume16/issue3/art-03.html (accessed on 20 August 2013).

24. Sas, W.; Gabryś, K. Laboratory measurement of shear stiffness in resonant column apparatus. *ACTA Sci. Pol. Ser. Arch.* **2012**, *11*, 29–39.

25. Gabryś, K.; Sas, W.; Soból, E. Small-strain dynamic characterization of clayey soil from Warsaw. *ACTA Sci. Pol. Ser. Arch.* **2015**, *14*, 55–65.

26. GDS Instruments Ltd. Bender Elements. In *The GDS Bender Elements System Hardware Handbook for Vertical and Horizontal Elements*; GDS Instruments Ltd.: Hampshire, UK, 2005.

27. Cheng, Z.; Leong, E.C. A Hybrid Bender Element-Ultrasonic System for Measurement of Wave Velocity in Soils. *Geotech. Test. J.* **2014**, *37*, 377–388.

28. Chan, C.M. Bender Element Test in Soil Specimens: Identifying the Shear Wave Arrival Time. *EJGE* **2010**, *15*, 1263–1276.

29. Clayton, C.R.I.; Theron, M.; Best, A.I. The measurement of vertical shear-wave velocity using side-mounted bender elements in the triaxial apparatus. *Géotechnique* **2004**, *54*, 495–498.

30. Brocanelli, D.; Rinaldi, V. Measurement of low-strain material damping and wave velocity with bender elements in the frequency domain. *Can. Geotech. J.* **1998**, *35*, 1032–1040.

31. Arroyo, M.; Greening, P.; Wood, D.M. An estimate of uncertainty in current pulse test practice. *Riv. Ital. Geotec.* **2001**, *37*, 17–35.

32. Mitaritonna, G.; Amorosi, A.; Cotecchia, F. Multidirectional bender element measurements in the triaxial cell: Equipment set-up and signal interpretation. *Riv. Ital. Geotec.* **2010**, *1*, 50–69.

33. Jovičić, V.; Coop, M.R.; Simic, M. Objective criteria for determining G_{max} from bender element tests. *Géotechnique* **1996**, *46*, 357–362.

34. Polish Committee for Standardization. *Geotechnical Investigations—Soil Classification—Part 2: Classification Rules*; PN–EN ISO 14688-2:2004; Polish Committee for Standardization: Warsaw, Poland, 2004. (In Polish)

35. Sas, W.; Gabryś, K.; Szymański, A. Effect of time on dynamic shear modulus of selected cohesive soil of one section of Express Way No S2 in Warsaw. *Acta Geophys.* **2015**, *63*, 398–413.

36. Head, K.H. *Manual of Soil Laboratory Testing. Volume 3 Effective Stress Tests*; J. Wiley & Sons Ltd.: New York, NY, USA, 1998.

37. Viggiani, G. Panellist discussion: Recent advances in the interpretation of bender element tests. In *Pre-Failure Deformation of Geomaterials*; Satoru, S., Toshiyuki, M., Seiichi, M., Eds.; Balkema: Rotterdam, The Netherland, 1995; Volume 2, pp. 1099–1104.

38. Brignoli, E.; Gotti, M.; Stokoe, K.H., II. Measurement of shear waves in laboratory specimens by means of piezoelectric transducers. *ASTM Geotech. Test. J.* **1996**, *19*, 384–397.

39. Kawaguchi, T.; Mitachi, T.; Shibuya, S. Evaluation of shear wave travel time in laboratory bender element test. In Proceedings of 15th International Conference on Soil Mechanics and Geotechnical Engineering, Istambul, Turkey, 28–31 August 2001; Volume 1, pp. 155–158.

40. Leong, E.C.; Yeo, S.H.; Rahardjo, H. Measuring Shear Wave Velocity Using Bender Elements. *Geotech. Test. J.* **2003**.

41. Hasan, A.M.; Wheeler, S.J. Measuring travel time in bender/extender element tests. In Proceedings of the 16th ECSMGE Geotechnical Engineering for Infrastructure and Development, Edinburg, UK, 13–17 September 2015; pp. 3171–3176.

42. Choo, H.; Larrahondo, J.; Burns, S.E. Coating Effects of Nano-Sized Particles onto Sand Surfaces: Small Strain Stiffness and Contact Mode of Iron Oxide-Coated Sands. *J. Geotech. Geoenviron. Eng.* **2015**, *141*, 04014077:1–04014077:10.

43. Lee, C.-J.; Hung, W.-Y.; Tsai, C.-H.; Chen, T.; Tu, Y.; Huang, C.-C. Shear wave velocity measurements and soil-pile system identifications in dynamic centrifuge tests. *Bull. Earthq. Eng.* **2014**, *12*, 717–734.

44. Fedrizzi, F.; Raviolo, P.L.; Vigano, A. Resonant column and cyclic torsional shear experiments on soils of the Trentino valleys. In Proceedings of the 16th ECSMGE Geotechnical Engineering for Infrastructure and Development, Edinburg, UK, 13–17 September 2015; pp. 3437–3442.

45. Gabryś, K.; Szymański, A. Badania parametrów odkształceniowych gruntów spoistych w kolumnie rezonansowej [In Polish] Research of deformation parameters of cohesive soils in resonant column. *Inżynieria Morska i Geotech.* **2012**, *4*, 324–327.

46. Amat, A.S. Elastic Stiffness Moduli of Huston Sand. Master's Thesis, University of Bristol, Bristol, UK, 2007.

47. Ali, H.; Mahbaz, S.B.; Cascante, G.; Garbinsky, M. Low strain measurement of shear modulus with resonant column and bender element tests—Frequency effects. In Proceedings of the GeoMontreal 2013: Geoscience for Sustainability, Montreal, QC, Canada, 29 September–3 October 2013.

48. Das, B.M.; Ramana, G.V. *Principles of Soil Dynamics*, 2nd ed.; United States of America: Stamford, CT, USA, 2011.

49. Mavko, G. *Parameters that Influence Seismic Velocity. Conceptual Overview of Rock and Fluid Factors that Impact Seismic Velocity and Impedance*; Stanford Rock Physics Laboratory: Standford, CA, USA, 2009.

50. Campanella, R.G.; Stewart, W.P. Seismic cone analysis using digital signal processing for dynamic site characterization. *Can. Geotech. J.* **1992**, *29*, 477–486.

51. Poisson's ratio. Available online: http://www.engineeringtoolbox.com (accessed on 21 September 2015).

52. Otsubo, M.; O'Sullivan, C.; Sim, W.W.; Ibraim, E. Quantitative assessment of the influence of surface roughness on soil stiffness. *Géotechnique* **2015**, *65*, 694–700.

53. Wichtmann, T.; Triantafyllidis, T. Stiffness and Damping of Clean Quartz Sand with Various Grain-Size Distribution Curves. *J. Geotech. Geoenviron. Eng.* **2014**, *140*, 06013003:1–06013003:4.

54. Qiu, T.; Huang, Y.; Guadalupe-Torres, Y.; Baxter, C.D.P.; Fox, P.J. Effective Soil Density for Small-Strain Shear Waves in Saturated Granular Materials. *J. Geotech. Geoenviron. Eng.* **2015**, *141*, 04015036:1–04015036:11.

55. Aggelis, D.G.; Tsinopoulos, S.V.; Polyzos, D. An iterative effective medium approximation (IEMA) for wave dispersion and attenuation predictions in particulate composites, suspensions and emulsions. *J. Acoust. Soc. Am.* **2004**, *116*, 3443–3452.

Design of a Stability Augmentation System for an Unmanned Helicopter Based on Adaptive Control Techniques

Shouzhao Sheng and Chenwu Sun

Abstract: The task of control of unmanned helicopters is rather complicated in the presence of parametric uncertainties and measurement noises. This paper presents an adaptive model feedback control algorithm for an unmanned helicopter stability augmentation system. The proposed algorithm can achieve a guaranteed model reference tracking performance and speed up the convergence rates of adjustable parameters, even when the plant parameters vary rapidly. Moreover, the model feedback strategy in the algorithm further contributes to the improvement in the control quality of the stability augmentation system in the case of low signal to noise ratios, mainly because the model feedback path is noise free. The effectiveness and superiority of the proposed algorithm are demonstrated through a series of tests.

Reprinted from *Appl. Sci.* Cite as: Sheng, S.; Sun, C. Design of a Stability Augmentation System for an Unmanned Helicopter Based on Adaptive Control Techniques. *Appl. Sci.* **2015**, *5*, 575–586.

1. Introduction

It is essential that the flight control system of an unmanned helicopter (UH) should be endowed with well-suited automatic capabilities to carry out flight missions. However, the flight performance of an UH is intimately dependent on the stability and control characteristics of the UH [1,2]. Unlike some mechanical systems with desirable structural properties, UH is normally an inherently unstable system without stability augmentation control strategy. Furthermore, some of the aerodynamic parameters vary with flight environments or system conditions, with the result that it is very difficult to design the stability augmentation system for an UH using the conventional control methods [3].

UH is also a complicated nonlinear dynamic system described by nonlinear differential equations. However, for simplicity in the design of controllers for an UH, the linearized models are normally derived from nonlinear differential equations if the UH operates around an operating point. Many linear control techniques for application to UH flight control systems have been proposed in literature, among which single-input, single-output (SISO) feedback control methods are by far the most common choices with few dependencies on dynamic models. In [4], a SISO PD control law is adopted and further optimized for both hovering and forward flight of the CMU-R50 UH. In [5], a SISO PID control law is implemented for automatic hovering of the Ursa Major 3 UH. The above SISO methods have the advantages of conceptual and computational simplicity. However they may decrease the stability and control qualities of UHs without considering parametric uncertainties and cross-couplings among axes.

Therefore, in order to improve the flight performance, a lot of research effort has been devoted to the design of advanced stability augmentation systems. Previous research reported in the literature includes gain scheduling [6], linear-quadratic regulation (LQR) or linear-quadratic Gaussian (LQG) approach [7], decentralized decoupled model predictive approach [8] and intelligent control methods like neural network [9] and fuzzy logic approach [10], *etc.* Several flight control systems using H$_\infty$ control methods, which can provide the robust stability and performance for the systems subject to uncertainties and disturbances, have been designed for mini UHs. In [11], a H$_\infty$ loop shaping technique is utilized for the stability augmentation system of Bell-205 helicopter. Mixed-norm optimization and weighted H$_\infty$ mixed sensitivity optimization methods are respectively designed to improve the stability and maneuverability characteristics of UHs [12,13]. Although these methods have achieved acceptable flight performance, they still rely heavily on the plant model. More importantly, the above-mentioned methods fail to consider the adverse effects of parametric uncertainties and measurement noises on the flying qualities. Some existing adaptive techniques can accommodate the parametric uncertainties more effectively without considering measurement noises [14–18]. A novel modified model reference adaptive control (MRAC) strategy is developed with added noise [19]. However, this method only reduces the noise disturbance using a low-pass filter. Usually in practical use, those existing adaptive control methods can hardly minimize the adverse effect of measurement noises on the flying qualities. The NASA Marshall Space Flight Center has developed an adaptive augmenting control (AAC) algorithm for launch vehicles by adapting a well tuned classical control algorithm to unexpected environments or variations in vehicle dynamics. The AAC algorithm has been successfully tested in a relevant environment. However it needs to be further evaluated by the flight tests [20].

The essential parameter regulation schemes can reduce the complexity of a high performance control system design problem in the presence of parametric uncertainties and measurement noises. On this basis, this study aims to develop an adaptive model feedback control algorithm for a prototype unmanned helicopter stability augmentation system. The proposed adaptive algorithm can achieve a guaranteed model reference tracking performance and speed up the convergence rates of adjustable parameters, even when the plant parameters vary rapidly. Moreover, the model feedback control strategy in the algorithm can further improve the control quality of the stability augmentation system because the model feedback path is noise free. The experimental setups and the actual flight test results using the proposed algorithm are shown and the results are discussed.

This paper describes the control problem for the prototype UH stability augmentation system in Section 2. The adaptive model feedback control algorithm for the prototype UH stability augmentation system is presented in Section 3, and Section 4 provides the improvement of the algorithm. The flight test results are shown in Section 5. Finally, conclusions are drawn in Section 6.

2. The Control Problem for the Prototype UH Stability Augmentation System

The prototype UH, with net weight 180 Kg and height 1.9 m, is a vertical takeoff and landing aircraft which includes two coaxial rotors and a fuselage with toroidal portion, as shown in Figure 1. A duct is formed through the fuselage. A propeller assembly is mounted to the top portion of the fuselage with a main rotor, 4.4 m in diameter, above the fuselage. A ducted rotor assembly in fuselage is used to compensate the propeller antitorque as well as providing some fraction of lift. The coaxial rotors, main and ducted, rotate at 800 rpm in the opposite directions with the main rotor providing about 70% of lift, drag, and pitch and roll movements of UH and the ducted rotor providing about 30% of lift and yaw movement.

Figure 1. The prototype unmanned helicopter.

The stability augmentation system design for the prototype UH is a challenging task because the UH dynamics are highly nonlinear and subject to parametric uncertainties. In addition, the plant parameters change with the flight environments (e.g., the aerodynamic constants) or the system conditions (e.g., lift curve slopes). The adaptive control strategy is adopted to solve the above problems in this study.

By assuming that the fuselage is a rigid body and the main rotor speed is constant, we can describe the nonlinear kinematic equations associated with the six degrees of freedom (6-DOF) as an equivalent block in the following manner:

$$\widehat{X} = f(\widehat{X}, \widehat{U}, \Theta) \tag{1}$$

258

where $\widehat{X} = \left[\widehat{u},\widehat{v},\widehat{w},\widehat{\theta},\widehat{\phi},\widehat{\psi},\widehat{q},\widehat{p},\widehat{r}\right]^T$ and $\widehat{U} = \left[\widehat{\delta}_e,\widehat{\delta}_a,\widehat{\delta}_r,\widehat{\delta}_c\right]^T$ represent the state vector and the control vector, respectively; Θ represents the unsteady aerodynamic parameters set, which is difficult and expensive to measure and is, therefore, not available in most cases. Of the state variables in $\widehat{X},\widehat{U}$, here $\widehat{u},\widehat{v},\widehat{w}$ are forward velocity, lateral velocity and vertical velocity, respectively; $\widehat{\theta},\widehat{\phi},\widehat{\psi}$ are pitch angle, roll angle and yaw angle, respectively; $\widehat{q},\widehat{p},\widehat{r}$ are pitch rate, roll rate and yaw rate, respectively; $\widehat{\delta}_e,\widehat{\delta}_a,\widehat{\delta}_r,\widehat{\delta}_c$ are pitch cyclic, roll cyclic, ducted rotor collective and main rotor collective, respectively.

One common method to solve the control problem for the nonlinear system in Equation (1) is through linearization. The linearized model about a trim condition of the nonlinear dynamics is then represented as

$$\dot{X} = AX + BU \tag{2}$$

where $X = \widehat{X} - X_e = [u,v,w,\theta,\phi,\psi,q,p,r]^T$ and $U = \widehat{U} - U_e = [\delta_e,\delta_a,\delta_r,\delta_c]^T$ are the increments at a specified trim condition; A and B are the system matrix and the control-input matrix, respectively. Accordingly, X_e and U_e are respectively the trim state and the trim input with respect to hovering, lifting or forward flight, which must satisfy the equation

$$f(X_e, U_e, \Theta) = 0 \tag{3}$$

Note that X_e and U_e, obtained by solving Equation (3), are unknown as well. Therefore, we define

$$X_0 = X_e + E_x, \quad U_0 = U_e + E_u \tag{4}$$

X_0 and U_0 can also be regarded as the estimates of X_e and U_e, respectively. Equation (2) can then be rewritten as:

$$\dot{X} = AX + BU + E \tag{5}$$

where $X = \widehat{X} - X_0$, $U = \widehat{U} - U_0$ and $E = AE_x + BE_u$. E is equivalent to an unknown input disturbance at the trim condition.

To avoid repetition, the present study is only focused on the stability augmentation system of the longitudinal axis to demonstrate the proposed adaptive algorithm, which, without loss of generality, can apply to other axes. Thus, referring to Equation (5), the linearized model of the longitudinal axis can be denoted by:

$$\dot{q} = M_u^q u + M_v^q v + M_w^q w + M_\theta^q \theta + M_\phi^q \phi + M_\psi^q \psi + M_q^q q + M_p^q p + M_r^q r + M_{\delta_e}^q \delta_e + M_{\delta_a}^q \delta_a + M_{\delta_r}^q \delta_r + M_{\delta_c}^q \delta_c + M_t^q \tag{6}$$

where $M_u^{\dot{q}}, M_v^{\dot{q}}, M_w^{\dot{q}}, M_\theta^{\dot{q}}, M_\phi^{\dot{q}}, M_\psi^{\dot{q}}, M_q^{\dot{q}}, M_p^{\dot{q}}, M_r^{\dot{q}}, M_{\delta_e}^{\dot{q}}, M_{\delta_a}^{\dot{q}}, M_{\delta_r}^{\dot{q}}$ and $M_{\delta_c}^{\dot{q}}$ represent the unknown time-varying aerodynamic parameters; $M_t^{\dot{q}}$ is the corresponding component of the vector E.

Given the corresponding wind tunnel test data, the effect of θ, ϕ, ψ and δ_r on q is negligible and can therefore be ignored, the linearized model of longitudinal axis can then be simplified as follows:

$$\dot{q} = M_u^{\dot{q}}u + M_v^{\dot{q}}v + M_w^{\dot{q}}w + M_q^{\dot{q}}q + M_p^{\dot{q}}p + M_r^{\dot{q}}r + M_{\delta_e}^{\dot{q}}\delta_e + M_{\delta_a}^{\dot{q}}\delta_a + M_{\delta_c}^{\dot{q}}\delta_c + M_t^{\dot{q}} \quad (7)$$

Overall, the complexity of the stability augmentation system design originates mainly from the parametric uncertainties together with measurement noises.

3. Design of the Stability Augmentation System Based on an Adaptive Model Feedback Control Algorithm

3.1. Theorem 1

Consider the plant of the form of Equation (7), with $M_q^{\dot{q}} < 0$ and $M_{\delta_e}^{\dot{q}} > 0$. Assume that

$$\delta_e = k_u^{\dot{q}}u + k_v^{\dot{q}}v + k_w^{\dot{q}}w + k_p^{\dot{q}}p + k_r^{\dot{q}}r + k_{\delta_a}^{\dot{q}}\delta_a + k_{\delta_c}^{\dot{q}}\delta_c + k_t^{\dot{q}} + k_{q_m}^{\dot{q}}q_m + k_{\delta_{em}}^{\dot{q}}\delta_{em} \quad (8)$$

where

$$k_{q_m}^{\dot{q}} = \left(k_{q_m}^{\dot{q}_m} - k_q^{\dot{q}} \right) / k_{\delta_e}^{\dot{q}}, \quad k_{\delta_{em}}^{\dot{q}} = k_{\delta_{em}}^{\dot{q}_m} / k_{\delta_e}^{\dot{q}} \quad (9)$$

and where $k_u^{\dot{q}}, k_v^{\dot{q}}, k_w^{\dot{q}}, k_q^{\dot{q}}, k_p^{\dot{q}}, k_r^{\dot{q}}, k_{\delta_e}^{\dot{q}}, k_{\delta_a}^{\dot{q}}, k_{\delta_c}^{\dot{q}}$ and $k_t^{\dot{q}}$ are adjustable parameters; q_m is the output of the ideal decoupled model given by:

$$\dot{q}_m = k_{q_m}^{\dot{q}_m}q_m + k_{\delta_{em}}^{\dot{q}_m}\delta_{em} \quad (10)$$

$k_{q_m}^{\dot{q}_m}$ and $k_{\delta_{em}}^{\dot{q}_m}$ denote the model parameters determined according to ADS-33, δ_{em} the manipulated input. Thus, q asymptotically converge to q_m as the adaptive laws are given by:

$$\begin{cases} \dot{k}_q^{\dot{q}} = -\rho_q \left(\kappa + e \right) q \text{ and } k_q^{\dot{q}} < 0, \\ \dot{k}_{\delta_e}^{\dot{q}} = -\rho_{\delta_e} \left(\kappa + e \right) \delta_{ei}, \\ \dot{k}_\chi^{\dot{q}} = \rho_\chi \left(\kappa + e \right) \chi, \\ \dot{k}_t^{\dot{q}} = \rho_t \left(\kappa + e \right), \end{cases} \quad (11)$$

where $\rho_q, \rho_{\delta_e}, \rho_\chi$ and ρ_t are greater than zero; χ represents u, v, w, p, r, δ_a and δ_c, respectively; the tracking error is defined as:

$$e = q_m - q \tag{12}$$

the time-varying parameter κ should satisfy the following constraint:

$$\text{sgn}(\kappa) = \text{sgn}\left(\dot{e} - k_q^{\dot{q}} e\right) \tag{13}$$

the generalized input is denoted by:

$$\delta_{ei} = k_{q_m}^{\dot{q}} q_m + k_{\delta_{em}}^{\dot{q}} \delta_{em} \tag{14}$$

3.2. Proof

Substituting Equation (14) into Equation (8) yields

$$\delta_e = k_u^{\dot{q}} u + k_v^{\dot{q}} v + k_w^{\dot{q}} w + k_p^{\dot{q}} p + k_r^{\dot{q}} r + k_{\delta_a}^{\dot{q}} \delta_a + k_{\delta_c}^{\dot{q}} \delta_c + k_t^{\dot{q}} + \delta_{ei} \tag{15}$$

Again the generalized plant shown in Equation (16) is derived by substituting Equation (15) into Equation (7).

$$\dot{q} = M_q^{\dot{q}} q + M_{\delta_e}^{\dot{q}} \delta_{ei} + \hat{M}_u^{\dot{q}} u + \hat{M}_v^{\dot{q}} v + \hat{M}_w^{\dot{q}} w + \hat{M}_p^{\dot{q}} p + \hat{M}_r^{\dot{q}} r + \hat{M}_{\delta_a}^{\dot{q}} \delta_a + \hat{M}_{\delta_c}^{\dot{q}} \delta_c + \hat{M}_t^{\dot{q}} \tag{16}$$

where $\hat{M}_\bullet^{\dot{q}} = M_\bullet^{\dot{q}} + M_{\delta_e}^{\dot{q}} k_\bullet^{\dot{q}}$.

Similarly, we have

$$\dot{q}_m = k_q^{\dot{q}} q_m + k_{\delta_e}^{\dot{q}} \delta_{ei} \tag{17}$$

Then, substituting Equations (16) and (17) into Equation (12) yields:

$$\dot{e} = k_q^{\dot{q}} e + b_q q + b_{\delta_{ei}} \delta_{ei} + b_u u + b_v v + b_w w + b_p p + b_r r + b_{\delta_a} \delta_a + b_{\delta_c} \delta_c + b_t \tag{18}$$

where

$$\begin{cases} b_q = k_q^{\dot{q}} - M_q^{\dot{q}}, \\ b_{\delta_{ei}} = k_{\delta_e}^{\dot{q}} - M_{\delta_e}^{\dot{q}}, \\ b_\bullet = -M_{\delta_e}^{\dot{q}} k_\bullet^{\dot{q}} - M_\bullet^{\dot{q}}, \end{cases} \tag{19}$$

Furthermore, the adaptive laws are chosen so that certain stability conditions based on Lyapunov theory are satisfied. Consider the Lyapunov function candidate

$$V = e^2 + \lambda_q b_q^2 + \lambda_{\delta_{ei}} b_{\delta_{ei}}^2 + \lambda_u b_u^2 + \lambda_v b_v^2 + \lambda_w b_w^2 + \lambda_p b_p^2 + \lambda_r b_r^2 + \lambda_{\delta_a} b_{\delta_a}^2 + \lambda_{\delta_c} b_{\delta_c}^2 + \lambda_t b_t^2 \tag{20}$$

where $\lambda_q, \lambda_{\delta_{ei}}, \lambda_u, \lambda_v, \lambda_w, \lambda_p, \lambda_r, \lambda_{\delta_a}, \lambda_{\delta_c}$ and λ_t are greater than zero. We can therefore conclude that $\dot{V}$ is negative definite as

$$\begin{cases} k_q^{\dot{q}} < 0, \\ \lambda_\chi \dot{b}_\chi = -\kappa\chi - e\chi, \\ \lambda_t \dot{b}_t = -\kappa - e, \end{cases} \tag{21}$$

in which χ represents $q, \delta_{ei}, u, v, w, p, r, \delta_a$ and δ_c, respectively. From Equations (19) and (21), we can easily deduce the adaptive laws given by Equation (11).

3.3. Remarks

i. The direct addition of an external feedback term like $a \cdot q$ to δ_e is necessary without any modification of the adaptive laws if $M_q^{\dot{q}}$ cannot satisfy $M_q^{\dot{q}} < 0$ at a certain flight condition, where a is used to guarantee the stability of the plant, that is, $M_q^{\dot{q}} + M_{\delta_e}^{\dot{q}} a < 0$ in this case.

ii. In consideration of the adverse effect of measurement noises, the variance of the noise component of δ_e can be reduced while using the proposed adaptive algorithm, mainly because the model feedback path is noise free, which can result in the improvement in the control quality of the stability augmentation system, especially in the case of low signal to noise ratios (SNR).

iii. Compared to the model reference adaptive control (MRAC) method, the additional time-varying parameter κ in the proposed algorithm can contribute to the convergence rates of adjustable parameters when the plant parameters vary rapidly.

4. The Improvement and Application of the Adaptive Laws

The adaptive laws are difficult to apply in practice because $\dot{q}$ is non-measurable in most cases. According to linear system theory, Equation (18) can be rewritten as follows:

$$\dot{e}_f = k_q^{\dot{q}} e_f + b_q q_f + b_{\delta_{ei}} \delta_{eif} + b_u u_f + b_v v_f + b_w w_f + b_p p_f + b_r r_f + b_{\delta_a} \delta_{af} + b_{\delta_c} \delta_{cf} + b_t t_f \tag{22}$$

where $e_f, q_f, \delta_{eif}, u_f, v_f, w_f, p_f, r_f, \delta_{af}, \delta_{cf}$ and t_f are respectively served as the outputs of an arbitrarily chosen filter $G_f(s)$ in response to $e, q, \delta_{ei}, u, v, w, p, r, \delta_a, \delta_c$

and $1(t)$. The adaptive laws, derived by using the similar Lyapunov function described in Equation (20), are rewritten as:

$$
\begin{cases}
\dot{k}_q^{\dot{q}} = -\rho_q \left(\kappa_f + e_f \right) q_f \text{ and } k_q^{\dot{q}} < 0, \\
\dot{k}_{\delta_e}^{\dot{q}} = -\rho_{\delta_e} \left(\kappa_f + e_f \right) \delta_{eif}, \\
\dot{k}_\chi^{\dot{q}} = \rho_\chi \left(\kappa_f + e_f \right) \chi_f, \\
\dot{k}_t^{\dot{q}} = \rho_t \left(\kappa_f + e_f \right),
\end{cases}
\tag{23}
$$

where κ_f satisfies the following constraint:

$$
\text{sgn} \left(\kappa_f \right) = \text{sgn} \left(\dot{e}_f - k_q^{\dot{q}} e_f \right)
\tag{24}
$$

The improved adaptive laws are therefore subject to a certain form of the filter. Assume that the filter transfer function is given by:

$$
G_f(s) = \frac{1}{Ts + 1}
\tag{25}
$$

where T is the time constant. In this case, Equation (24) can be written as:

$$
\text{sgn} \left(\kappa_f \right) = \text{sgn} \left\{ \frac{1}{T} \left(e - e_f \right) - k_q^{\dot{q}} e_f \right\}
\tag{26}
$$

which indicates that κ_f is available. Figure 2 shows the schematic diagram of the adaptive stability augmentation system with model feedback. Note that the proper choice of κ_f can speed up the convergence of adjustable parameters.

Note that $k_{q_m}^{\dot{q}}$ and $k_{\delta_{em}}^{\dot{q}}$ are completely dominant over other adjustable parameters in Equation (8), however the noise component of q_m can be ignored on the basis of the assumption that the manipulated input δ_{em} is noise free. For δ_e, the adaptive model feedback strategy can therefore lead to a marked loss in noises.

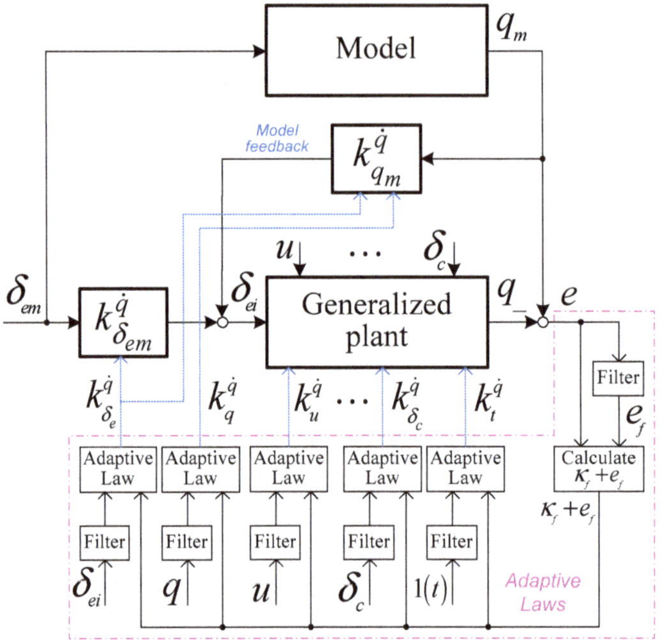

Figure 2. The schematic diagram of the adaptive stability augmentation system.

5. Flight Tests

In this section, the performance of the proposed algorithm is demonstrated for the stability augmentation system of the longitudinal axis. All pre-chosen parameters are given as follows:

(1)   The model parameters: $k_{q_m}^{\dot{q}_m} = -1.5$ and $k_{\delta_{em}}^{\dot{q}_m} = 2.25$;

(2)   The initial values of adjustable parameters: $k_u^{\dot{q}}(0), k_v^{\dot{q}}(0), k_w^{\dot{q}}(0), k_q^{\dot{q}}(0), k_r^{\dot{q}}(0),$ $k_{\delta_a}^{\dot{q}}(0), k_{\delta_c}^{\dot{q}}(0)$ and $k_t^{\dot{q}}(0)$ are assumed to equal zero, and $k_p^{\dot{q}}(0) = 0, k_{\delta_e}^{\dot{q}}(0) = 1.0$;

(3)   The time constant of the filter: $T = 0.02$;

(4)   The parameters of the adaptive control laws: $\rho_\bullet = 50$; $\kappa_f = 2.0 \left[\frac{1}{T} \left(e - e_f \right) - k_q^{\dot{q}} e_f \right]$; $a = -0.2$.

The proposed algorithm is first implemented in a numerical simulation based on the nonlinear helicopter model. It is then applied to the prototype UH to evaluate the flight performance of the stability augmentation system.

5.1. Task 1: Numerical Simulation

The numerical simulation test, shown in Figure 3, is conducted using the nonlinear dynamics model of the prototype unmanned helicopter, together with a simplified closed-form trim calculation. The manipulated input signal is generated from a joystick device. Various moderately aggressive maneuvers are conducted during the simulation to evaluate the performance of the proposed algorithm at different operate points of the flight envelope. A comparison is made between the output of the plant and that of the model, as shown in Figure 3a,b, which demonstrates that the tracking error can rapidly approach zero in 0.9 s and remain in a bounded range. We can therefore conclude that the proposed adaptive algorithm achieves a guaranteed model reference tracking performance and has a good convergence property. The tracking error remains in a bounded range even though the adjustable parameters vary rapidly during the test. However the conventional MRAC approach is not applicable to the case where the variations of the plant parameters are significant.

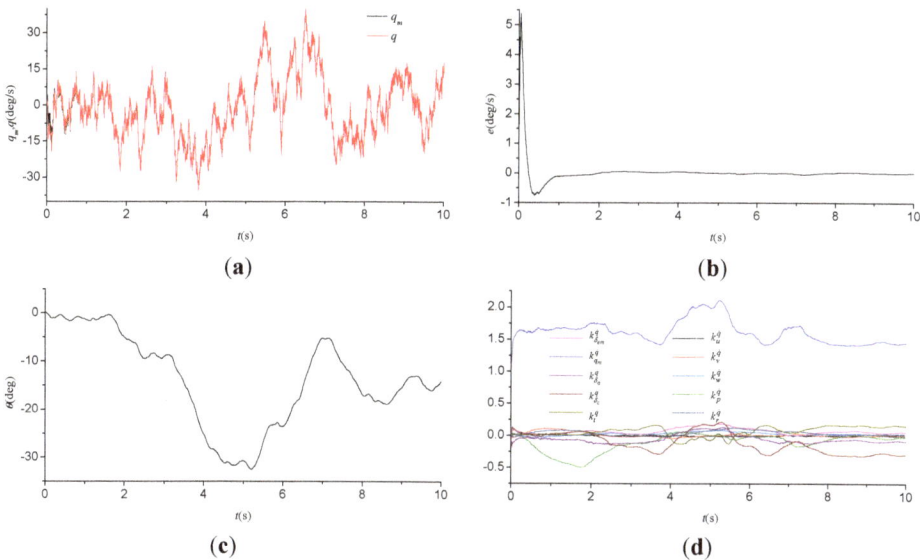

Figure 3. Numerical simulation results of the proposed algorithm. (**a**) q_m and q; (**b**) e; (**c**) θ; (**d**) Adjustable parameters.

5.2. Task 2: Flight Tests

The flight tests are conducted to compare the performance of the proposed algorithm with that of the conventional PID while those two approaches applies to the prototype unmanned helicopter stability augmentation system, respectively. For a fair comparison, the attitude and position controllers remain unchanged. Our control software system, which runs on a DSP-based hardware platform with a control period of 10 ms, can provide reliable support for high precision timer and synchronization operations. The sensors (including an AHRS unit and a DGPS unit)

265

are used to provide the information related to the flight status, such as the angles, angular rates, velocity and position. During flight, Wi-Fi and serial-based data links provide a link to the ground station computer that allows monitoring the real-time flight information and uploading remote control commands such as way points. Testing of the proposed adaptive algorithm begins with hover, followed by simple way-point navigation due to the limitations of the test conditions such as flight safety. In consideration of measurement noises of the sensors, the results from these flight tests are provided in Figures 4 and 5.

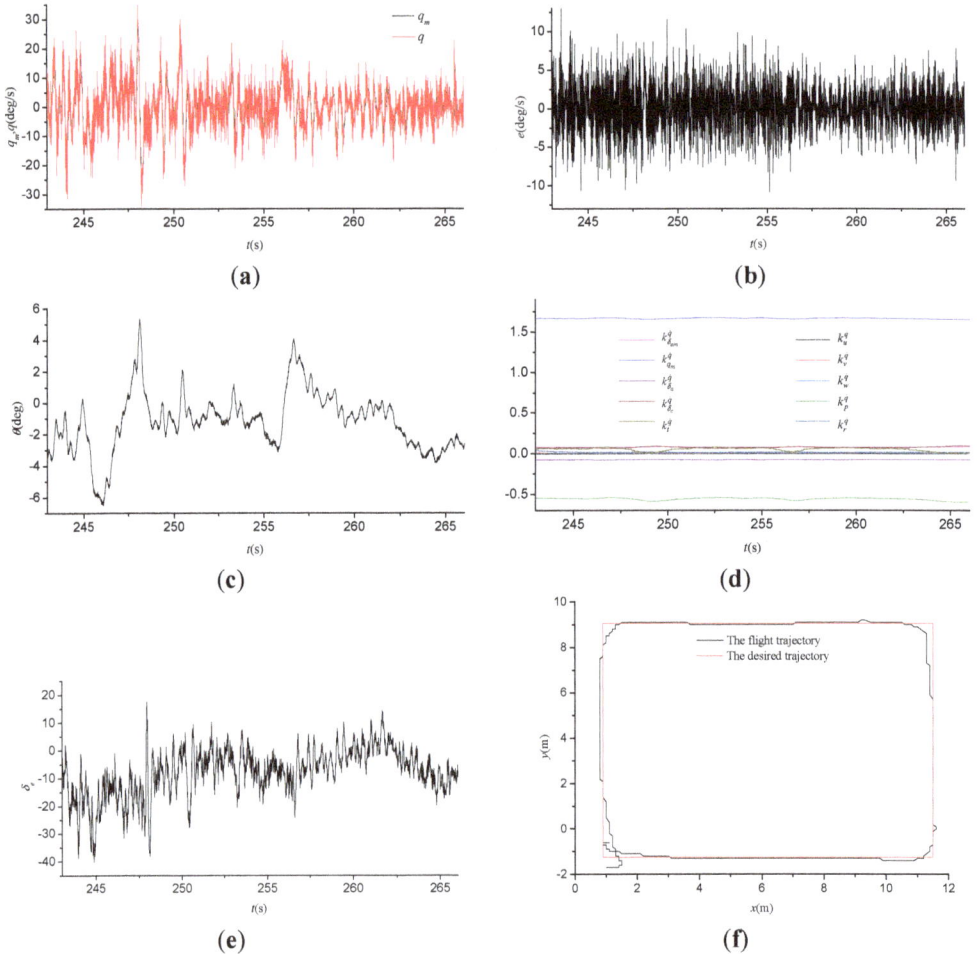

Figure 4. Flight test results of the proposed algorithm. (**a**) q_m and q; (**b**) e; (**c**) θ; (**d**) Adjustable parameters; (**e**) δ_e; (**f**) The ground track view of trajectory.

The performance of the adaptive stability augmentation system is first evaluated at low speeds where a square pattern is flown, as shown in Figure 4. The noisy tracking error, shown in Figure 4b, is also within an acceptable range. Moreover, the model feedback strategy in the proposed algorithm contributes to the restricted variation of the control signal δ_e shown in Figure 4e, which explicitly improves the external command position tracking performance, and implicitly demonstrates the improvement in the control quality of the stability augmentation system.

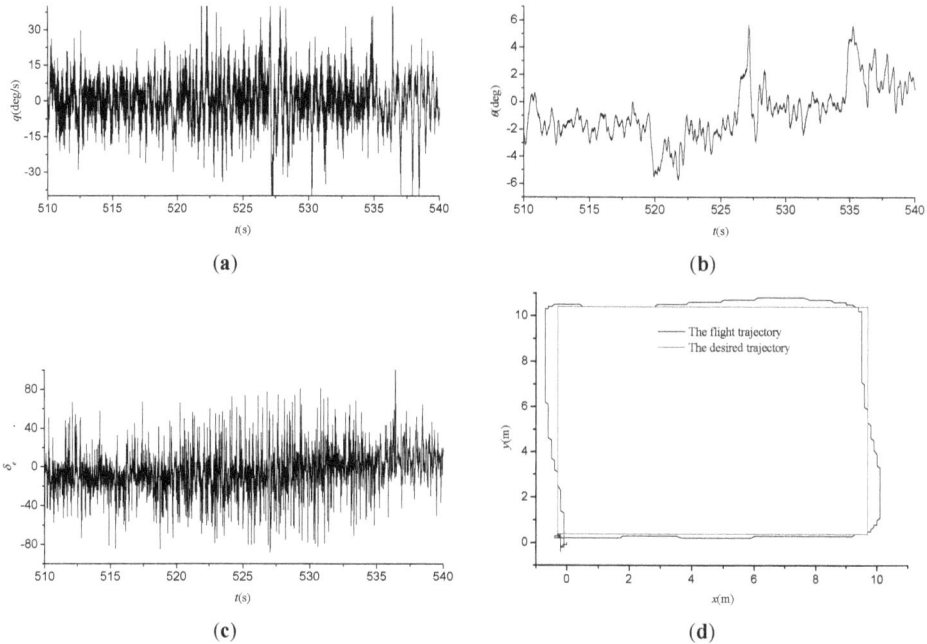

Figure 5. Flight test results of the PID controller. (**a**) q; (**b**) θ; (**c**) δ_e; (**d**) The ground track view of trajectory.

By comparison with the above-mentioned results, the flight test results of the conventional PID controller implemented through the widely used Pixhawk autopilot module, shown in Figure 5, appear to deteriorate. For instance, the variation of the control signal δ_e, together with the external command position tracking errors, increases when the prototype UH is commanded to perform the same flight mission.

Based on the flight test results, we can conclude that the proposed adaptive model feedback control algorithm is able to adapt to rapidly changing flight conditions and effectively enhance the performance of the prototype UH stability augmentation system in the presence of parametric uncertainties and measurement noises, which results in the improvement in flying qualities. More significantly, most existing designs would require accurate models at each point, while the proposed design does not.

6. Conclusions

This paper presents the adaptive model feedback control algorithm for the prototype UH stability augmentation system in the presence of parametric uncertainties and measurement noises. The proposed adaptive algorithm is able to achieve a guaranteed model reference tracking performance and speed up the convergence rates of adjustable parameters, even when the plant parameters vary rapidly. Moreover, the model feedback strategy in the proposed algorithm further contributes to the improvement in the control quality of the stability augmentation system in the case of low SNR. The flight test results have shown that the proposed algorithm can considerably improve the flying qualities of the prototype UH.

Acknowledgments: This study was supported in part by National Natural Science Foundation of China (NSFC) (Under Grant No. 61374188), Aeronautical Science Foundation of China (Under Grant No. 2013ZC52033), Natural Science Foundation of Jiangsu Province of China (Under Grant No. BK20141412).

Author Contributions: All authors discussed the contents of the manuscript. Shouzhao Sheng contributed to the research idea and the framework of this study. Chenwu Sun performed the experimental work.

Conflicts of Interest: The authors declare no conflict of interest.

References

1. Cai, G.; Chen, B.M.; Lee, T.H. *Unmanned Rotorcraft Systems*, 1st ed.; Springer: New York, NY, USA, 2011.
2. Ren, B.; Ge, S.S.; Chen, C.; Fua, C.H.; Lee, T.H. *Modeling, Control and Coordination of Helicopter Systems*, 1st ed.; Springer: New York, NY, USA, 2012.
3. Cook, M.V. *Flight Dynamics Principles: A Linear Systems Approach to Aircraft Stability and Control*, 3rd ed.; Butterworth-heinemann: Oxford, UK, 2012.
4. Mettler, B. *Identification Modeling and Characteristics of Miniature Rotorcraft*, 1st ed.; Springer: New York, NY, USA, 2011.
5. Shim, D.H.; Kim, H.J.; Sastry, S. Control System Design for Rotorcraft-Based Unmanned Aerial Vehicles Using Time-Domain System Identification. In Control Applications, Proceedings of the 2000 IEEE International Conference on, Anchorage, AK, USA, 25–27 September 2000; pp. 808–813.
6. Takahashi, M.; Schulein, G.; Whalley, M. Flight Control Law Design and Development for an Autonomous Rotorcraft. In Proceedings of the 64th Annual Forum of the American Helicopter Society, Montreal, Quebec, Canada, 29 April–1 May 2008; pp. 1652–1671.
7. How, J.; Bethke, B.; Frank, A.; Dale, D.; Vian, J. Real time indoor autonomous vehicle test environment. *IEEE Control Syst. Mag.* **2008**, *28*, 51–64.

8. Shim, D.H.; Kim, H.J.; Sastry, S. Decentralized Nonlinear Model Predictive Control of Multiple Flying Robots. In Decision and Control, Proceedings of 42nd IEEE Conference on, Maui, HI, USA, December 2003; pp. 3621–3626.

9. Buskey, G.; Wyeth, G.; Roberts, J. Autonomous Helicopter Hover Using an Artificial Neural Network. In Proceedings of the IEEE Conference on Robotics and Automation, Seoul, Korea, 21–26 May 2001; pp. 1635–1640.

10. Garcia, R.; Valavanis, K. The implementation of an autonomous helicopter testbed. *J. Intell. Robot. Syst.* **2009**, *54*, 423–454.

11. Smerlas, A.J.; Walker, D.J.; Postlethwaite, I.; Strange, M.E.; Howitt, J.; Gubbels, A.W. Evaluating H_∞ controllers on the NRC Bell 205 fly-by-wire helicopter. *Control Eng. Pract.* **2001**, *9*, 1–10.

12. Walker, D.J. Multivariable control of the longitudinal and lateral dynamics of a fly-by-wire helicopter. *Control Eng. Pract.* **2003**, *11*, 781–795.

13. Gadewadikar, J.; Lewis, F.; Subbarao, K.; Chen, B.M. Structured H-infinity command and control-loop design for unmanned helicopters. *J. Guid. Control Dynam.* **2008**, *31*, 1093–1102.

14. Dytek, Z.T. Adaptive Control of Unmanned Aerial Systems. Ph.D. Thesis, Massachusetts Institute of Technology, Cambridge, MA, USA, September 2010.

15. Krupadanam, A.S.; Annaswamy, A.M.; Mangoubi, R.S. Multivariable adaptive control design with applications to autonomous helicopters. *J. Guid. Control Dynam.* **2002**, *25*, 843–851.

16. Liu, Y.; Tao, G. Modeling and model reference adaptive control of aircraft with asymmetric damage. *J. Guid. Control Dynam.* **2010**, *33*, 1500–1517.

17. Dauer, J.C.; Faulwasser, T.; Lorenz, S.; Findeisen, R. Optimization-Based Feed Forward Path Following for Model Reference Adaptive Control of an Unmanned Helicopter. In Proceedings of the AIAA Guidance, Navigation, and Control Conference, Boston, MA, USA, 19–22 August 2013.

18. Rusnak, I.; Weiss, H.; Barkana, I. Improving the performance of existing missile autopilot using simple adaptive control. *Int. J. Adapt Control Signal Process.* **2014**, *28*, 732–749.

19. Neild, S.A.; Yang, L.; Wagg, D.J. A modified model reference adaptive control approach for systems with noise or unmodelled dynamics. *Proc. Inst. Mech. Eng., Part I: J. Syst. Control Eng.* **2008**, *222*, 197–208.

20. VanZwieten, T.; Gilligan, E.; Wall, J.; Orr, J.; Miller, C.; Hanson, C. Adaptive Augmenting Control Flight Characterization Experiment on an F/A-18. In Proceedings of American Astronautical Society Guidance & Control Conference, Breckenridge, CO, USA, 31 January–5 February 2014; Volume 1, pp. 1–17.

Numerical and Experimental Characterization of Fiber-Reinforced Thermoplastic Composite Structures with Embedded Piezoelectric Sensor-Actuator Arrays for Ultrasonic Applications

Klaudiusz Holeczek, Eric Starke, Anja Winkler, Martin Dannemann and Niels Modler

Abstract: The paper presents preliminary numerical and experimental studies of active textile-reinforced thermoplastic composites with embedded sensor-actuator arrays. The goal of the investigations was the assessment of directional sound wave generation capability using embedded sensor-actuator arrays and developed a wave excitation procedure for ultrasound measurement tasks. The feasibility of the proposed approach was initially confirmed in numerical investigations assuming idealized mechanical and geometrical conditions. The findings were validated in real-life conditions on specimens of elementary geometry. Herein, the technological aspects of unique automated assembly of thermoplastic films containing adapted thermoplastic-compatible piezoceramic modules and conducting paths were described.

Reprinted from *Appl. Sci.* Cite as: Holeczek, K.; Starke, E.; Winkler, A.; Dannemann, M.; Modler, N. Numerical and Experimental Characterization of Fiber-Reinforced Thermoplastic Composite Structures with Embedded Piezoelectric Sensor-Actuator Arrays for Ultrasonic Applications. *Appl. Sci.* **2016**, *6*, 55.

1. Introduction

Textile-reinforced thermoplastic composites (TRTC) show a high potential for serial manufacturing of innovative function-integrating lightweight constructions. Due to the textile structure, the layered construction, and the associated specific production processes, such materials enable the possibility for a matrix-homogeneous integration of functional elements such as sensors, actuators, or even electronic circuit boards [1–7]. The matrix-homogeneity is achieved by the utilization of identical thermoplastic materials both for the functional elements and the matrix of the composite component. Signals from integrated systems can be favorably utilized to broaden the application scope of TRTC through the implementation of auxiliary functions, e.g., active vibration damping, condition, or structural health monitoring [8,9]. State of the art solutions utilize conventional piezoelectric transducers, e.g., macro-fiber composites, or active fiber composites, which are mainly adhesively bonded to the structural components [10,11]. The associated assembly and bonding processes are characterized by several labor-extensive work steps. To successfully transform the current manual production process of active textile-reinforced thermoplastic composites into a fully automated mass production process, novel piezoelectric modules and adapted manufacturing technologies are necessary [12]. In the following studies an assembly method is presented, which is based on an automated thermal or ultrasonic fixing of functional elements on thermoplastic carrier films. The assembly process has been named *ePreforming* and the respective outcome—a functionalized film—is called the *ePreform* [12,13]. The *ePreforming* technology enables a direct coupling of the piezoceramic transducer to the host structure, which reduces deformation losses compared to adhesively-bonded transducers [10]. Therefore, the coupling efficiency should be higher for integrated than bonded piezoceramic actuators.

A vital prerequisite for this process is the use of thermoplastic-compatible piezoceramic modules (TPMs). These modules are based on a piezoceramic functional layer (wafer or fiber composite) enclosed between two thermoplastic carrier films that are metallized with electrode structures. Thereby, the TPM-carrier films and the matrix of the fiber-reinforced composite components are made from identical materials. The possibility for *ePreforming* the whole arrays of actuators and sensors enables new application fields of these active structures with a high potential for ultrasonic measuring tasks, like the measurement of flow rate or distance.

2. Phenomenological Description

Embedded piezoelectric transducer arrays are suitable for transmission of mechanical waves into adjacent media (radiation of sound) and reception of sound waves. This property can be used for ultrasound-based measuring tasks, condition monitoring, or structural health monitoring (SHM) applications [14–16]. For condition monitoring or SHM the influence of flaws, defects, or damage on the plate wave propagation characteristics is applied [17,18]. For ultrasound measurement tasks the plate waves are used for an efficient radiation of sound waves into the adjacent media, e.g., water or air [19]. This effect can also be reversed in order to receive the sound waves [20].

2.1. Ultrasound Radiation and Reception

The sound radiation and reception is based on the interaction of fluid and solid waves. Especially the flexural wave, also known as the first asymmetric Lamb wave mode (A_0-mode), can be efficiently used for sound radiation due to the high out-of-plane displacement amplitudes. An exemplary interaction between the waves in the plate and acoustic waves is shown in Figure 1.

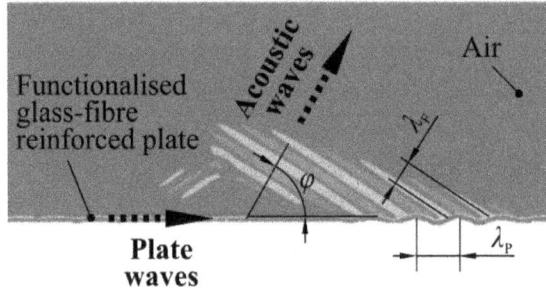

Figure 1. Radiation of the ultrasound wave from the functionalized Textile-reinforced thermoplastic composites (TRTC) plate at an angle φ.

The radiation of the Lamb waves into the adjacent fluid takes place at an angle φ which results from the ratio of the wavelength in the fluid λ_F to the wavelength of the Lamb wave λ_P. This angle can be calculated either from the wavelengths or from the velocities of the waves using the following equation:

$$\phi = \arccos\frac{\lambda_F}{\lambda_P} = \arccos\frac{c_F}{c_P} \tag{1}$$

wherein c_F is the speed of sound in the fluid and c_P is the phase velocity of the flexural wave in the plate. Since the phase velocity is not only a function of the mechanical material properties but also of the excitation frequency, the radiation angle can be set by modifying the excitation signal characteristics of the structure-integrated actuators. In order to scrutinize the procedure to set the radiation angle a case study employing the mechanical properties of the analyzed material is presented below.

Glass fiber-reinforced polyamide 6 (GF/PA6) plate with the thickness of 2 mm is studied. The dispersion curves—describing the relation between the wave speed and the excitation frequency—can be calculated analytically based on mathematical relation presented in various textbooks (see e.g., [21]) by assuming isotropic material properties (Young's modulus E = 20 GPa, mass density ρ = 1800 kg/m³, Poisson's ratio ν = 0.3). Figure 2 shows the dispersion curves with the phase velocities and the group velocities of the A_0- and the S_0-modes.

The wavelength calculated from the phase velocity of the A_0-mode is shown in Figure 3A as a function of the excitation frequency. For the radiation in the air (c_F = 340 m/s), the radiation angle is shown in Figure 3B.

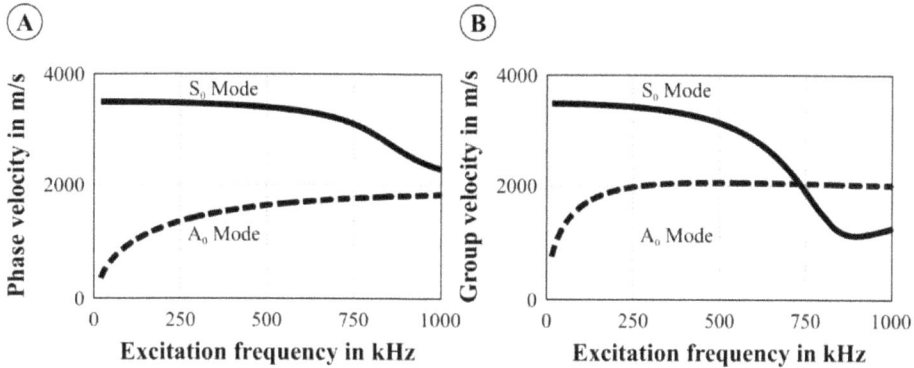

Figure 2. Dispersion diagrams for a GF/PA6 plate. (A) Phase velocity; (B) Group velocity.

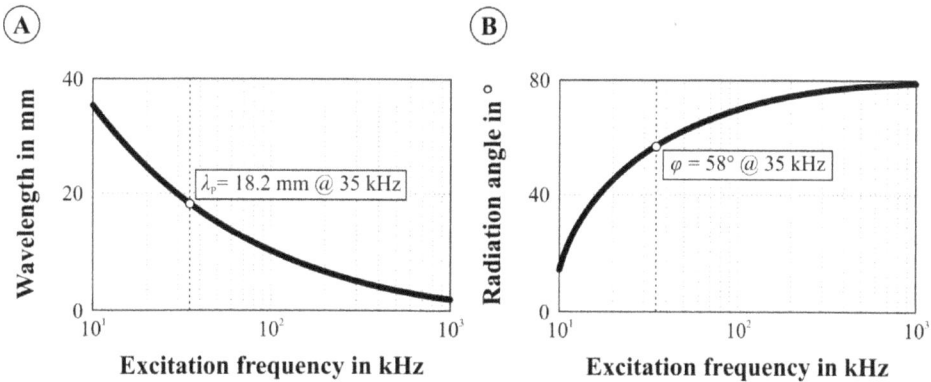

Figure 3. Determination of the radiation angle. (A) Wavelength of the A_0-mode in a 2 mm thick GF/PA6 plate; (B) Radiation angle of sound waves in air.

For example, at a frequency of 35 kHz the A_0 Lamb wave phase velocity is 640 m/s and the resulting wavelength of the flexural wave is 18.2 mm. The wavelength in air is 9.7 mm which leads to a radiation angle of 58°. As Figure 3 shows, this angle increases with increasing frequency. The angle can be adjusted to specific applications either by adapting the material and the thickness of the plate or by changing the excitation frequency.

2.2. Transducer Arrays for the Generation of Directional Waves in a Plate

For some measurement applications, especially when localization of obstacles is of key interest, it is advantageous to directionally send or receive sound waves. The first step to obtain directional ultrasound radiation is the generation of directional flexural waves in the plate. This can be achieved by using transducer arrays instead of a single transducer. Through adaptation of the actuator array configuration, the mainly excited or received wave mode can be precisely controlled [22]. Furthermore, the directivity of the waves can also be adjusted [23,24]. The simplest setup is a one-dimensional array on the surface of the plate. Depending on the size and the distance between the individual transducers and the time-delays in the electrical driving signal, the directional characteristics can be adjusted, for example, to amplify the wave generation in one direction. It is also possible to use electronic beam-steering (see e.g., [25]) in order to vary the direction of the dominant wave generation.

For a one-dimensional array the optimal delay time for an additive superposition in the array longitudinal direction can be analytically calculated using the transducer offset Δl and the phase velocity c_p. In the case of a GF/PA6 plate and a 15 mm offset between the actuators the optimal time delay yields [26]:

$$\Delta t = \frac{\Delta l}{c_p} = \frac{15 \text{ mm}}{640 \text{ ms}^{-1}} = 23.4 \text{ µs} \tag{2}$$

3. Application of Integrated Transducer Arrays for Generation of Sound Waves

In order to confirm the capability to apply the integrated transducer arrays for effective generation of sound waves, firstly, a numerical model has been elaborated, where the possibility to generate a directional wave is analyzed. These investigations were followed by the experimental investigations where, firstly, the manufacturing process of the active composite is described, followed by the characterization of the waves in the plate using a laser scanning vibrometer and the measurement of the resulting sound waves in the air using a microphone.

3.1. Numerical Investigations

A three-dimensional parametrical model consisting of two sub-models: a free-free supported plate and attached to it four piezoelectric transducers (transducer array), was elaborated (see Figure 4). The model has been created using commercially available finite-element software [27]. To realistically simulate the interaction between the electrical driving signal of the transducers and the resulting solid waves, a coupled electromechanical problem has to be solved. Therefore the transducers were modeled as 3-D, 20-node, coupled-field solid elements (SOLID 226). The elastic base structure has been meshed with 116,568 elements of type SOLID 186 with a maximum element size of 0.2 mm.

In order to capture the time- and space-dependent phenomena connected with the propagation of mechanical waves, a transient analysis with the time step of 2.5 µs has been performed. Herein, the termination time was set to 500 µs which results in

200 load steps per one simulation. The mechanical properties of the base structure are presented in Table A1.

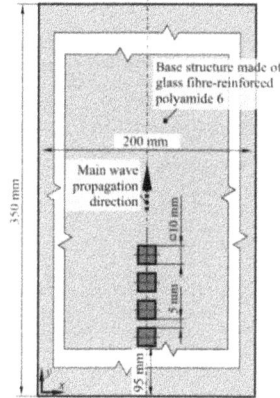

Figure 4. General overview of the elaborated model with geometrical interpretation of the modifiable parameters.

The transducers were driven by a Hann-windowed sine signal with 35 kHz center frequency and amplitude A equal to 6 V (Figure 5). The signal characteristics have been selected based on the technical specifications of the existing devices, planned to be used in the experimental investigations.

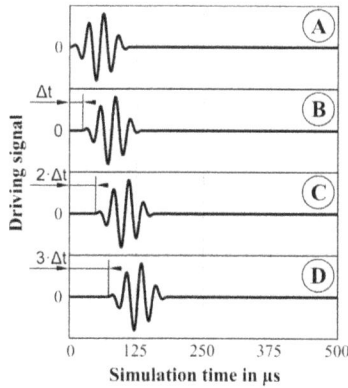

Figure 5. Transducer driving signals. (**A**) For the first transducer within the array. (**B**) For the second transducer within the array. (**C**) For the third transducer within the array. (**D**) For the fourth transducer within the array.

In order to generate directional waves in the plate, the transducers have to be activated in predefined moments in order to amplify the desired plate mode as well

275

as to shape the directional characteristics of the propagating wave. The geometrical configuration together with the material properties of the base structure governs the optimal time delay (Δt) which in the presented study equals 23.4 µs (see Section 2).

Since the analyzed phenomena are time- and location-dependent, a simple and informative presentation of the results is greatly limited. In order to visualize the wave field, snapshots at some time step have been created. For the results presentation, the wave field for one time step namely at 20 µs has been selected based on the following criteria:

- the wave has to propagate through some distance in order to reveal its directivity and plate mode character; and
- the reflections of higher Lamb modes—if any present—from the plate edges should not interfere with the main wave front;

The obtained results (Figure 6) confirmed that through application deliberately-activated structure-integrated transducers organized into a linear array the generation of directional wave is possible. It is clearly visible that the wave amplitudes along the main wave propagation direction are amplified in comparison with those propagating in the other direction. It can be observed that some side lobes are also present. In order to confirm these observations experimental investigations on nominally-identical structures have been conducted.

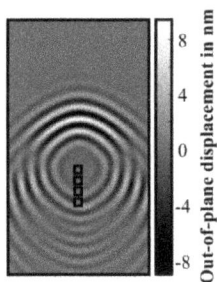

Figure 6. Out-of-plane displacements of the plate wave generated by a transducer array at t = 20 µs. For the animation of the wave field please refer to supplementary material.

3.2. Manufacturing Process

For investigations regarding the application of structure-integrated transducers for sound wave generation, manufacturing of the active textile-reinforced thermoplastic composites (TRTC), the application of thermoplastic-compatible piezoceramic modules (TPMs) on the surface of a fiber-reinforced semi-finished plate (organic sheet) was necessary. The organic sheet consists of a glass fiber-reinforced polyamide 6 (type: 102-RG600(x)/47% Roving Glass—PA 6 Consolidated Composite

Laminate produced by Bond-Laminates GmbH). The *ePreform* consists of a polyamide 6 (PA 6) film with a thickness of 100 µm. In regard to the built up of the TPM, commercially available monolithic wafers (type: PZT 5A1, electromechanical properties can be found in Table A2) with a square area of 100 mm² and a thickness of 0.2 mm were used (see Figure 7B). The outer surfaces of the wafers were metallized by silver printed electrodes and contacted by conductive copper tapes. This layup was embedded into two PA 6 carrier films with a thickness of 100 µm (see Figure 7A). The TPMs are polarized in the thickness direction and work in the d_{31} mode, which means a principal deflection normal to the polarization direction.

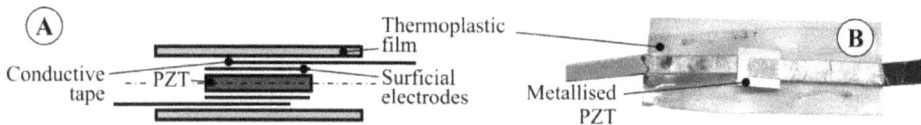

Figure 7. Prototypic thermoplastic-compatible piezoceramic modules (TPM) configuration. (**A**) Built up; (**B**) Consolidated TPM (d_{31} mode).

A major precondition for the embedding of the TPMs into the composite structures is the use of identical thermoplastic materials for the TPM carrier films and the matrix of the fiber-reinforced structure. During the consolidation, the module will be matrix homogeneous integrated into the composite structure. Compared to adhesively-bonded modules the integration of TPM enables an efficient coupling of the piezoceramic layer to the reinforcement as shown in Figure 8.

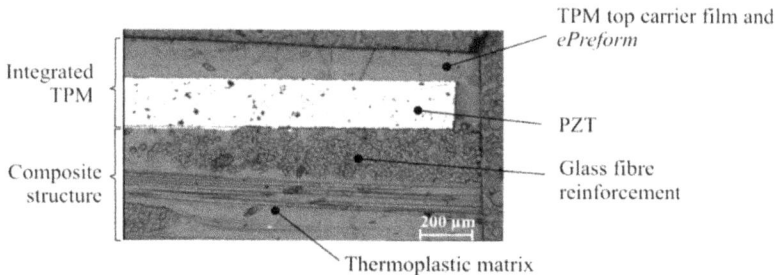

Figure 8. Micrograph of a matrix homogeneous integrated TPM.

The conceptual manufacturing process for active TRTC, which bases on a press technology, starts with the first process step, the so-called *ePreforming process*. It comprises the rollup and cutting of a thermoplastic film, the precise positioning, and fixing of TPMs, the application of conductive paths and the accompanied electrical contacting (see Figure 9).

277

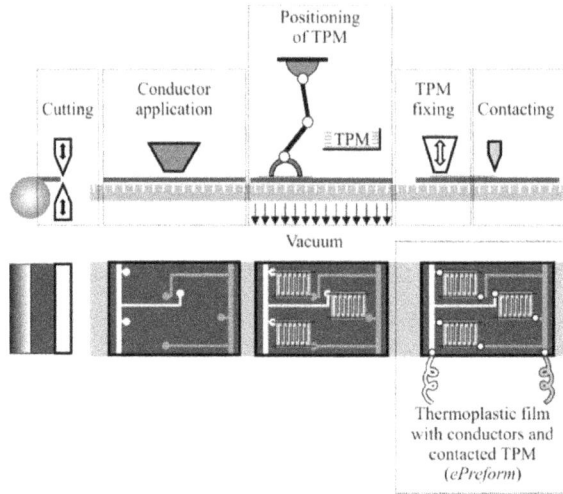

Figure 9. Schematic of the *ePreforming* process.

In the developed manufacturing process (see Figure 10) the *ePreform* is positioned in the pressing die and covered by a preheated and melted organic sheet and, subsequently, the press closes. During the consolidation the TPMs are simultaneously polarized [28].

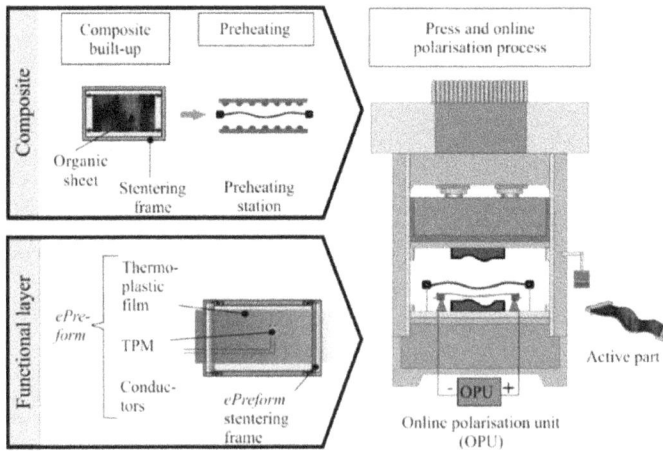

Figure 10. Schematic manufacturing process of active TRTC.

In regard to the manufacturing of the *ePreform* an especially-developed *ePreforming* unit was used in order to assure the automated assembly of thermoplastic films with TPM and the necessary conductors. The transfer of TPM from storage to the predefined position is realized by a vacuum gripper, whereas the fixation to the thermoplastic film can be realized by thermal stapling or ultrasonic welding. In this investigation the TPM were fixed by thermal stapling. The leads are automatically rolled up from the wire coil, fixed by thermal welding points, and cut at the end of the conductive paths. In these studies the leads consists of tin-coated copper wires with a diameter of 0.21 mm. Furthermore the TPMs, shown in Figure 11, are arranged to a linear pattern of six elements which have an offset of the piezoceramic wafers of 15 mm (see Figure 11).

For the initial prototypic tests the *ePreform* was positioned and fixed at the organic sheet plate manually using thermal resistant polyimide tape. The plate had dimensions of 1000 mm length, 600 mm width, and 2 mm thickness, and the TPM pattern was positioned 250 mm from the short side in the middle of the plate. In contrast to the introduced manufacturing process of such active parts, the consolidation of the investigated plate was performed by an autoclave process because of its prototypic configuration. The main process parameters are a maximum temperature of 230 °C, a dwell time at maximum temperature for 2 min, a consolidation pressure of 5 bar, and a vacuum of 20 mbar.

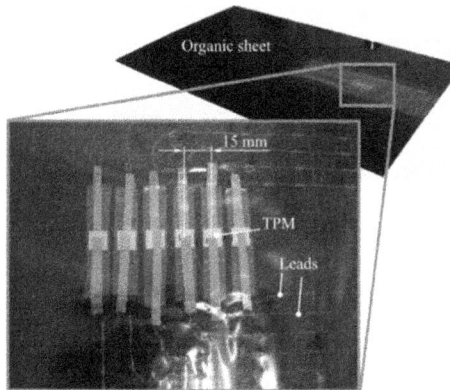

Figure 11. Organic sheet, *ePreform* with TPM pattern, and conductors (thermally stapled).

3.3. Experimental Investigations

The aim of the experimental studies was to characterize the generation of a directed acoustic wave using the integrated actuator array. Herein, two experimental techniques were utilized, *i.e.* laser Doppler vibrometry (LDV) and microphone measurements in order to identify the actuator-induced wave propagation of the investigated plate and the corresponding acoustic wave.

3.3.1. Laser Doppler Vibrometry

To assess the mechanical wave induced by the actuators, a series of experiments using the scanning laser Doppler vibrometer (type PSV-400 produced by Polytec GmbH, Waldbronn, Germany) were conducted. The application of a contactless measurement system guarantees that the mechanical properties of the investigated object are not distorted by the additional mass of typical vibration sensors.

Since the output signal of the laser scanning vibrometer is directly proportional to the velocity of the targeted surface along the laser beam direction, the LDV has been positioned normal to the analyzed plate. In order to assure to high reflection of the laser light and, hence, a high signal-to-noise ratio, a large section on the plate surface was covered with a reflective spray paint. A regular spatial distribution of 55×97 discrete measuring points in approximately 2 mm distance has been selected to guarantee reliable capturing of the wave's spatial and time development. The investigated plate was hung vertically using a thin light rope (Figure 12).

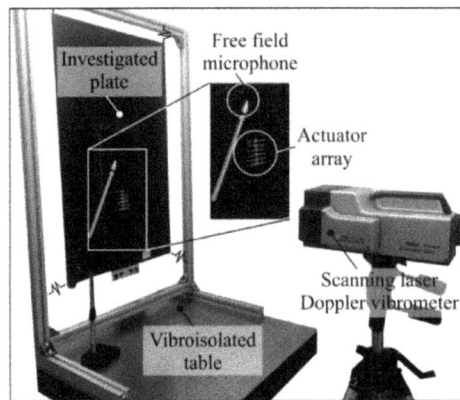

Figure 12. Experimental setup for the determination of propagation of induced mechanical and acoustical waves.

The actuators were excited with a typical burst signal as it can be used for different measurement tasks. Based on the plate dimensions and the array setup a center frequency of 35 kHz was chosen. The signal consists of a four-periods-long sinusoidal signal which was windowed with a Hann window to limit the bandwidth. The triggered 2.5 ms long time series of out-of-plane velocities were recorded 20 times in every point and subsequently averaged to reduce the noise content in the acquired signals. The time resolution of the LDV was set to 4.883 μs to assure at least five samples per measured signal period.

3.3.2. Microphone Measurements

Due to the excitation of the integrated actuator-array mechanical flexural plate waves were generated. In regard to the achievement of a directed radiation of plate waves, the excitations of the transducers were delayed 24 μs relatively to the previous transducer. In regard to the directional characteristic of the acoustic wave and the defined radiation angle, the spatial distribution of sound pressure was recorded. For this purpose a ¼″ free-field microphone (MK301, Microtech Gefell, Gefell, Germany) was used and the sound pressure was determined at different heights (10 mm, 30 mm, 70 mm) over the plate. The raster of the discrete measuring points in each height was set to 20 mm. Figure 13 shows the planar coordinate system in the plate plane, whereupon the origin was set to the end of the first transducer element.

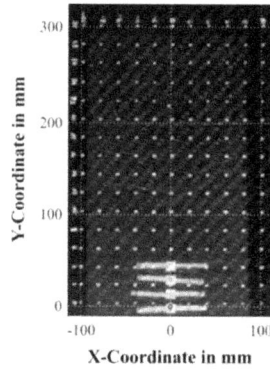

Figure 13. Coordinate system for the microphone measurement (origin at the first transducer of the array).

The recorded time series of out-of-plane velocities and sound pressures obtained using the laser Doppler velocimetry technique and microphone provide a basis for the subsequent analysis regarding the mechanical and acoustic wave propagation. The results of this analysis are presented in this section.

3.3.3. Propagation of Structural Mechanical Waves

Mainly, the impact of the time delay between the actuator initialization has been studied. The wavefront for cases with zero time delay and optimal time delay are presented in Figure 14. In this context, an optimal time delay of 24 μs was experimentally determined. The difference of 0.6 μs between theoretical and experimental delay times (see Section 2) is affected by uncertainties and tolerances of the geometry and the material properties of the experimental setup.

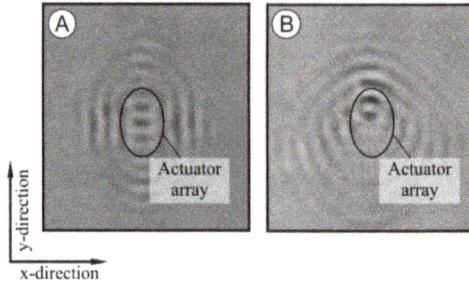

Figure 14. Out-of-plane velocities at $t = 20$ µs for different time delays. (**A**) for zero seconds; (**B**) for 24 ms.

It is clearly observable that, while for the zero time delay the wave propagates equally in positive and negative y-direction, for the case with the optimized time delay, the wave propagates mainly in the positive y-direction. Additionally, the wave amplitude of the latter case was twice as much as in the first case. In order to generate evidence, that such wavefronts cause a directed sound wave, the signals recorded using the microphone were analyzed.

3.3.4. Propagation of Acoustic Waves

Figure 15 shows the maximum sound pressure of the received bursts for the scanned heights. Herein, the results for the plane 10 mm above the structure show an obvious maximum of about 90 dB at the end of the array. In a height of 70 mm the sound pressure level is about 82 dB to 84 dB and extended over a larger area. The measured data fits quite well to the expected radiation angle of approximately 60°.

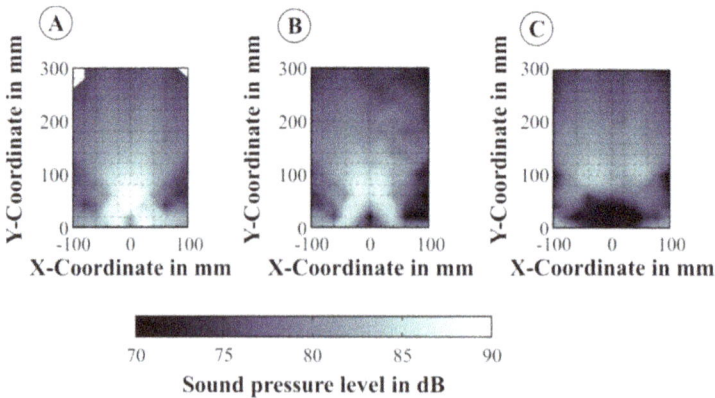

Figure 15. Measured maximum sound pressure level of the radiated ultrasound burst at different heights. (**A**) 10 mm above the plate; (**B**) 30 mm above the plate; (**C**) 70 mm above the plate.

282

On closer examination of the results, the lateral emission of ultrasound (side lobes) is recognizable. This is caused by the side lobes of the plate waves discussed in the previous section. These side lobes can be suppressed or prevented by a modified array design.

4. Conclusions and Outlook

The integration of sensors and actuators like piezoceramic modules exhibits a high potential to functionalize composite materials. Especially, a manufacturing technology ready for serial production can reduce the costs of active structures and, thus, lead to a wider range of possible applications. The developed *ePreforming* technology gives the possibility to integrate sensors and actuators into composites using established manufacturing technologies, like press processes. Furthermore, the process enables the integration of sensor-actuator arrays with reproducible positioning. Both possibilities could be validated in the performed manufacturing studies. Moreover, the robustness of the *ePreforming* technique, in terms of fatigue behavior in comparison with conventionally-bonded functional elements, will be analyzed in further studies.

Considering the functional testing, experimental studies were done, showing the high potential of piezoelectric transducer arrays for an angular radiation of ultrasound waves. In the future, this effect can be used to realize new material-integrated concepts for measuring distances or flow rates. Additionally, the developed concept can be applied for SHM applications based on the analysis of Lamb wave scatter on discontinuities or cracks.

Supplementary Materials: The following are available online at www.mdpi.com/2076-3417/6/3/55/s1, Video S1: Animation of the wave field for analyzed structure.

Acknowledgments: The presented work is part of the research within the context of the Collaborative Research Centre/Transregio (SFB/TR) 39 PT-PIESA, subproject B04, and the Collaborative Research Centre (SFB) 639, subproject D3. The transducers and transducer arrays were kindly provided by the subprojects A05 and T03 of the SFB/TR39. The authors are grateful to the Deutsche Forschungsgemeinschaft (DFG; eng. German Research Foundation) for the financial support of the SFB/TR39 and the SFB639.

Author Contributions: Klaudiusz Holeczek conceived and wrote this article as well as analyzed the experimental data; Eric Starke setup the numerical model as well as performed the simulations; Anja Winkler designed and conducted the manufacturing of all structures used throughout the studies and performed the mechanical characterization; Martin Dannemann and Niels Modler supervised and coordinated the investigations as well as performed checkup of the manuscript's logical structure.

Conflicts of Interest: The authors declare no conflict of interest.

Appendix A

Table A1. Mechanical material parameters of the TEPEX® 102-RG600(x)/47% tested according to DIN EN ISO 527-4 and DIN EN ISO 14129 standards or provided by the manufacturer in [29].

Property	Symbol	Value and Unit
Young's Modulus	$E_\parallel$	19.83 GPa
	$E_\perp$	19.37 GPa
Shear Modulus	$G_\#$	7.3 GPa
Poisson's ratio	$\upsilon_{\parallel\perp}$	0.17
Density	ρ	1.8 g/cm^3

Table A2. Material parameters of the PIC181 calculated on the basis of the manufacturer data [30].

Property	Symbol	Value and Unit
Density	ρ	7.8 g/cm^3
Young's Modulus	E_{11}	122 GPa
	E_{21}	57 GPa
	E_{31}	54 GPa
	E_{22}	122 GPa
	E_{23}	54 GPa
	E_{33}	103 GPa
	E_{44}	34 GPa
	E_{55}	33 GPa
	E_{66}	33 GPa
Permittivity in vacuum	ε_0	8.85×10^{-12} F/m
Relative permittivity	ε^T_{11}	413
	ε^T_{33}	877
Piezoelectric stress coefficients	e_{31}	-7.1 C/m^2
	e_{33}	14.4 C/m^2
	e_{15}	15.2 C/m^2

References

1. Dumstorff, G.; Paul, S.; Lang, W. Integration without disruption: The basic challenge of sensor integration. *Sens. J. IEEE* **2014**, *14*, 2102–2111.
2. Kinet, D.; Mégret, P.; Goossen, K.W.; Qiu, L.; Heider, D.; Caucheteur, C. Fiber Bragg grating sensors toward structural health monitoring in composite materials: Challenges and solutions. *Sensors* **2014**, *14*, 7394–7419.

3. Reddy, J.N. On laminated composite plates with integrated sensors and actuators. *Eng. Struct.* **1999**, *21*, 568–593.

4. Zysset, C.; Kinkeldei, T.W.; Münzenrieder, N.; Cherenack, K.; Tröster, G. Integration method for electronics in woven textiles. *Compon. Packag. Manuf. Technol. IEEE Trans.* **2012**, *2*, 1107–1117.

5. Pfeifer, G.; Starke, E.; Fischer, W.J.; Roscher, K.U.; Landgraf, J. Sensornetzwerke in faserverstärkten Verbundwerkstoffen mit drahtloser Signalübertragung. *Z. Kunststofftechnik/J. Plast. Technol.* **2007**, *5*, 1–15. (In German).

6. Roscher, K.-U.; Grätz, H.; Heinig, A.; Fischer, W.-J.; Pfeifer, G.; Starke, E. Integrated sensor network with event-driven activation for recording impact events in textile-reinforced composites. In Proceedings of the 2007 IEEE Sensors, Atlanta, GA, USA, 28–31 October 2007; pp. 128–131.

7. Hufenbach, W.; Adam, F.; Körner, I.; Winkler, A.; Weck, D. Combined joining technique for thermoplastic composites with embedded sensor networks. Available online: http://jim.sagepub.com/ content/early/2013/01/03/1045389X12471870 (accessed on 13 February 2016).

8. Kostka, P.; Holeczek, K.; Filippatos, A.; Langkamp, A.; Hufenbach, W. *In situ* integrity assessment of a smart structure based on the local material damping. *J. Intel. Mater. Syst. and Struct.* **2013**, *24*, 299–309.

9. Kostka, P.; Holeczek, K.; Hufenbach, W. Structure-integrated active damping system: Integral strain-based design strategy for the optimal placement of functional elements. *Int. J. Compos. Mater.* **2013**, *3*, 53–58.

10. Hufenbach, W.; Gude, M.; Heber, T. Embedding *versus* adhesive bonding of adapted piezoceramic modules for function-integrative thermoplastic composite structures. *Compos. Sci. Technol.* **2011**, *71*, 1132–1137.

11. Moharana, S.; Bhalla, S. Influence of adhesive bond layer on power and energy transduction efficiency of piezo-impedance transducer. *J. Intell. Mater. Syst. Struct.* **2014**, *26*, 247–259.

12. Hufenbach, W.; Gude, M.; Heber, T. Development of novel piezoceramic modules for adaptive thermoplastic composite structures capable for series production. In 22nd Eurosensors Conference, Dersden, Germany, 7–10 September 2008; pp. 22–27.

13. Hufenbach, W.; Modler, N.; Winkler, A. Sensitivity analysis for the manufacturing of thermoplastic e-preforms for active textile reinforced thermoplastic composites. *Procedia Mater. Sci.* **2013**, *2*, 1–9.

14. Raghavan, A.; Cesnik, C.E.S. Review of guided-wave structural health monitoring. *Shock. Vib. Dig.* **2007**, *39*, 91–114.

15. Staszewski, W.J.; Lee, B.C.; Mallet, L.; Scarpa, F. Structural health monitoring using scanning laser vibrometry: I. Lamb wave sensing. *Smart Mater. Struct.* **2004**, *13*, 251–260.

16. Holeczek, K.; Kostka, P.; Modler, N. Dry friction contribution to damage-caused increase of damping in fiber-reinforced polymer-based composites. *Adv. Eng. Mater.* **2014**, *16*, 1284–1292.

17. Kostka, P.; Holeczek, K.; Hufenbach, W. A new methodology for the determination of material damping distribution based on tuning the interference of solid waves. *Eng. Struct.* **2015**, *83*, 1–6.

18. Holeczek, K.; Dannemann, M.; Modler, N. Auslegung reflexionsfreier Dämpfungsmaßnahmen für mechanische Festkörperwellen. In Proceedings of the VDI-Fachtagung Schwingungsdämpfun, Leonberg, Germany, 22–23 September 2015; pp. 155–170. (In German).

19. Brekhovskikh, L. *Waves in Layered Media*, 2nd ed.; Elsevier Science: Oxford, UK, 1980.

20. Kunadt, A.; Pfeifer, G.; Fischer, W.J. Ultrasound flow sensor based on arrays of piezoelectric transducers integrated in a composite: Materials Science Engineering, Symposium B6—Hybrid Structures. *Procedia Mater. Sci.* **2013**, *2*, 160–165.

21. Rose, J.L. *Ultrasonic Waves in Solid Media*; Cambridge Univ. Press: Cambridge, UK, 2004.

22. Li, J.; Rose, J.L. Implementing guided wave mode control by use of a phased transducer array. *Ultrason. Ferroelectr. Freq. Control, IEEE Trans.* **2001**, *48*, 761–768.

23. Yu, L.; Giurgiutiu, V. *In-situ* optimized PWAS phased arrays for Lamb wave structural health monitoring. *J. Mech. Mater. Struct.* **2007**, *2*, 459–487.

24. Giurgiutiu, V. Tuned Lamb Wave Excitation and Detection with Piezoelectric Wafer Active Sensors for Structural Health Monitoring. *J. Intell. Mater. Syst. Struct.* **2005**, *16*, 291–305.

25. Moulin, E.; Bourasseau, N.; Assaad, J.; Delebarre, C. Lamb-wave beam-steering for integrated health monitoring applications. In Proceedings of the Nondestructive Evaluation and Health Monitoring of Aerospace Materials and Composites, Gyekenyesi, San Diego, CA, USA, March 2003; pp. 124–131.

26. Giurgiutiu, V. *Structural Health Monitoring with Piezoelectric Wafer Active Sensors*; Elsevier/Academic Press: Amsterdam, The Netherlands, 2008.

27. ANSYS, Release 15.0; Software for Engineering Analysis. ANSYS® Academic Research, Pittsburgh, Pennsylvania, USA, 2013.

28. Winkler, A.; Modler, N. Online poling of thermoplastic-compatible piezoceramic modules during the manufacturing process of active fiber-reinforced composites. *Mater. Sci. Forum*, **2015**, *825–826*, 787–794.

29. Bond-Laminates GmbH. Material Data Sheet: Tepex® dynalite 102-RG600(x)/47% Roving Glass–PA 6 Consolidated Composite Laminate, 2014. MatWeb: Material property data. Available online: http://bond -laminates.com/uploads/tx_lxsmatrix/ MDS_102-RG600_x_-47_.pdf (accessed on 13 February 2016).

30. Physik Instrumente (PI) GmbH & Co. KG. Available online: http://www.piceramic.com/ download/ PI_Ceramic_Material_Data.pdf (accessed on 13 February 2016).

Acoustic Emissions to Measure Drought-Induced Cavitation in Plants

Linus De Roo, Lidewei L. Vergeynst, Niels J.F. De Baerdemaeker and Kathy Steppe

Abstract: Acoustic emissions are frequently used in material sciences and engineering applications for structural health monitoring. It is known that plants also emit acoustic emissions, and their application in plant sciences is rapidly increasing, especially to investigate drought-induced plant stress. Vulnerability to drought-induced cavitation is a key trait of plant water relations, and contains valuable information about how plants may cope with drought stress. There is, however, no consensus in literature about how this is best measured. Here, we discuss detection of acoustic emissions as a measure for drought-induced cavitation. Past research and the current state of the art are reviewed. We also discuss how the acoustic emission technique can help solve some of the main issues regarding quantification of the degree of cavitation, and how it can contribute to our knowledge about plant behavior during drought stress. So far, crossbreeding in the field of material sciences proved very successful, and we therefore recommend continuing in this direction in future research.

Reprinted from *Appl. Sci.* Cite as: De Roo, L.; Vergeynst, L.L.; De Baerdemaeker, N.J.F.; Steppe, K. Acoustic Emissions to Measure Drought-Induced Cavitation in Plants. *Appl. Sci.* **2016**, *6*, 71.

1. Introduction

Stress in materials or structures is often accompanied by the built up of mechanical pressures, which, upon release, lead to elastic wave propagation away from the stressed zone [1]. These waves are called acoustic emissions. Today, the acoustic emission technique is widely applied for material testing and structural health monitoring on engineering materials such as concrete [2–4], metal alloys [5] and fiber composite materials [6]. However, the oldest reported scientifically planned acoustic emission experiment avant la lettre was performed on wood in 1933 by Fuyuhiko Kishinouye [3,7,8], long before the term acoustic emission (AE) was introduced by Schofield in 1961 [3]. In 1966, Milburn and Johnson used a similar measurement set-up as Kishinouye to detect for the first time AE signals in plants, when they were subjected to dehydration.

Nowadays, drought associated with global warming gains increasing interest. How plants cope with drought stress is a topic of an intense debate [9–16] and urges the need for a good measure of drought stress. In this paper, the relevant literature

287

contributing to the development of the AE technique is reviewed in order to propose it as a promising method to measure drought-induced cavitation in plants.

Before starting, we want to point out the different use of the concept stress between physicists and plant scientists. As described by Lichtenthaler [17] and according to physics, stress in plants means the state of a plant under the condition of a force applied. The response of the plant to this stress is called strain as long as no damage occurs. This distinction is not often made in plant sciences, where stress and strain are mostly used interchangeably, but which can be confusing in other research fields. Therefore, in what follows, we will use both physically defined terms as an attempt to also introduce this stress concept in plant sciences.

2. The Importance of Water

On a daily basis, plants extract water from the soil via the roots from where it is transported in the stem towards the leaves, where it eventually transpires into the atmosphere [18]. This seemingly wasteful process is vital for plant survival. Water is the transport medium that carries nutrients from the soil towards the plant organs and distributes generated carbohydrates throughout the plant. The evaporation of water in the stomata of leaves provides a cooling function, which is necessary to prevent overheating of leaves during sunny days. The water in living cells also provides a crucial role in the firmness and elasticity of soft tissues. The positive pressure that is exerted on the cell walls, called turgor pressure, is essential for growth (through cell growth and cell division) and fulfills the role of backbone in non-woody tissues such as leaves and petioles. Only a minor fraction of the transported water (<1%) is used to make new plant material through photosynthesis [19,20]. Given this multitude of functions, it is hence no wonder that water shortage is one of the main causes of plant mortality [21,22].

Water is transported in the xylem tissue of the plant following a gradient in water potential. Xylem is a porous structure of dead cells containing a network of parallel conduits, interconnected by pits [23]. The xylem conduits operate under negative pressure, or tension. According to the cohesion-tension theory [18], the origin of this tension is the evaporation of water in the stomatal region. When the stomata in the leaves are open, water is transpired due to the difference in water vapour pressure between the atmosphere and the substomatal cavity. As water evaporates into the air spaces in the leaf, water menisci in the small capillaries (nanometer scale) in adjacent cell walls are retracted and capillary forces (due to strong adhesion) pull the menisci back towards the surface. The network of many small capillaries in the cell wall acts thus as a wick for water rise [24]. Thanks to the strong cohesion between water molecules, the tension is transmitted downwards and water can be drawn towards the leaves. The tension in the xylem conduits may increase enormously when faced with a dry soil or with a great transpirational demand that exceeds the rate of water supply from the roots or from internal water reserves. This involves a risk of gas bubbles entering the conduits, which may expand and quickly fill the whole conduit. The water released from the conduits during this process may contribute to the transpiration stream, but on the other hand, the water conducting system will be locally interrupted. Because adjacent conduits are interconnected via pits in the conduit walls, the sap may circumvent the embolized conduit. However, when too many conduits are embolized, this will impair plant functioning. The formation of air emboli that block sap flow in xylem conduits is currently of high interest because it is one of the key processes leading to plant mortality during drought [22]. The phenomenon is called cavitation, which is the mechanical breakage of the continuous xylem water column and occurs when the tensile strength of the column is exceeded [25]. According to the current knowledge, the main cause of cavitation is the failure of a pit in the conduit wall to prevent gas from entering the conduit at strong tension [26], known as the air-seeding hypothesis [27]. Recently, Schenk *et al.* [28] postulated that nanobubbles are snapped off during air-seeding. These nanobubbles are stabilized by surfactants and may exist in plant sap under tension. They may eventually result in an embolism when the size of the nanobubble exceeds a critical threshold due to increasing tension or when many nanobubbles coalesce.

A plant's vulnerability to cavitation is often used as a key feature of its drought resistance [29], and has been defined by plotting the percentage loss of hydraulic conductivity (PLC, %) against decreasing xylem water potential (ψ, MPa), which results in a vulnerability curve (VC) (Figure 1) [30]. The xylem water potential at 50% PLC (P50) is the most common parameter to describe a species "drought resistance". The cumulative number of AE, originating from cavitation events, are a good estimate for conductivity loss [31–35] and thus has the potential to be used as an indirect and non-destructive method to construct VCs and to determine drought resistance of plants.

Because of prevailing tensions in plants, xylem is under a metastable state [37,38], which makes it a difficult object to study. A small intrusion in the tissue will cause the formation of emboli in the conduits, hereby influencing the very mechanism that is about to be studied. Therefore, the current destructive methods are under intense debate [31,39], triggered as well by recent evidence for possible artefacts in these techniques [40,41]. The use of the non-destructive AE technique has therefore gained renewed interest in plant sciences [31,33]. In particular, the question about how plants cope with drought stress in a changing climate, and which mechanisms are underlying drought resistance are gaining increased importance [9,11,29]. Application of AE in plant research can help with answering these fundamental questions to increase our knowledge on drought tolerance of plants and their ability to recover from and adapt to predicted future droughts.

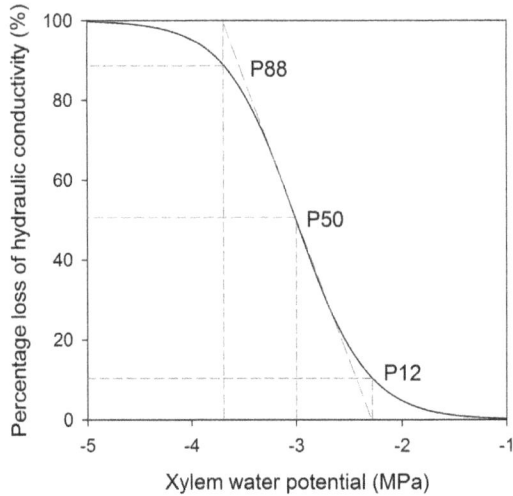

Figure 1. Typical vulnerability curve with P12, P50 and P88 representing the xylem water potential at which, respectively, 12%, 50% and 88% of xylem hydraulic conductivity is lost, adapted from Domec and Gartner [30] and Fichot *et al.* [36].

3. Acoustic Emission (AE) Application to Measure Drought-Induced Cavitation: From Past to Present

As stated in the introduction, Milburn and Johnson [42] recorded sounds in plants when they were subjected to dehydration. They detected audible vibrations (<20 kHz) in petioles of dehydrating leaves of diverse plant species by fixing the petiole on a phonograph pick-up needle (Figure 2). Because measurements were often disturbed by environmental noise, the step was made towards AE detection in the ultrasonic frequency range (>20 kHz) by Tyree and Dixon [43]. This has facilitated various experiments because the problematic ambient noise in the audible range could be electronically filtered. Based on similar sound production by the rupture of plant sap under tension in glass tubes [18,44], it was hypothesized that the rupture of sap in plant conduits produced the observed sounds during drying. Although extensive circumstantial evidence was provided to support this hypothesis [45–48], it was realized that sounds may be produced by other mechanisms too. Sounds were observed during drying of plant tissues that do not contain conduits [45,49], during re-watering [48], and also during freezing [50–52] and thawing [50,51]. The AE technique is also applied to monitor internal cracking of wood during drying [53,54] and soaking [53]. Moreover, Gagliano *et al.* [55] speculated that plants may actively produce sounds for short-distance communication. AE detection is thus an interesting tool, applicable in a wide range of domains. Further scope of this review is, however, on its application in the detection of drought-induced cavitation in plants.

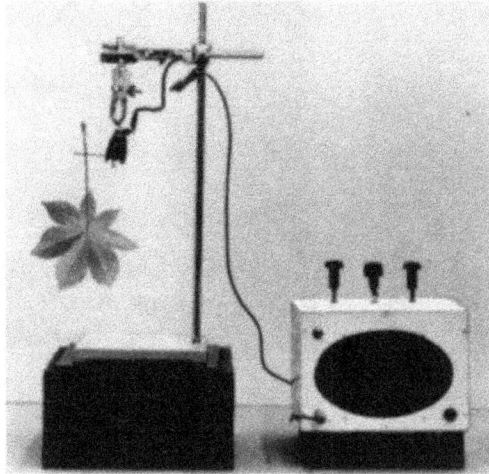

Figure 2. Picture of the vibration detector used by Milburn and Johnson [42] to detect drying-induced sounds in leaf petioles.

During dehydration of fresh plant material, AE signal detection was found to be a valid method to measure cavitation in woody branches [35], leaves [56], herbaceous stems [57] and sap wood sections [49], while others found a poor correspondence between hydraulic and acoustic measurements [58,59]. Especially in angiosperm species, the continued AE activity after loss of most of the hydraulic conductivity was a great cause of concern [31,60]. Other processes than cavitation in sap-conducting elements that could cause extra AE signals during drought stress have been suggested by various authors: cavitation of fibers, wood tracheids or ray cells [58,59,61,62], mechanical strains [25,48], and microscopic failure [43,63,64]. Moreover, the actual mechanism that causes cavitation-related AE signals is not exactly known. Different processes have also been suggested: vibration of the conduit wall after sudden pressure release due to cavitation [42,43,60], oscillation of hydrogen bounds in water after pressure release [43], conversion from liquid water to vapour during cavitation of the water [25], pit membrane rupture [65], torus aspiration (in gymnosperms [43]), the entry of a gas bubble through a pore in the pit membrane [25], and subsequent bubble oscillation [23]. Moreover, as the number of detected cavitation related AE signals may be larger than the number of cavitated conduits [56,66,67], different AE-inducing processes might be involved during cavitation.

Attempts have been made to use signal features in order to get more insights into the underlying processes. Ritman and Milburn [61] suggested that vessel length has an influence on the cut-off frequency of the detected signals. They suggested that cavitation in long vessels produced broadband signals with frequencies down to 500 Hz, whereas cavitation in fibers and wood tracheids only produced higher frequencies (>100 kHz). Tyree and Sperry [60] observed that, when detecting AE signals in the frequency range 50–1000 kHz, the frequency spectra changed towards higher frequencies when the degree of cavitation increased in *Thuja*, *Pinus*, and *Acer* stems. However, they stayed indecisive about the possible origin of the signals. Based on the waveforms, Laschimke *et al.* [68] distinguished two types of AE signals in dehydrating leaves of *Ulmus glabra*: the abrupt disruption of the water column and an oscillating source that was related to the vibration of gas bubbles during sap flow. It was also found that AE energy is a function of conduit size and xylem tension [34] and Rosner *et al.* [66] found that this parameter might be a better measure for hydraulic conductivity loss than AE counts.

Although further research is necessary to develop reliable methods for AE data interpretation, this technique has a high potential for continuous long-term cavitation monitoring in the field [69–73]. Already in 1983, Tyree and Dixon [43] (page 1099) were looking forward to the scientific progress that could be realized with the AE method:

"If it can be proved that most ultrasonic AE are a result of cavitation events, then we will have a powerful diagnostic tool that may give us new insight into the water relations of trees and other plants."

However, Quarles [74], who was especially interested in AE detection of fractures in wood, realized that deeper insights in the AE method would be essential for successful implementation:

"For successful implementation, it will be essential to understand how the propagating acoustic wave changes as a function of factors such as distance and propagating direction between the acoustic source and the receiving transducer."

During the past couple of decades, rapid developments in microelectronics have resulted in great advances in the AE technique. Currently, high performance acquisition systems are available that are able to record and store waveforms from multiple channels at high sample frequencies [64,75]. Despite the large amount of research concerning AE application to measure drought-induced cavitation, the full potential of these state of the art measurement systems has not been used so far. In what follows, current state-of-the-art and applications of AE in cavitation research will be illustrated as well as future opportunities. These deliver important findings that contribute to the development of a powerful diagnostic AE-tool for online and non-destructive measurement of cavitation.

4. Current State-of-the-Art and Application of AE in Plant Sciences

4.1. Endpoint Selection

Due to the many different AE sources acting during plant dehydration, AE activity can be detected long after most of the hydraulic conductivity is lost [31,76].

This makes it difficult to determine the endpoint (*i.e.*, 100% PLC) of the VC, which physiologically corresponds with complete cavitation of the xylem vessels and, thus, the full breakdown of the plants hydraulic pathway. As a possible solution, many studies make use of arbitrary methods. In addition, they focus more on gymnosperms, because gymnosperms have a more uniform and plain anatomy, and, thus, a more straightforward AE pattern compared to angiosperms [34]. Wolkerstorfer, Rosner *et al.* [64] filtered AE measured on dehydrating *Pinus* by drawing a straight line "by eye" through the point cloud of amplitude *versus* time. If no clear groups can be distinguished in the point cloud, they suggest using other AE features such as cumulative amplitude or energy until plausible VCs can be constructed. However, one has to be cautious when using this method, because the filtering has a strong effect on the estimated vulnerability. Other arbitrary methods include setting the endpoint at a water potential value that equals (i) the endpoint taken from parallel testing with another method [77]; (ii) a value taken from literature; or (iii) the turgor loss point of the leaves [78,79].

Vergeynst *et al.* [80] developed a more mathematical approach to determine the VC endpoint. Given that PLC should by definition be linked to cumulative AE, and that, according to Aggelis *et al.* [81], a specific AE-inducing mechanism results in an AE activity, Vergeynst *et al.* [80] recommended determining the VC endpoint by the end of the AE activity peak, which mathematically corresponds to the local maximum of the third derivative of the curve of cumulative AE *versus* time (Figure 3).

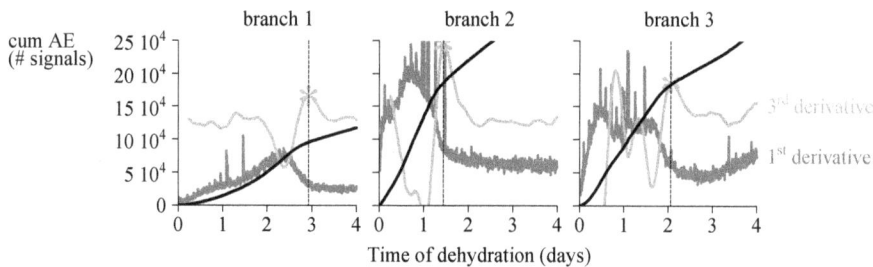

Figure 3. Vergeynst *et al.* [80] calculated the point of 100% loss of hydraulic conductivity (P100) as the endpoint of the acoustic emission (AE) activity (first derivative of the curve of cumulative AE) peak, where the third derivative of the curve of cumulative AE signals in time reached a local maximum (indicated by the vertical dashed line and the "*" symbol). The black line shows the curve of cumulative AE. Results of measurements on three different grapevine branches are shown.

Although the AE activity may continue after the selected endpoint (Figure 3), the obtained P50 values (-2.30 to -2.73 MPa) corresponded well with values found in literature for grapevine (-2.17 to -2.97 MPa; [82,83]), from which was concluded that the endpoint was accurate, and that the remaining AE activity was related to other sources than cavitation. Strong evidence for their approach was given by validation with X-ray micro-computed tomography (μCT), which showed quite similar patterns for both visually and acoustically detected cavitation [31] (Figure 4).

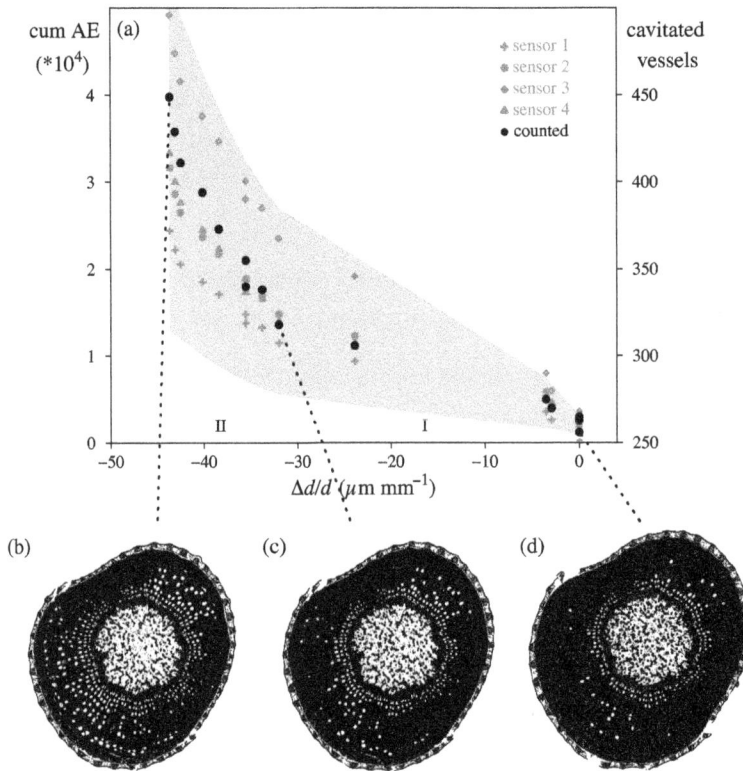

Figure 4. (**a**) Cumulative AE (cum AE, grey symbols, left y-axis) detected by four different AE sensors (1–4) showed a pattern similar to the number of cavitated vessels (black symbols, right y-axis) counted on μCT images when plotted against relative radial diameter shrinkage ($\Delta d/d$), which is a measure for decreasing xylem water potential. The μCT cross-sections of the grapevine branch are shown for the beginning (**d**) and end (**b**) of the dehydration experiment and at the breakpoint between Phases I and II (**c**). The grey zone in (**a**) delimits the 99.7% confidence interval around the mean cumulative AE curve [33].

The difficulty in determining the endpoint of VCs based on AEs has also been addressed by Nolf *et al.* [84]. To tackle this problem, these authors hypothesized that the highest acoustic activity should occur near the steepest part of the VC, which is the inflection point, reflecting P50. They obtained good similarity when comparing their method with hydraulic measurements of 16 species. The major drawback of this approach is that the VC has to have a perfect sigmoidal S-shape. Any deviation from this ideal curve might cause the steepest part to deviate from the targeted P50 value.

4.2. AE Feature Extraction

A major challenge in using AEs as an indirect measure for cavitation is determination of the source of the detected signal [31]. The AE signal is shaped by the followed path from source to sensor, which makes signal interpretation not straightforward because of the wood's anisotropic characteristics [85]. These include differences in sound propagation in the three wood directions, wood density, xylem water content and wood anatomy [34]. Previous AE application in cavitation research mainly focused on the cumulative AE signal because of the assumption that the majority of the signals correspond to a loss in hydraulic conductivity [43,46,86,87]. Given the many different AE sources during dehydration, this assumption is, however, not always valid [64,87]. Extraction of the AE signals caused by cavitation is therefore necessary. A method that has already proven its success is extraction of the signals based on corresponding wave features. In industrial lumber drying, for example, the amplitude and energy of AE signals were used to pinpointing wood checking [88–92].

Rosner *et al.* [66] measured AEs on juvenile and mature wood samples, taken from living *Picea abies* trees, with a broadband sensor in the frequency range 100–1000 kHz. After extraction of the waveform features amplitude, rise time, rise angle and absolute energy (Figure 5), cumulative AE energy appeared to be a good measure for PLC and, thus, for VC development. This was confirmed in other studies on leaves of *P. ponderosa*, *P. nigra*, *C. chrysophylla* and *P. japonica* [93], and on *P. abies* wood samples [94]. Mayr and Rosner [34] found a correlation between tracheid lumen area and mean AE energy in *P. abies* wood samples using a 150 kHz resonant sensor, which responds most strongly to acoustic waves in the 50–200 kHz frequency range. A comparison between samples with different early-latewood distribution revealed that the mean energy of AEs during wood drying increased in earlywood but decreased in samples containing more latewood. This was attributed to the homogeneous structure of earlywood and its larger tracheid diameters, causing AE energy to increase with increasing xylem tension. More resistant latewood tracheids, which cavitate at a more negative xylem tension, emitted lower energy because of their smaller diameters and more heterogeneous structure [34,64]. Ponomarenko *et al.* [23] found a similar relationship, confirming that AE energy is linked to mechanical elastic energy, which is released during cavitation in conifer tracheids. Whereas AE feature extraction to determine cavitation-related AE has been successfully applied in conifers, it remains a challenge for the more heterogeneous angiosperms [31,34].

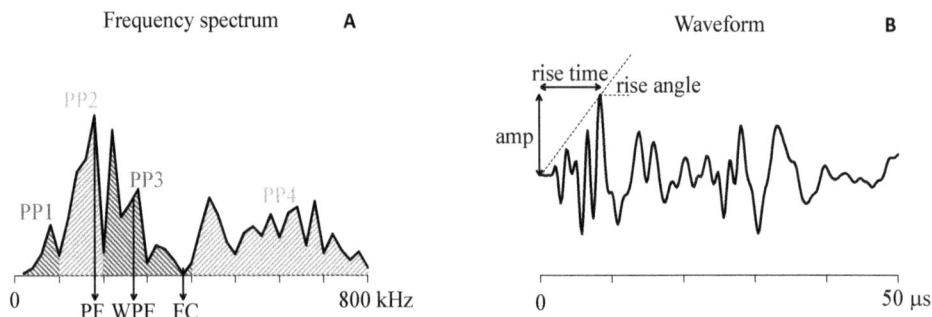

Figure 5. (A) Frequency features peak frequency (PF), weighted peak frequency (WPF), frequency centroid (FC) and four partial powers (PP) were used for the automated clustering algorithm; and (B) Waveform features rise time, amplitude and rise angle describe the shape of the AE signal [80].

Vergeynst *et al.* [80] proposed clustering to determine cavitation-related AE in angiosperms. By using state-of-the-art techniques from material sciences, finite element modelling and an automated clustering algorithm [85,95,96], in combination with broadband point-contact sensors, they were able to extract the AE that originated from cavitation during dehydration of grapevine branches. The flat frequency response of these sensors in a wide frequency range (20–1000 kHz) allows differentiation between different AE sources. Instead of relating waveform features of AE signals with cavitation, Vergeynst *et al.* [80] recommended to use the frequency features of AE signals. Following Sause and Horn [97], they extracted from the frequency spectrum, peak frequency (PF), frequency centroid (FC), weighted PF (WPF, geometric mean of PF and FC) and the partial powers of the following frequency ranges: 0–100 kHz, 100–200 kHz, 200–400 kHz and 400–800 kHz (Figure 5). The signals that typified the cavitation phase were characterized by a high intensity at 100–200 kHz. Another signal type, with high intensity in the range 400–800 kHz, was strongly related with branch shrinkage, and probably originated from micro-fractures, which are small fissures in the stretched cell wall or pith membrane due to an initial volume change. Finally, a low-frequency signal type (high intensity below 100 kHz) was identified that was attributed to macro-fractures or free drainage of water through the porous wood medium. This clustering approach to identify cavitation-related AE signals is a great advance, and may lead to the development of a powerful tool to also investigate cavitation in plants in their native environment. AE sources, such as micro- and macro-fractures, will no longer disturb cavitation measurements but may provide additional information on, for instance, use of elastic water reserves or mechanical strains.

It has often been assumed that one AE signal represents cavitation of a single water column [98]. However, except from the observed one-to-one relationship by

Tyree and Dixon [47] and Lewis [99] on small sapwood samples of two gymnosperm species, the number of AE signals has been either lower [23,100–102] or higher [56,66] than the number of cavitating conduits in further experiments. In measurements on grapevines, the number of cavitation-related AE signals exceeded the number of vessels in the branch by one or two size orders [33]. Moreover, the amplitude distribution with maximum signal density near the detection threshold of 28 dB_{AE} suggested that a considerable part of the signals could not be detected because their amplitude fell below the detection threshold, and thus probably even more AE signals are produced during xylem cavitation [103]. The use of less sensitive measurement systems, or higher detection threshold settings, might explain cases where less AE signals than cavitated conduits were observed.

One reason for the higher number of signals observed by Vergeynst *et al.* [80] could be overlapped between adjacent cluster types, so that part of the cavitation-related signal type possibly originated from other co-occurring AE sources, such as micro-fractures or water drainage. This overlap might be caused by different attenuation of the AE signals dependent on their frequency. The effect of attenuation was also observed during a freeze-thaw experiment, where it decreased in frozen samples [104]. A second hypothesis is that the cavitation process generates many AE signals. According to the nanobubble theory of Schenk *et al.* [28], many nanobubbles may be formed and exist in the xylem before they coalesce and form an embolism. Coalescence of these nanobubbles and subsequent bubble collapse may result in much more AE signals than cavitated conduits. Alternatively, fissures in the stretched pith membrane or rearrangements of the cell wall layers due to pressure redistribution after cavitation may cause many AE signals per cavitated conduit. Further research and more detailed modelling of the actual micro-mechanical processes at the AE source, such as bubble formation and coalescence, may hopefully throw light on cavitation-related AE sources and, accordingly, on the processes behind emboli formation.

4.3. In Vivo Measurements

Dealing with drought stress is a dynamic process and the resistance of a plant against drought depends on both physiological and anatomical characteristics [21]. To obtain a good understanding of a plant's behavior during drought stress, *in vivo* measurements on plants are necessary. Conventional methods to determine a plants' vulnerability to cavitation are destructive, and their results are recently under intense debate because several artefacts are reported to play a role [40,41,105]. In addition, these methods are labor-intensive, which hampers their applicability in the field. The call for non-destructive methods is therefore now more urgent than ever [39]. Cochard *et al.* [41] recently recommended μCT as the standard technique to measure hydraulic vulnerability, but this method is not suitable for field applications. Today, the AE technique is used in a destructive way and, thus, subject to similar artefacts as the conventional methods, but in contrast to these established methods, the AE technique has the potential to measure non-destructively, enabling automated and continuous measurements of cavitation in the field. Because of the difficult signal interpretation of AEs, only a few studies have focused on the possibility of *in vivo* measurements of cavitation on actively growing trees with acoustic sensors. Field measurements on *P. sylvestris* with a 150 kHz resonant sensor during the growing season showed good relationships between AE activity and stem diameter variation [73] or sap flow [106,107]. In addition, such continuous measurements will provide valuable information on plant behavior during drought and, more specifically, contribute to the debated process of cavitation recovery [36,108–111].

When using the AE technique on living plants in lab or field conditions, Vergeynst *et al.* [80] use a broadband point-contact sensor (20–1000 kHz) with flat frequency response in order to differentiate between the different AE sources. Based on preliminary analysis and clustering, a minimum set of signal frequency features may be selected (weighted peak frequency, frequency centroid and partial powers) that enable discrimination between different AE sources. However, when it is sufficiently demonstrated that the resonant sensor shows good results, its use might be preferred in further practical applications because of the more straightforward data processing and interpretation. A second point of attention for *in vivo* measurements is the particular nature of cavitation detection. Contrary to the hydraulic method and visualization methods, the AE technique measures changes in the degree of embolism rather than the degree of embolism itself. This principle should be kept in mind when preparing and interpreting AE measurements of cavitation. If the initial state of xylem embolism is known, cumulative AE measurements might result in the degree of embolism if cavitation has not been repaired in the measurement period. This will require a reference number of AE signals at 100% embolism, which may be obtained in a VC or at the end of a drought experiment. Combination of broadband AE measurements with other non-destructive measurements of sap flow, water potential, water content and xylem/phloem shrinkage [20] may yield the largest insight into plant water dynamics during drought. A better knowledge of these dynamics is crucial to feed mechanistic plant and climate models and will help with guiding mitigation of climate change impacts on plants in natural and agricultural communities [112].

5. Conclusions and Future Perspectives

How plants cope with drought in a changing climate is an active area of research, but a lot of questions remain unanswered. Mechanisms and strategies that underlie plant survival and mortality during drought are the subject of an intense debate [9–16]. The main drivers of this discussion are, however, the used methods instead of the actual mechanisms that occur in plants [39,41,110,113]. Development of a universal technique to measure cavitation, and vulnerability to it, will be essential to bring the debate to a next level, which will change the focus towards physiology behind drought resistance. In this review, we showed that revival in the use of AEs for detection of plants' vulnerability to drought is very promising. It might become a powerful non-destructive, readily automated and online method. Although great advances were recently made in dealing with the main criticisms of the technique, such as being indirect and endpoint selection difficulties, further research is needed. The behavior of AEs in plant material has to be studied, and questions regarding wave propagation through wood, the behavior of dehydrating wood, and the mechanisms behind cavitation need to be answered in order to fundamentally link AEs and cavitation events. In order to achieve this, we recommend:

- A combination of cavitation measurements with AE broadband sensors and μCT in a broad range of plant species: interpretation of the acoustic signals in combination with visuals will govern the largest insights in the mechanisms underlying cavitation.

- Feature extraction from the AE signals: this will allow a comprehensive analysis of the AE sources, and will deliver valuable information on cavitation and other AE producing processes that occur in drought-stressed plants.

- Detailed study of wave propagation and attenuation in plants, both in dehydrating and in frozen samples.

- Investigation of the effects of debarking prior to acoustic measurements: as the bark can be an additional AE source, this might influence the captured signals.

- Further validation of the use of broadband point-contact AE sensors in the field of plant hydraulics *versus* the 150 kHz resonance AE sensor.

- Development of an *in situ* AE measurement protocol for living plants, and its translation to drought sensitivity.

To conclude, the use of AEs in plant sciences to measure cavitation is promising and is gaining increasing interest. As implementation of some of the state-of-the-art techniques from material sciences already pushed frontiers in cavitation research, we believe that cross-fertilization between these different scientific domains will also be beneficial for both in the future.

Acknowledgments: The authors thank the Research Foundation—Flanders (FWO) for funding (research program G.0941.15N granted to Kathy Steppe, and PhD Fellowship granted to Lidewei L. Vergeynst).

Author Contributions: Lidewei L. Vergeynst, Linus De Roo and Niels J. F. De Baerdemaeker reviewed the literature. Kathy Steppe supervised the work. Linus De Roo wrote most of the manuscript with the input of Lidewei L. Vergeynst, and all authors contributed by discussing and reviewing drafts and the final version of the manuscript. Linus De Roo and Lidewei L. Vergeynst contributed equally to the manuscript.

Conflicts of Interest: The authors declare no conflict of interest.

References

1. Ernst, R.; Zwimpfer, F.; Dual, J. One sensor acoustic emission localization in plates. *Ultrasonics* **2016**, *64*, 139–150.
2. Robeyst, N.; Grosse, C.U.; de Belie, N. Measuring the change in ultrasonic p-wave energy transmitted in fresh mortar with additives to monitor the setting. *Cem. Concrete Res.* **2009**, *39*, 868–875.
3. Ohtsu, M. History and fundamentals. In *Acoustic Emission Testing*; Springer: Heidelberg, Germany, 2008; pp. 11–18.
4. Ohtsu, M. Fundamentals and applications. In *Acoustic Emission and Related Non-Destructive Evaluation Techniques in the Fracture Mechanics of Concrete*; Woodhead Publishing: Cambridge, UK, 2015; Volume 1, pp. 11–18.
5. Jayakumar, T.; Mukhopadhyay, C.K.; Venugopal, S.; Mannan, S.L.; Raj, B. A review of the application of acoustic emission techniques for monitoring forming and grinding processes. *J. Mater. Process. Technol.* **2005**, *159*, 48–61.
6. Hamstad, M.A. Thirty years of advances and some remaining challenges in the application of acoustic emission to composite materials. In *Acoustic Emission beyond the Millennium*; Kishi, T., Ohtsu, M., Yuyama, S., Eds.; Elsevier: Oxford, UK, 2000; pp. 77–91.
7. Drouillard, T.F. Anecdotal history of acoustic emission from wood. *J. Acoust. Emiss.* **1990**, *9*, 155–176.
8. Kishinouye, F. An experiment on the progression of fracture (a preliminary report). *J. Acoust. Emiss.* **1990**, *9*, 177–180.
9. Anderegg, W.R.; Berry, J.A.; Smith, D.D.; Sperry, J.S.; Anderegg, L.D.; Field, C.B. The roles of hydraulic and carbon stress in a widespread climate-induced forest die-off. *Proc. Natl. Acad. Sci. USA* **2012**, *109*, 233–237.
10. Allen, C.D.; Macalady, A.K.; Chenchouni, H.; Bachelet, D.; McDowell, N.; Vennetier, M.; Kitzberger, T.; Rigling, A.; Breshears, D.D.; Hogg, E.T. A global overview of drought and heat-induced tree mortality reveals emerging climate change risks for forests. *For. Ecol. Manag.* **2010**, *259*, 660–684.
11. Anderegg, W.R.L.; Flint, A.; Huang, C.-Y.; Flint, L.; Berry, J.A.; Davis, F.W.; Sperry, J.S.; Field, C.B. Tree mortality predicted from drought-induced vascular damage. *Nat. Geosci.* **2015**, *8*, 367–371.
12. McDowell, N.G.; Ryan, M.G.; Zeppel, M.J.; Tissue, D.T. Feature: Improving our knowledge of drought-induced forest mortality through experiments, observations, and modeling. *New Phytol.* **2013**, *200*, 289–293.
13. Martínez-Vilalta, J.; Lloret, F.; Breshears, D.D. Drought-induced forest decline: Causes, scope and implications. *Biol. Lett.* **2012**, *8*, 689–691.
14. Zeppel, M.J.; Anderegg, W.R.; Adams, H.D. Forest mortality due to drought: Latest insights, evidence and unresolved questions on physiological pathways and consequences of tree death. *New Phytol.* **2013**, *197*, 372–374.

15. Doughty, C.E.; Metcalfe, D.; Girardin, C.; Amézquita, F.F.; Cabrera, D.G.; Huasco, W.H.; Silva-Espejo, J.; Araujo-Murakami, A.; da Costa, M.; Rocha, W. Drought impact on forest carbon dynamics and fluxes in amazonia. *Nature* **2015**, *519*, 78–82.

16. Hartmann, H.; Adams, H.D.; Anderegg, W.R.; Jansen, S.; Zeppel, M.J. Research frontiers in drought-induced tree mortality: Crossing scales and disciplines. *New Phytol.* **2015**, *205*, 965–969.

17. Lichtenthaler, H.K. Vegetation stress: An introduction to the stress concept in plants. *J. Plant Physiol.* **1996**, *148*, 4–14.

18. Dixon, H.H. *Transpiration and the Ascent of Sap in Plants*; Macmillan and Co.: London, UK, 1914.

19. Kozlowski, T.T.; Pallardy, S.G. *Growth Control in Woody Plants*; Elsevier: San Diego, CA, USA, 1997.

20. Steppe, K.; Sterck, F.; Deslauriers, A. Diel growth dynamics in tree stems: Linking anatomy and ecophysiology. *Trends Plant Sci.* **2015**, *20*, 335–343.

21. Sevanto, S.; McDowell, N.G.; Dickman, L.T.; Pangle, R.; Pockman, W.T. How do trees die? A test of the hydraulic failure and carbon starvation hypotheses. *Plant Cell Environ.* **2014**, *37*, 153–161.

22. McDowell, N.; Pockman, W.T.; Allen, C.D.; Breshears, D.D.; Cobb, N.; Kolb, T.; Plaut, J.; Sperry, J.; West, A.; Williams, D.G. Mechanisms of plant survival and mortality during drought: Why do some plants survive while others succumb to drought? *New Phytol.* **2008**, *178*, 719–739.

23. Ponomarenko, A.; Vincent, O.; Pietriga, A.; Cochard, H.; Badel, E.; Marmottant, P. Ultrasonic emissions reveal individual cavitation bubbles in water-stressed wood. *J. R. Soc. Interface* **2014**, *11*.

24. Nobel, P.S. *Physicochemical and Environmental Plant Physiology*; Academic Press: San Diego, CA, USA, 1999.

25. Jackson, G.; Grace, J. Cavitation and water transport in trees. *Endeavour* **1994**, *18*, 50–54.

26. Rockwell, F.E.; Wheeler, J.K.; Holbrook, N.M. Cavitation and its discontents: Opportunities for resolving current controversies. *Plant Physiol.* **2014**, *164*, 1649–1660.

27. Zimmermann, M. *Xylem structure and the ascent of sap. Xylem Structure and the Ascent of Sap*; Springer-Verlag: Berlin, Germany, 1983.

28. Schenk, H.J.; Steppe, K.; Jansen, S. Nanobubbles: A new paradigm for air-seeding in xylem. *Trends Plant Sci.* **2015**, *20*, 199–205.

29. Choat, B.; Jansen, S.; Brodribb, T.J.; Cochard, H.; Delzon, S.; Bhaskar, R.; Bucci, S.J.; Feild, T.S.; Gleason, S.M.; Hacke, U.G.; *et al.* Global convergence in the vulnerability of forests to drought. *Nature* **2012**, *491*, 752–755.

30. Domec, J.-C.; Gartner, B.L. Cavitation and water storage capacity in bole xylem segments of mature and young douglas-fir trees. *Trees* **2001**, *15*, 204–214.

31. Rosner, S. A new type of vulnerability curve: Is there truth in vine? *Tree Physiol.* **2015**, *35*, 410–414.

32. Cochard, H.; Badel, E.; Herbette, S.; Delzon, S.; Choat, B.; Jansen, S. Methods for measuring plant vulnerability to cavitation: A critical review. *J. Exp. Bot.* **2013**, *64*, 4779–4791.

33. Vergeynst, L.L.; Dierick, M.; Bogaerts, J.A.; Cnudde, V.; Steppe, K. Cavitation: A blessing in disguise? New method to establish vulnerability curves and assess hydraulic capacitance of woody tissues. *Tree Physiol.* **2015**, *35*, 400–409.

34. Mayr, S.; Rosner, S. Cavitation in dehydrating xylem of picea abies: Energy properties of ultrasonic emissions reflect tracheid dimensions. *Tree Physiol.* **2011**, *31*, 59–67.

35. Gullo, M.L.; Salleo, S. Different vulnerabilities of *Quercus ilex* L. To freeze-and summer drought-induced xylem embolism: An ecological interpretation. *Plant Cell Environ.* **1993**, *16*, 511–519.

36. Fichot, R.; Brignolas, F.; Cochard, H.; Ceulemans, R. Vulnerability to drought-induced cavitation in poplars: Synthesis and future opportunities. *Plant Cell Environ.* **2015**, *38*, 1233–1251.

37. Dixon, H.H.; Joly, J. On the ascent of sap. *Philos. Trans. R. Soc. Lond. B* **1895**.

38. Steudle, E. The cohesion-tension mechanism and the acquisition of water by plant roots. *Annu. Rev. Plant Biol.* **2001**, *52*, 847–875.

39. Jansen, S.; Schuldt, B.; Choat, B. Current controversies and challenges in applying plant hydraulic techniques. *New Phytol.* **2015**, *205*, 961–964.

40. Wheeler, J.K.; Huggett, B.A.; Tofte, A.N.; Rockwell, F.E.; Holbrook, N.M. Cutting xylem under tension or supersaturated with gas can generate plc and the appearance of rapid recovery from embolism. *Plant Cell Environ.* **2013**, *36*, 1938–1949.

41. Cochard, H.; Delzon, S.; Badel, E. X-ray microtomography (micro-CT): A reference technology for high-resolution quantification of xylem embolism in trees. *Plant Cell Environ.* **2015**, *38*, 201–206.

42. Milburn, J.; Johnson, R. The conduction of sap: II. Detection of vibrations produced by sap cavitation in ricinus xylem. *Planta* **1966**, *69*, 43–52.

43. Tyree, M.T.; Dixon, M.A. Cavitation events in *Thuja occidentalis* L.? Utrasonic acoustic emissions from the sapwood can be measured. *Plant Physiol.* **1983**, *72*, 1094–1099.

44. Temperley, H.; Chambers, L.G. The behaviour of water under hydrostatic tension: I. *Proc. Phys. Soc.* **1946**, *58*, 420.

45. Milburn, J.A. Cavitation in *Ricinus* by acoustic detection: Induction in excised leaves by various factors. *Planta* **1973**, *110*, 253–265.

46. Tyree, M.T.; Dixon, M.A.; Thompson, R.G. Ultrasonic acoustic emissions from the sapwood of *Thuja occidentalis* measured inside a pressure bomb. *Plant Physiol.* **1984**, *74*, 1046–1049.

47. Tyree, M.T.; Dixon, M.A.; Tyree, E.L.; Johnson, R. Ultrasonic acoustic emissions from the sapwood of cedar and hemlock an examination of three hypotheses regarding cavitations. *Plant Physiol.* **1984**, *75*, 988–992.

48. Milburn, J.A. Cavitation studies on whole ricinus plants by acoustic detection. *Planta* **1973**, *112*, 333–342.

49. Kikuta, S.B. Ultrasound acoustic emissions from bark samples differing in anatomical characteristics. *Phyton* **2003**, *43*, 161–178.

50. Raschi, A.; Mugnozza, G.S.; Surace, R.; Valentini, R.; Vazzana, C. The use of ultrasound technique to monitor freezing and thawing of water in plants. *Agric. Ecosyst. Environ.* **1989**, *27*, 411–418.

51. Mayr, S.; Sperry, J.S. Freeze-thaw-induced embolism in pinus contorta: Centrifuge experiments validate the "thaw-expansion hypothesis" but conflict with ultrasonic emission data. *New Phytol.* **2010**, *185*, 1016–1024.

52. Charrier, G.; Charra-Vaskou, K.; Kasuga, J.; Cochard, H.; Mayr, S.; Ameglio, T. Freeze-thaw stress: Effects of temperature on hydraulic conductivity and ultrasonic activity in ten woody angiosperms. *Plant Physiol.* **2014**, *164*, 992–998.

53. Moliński, W.; Raczkowski, J.; Poliszko, S.; Ranachowski, Z. Mechanism of acoustic emission in wood soaked in water. *Holzforschung* **1991**, *45*, 13–17.

54. Rosner, S. Characteristics of acoustic emissions from dehydrat-ing wood related to shrinkage processes. *J. Acoustic Emission* **2007**, *25*, 149–157.

55. Gagliano, M.; Mancuso, S.; Robert, D. Towards understanding plant bioacoustics. *Trends Plant Sci.* **2012**, *17*, 323–325.

56. Kikuta, S.; Gullo, M.; Nardini, A.; Richter, H.; Salleo, S. Ultrasound acoustic emissions from dehydrating leaves of deciduous and evergreen trees. *Plant Cell Environ.* **1997**, *20*, 1381–1390.

57. Hacke, U.G.; Sperry, J.S.; Pittermann, J. Drought experience and cavitation resistance in six shrubs from the great basin, utah. *Basic Appl. Ecol.* **2000**, *1*, 31–41.

58. Sperry, J.S.; Tyree, M.T.; Donnelly, J.R. Vulnerability of xylem to embolism in a mangrove *vs.* an inland species of rhizophoraceae. *Physiol. Plant.* **1988**, *74*, 276–283.

59. Cochard, H.; Tyree, M.T. Xylem dysfunction in quercus: Vessel sizes, tyloses, cavitation and seasonal changes in embolism. *Tree Physiol.* **1990**, *6*, 393–407.

60. Tyree, M.; Sperry, J. Characterization and propagation of acoustic emission signals in woody plants: Towards an improved acoustic emission counter. *Plant Cell Environ.* **1989**, *12*, 371–382.

61. Ritman, K.; Milburn, J. Acoustic emissions from plants: Ultrasonic and audible compared. *J. Exp. Bot.* **1988**, *39*, 1237–1248.

62. Kasuga, J.; Charrier, G.; Uemura, M.; Améglio, T. Characteristics of ultrasonic acoustic emissions from walnut branches during freeze-thaw-induced embolism formation. *J. Exp. Bot.* **2015**, *66*, 1965–1975.

63. Rosner, S.; Konnerth, J.; Plank, B.; Salaberger, D.; Hansmann, C. Radial shrinkage and ultrasound acoustic emissions of fresh *versus* pre-dried norway spruce sapwood. *Trees* **2010**, *24*, 931–940.

64. Wolkerstorfer, S.V.; Rosner, S.; Hietz, P. An improved method and data analysis for ultrasound acoustic emissions and xylem vulnerability in conifer wood. *Physiol. Plant* **2012**, *146*, 184–191.

65. Rosner, S. Waveform features of acoustic emission provide information about reversible and irreversible processes during spruce sapwood drying. *BioResources* **2012**, *7*, 1253–1263.

66. Rosner, S.; Klein, A.; Wimmer, R.; Karlsson, B. Extraction of features from ultrasound acoustic emissions: A tool to assess the hydraulic vulnerability of norway spruce trunkwood? *New Phytol.* **2006**, *171*, 105–116.

67. Mayr, S.; Schmid, P.; Laur, J.; Rosner, S.; Charra-Vaskou, K.; Dämon, B.; Hacke, U.G. Uptake of water via branches helps timberline conifers refill embolized xylem in late winter. *Plant Physiol.* **2014**, *164*, 1731–1740.

68. Laschimke, R.; Burger, M.; Vallen, H. Acoustic emission analysis and experiments with physical model systems reveal a peculiar nature of the xylem tension. *J. Plant Physiol.* **2006**, *163*, 996–1007.

69. Tyree, M.T.; Fiscus, E.L.; Wullschleger, S.; Dixon, M. Detection of xylem cavitation in corn under field conditions. *Plant Physiol.* **1986**, *82*, 597–599.

70. Ikeda, T.; Ohtsu, M. Detection of xylem cavitation in field-grown pine trees using the acoustic emission technique. *Ecol. Res.* **1992**, *7*, 391–395.

71. Jackson, G.; Grace, J. Field measurements of xylem cavitation: Are acoustic emissions useful? *J. Exp. Bot.* **1996**, *47*, 1643–1650.

72. Perks, M.P.; Irvine, J.; Grace, J. Xylem acoustic signals from mature pinus sylvestris during an extended drought. *Ann. For. Sci.* **2004**, *61*, 1–8.

73. Hölttä, T.; Vesala, T.; Nikinmaa, E.; Perämäki, M.; Siivola, E.; Mencuccini, M. Field measurements of ultrasonic acoustic emissions and stem diameter variations. New insight into the relationship between xylem tensions and embolism. *Tree Physiol.* **2005**, *25*, 237–243.

74. Quarles, S.L.; Association, W.D.K.; Association, W.D.K. Acoustic Emission Generated during Drying. In Proceedings of the 41st Western Dry Kiln Association Meeting, Corvallis, OR, USA, 1990; p. 53.

75. Michlmayr, G.; Cohen, D.; Or, D. Sources and characteristics of acoustic emissions from mechanically stressed geologic granular media—A review. *Earth Sci. Rev.* **2012**, *112*, 97–114.

76. Tyree, M.T.; Sperry, J.S. Vulnerability of xylem to cavitation and embolism. *Annu. Rev. Plant Biol.* **1989**, *40*, 19–36.

77. Hacke, U.; Sauter, J. Xylem dysfunction during winter and recovery of hydraulic conductivity in diffuse-porous and ring-porous trees. *Oecologia* **1996**, *105*, 435–439.

78. Nardini, A.; Tyree, M.T.; Salleo, S. Xylem cavitation in the leaf of prunus laurocerasusand its impact on leaf hydraulics. *Plant Physiol.* **2001**, *125*, 1700–1709.

79. Salleo, S.; Gullo, L.; Raimondo, F.; Nardini, A. Vulnerability to cavitation of leaf minor veins: Any impact on leaf gas exchange? *Plant Cell Environ.* **2001**, *24*, 851–859.

80. Vergeynst, L.L.; Sause, M.G.R.; Steppe, K. Clustering reveals cavitation-related acoustic emission signals from dehydrating branches. *Tree Physiol.* **2016**. in press.

81. Aggelis, D.; Matikas, T.; Shiotani, T. Advanced acoustic techniques for health monitoring of concrete structures. In *The Song's Handbook of Concrete Durability*; Kim, S.H., Ann, K.Y., Eds.; Middleton Publishing Inc: Middleton, WI, USA, 2010; pp. 331–378.

82. Choat, B.; Drayton, W.M.; Brodersen, C.; Matthews, M.A.; Shackel, K.A.; Wada, H.; Mcelrone, A.J. Measurement of vulnerability to water stress-induced cavitation in grapevine: A comparison of four techniques applied to a long-vesseled species. *Plant Cell Environ.* **2010**, *33*, 1502–1512.

83. Brodersen, C.R.; McElrone, A.J.; Choat, B.; Lee, E.F.; Shackel, K.A.; Matthews, M.A. *In vivo* visualizations of drought-induced embolism spread in vitis vinifera. *Plant Physiol.* **2013**, *161*, 1820–1829.

84. Nolf, M.; Beikircher, B.; Rosner, S.; Nolf, A.; Mayr, S. Xylem cavitation resistance can be estimated based on time-dependent rate of acoustic emissions. *New Phytol.* **2015**, *208*, 625–632.

85. Vergeynst, L.L.; Sause, M.G.; Hamstad, M.A.; Steppe, K. Deciphering acoustic emission signals in drought stressed branches: The missing link between source and sensor. *Front. Plant Sci.* **2015**, *6*.

86. Lo, G.M.; Salleo, S. Three different methods for measuring xylem cavitation and embolism: A comparison. *Ann. Bot.* **1991**, *67*, 417–424.

87. Cochard, H. Vulnerability of several conifers to air embolism. *Tree Physiol.* **1992**, *11*, 73–83.

88. Niemz, P.; Emmler, R.; Pridöhl, E.; Fröhlich, J.; Lühmann, A. Vergleichende untersuchungen zur Anwendung von piezoelektrischen und Schallemissionssignalen bei der trocknung von Holz. *Holz als Roh und Werkstoff* **1994**, *52*, 162–168.

89. Lee, S.-H.; Quarles, S.; Schniewind, A. Wood fracture, acoustic emission, and the drying process part 2. Acoustic emission pattern recognition analysis. *Wood Sci. Technol.* **1996**, *30*, 283–292.

90. Beall, F. Overview of the use of ultrasonic technologies in research on wood properties. *Wood Sci. Technol.* **2002**, *36*, 197–212.

91. Kawamoto, S.; Williams, R.S. *Acoustic Emission and Acousto-Ultrasonic Techniques for Wood and Wood-Based Composites: A Review*; US Department of Agriculture: Madison, WI, USA, 2002.

92. Beall, F.C.; Breiner, T.A.; Wang, J. Closed-loop control of lumber drying based on acoustic emission peak amplitude. *For. Prod. J.* **2005**, *55*, 167–174.

93. Johnson, D.M.; Meinzer, F.C.; Woodruff, D.R.; McCulloh, K.A. Leaf xylem embolism, detected acoustically and by cryo-sem, corresponds to decreases in leaf hydraulic conductance in four evergreen species. *Plant Cell Environ.* **2009**, *32*, 828–836.

94. Rosner, S.; Karlsson, B.; Konnerth, J.; Hansmann, C. Shrinkage processes in standard-size norway spruce wood specimens with different vulnerability to cavitation. *Tree Physiol.* **2009**, *29*, 1419–1431.

95. Sause, M.G.R.; Gribov, A.; Unwin, A.R.; Horn, S. Pattern recognition approach to identify natural clusters of acoustic emission signals. *Pattern Recognit. Lett.* **2012**, *33*, 17–23.

96. Zelenyak, A.M.; Hamstad, M.A.; Sause, M.G. Modeling of acoustic emission signal propagation in waveguides. *Sensors* **2015**, *15*, 11805–11822.

97. Sause, M.; Horn, S. Simulation of acoustic emission in planar carbon fiber reinforced plastic specimens. *J. Nondestruct. Eval.* **2010**, *29*, 123–142.

98. Haacs, R.; Blank, R.W. Acoustic emission from drought-stressed red pine (*Pinus resinosa*). *J. Acoust. Emiss.* **1990**, *9*, 181–187.

99. Lewis, A.M. *Two Mechanisms for the Initiation of Embolism in Tracheary Elements and Other Dead Plant Cells under Water Stress*; Harvard University: Cambridge, MA, USA, 1987.

100. Sandford, A.; Grace, J. The measurement and interpretation of ultrasound from woody stems. *J. Exp. Bot.* **1985**, *36*, 298–311.

101. Ritman, K.; Milburn, J. Monitoring of ultrasonic and audible emissions from plants with or without vessels. *J. Exp. Bot.* **1991**, *42*, 123–130.

102. Kikuta, S.; Lo Gullo, M.; Kartusch, B.; Rosner, S.; Richter, H. Ultrasound Acoustic Emissions from Conifer Sapwood Sections: Relationship between Number of Events Detected and Number of Tracheids. In Proceedings of the IAWA International Symposium on Wood Sciences, Montpellier, France, 24–29 October 2004; p. 29.

103. Vergeynst, L.L. *Investigation and Application of the Acoustic Emission Technique to Measure Drought-Induced Cavitation in Woody Plants*; Ghent University: Ghent, Belgium, 2015.

104. Charrier, G.; Pramsohler, M.; Charra-Vaskou, K.; Saudreau, M.; Ameglio, T.; Neuner, G.; Mayr, S. Ultrasonic emissions during ice nucleation and propagation in plant xylem. *New Phytol.* **2015**, *207*, 570–578.

105. Torres-Ruiz, J.M.; Jansen, S.; Choat, B.; McElrone, A.J.; Cochard, H.; Brodribb, T.J.; Badel, E.; Burlett, R.; Bouche, P.S.; Brodersen, C.R.; *et al.* Direct X-ray microtomography observation confirms the induction of embolism upon xylem cutting under tension. *Plant Physiol.* **2015**, *167*, 40–43.

106. Zweifel, R.; Zeugin, F. Ultrasonic acoustic emissions in drought-stressed trees—More than signals from cavitation? *New Phytol.* **2008**, *179*, 1070–1079.

107. Steppe, K.; Zeugin, F.; Zweifel, R. Low-decibel ultrasonic acoustic emissions are temperature-induced and probably have no biotic origin. *New Phytol.* **2009**, *183*, 928–931.

108. Brodersen, C.R.; McElrone, A.J. Maintenance of xylem network transport capacity: A review of embolism repair in vascular plants. *Front. Plant Sci.* **2013**, *4*.

109. Zwieniecki, M.A.; Holbrook, N.M. Confronting maxwell's demon: Biophysics of xylem embolism repair. *Trends Plant Sci.* **2009**, *14*, 530–534.

110. McCulloh, K.A.; Meinzer, F.C. Further evidence that some plants can lose and regain hydraulic function daily. *Tree Physiol.* **2015**, *35*, 691–693.

111. Meinzer, F.C.; McCulloh, K.A. Xylem recovery from drought-induced embolism: Where is the hydraulic point of no return? *Tree Physiol.* **2013**, *33*, 331–334.

112. Zwieniecki, M.A.; Secchi, F. Threats to xylem hydraulic function of trees under "new climate normal" conditions. *Plant Cell Environ.* **2015**, *38*, 1713–1724.

113. McElrone, A.; Brodersen, C.; Alsina, M.; Drayton, W.; Matthews, M.; Shackel, K.; Wada, H.; Zufferey, V.; Choat, B. Centrifuge technique consistently overestimates vulnerability to water stress-induced cavitation in grapevines as confirmed with high-resolution computed tomography. *New Phytol.* **2012**, *196*, 661–665.

The Stiffness and Damping Characteristics of a Dual-Chamber Air Spring Device Applied to Motion Suppression of Marine Structures

Xiaohui Zeng, Liang Zhang, Yang Yu, Min Shi and Jifu Zhou

Abstract: Dual-chamber air springs are used as a key component for vibration isolation in some industrial applications. The working principle of the dual-chamber air spring device as applied to motion suppression of marine structures is similar to that of the traditional air spring, but they differ in their specific characteristics. The stiffness and damping of the dual-chamber air spring device determine the extent of motion suppression. In this article, we investigate the stiffness and damping characteristics of a dual-chamber air spring device applied to marine structure motion suppression using orthogonal analysis and an experimental method. We measure the effects of volume ratio, orifice ratio, excitation amplitude, and frequency on the stiffness and damping of the dual-chamber vibration absorber. Based on the experimental results, a higher-order non-linear regression method is obtained. We achieve a rapid calculation model for dual-chamber air spring stiffness and damping, which can provide guidance to project design.

Reprinted from *Appl. Sci.* Cite as: Zeng, X.; Zhang, L.; Yu, Y.; Shi, M.; Zhou, J. The Stiffness and Damping Characteristics of a Dual-Chamber Air Spring Device Applied to Motion Suppression of Marine Structures. *Appl. Sci.* **2016**, *6*, 74.

1. Introduction

Marine structure safety, especially in the deep sea, is an important concern. The deep-sea floating platform is one of the most important large structures used in ocean energy exploitation. Dynamic responses due to the wind and wave effects must be considered to ensure a safe design [1]. The combined action of complex multiple loads could cause a large vibration amplitude in marine structures [2,3]. Large amplitude and alternating vibration are the main detrimental effects that reduce the safety and fatigue life of a deep-sea platform. In the field of civil engineering, engineers use a wide variety of energy-absorbing devices to reduce the vibration of building structures. These energy-absorbing devices may also be applied to the problem of vibration suppression of deep-sea marine structures.

The vibration of a floating platform could be reduced using active or passive controls [4,5] and various devices have been proposed. A Tuned Liquid Column Damper (TLCD), a kind of passive vibration control device to suppress movement, was proposed by Sakai [6] in 1989. Lee *et al.* [7,8] first used the TLCD to suppress the surge and sway motions vibration of a Tension Leg Platform (TLP). Taflanidis *et al.* [9] further developed a simulation-based method for the design of mass dampers applied for the response mitigation of tension lag platforms. Lee *et al.* [10] experimentally studied the harmonic responses of the TLCD for wind excitations. Tanmoy *et al.* [11] studied the effective performance of TLCD and another passive vibration-mitigating device, a tuned liquid column ball damper for the control of wave-induced vibration. Zeng *et al.* [12] invented a new type of energy-absorbing device, the S-shaped TLCD. The S-shaped TLCD could effectively suppress the horizontal movement and vertical in-plane rotation of a TLP. Although these devices were effective for the suppression of surge and sway motion, TLCD was ineffective at the suppression of vertical movement, and sometime even enhanced this movement [10]. The vertical movement of the floating structures significantly impacts the strength and fatigue of the tension leg and mooring system, and the inability of these devices to suppress this movement could decrease the safety and service life of the structure.

Vibration-induced vertical motion of TLPs could be balanced by using absorbers with air springs. Dual-chamber air springs are an effective energy-absorbing device used as a key component for vibration isolation. Rijken, Bian, and Spillane *et al.* [13–15] applied a system of vibration absorbers using dual-chamber air springs and water columns to suppress resonant motions and studied the effects of the orifice ratio for structural damping. Bachrach and Rivin [16] studied the complex dynamic stiffness of the damper spring, a function of the excitation frequency. The experimental results measured by Kim and Lee [17] for a dual-chamber pneumatic spring exhibited significant amplitude-dependent nonlinear behavior. Jing *et al.* [18–20] proposed a characteristic output spectrum nonlinear (nCOS) method. The nCOS method is based on the theory of Volterra series expansion. They used the nCOS method for analysis and design of an air spring in a nonlinear vehicle suspension system. These studies focused on traditional forms of a dual-chamber air spring.

A dual-chamber air spring device applied to marine structure motion suppression requires a gas-liquid coupling air spring structure. This differs from the traditional air spring both in structure type and in the vibration characteristics. The compression of the gas in the traditional air spring was controlled by a piston, which acted on the pressure surface of the air spring [21–23]. As shown in Figure 1a, the displacement of the piston, x_0, causes a compression of the air in the upper chamber. The pressure variation of the air spring chamber is of the same phase and amplitude as the external load. For the gas-liquid coupling air spring device applied to motion suppression of marine structures, a liquid column can also move in an independent manner. As shown in Figure 1b, the amount of compressed air in the air chamber is the relative displacement between the liquid column and the top of the air spring, *i.e.* $x_1 - x_0$. The movement of the liquid column (similar to the piston) and the external excitation can differ in phases and amplitude. The stiffness and damping characteristics of this kind of dual-chamber air spring have not been investigated.

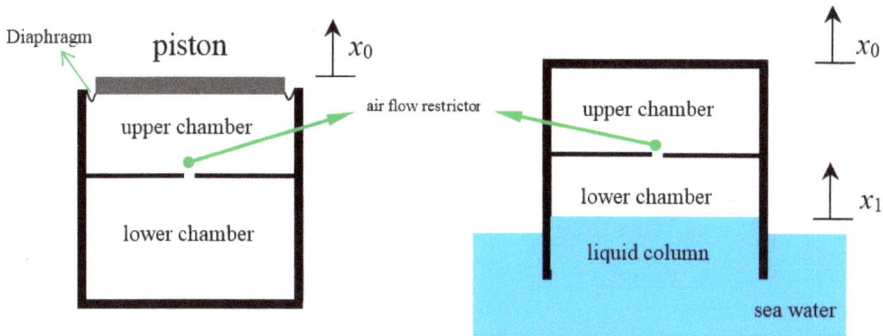

Figure 1. Schematic diagram of two kinds of dual-chamber air spring: (**a**) the traditional device; (**b**) a gas-liquid coupling device.

The two main determinants of structure movement suppression effect are the stiffness and damping of the energy-absorbing device. Previous studies of damping and stiffness focused on the traditional air spring, and used an approximation to simulate air flow through the orifice plates [24–26]. These results may not apply to the gas-liquid coupling dual-chamber air spring device. Moreover, few factors were investigated and analyzed. There is not a comprehensive study on which factors affect the stiffness and damping characteristics of dual-chamber air spring. Further studies must be carried out on the gas-liquid coupling dual-chamber air spring device.

Here, we investigate the damping and stiffness of a dual-chamber air spring device applied to motion suppression of marine structures. We used orthogonal analysis and an experimental method and found that stiffness and damping are a function of the volume ratio, orifice ratio, excitation amplitude, and frequency. We then measured the effects of these four factors on dual-chamber vibration absorber stiffness and damping. Based on the experimental results, we used higher-order non-linear regression method and generated a rapid calculation model for dual-chamber air spring stiffness and damping, which can provide guidance for future project design.

2. Theoretical Analysis

The gas-liquid coupling, dual-chamber, air spring energy-absorbing device has two chambers. The upper chamber functions for air storage and has a fixed volume, while the volume of the lower chamber, in contact with the water, varies with the motion of the water column. The gas and the oscillating liquid column are coupled and this device uses the interaction of oscillations of the liquid column and gas to achieve floating structure motion suppression.

The performance and design of the dual-chamber air spring device are usually modeled after a spring damper system. The characteristics of the damping device are described by the spring stiffness and damping coefficient [26–30]. In this work, the dual-chamber air spring device is a complex fluid-gas coupling system, but its dynamic behavior can also be characterized by the spring stiffness and damping coefficient. The physical model of the common spring damping system can be described using a parallel model or as a series model. Here, we explore the advantages and disadvantages of the two models to characterize the dynamic characteristics of the dual-chamber air spring device applied to motion suppression of marine structures.

In this section, three kinds of energy dissipation expression are derived: a series model, a parallel model, and work done by excitation force.

2.1. Parallel Model

As shown in Figure 2, the standard equation of the model is:

$$M\ddot{X} + C\dot{X} + KX = F \tag{1}$$

where M represents the overall mass of the vibration absorber structure, C represents the overall damping of the structure, and K refers to the overall stiffness of the structure.

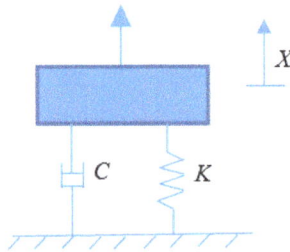

Figure 2. Parallel model.

The structure stiffness in the above equation changes with the air pressure in the vibration absorber air chamber. The change in air pressure is caused by the relative motion of the liquid column in the air chamber. The damping has two components. Some damping is due to fluid vortex shedding at the bottom of the vibration absorber and the friction between the outer wall and water and this damping is related to the motion of the vibration absorber structure. Damping is also caused by the effect of the air in the air chamber and the orifice as well as the friction between inner wall and water, and this damping is related to the motion of the liquid column in the vibration absorber to the vibration absorber.

For harmonic excitation $F = F_0 e^{i\omega t}$, the structural response X is also a harmonic motion which can be represented as $X = X_0 e^{i(\omega t - \varphi)}$, substituted into Equation (1) and using Euler's Formula, $e^{i\varphi} = \cos\varphi + i\sin\varphi$, the stiffness and damping of the vibration system can be written as:

$$K = \frac{F_0}{X_0}\cos\varphi + \omega^2 M \tag{2}$$

$$C = \frac{1}{\omega}\frac{F_0}{X_0}\sin\varphi \tag{3}$$

The mass matrix can be $M = [m_1, m_2, m_3, m_4, m_5, m_6, m_7]^T$. In the experiment, the mass of each different air chamber length structure is shown as follows:
$m_1 = 5.769, m_2 = 5.967, m_3 = 6.178, m_4 = 6.386, m_5 = 6.593, m_6 = 6.773,$
$m_7 = 7.001$, in kg.

The periodical energy consumption of damping is:

$$Q = \int C\dot{X}dX = \int_0^T \left(C\dot{X}\right)\dot{X}dt = \int_0^T C\dot{X}^2 dt = C\int_0^T \dot{X}^2 dt = \frac{1}{\omega}\frac{F_0}{X_0}\sin\varphi \sum_{i=1}^{i=f_s \cdot T} \dot{X}_i^2 t_i \tag{4}$$

where f_s refers to the sampling frequency of the experiment, f_s = 1024 Hz. Due to the equal-interval sampling of the experiment data, $t_i = \frac{1}{1024}$.

2.2. Series Model

As shown in Figure 3, the standard equation of the model is:

$$\begin{cases} M\ddot{X} + k(X - x) = F \\ k(x - X) + c\dot{x} = 0 \end{cases} \tag{5}$$

Substitute $F = F_0 e^{i\omega t}$ and $X = X_0 e^{i(\omega t - \varphi)}$ into the equation and apply Euler's Formula, $e^{i\varphi} = \cos\varphi + i\sin\varphi$, to determine the damping and stiffness:

$$c = \frac{F_0}{\omega X_0}\sin\varphi + \frac{F_0}{\omega X_0}\cot\varphi\cos\varphi + 2\omega M\cot\varphi + \frac{\omega^3 M^2 X_0}{F_0 \sin\varphi} \tag{6}$$

$$k = \frac{c}{\frac{1}{\omega}\cot\varphi + \frac{\omega M X_0}{F_0 \sin\varphi}} \tag{7}$$

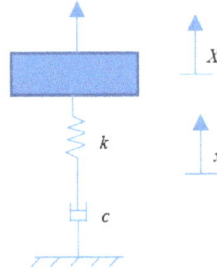

Figure 3. Series model.

The periodical energy consumption of damping is:

$$q = \int c\dot{x}\mathrm{d}x = \int_0^T c\dot{x}^2\mathrm{d}t = \sum_{i=1}^{i=f_s \cdot T} c\dot{x}_i^2 t_i \tag{8}$$

2.3. Work Done by Excitation Force

The periodical work done by excitation force of the system is:

$$W = \int F\mathrm{d}X = \int_0^T F\dot{X}\mathrm{d}t = \sum_{i=1}^{i=f_s \cdot T} F_i \dot{X}_i t_i \tag{9}$$

where F refers to the excitation load, and X refers to the structural response.

3. Experimental Method

3.1. Design of Experimental Device

The dual-chamber air spring vibration absorber described here is mainly used to restrain the large vibration of an offshore structure. In the experiment, a large water container was designed to simulate the action of sea water. The container is a cylindrical device with a diameter of 1500 mm (much larger than the diameter of 100 mm of the air chamber). As shown in Figure 4, the experimental device consists of: 1. Hydraulic pressure loader; 2. Force sensor; 3. Dual-chamber air spring; 4. Orifice plate; and 5. Air pressure sensor.

As shown in Figure 5, in order to study the effects of different upper and lower air chamber ratios on the vibrating characteristics, the length of the lower air chamber (200 mm) was kept constant while the upper air chamber was changed (100 mm, 150 mm, 200 mm, 250 mm, 300 mm, 350 mm, 400 mm) to study the effect of different volume ratios. The upper and lower air chambers are connected by an orifice plate and the air flows back and forth through an orifice. The orifice changes the air distribution of the chambers greatly and can influence the vibrating characteristics of the structure. In this experiment, the orifice diameters used to study the effect of different orifice ratios on the air spring were 0, 10, 20, 30, 40, 50, and 60 mm.

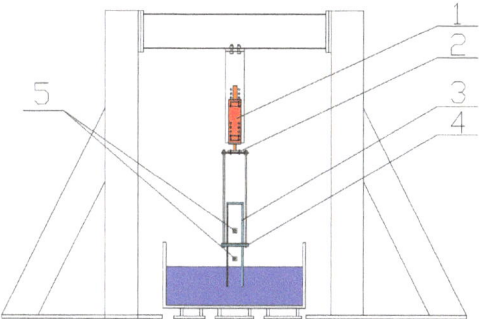

Figure 4. Schematic diagram for the experimental device.

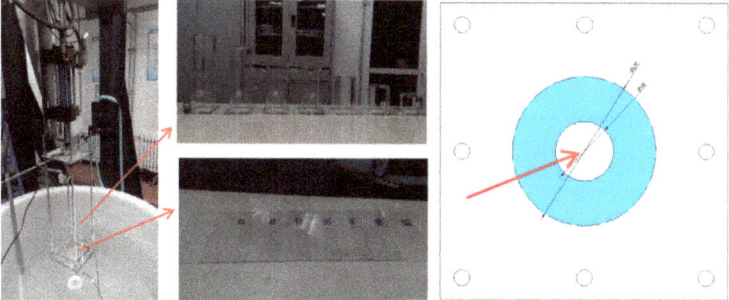

Figure 5. Upper chamber and orifice plates of the dual-chamber air spring.

As shown in Figure 6, a hydraulic servo fatigue machine is used as the loading system of the experiment. The fatigue machine has excellent performance and can be used for large structure experiments. By controlling the loading frequency and amplitude, we can test the effect of external load on the vibrating characteristics of the air spring. In the experiment, the amplitude range was 2, 4, 6, 8, 10, 12, and 14 mm, and the frequency range was 0.497–5.474 Hz.

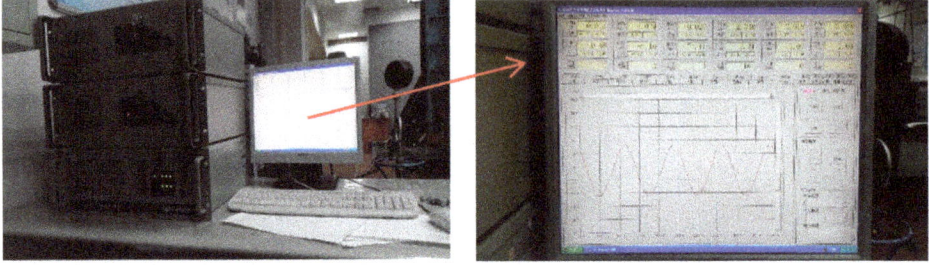

Figure 6. Loading system for experimental device.

3.2. Setting of Test Parameters

We studied the influences of each factor on dual-chamber air spring dynamic stiffness and damping. The experiment assumes the cross section of the upper air chamber equals that of the lower air chamber.

Variables to be considered: Volume of upper air chamber V_2, Volume of lower air chamber V_1, opening diameter d, amplitude of outer excitation A, and frequency of outer excitation ω.

The length of the lower air chamber, L_1, and the frequency of outer excitation, ω, were used as the basic variables with the following three dimensionless numbers:

Height of lower air chamber $L_1 = 0.2$ m

Diameter of air chamber $D = 0.1$ m

$\xi_1 = \dfrac{L_2}{L_1}$, volume ratio, values = 0.5, 0.75, 1, 1.25, 1.5, 1.75, 2

$\xi_2 = \dfrac{d}{D}$, orifice ratio, values = 0, 0.1, 0.2, 0.3, 0.4, 0.5, 0.6

$\xi_3 = \dfrac{A}{L_1}$, ratio of the amplitude of outer excitation and the height of air chamber, values = 0.01, 0.02, 0.03, 0.04, 0.05, 0.06, 0.07.

Frequency of outer excitation. Log value $\ln\omega$ was used to investigate the vibrating characteristics of the air spring at low frequency (values = $-0.7, -0.3, 0.1, 0.5, 0.9, 1.3, 1.7$).

In total, four variables and seven values were tested, as shown in Table 1:

Table 1. Four factors and seven levels chart.

Factor / Level	Volume Ratio (ξ_1)	Orifice Ratio (ξ_2)	Loading Amplitude Ratio (ξ_3)	Loading Frequency $\ln\omega$	Real Loading Frequency (Hz)
1	0.5	0.0	0.01	−0.7	0.497
2	0.75	0.1	0.02	−0.3	0.741
3	1.0	0.2	0.03	0.1	1.105
4	1.25	0.3	0.04	0.5	1.649
5	1.5	0.4	0.05	0.9	2.460
6	1.75	0.5	0.06	1.3	3.669
7	2.0	0.6	0.07	1.7	5.474

314

3.3. Orthogonal Experiment and Orthogonal Array

For experimental designs with multiple factors and multiple levels, a large number of experiments are required. For example, for the experimental scheme with four variables and seven levels in this paper, 2401 (7^4) experiments need to be carried out. Hence, a method with the minimum numbers of experiments is needed to find out the effect of each parameter on the vibrating characteristics of the air spring.

In 1947, Rao [31] developed factorial experiments, an optimization experimental method (an efficient testing strategy) that can determine the relation between factors and a test index and specify the primary and secondary factors using fewer trials. Taguchi [32] effectively used orthogonal arrays in his research. Azouzi and Guillot [33] examined the feasibility for an intelligent sensor fusion to estimate online surface finish and dimensional deviations using orthogonal arrays. Green, Krieger, and Wind [34] applied orthogonal arrays to conjoint analysis. The use of the orthogonal test method minimizes the number of tests but still allows determination of the changing rules of all factors. We applied the orthogonal experiment method to investigate the effect of upper and lower air chamber volume ratio, orifice ratio, loading amplitude, and loading frequency of outer excitation.

As an effective method of solving multi-factor experiment problems, orthogonal experimental design selects typical points from the overall experiment and tests them. These typical points are "even" and "regular": (1) in the experiment, each factor has the same occurrence number at each different level; (2) each combination of two factors at each different level occurred in the experiment and had the same occurrence number. Meeting these two criteria indicates that the experiment scheme designed by orthogonal experiment method is typical. This method reduces the number of experiments and can generally reflect the overall effect of each factor at each level on the index.

Range analysis is commonly used in scientific experiments. Change in experimental results often occurs in scientific experiments due to two types of factors. One is variations arising from random effects; such effects are controllable in experiments, and are therefore inevitable. The other is artificial control produces changes in experimental results. When such factors have a significant impact on the experiment, they are bound to significantly alter the results, accompanied by random factors. Conversely, when such factors have no significant effect on the experimental results, corresponding changes will not be manifested obviously, so changes in experimental results have been substantially ascribed to the effects of random factors. The purpose of conducting scientific experiments is often to determine whether these artificially controllable factors have an impact on the experimental results and what the effect is.

Range analysis is an effective tool to judge the above matters through analyzing the data variation in experimental results, because it can separate random variation from non-random variation in a hybrid state to help determine the source of dominant variations. Using range analysis, we analyzed the results of orthogonal design to explore the effect of various factors on air spring characteristics. In addition, we investigated the effects of four factors (upper and lower air chamber volume ratio, orifice ratio, loading amplitude, and loading frequency of outer excitation) on the air vibration characteristics of the dual-chamber. Seven variation levels of each factor were examined. The tests were performed 49 times, accounting for only 2% of the traditional number of trials. The arrangements and results of the orthogonal analysis scheme are shown in Appendix 1.

4. Experimental Results and Analysis of a Dual-Chamber Air Spring

We used an intuitive method in orthogonal analysis that looks at means and ranges. The mean average is the average value of different factors at the same level, and this can reveal the effect of different levels of single factors on the indicators. The range indicates the maximum value of the average numerical difference for each factor at each level, and reflects the impact of each column factor at different levels, from which the primary and secondary sequence of factors can be determined. If each factor has the same number of level, the influence extent of each factor can be judged directly by comparing the range size.

For 49 sets of experimental schemes, the comparison of energy dissipation is shown in Figure 7.

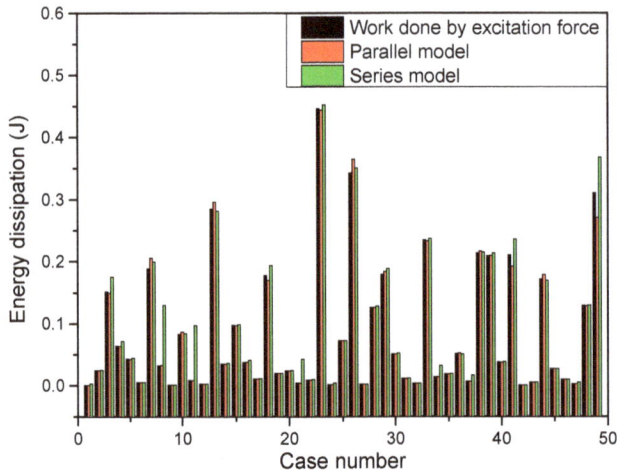

Figure 7. Energy dissipation test program comparison chart.

Figure 7 shows that under most operating conditions, there is a good fit between the parallel model and work done by excitation force, but under individual conditions, the series model shows better goodness of fit. From the relationship of energy dissipation mean of the 49 sets of orthogonal design schemes,

the goodness of fit of the parallel model $\alpha_p = 1 - \dfrac{\displaystyle\sum_0^{49} \dfrac{\left| Q_{parallel} - Q_{structure} \right|}{Q_{structure}}}{49} = 99.88\%.$

The goodness of fit of the series model $\alpha_s = 1 - \dfrac{\displaystyle\sum_0^{49} \dfrac{\left| Q_{series} - Q_{structure} \right|}{Q_{structure}}}{49} = 89.43\%.$

By the range relation of energy consumption, we can infer a higher goodness of fit between work done by excitation force and energy dissipation of the parallel model. Furthermore, the parallel model can be employed to make deeper analyses of the structure.

4.1. Influence of Various Parameters on Air Spring Stiffness

Stiffness is a function of the volume ratio, orifice ratio, excitation amplitude, and frequency. The influence law can be seen from the following figures:

As can be seen from Figure 8, the effect of the volume ratio on the dual-chamber vibration absorber stiffness is not monotonic. With an increase in the volume ratio, the stiffness of vibration absorber first slowly decreases and increases at a volume ratio of 1.25, then decreases and finally increases dramatically. As per Figure 9, the effect of the orifice ratio on the dual-chamber vibration absorber stiffness similarly is not monotonic. With an increase in the orifice ratio, the stiffness of vibration absorber first shows little change, then increases at a orifice ratio of 0.4, then decreases and finally increases to its highest level.

According to Figure 10, the effect of the loading amplitude ratio on the dual-chamber vibration absorber stiffness is basically monotonic. With an increase in the amplitude ratio, the stiffness of the dual-chamber air spring also increases and slightly protrudes at the amplitude ratio of 0.04. Figure 11 shows the effect of the loading frequency on the dual-chamber vibration absorber stiffness is also monotonic. The stiffness of the dual-chamber air spring monotonically increases with an increase in the loading frequency. As can be seen from Figure 12, the effect of each factor on the dual-chamber vibration absorber stiffness varies, where the volume ratio, orifice ratio, and loading amplitude ratio exert considerable influence but the loading frequency has the maximum impact. The results described in Figure 12 are only applicable when considering the volume ratio, the orifice ratio, load amplitude, and frequency in this case.

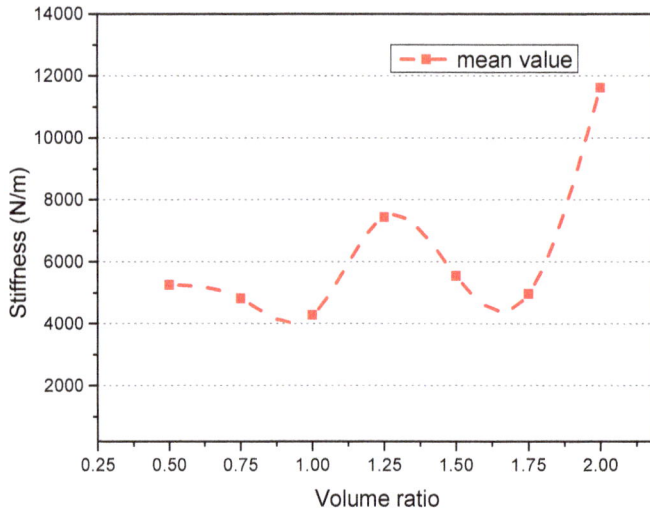

Figure 8. Effect of the volume ratio on the dual-chamber air spring stiffness.

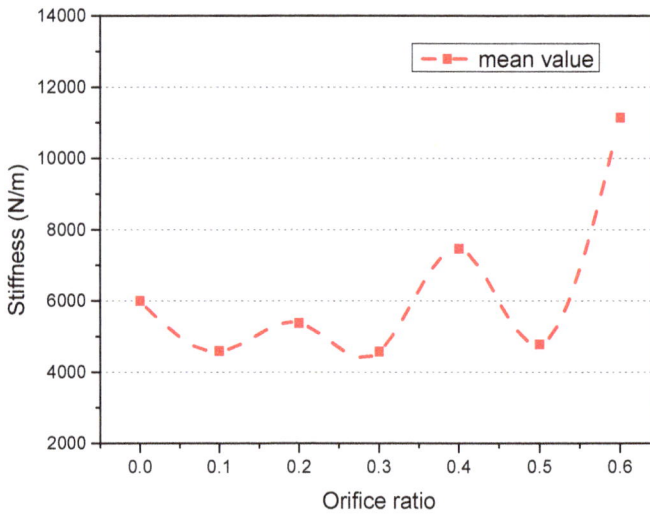

Figure 9. Effect of the orifice ratio on the dual-chamber air spring stiffness.

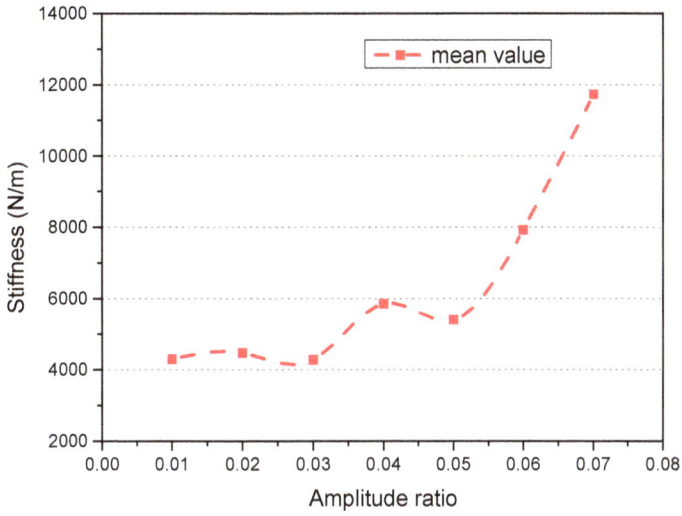

Figure 10. Effect of the amplitude ratio on the dual-chamber air spring stiffness.

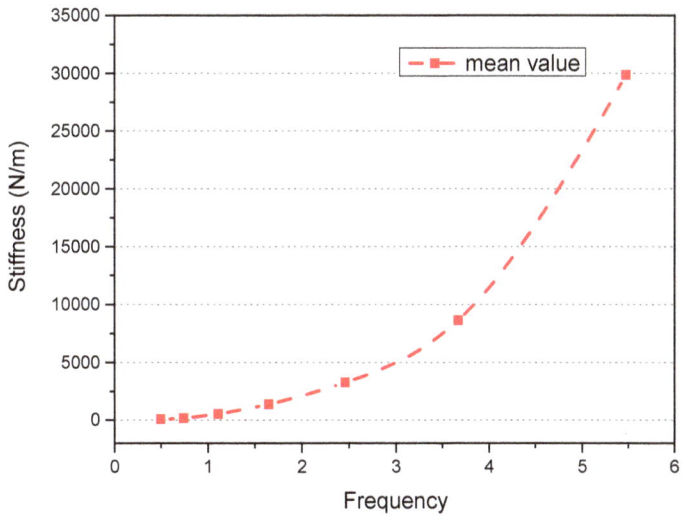

Figure 11. Effect of the frequency on the dual-chamber air spring stiffness.

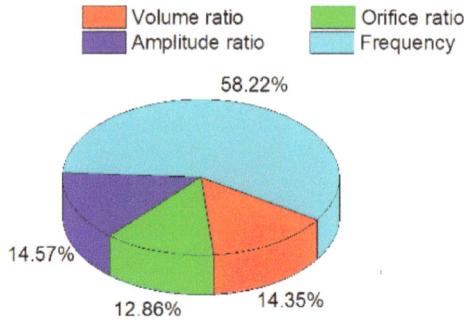

Figure 12. Summary of the effects of the four factors on the dual-chamber air spring stiffness.

4.2. Influence of Various Parameters on Air Spring Damping

Damping is a function of the volume ratio, orifice ratio, excitation amplitude, and frequency. In this section the four factors for the dual-chamber air spring damping effects are examined.

The influence law can be seen from the following figures:

Figure 13 shows that the effect of the volume ratio of the dual-chamber vibration absorber damping is not monotonic. With an increase in the volume ratio, the damping of the vibration absorber first slowly decreases and then increases at a volume ratio of 1.25, then decreases and then increases. As per Figure 14, the effect of the orifice ratio on the dual-chamber vibration absorber damping is also not monotonic. With an increase in the orifice ratio, the damping of vibration absorber first shows little change, then increases at an orifice ratio of 0.4, then decreases and then increases.

According to Figure 15, the effect of the loading amplitude ratio on the dual-chamber vibration absorber damping is not monotonic. With an increase in the amplitude ratio, the structural damping shows little change at first and then declines at 0.05, before eventually increasing. Figure 16 indicates that the effect of the loading frequency on the dual-chamber vibration absorber damping is basically monotonic. With an increase in the amplitude ratio, the vibration absorber damping increases until 3.669 Hz, and then increases dramatically. As summarized in Figure 17, the effect of each factor on the dual-chamber vibration absorber damping varies, where the volume ratio, orifice ratio, and loading amplitude ratio have strong effects but the loading frequency has the maximum impact. The results described in Figure 17 are only applicable when considering the volume ratio, the orifice ratio, load amplitude, and frequency in this case.

320

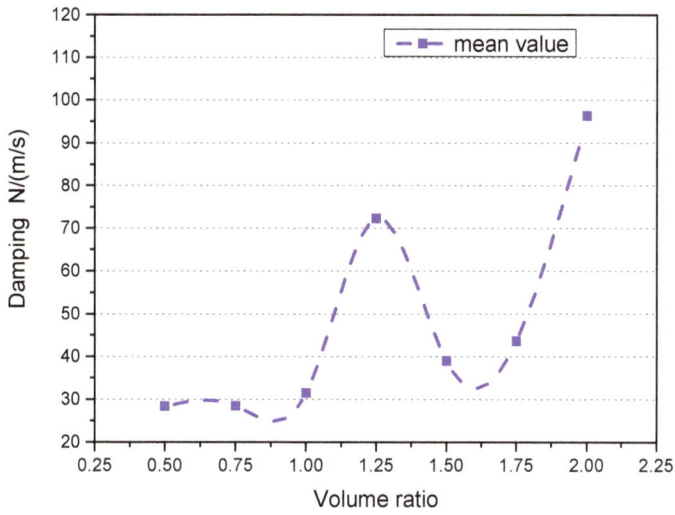

Figure 13. Effect of the volume ratio on the dual-chamber air spring damping.

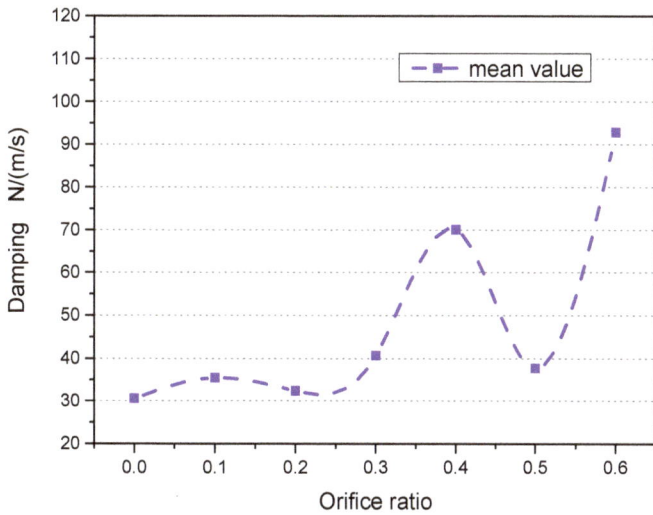

Figure 14. Effect of the orifice ratio on the dual-chamber air spring damping.

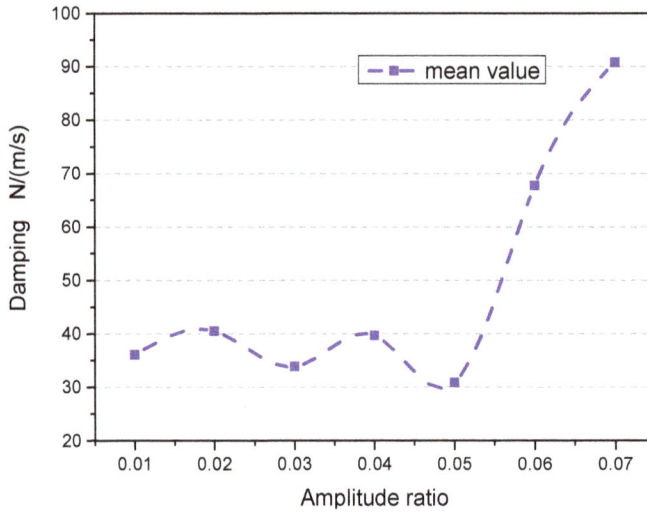

Figure 15. Effect of the amplitude ratio on the dual-chamber air spring damping.

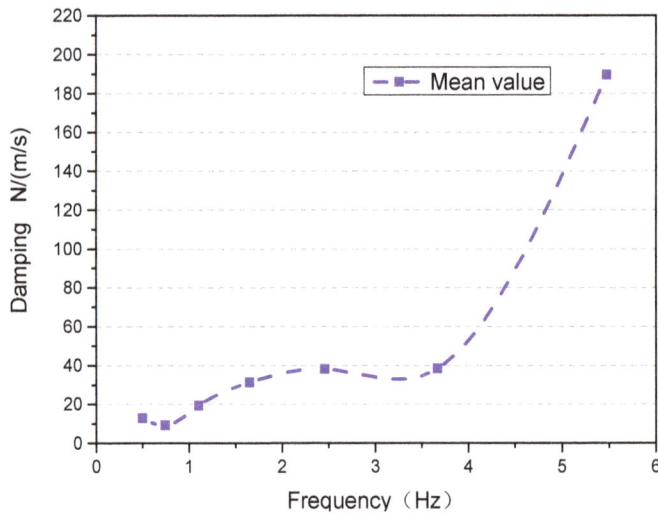

Figure 16. Effect of the frequency on the dual-chamber air spring damping.

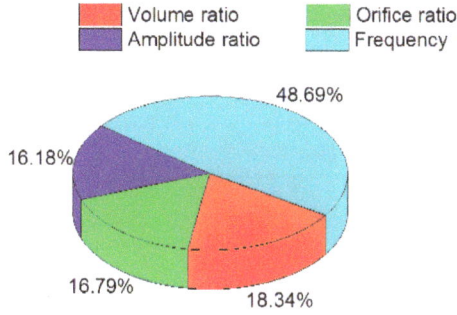

Figure 17. Summary of the effects of the four factors on dual-chamber air spring damping.

Thus, the four factors influence the stiffness and damping of the gas-liquid coupling dual-chamber air spring in a complex manner; there is not a simple linear relationship. Each factor has different effects on stiffness and damping. Prediction of the effects on stiffness and damping of the gas-liquid coupling dual-chamber air spring based on these four factors would be useful for efficient design, and in the next chapter, we propose an efficient model of dual-chamber air spring stiffness and damping to do this.

5. Rapid Calculation Model of Dual-Chamber Air Spring Stiffness and Damping

From the previous analyses, we can see that volume ratio, orifice ratio, loading amplitude, and loading frequency all have complex effects on the stiffness and damping of the dual-chamber air spring device and are not monotonic in most cases. Therefore complex models must be constructed using statistical analyses of the experimental data.

5.1. Normalization Process of Each Factor

In order to facilitate the analysis of each factor's effect, the normalization process of each factor is performed.

Normalized volume ratio $\varphi_1 = \dfrac{\xi_1}{2}$, values of 0.25, 0.375, 0.5, 0.625, 0.75, 0.875, 1.

Normalized orifice ratio $\varphi_2 = \dfrac{\xi_2}{0.6}$, values of 0, 0.167, 0.333, 0.5, 0.667, 0.833, 1.

Normalized loading amplitude ratio $\varphi_3 = \dfrac{\xi_3}{0.07}$, values of 0.142, 0.286, 0.429, 0.571, 0.714, 0.857, 1.

Normalized loading frequency $\varphi_4 = \dfrac{\ln\omega + 0.7}{2.4}$, values of 0, 0.167, 0.333, 0.5, 0.667, 0.833, 1.

The functional relationship between the stiffness and damping of the dual-chamber air spring and each factor is:

$$K_{modle} = f_1(\varphi_1, \varphi_2, \varphi_3, \varphi_4) \tag{10}$$

$$C_{modle} = f_2(\varphi_1, \varphi_2, \varphi_3, \varphi_4) \tag{11}$$

5.2. Determination of Rapid Calculation Model of Stiffness and Damping

From the above analysis, each factor has a nonlinear impact on the stiffness and damping of the dual-chamber air spring. The quadratic function can not accurately simulate the variation law of the stiffness and damping of the dual-chamber air spring. When the power of an independent variable is more than three, the regression function becomes very unstable [35], thus destabilizing the application of the regression model. In order to more accurately reflect the effect law of each factor on the stiffness and damping of the dual-chamber air spring, this paper uses cubic polynomials, which can usually be converted into ordinary multiple linear regression for processing. Experimental data are used to fit the expression of the stiffness and damping of the dual-chamber air spring.

Suppose the stiffness experimental value meets the following cubic polynomials:

$$
\begin{aligned}
f_1 &= \beta_0 + \sum_{j=1}^{34} \beta_j g_j(\varphi_1, \varphi_2, \varphi_3, \varphi_4) \\
&= \beta_0 + \beta_1\varphi_1 + \beta_2\varphi_2 + \beta_3\varphi_3 + \beta_4\varphi_4 + \beta_5\varphi_1^2 + \beta_6\varphi_2^2 + \beta_7\varphi_3^2 + \beta_8\varphi_4^2 + \beta_9\varphi_1\varphi_2 + \beta_{10}\varphi_1\varphi_3 + \beta_{11}\varphi_1\varphi_4 \\
&\quad + \beta_{12}\varphi_2\varphi_3 + \beta_{13}\varphi_2\varphi_4 + \beta_{14}\varphi_3\varphi_4 + \beta_{15}\varphi_1^3 + \beta_{16}\varphi_2^3 + \beta_{17}\varphi_3^3 + \beta_{18}\varphi_4^3 + \beta_{19}\varphi_1^2\varphi_2 + \beta_{20}\varphi_1^2\varphi_3 + \beta_{21}\varphi_1^2\varphi_4 \\
&\quad + \beta_{22}\varphi_1\varphi_2^2 + \beta_{23}\varphi_1\varphi_3^2 + \beta_{24}\varphi_1\varphi_4^2 + \beta_{25}\varphi_2^2\varphi_3 + \beta_{26}\varphi_2^2\varphi_4 + \beta_{27}\varphi_2\varphi_3^2 + \beta_{28}\varphi_2\varphi_4^2 + \beta_{29}\varphi_3^2\varphi_4 + \beta_{30}\varphi_3\varphi_4^2 \\
&\quad + \beta_{31}\varphi_1\varphi_2\varphi_3 + \beta_{32}\varphi_1\varphi_2\varphi_4 + \beta_{33}\varphi_1\varphi_3\varphi_4 + \beta_{34}\varphi_2\varphi_3\varphi_4 + \varepsilon_i(o^4)
\end{aligned} \tag{12}
$$

where $\beta_0, \beta_1, \beta_2, ..., \beta_{34}$ are undetermined coefficients; $\varphi_1, \varphi_2, \varphi_3, \varphi_4$ are controllable variables in the experiment; $g_j(\varphi_1, \varphi_2, \varphi_3, \varphi_4)$ is function expression corresponding to β_j; and f_1 is the stiffness of dual-chamber air spring.

This paper uses the least squares method to estimate the undetermined coefficients in the above formula to obtain the regression equation for stiffness:

$$
\begin{aligned}
\frac{\hat{f_1}}{E13} &= -0.0952 + 1.4139\varphi_1 - 0.6852\varphi_2 - 1.3314\varphi_3 + 1.2366\varphi_4 - 0.9296\varphi_1^2 + 0.5123\varphi_2^2 \\
&\quad + 0.3488\varphi_3^2 - 0.7531\varphi_4^2 - 0.3542\varphi_1\varphi_2 - 0.7975\varphi_1\varphi_3 - 1.5502\varphi_1\varphi_4 + 2.7841\varphi_2\varphi_3 \\
&\quad - 0.5386\varphi_2\varphi_4 + 0.2573\varphi_3\varphi_4 - 0.1684\varphi_1^3 - 0.5071\varphi_2^3 + 1.4815\varphi_3^3 - 0.2310\varphi_4^3 \\
&\quad + 0.8761\varphi_1^2\varphi_2 + 1.3092\varphi_1^2\varphi_3 + 0.1184\varphi_1^2\varphi_4 - 0.0996\varphi_1\varphi_2^2 - 0.3868\varphi_1\varphi_3^2 + 0.0888\varphi_1\varphi_4^2 \\
&\quad + 1.1500\varphi_2^2\varphi_3 - 0.6931\varphi_2^2\varphi_4 - 2.5784\varphi_2\varphi_3^2 + 0.1512\varphi_2\varphi_4^2 - 2.7151\varphi_3^2\varphi_4 + 1.6950\varphi_3\varphi_4^2 \\
&\quad - 2.1074\varphi_1\varphi_2\varphi_3 + 1.1258\varphi_1\varphi_2\varphi_4 + 1.3134\varphi_1\varphi_3\varphi_4 + 0.6595\varphi_2\varphi_3\varphi_4
\end{aligned} \tag{13}
$$

Similarly, the damping regression equation can be acquired based on the stiffness calculation method:

$$
\begin{aligned}
\frac{\hat{f}_2}{E11} =\ & -0.2648 + 3.0819\varphi_1 - 1.6038\varphi_2 - 2.5479\varphi_3 + 2.5604\varphi_4 - 2.1802\varphi_1^2 + 1.4020\varphi_2^2 \\
& +1.5283\varphi_3^2 - 1.6530\varphi_4^2 - 0.8512\varphi_1\varphi_2 - 1.5508\varphi_1\varphi_3 - 3.7226\varphi_1\varphi_4 + 5.4696\varphi_2\varphi_3 \\
& -0.8495\varphi_2\varphi_4 + 0.0787\varphi_3\varphi_4 - 0.1734\varphi_1^3 - 0.9562\varphi_2^3 + 1.4337\varphi_3^3 - 0.2797\varphi_4^3 \\
& +1.6602\varphi_1^2\varphi_2 + 2.1616\varphi_1^2\varphi_3 + 0.8800\varphi_1^2\varphi_4 - 0.1481\varphi_1\varphi_2^2 - 0.3982\varphi_1\varphi_3^2 + 0.6600\varphi_1\varphi_4^2 \\
& +0.9527\varphi_2^2\varphi_3 - 0.8391\varphi_2^2\varphi_4 - 3.9493\varphi_2\varphi_3^2 + 0.4225\varphi_2\varphi_4^2 - 3.5250\varphi_3^2\varphi_4 + 2.5354\varphi_3\varphi_4^2 \\
& -3.3547\varphi_1\varphi_2\varphi_3 + 1.6821\varphi_1\varphi_2\varphi_4 + 1.9625\varphi_1\varphi_3\varphi_4 + 0.3758\varphi_2\varphi_3\varphi_4
\end{aligned}
\tag{14}
$$

Then the stiffness and damping of the dual-chamber air spring can be calculated by the following formulas:

$$
K_{\text{modle}} = f_1\left(\varphi_1, \varphi_2, \varphi_3, \varphi_4\right) = \hat{f}_1 \cdot E13 \tag{15}
$$

$$
C_{\text{modle}} = f_2\left(\varphi_1, \varphi_2, \varphi_3, \varphi_4\right) = \hat{f}_2 \cdot E11 \tag{16}
$$

Equations (15) and (16) determine the cubic polynomial equation of stiffness and damping.

5.3. Test of Calculation Model

In statistics, regression analysis is performed using the variables. The coefficient of determination. R^2 is the ratio between the regression sum of squares and sum of squares for total. R^2 is between 0 and 1. The closer it is to 1, the better the fit of the regression. A goodness of fit of more than 0.8 is typically considered acceptable.

In multiple regression analysis, R^2 expression:

$$
R^2 = 1 - \frac{SSE}{SST} \tag{17}
$$

where SST is the sum of squares for total; and SSE is the sum of squares for error.

$$
SST = \sum_{i=1}^{N} \left(f_{1i} - \overline{f}_{1i}\right)^2 \tag{18}
$$

$$
SSE = \sum_{i=1}^{N} \left(f_{1i} - \hat{f}_{1i}\right)^2 \tag{19}
$$

The coefficient of determination R^2 is related to the number of independent variables. In order to accurately test the accuracy of the model, it is necessary to take into account the degrees of freedom and make adjustments of R^2 expression according to the size of the formula.

$$R_a^2 = 1 - \frac{SSE/(n-p-1)}{SST/(n-p)} = 1 - \left(1 - R^2\right) \cdot \left(\frac{n-1}{n-p-1}\right) \tag{20}$$

$R_a{}^2$ is denoted as the adjusted coefficient of determination; n is the total number of experiments, and p is the number of variables. The adjusted coefficient of determination more accurately reflects the degree of fit between the model and experimental data.

By making an analysis of the data ($n = 49$, $p = 35$), the following can be obtained:

1. For the stiffness model $R^2 = 0.99154$, $R_a{}^2 = 0.96378$, the goodness of fit is 96.378%, indicating that the model can well simulate and predict the experimental results. This indicates the effectiveness of the cubic polynomial regression equation fit.
2. For the damping model, $R^2 = 0.98970$, $R_a{}^2 = 0.95470$, the goodness of fit is 95.470%, again indicating that the model can well simulate and predict the experimental results. This again supports the effectiveness of the cubic polynomial regression equation fit.

5.4. Prediction of Experimental Results by the Rapid Calculation Model

To further validate the forecast accuracy of the stiffness and damping rapid calculation model, we conducted a series of experiments. To ensure the universality, these experiments consist of random combinations of four factors at different levels, different from the levels tested in the test programs in the orthogonal table.

The straight lines in Figures 18 and 19 are experimental values, while red dots are predictive values for the rapid calculation model. If a dot falls on the line, this indicates the predicted value is equivalent to the experimental value; deviations from the straight line indicate differences between the predicted and the experimental values. As seen in the figures, the predicted and experimental values fit well. For more horizontal level combination schemes, the predicted value can be calculated based on the rapid calculation model proposed in this paper, which can play a guiding role to engineering structural design and applications.

The experiments performed here were representative in the parameter range, and the polynomial prediction model is appropriate for these values. The application of the polynomial prediction model is as follows: upper and lower air chamber volume ratio, $\xi_1 \in [0.5 - 2.0]$, orifice ratio, $\xi_2 \in [0 - 0.6]$, loading amplitude ratio, $\xi_3 \in [0.01 - 0.07]$, and loading frequency $f \in [0.497 - 5.474]$.

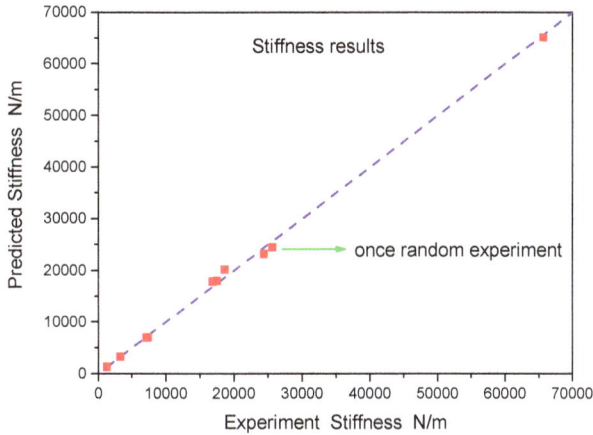

Figure 18. Comparison of experimental and predicted values on the dual-chamber air spring stiffness.

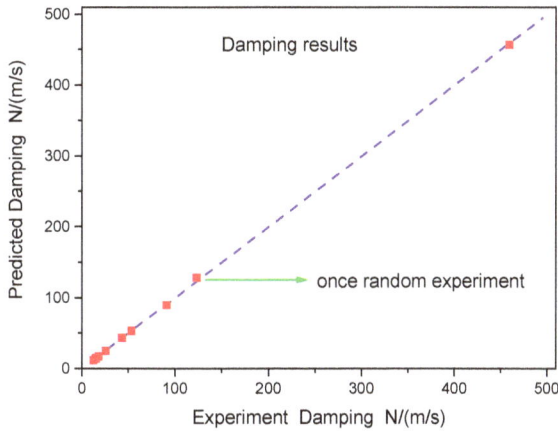

Figure 19. Comparison of experimental and predicted values on the dual-chamber air spring damping.

6. Conclusions

The mechanical differences between the gas-liquid coupling dual-chamber air spring device and the traditional one make the analysis of effects on stiffness and damping more challenging. The compression of the gas in the traditional air spring is directly on the external load of the air chamber by the piston. The pressure variation of the air spring chamber is the same phase and amplitude as the external load, but for the dual-chamber air spring device applied to motion suppression of marine structures, the movement of the liquid column and the external excitation differ in phases and amplitude.

327

The current study analyzes the characteristics of a dual-chamber air spring device. A parallel model and series model were used to simulate the dual-chamber air spring energy-absorbing device. An orthogonal test scheme was used to investigate the effect laws of four factors (upper and lower air chamber volume ratio, orifice ratio, loading amplitude, and loading frequency of outer excitation) on the air vibration characteristics of the dual-chamber. Based on the experimental results, a higher-order non-linear regression method was obtained, achieving a rapid calculation model for dual-chamber air spring stiffness and damping, and the reliability of the method was verified experimentally. Using the rapid calculation model, from the upper and lower air chamber volume, orifice ratio, the frequency, and amplitude of external load, we can determine the stiffness and damping of the dual-chamber air spring device applied to motion suppression of marine structures by the formulas (15) and (16). The main findings are:

1. Based on energy consumption results, the goodness of fit of the parallel model was 89.43%, and the goodness of fit of the series model was 99.88%. The parallel model is more consistent with the real physical model.
2. The effects of volume ratio and orifice ratio on dual-chamber vibration absorber stiffness were not monotonic, but the loading amplitude ratio and frequency tended toward monotonic increasing.
3. The effects of the volume ratio, orifice ratio, and loading amplitude ratio on the dual-chamber vibration absorber stiffness did not behave in a monotonic manner, but the loading frequency on damping tended toward monotonic increasing.
4. A polynomial rapid calculation model for stiffness and damping was constructed. The accuracy of the rapid calculation model results was verified by the experimental results, and the predicted values were in good agreement with the experimental values.

Acknowledgments: The authors would like to express their gratitude for the financial support of the National Natural Science Foundation of China (Grant Nos. 11072246, 51490673, 10702073, 11232012) and the National Basic Research Program of China (973 Program 2014CB046801)

Author Contributions: Xiaohui Zeng put forward the overall idea. Liang Zhang contributed to the theoretical and experimental analysis, and wrote the manuscript. Yang Yu contributed to the experimental study. Min Shi contributed to data analysis. Jifu Zhou contributed to data analysis.

Conflicts of Interest: The authors declare no conflict of interest.

Appendix

Table A1. Orthogonal table L49 and the experimental analysis results.

Case	Volume Ratio (ξ_1)	Orifice Ratio (ξ_2)	Loading Amplitude Ratio (ξ_3)	Loading Frequency $\ln\omega$	Real Loading Frequency (Hz)	Parallel Model K	Parallel Model C	Parallel Model Q	Series Model k	Series Model c	Series Model q	Work Done by Excitation Force W
1	0.5	0	0.01	-0.7	0.497	109.192	14.772	0.0004292	128.627	97.766	0.003081	0.0004297
2	0.5	0.1	0.03	0.5	1.649	1249.035	19.870	0.02411	1282.923	752.237	0.02483	0.02437
3	0.5	0.2	0.05	1.7	5.474	24512.047	58.106	0.1492	24674.823	8808.250	0.1752	0.1516
4	0.5	0.3	0.07	0.1	1.105	503.285	15.400	0.06352	525.984	356.854	0.07156	0.06384
5	0.5	0.4	0.02	1.3	3.669	7208.513	47.043	0.04246	7371.526	2127.292	0.04434	0.04304
6	0.5	0.5	0.04	-0.3	0.741	158.223	7.249	0.005029	165.412	166.801	0.005102	0.005048
7	0.5	0.6	0.06	0.9	2.460	3121.949	36.450	0.2058	3223.487	1157.179	0.1996	0.1887
8	0.75	0	0.07	1.3	3.669	10804.935	10.771	0.03249	10810.636	20423.049	0.1299	0.03209
9	0.75	0.1	0.02	-0.3	0.741	152.925	5.989	0.001248	158.002	186.403	0.001331	0.001287
10	0.75	0.2	0.04	0.9	2.460	3298.782	35.056	0.08677	3387.664	1336.123	0.08447	0.08329
11	0.75	0.3	0.06	-0.7	0.497	98.908	8.265	0.008236	105.624	129.970	0.09743	0.008478
12	0.75	0.4	0.01	0.5	1.649	1277.190	20.470	0.002832	1312.362	763.784	0.002948	0.002881
13	0.75	0.5	0.03	1.7	5.474	17619.902	102.952	0.2962	18330.771	2654.775	0.2812	0.2846
14	0.75	0.6	0.05	0.1	1.105	519.751	15.930	0.03474	543.270	367.970	0.03593	0.03468
15	1	0	0.06	0.5	1.649	1412.352	24.368	0.09738	1457.423	787.955	0.09871	0.09771
16	1	0.1	0.01	1.7	5.474	17009.600	98.264	0.03743	17680.432	2589.838	0.04081	0.0374
17	1	0.2	0.03	0.1	1.105	561.650	16.714	0.01086	585.609	408.524	0.01111	0.01092
18	1	0.3	0.05	1.3	3.669	7642.988	36.205	0.17	7734.053	3074.820	0.1937	0.178
19	1	0.4	0.07	-0.3	0.741	179.227	8.004	0.02	186.963	193.429	0.02007	0.02
20	1	0.5	0.02	0.9	2.460	3121.603	28.248	0.02371	3182.590	1474.095	0.02413	0.02384
21	1	0.6	0.04	-0.7	0.497	99.429	8.566	0.004021	106.606	127.237	0.04211	0.004046
22	1.25	0	0.05	-0.3	0.741	175.538	7.100	0.00899	181.754	207.608	0.009773	0.009058
23	1.25	0.1	0.07	0.9	2.460	3294.971	45.801	0.4436	3446.867	1039.328	0.452	0.446
24	1.25	0.2	0.02	-0.7	0.497	93.417	19.146	0.001581	131.580	66.013	0.004269	0.001549
25	1.25	0.3	0.04	0.5	1.649	1337.610	35.092	0.07292	1436.307	510.685	0.07293	0.07287
26	1.25	0.4	0.06	1.7	5.474	39581.430	324.946	0.365	42733.898	4404.865	0.3508	0.3427

329

Table 1. *Cont.*

Case	Volume Ratio (ξ_1)	Orifice Ratio (ξ_2)	Loading Amplitude Ratio (ξ_3)	Loading Frequency lnω	Real Loading Frequency (Hz)	Parallel Model			Series Model			Work Done by Excitation Force
						K	C	Q	k	c	q	W
27	1.25	0.5	0.01	0.1	1.105	507.804	26.241	0.002135	573.125	230.241	0.002108	0.002117
28	1.25	0.6	0.03	1.3	3.669	7085.175	47.500	0.1258	7254.263	2037.839	0.1285	0.1262
29	1.5	0	0.04	1.7	5.474	25749.741	114.565	0.184	26352.096	5012.048	0.1885	0.1795
30	1.5	0.1	0.06	0.1	1.105	578.651	17.741	0.05048	604.853	409.543	0.05246	0.05053
31	1.5	0.2	0.01	1.3	3.669	7492.063	34.781	0.01139	7577.801	3074.069	0.01196	0.0116
32	1.5	0.3	0.03	-0.3	0.741	191.848	8.807	0.003483	200.600	201.881	0.003591	0.003496
33	1.5	0.4	0.05	0.9	2.460	3380.530	47.534	0.2329	3540.001	1055.189	0.2364	0.2345
34	1.5	0.5	0.07	-0.7	0.497	83.222	14.337	0.01418	107.242	64.009	0.03212	0.01412
35	1.5	0.6	0.02	0.5	1.649	1376.694	34.954	0.01857	1471.837	540.732	0.01881	0.01865
36	1.75	0	0.03	0.9	2.460	3187.457	27.938	0.05243	3245.880	1552.170	0.05043	0.05119
37	1.75	0.1	0.05	-0.7	0.497	85.259	12.130	0.006464	102.043	73.747	0.01677	0.006484
38	1.75	0.2	0.07	0.5	1.649	1529.732	46.205	0.2168	1679.347	518.626	0.2152	0.2133
39	1.75	0.3	0.02	1.7	5.474	18778.734	133.334	0.2094	19897.493	2371.391	0.2137	0.2089
40	1.75	0.4	0.04	0.1	1.105	573.199	29.053	0.03713	644.131	263.825	0.03814	0.03742
41	1.75	0.5	0.06	1.3	3.669	10461.160	45.520	0.1921	10566.332	4573.206	0.2358	0.2106
42	1.75	0.6	0.01	-0.3	0.741	174.026	10.879	0.0004944	188.745	139.501	0.0004885	0.0004824
43	2	0	0.02	0.1	1.105	564.645	14.696	0.005067	583.071	465.062	0.005049	0.005031
44	2	0.1	0.04	1.3	3.669	9798.212	48.073	0.179	9923.452	3809.107	0.1693	0.1718
45	2	0.2	0.06	-0.3	0.741	198.357	16.858	0.02693	229.366	124.692	0.02689	0.02692
46	2	0.3	0.01	0.9	2.460	3509.969	47.140	0.009806	3661.021	1142.528	0.009924	0.009847
47	2	0.4	0.03	-0.7	0.497	83.851	13.458	0.002517	104.857	67.178	0.005397	0.002638
48	2	0.5	0.05	0.5	1.649	1490.134	39.147	0.1286	1600.387	568.245	0.1297	0.129
49	2	0.6	0.07	1.7	5.474	65716.150	495.144	0.2703	70124.859	7875.753	0.368	0.3109

References

1. Taylor, R.E.; Jefferys, E.R. Variability of hydrodynamic load predictions for a tension leg platform. *Ocean Eng.* **1986**, *13*, 449–490.
2. Zeng, X.H.; Shen, X.P.; Wu, Y.X. Governing equations and numerical solutions of tension leg platform with finite amplitude motion. *Appl. Math. Mech.* **2007**, *28*, 37–49.
3. Zeng, X.H.; Li, X.W.; Liu, Y.; Wu, Y.X. Nonlinear dynamic responses of tension leg platform with slack-taut tether. *China Ocean Eng.* **2009**, *23*, 37–48.
4. Ahmad, S.K.; Ahmad, S. Active control of non-linearly coupled TLP response under wind and wave environments. *Comput. Struct.* **1999**, *72*, 735–747.
5. Alves, R.M.; Batista, R.C. *Active/Passive Control of Heave Motion for TLP Type Offshore Platform*; Chung, J.S., Matsui, T., Koterayama, W., Eds.; International Society of Offshore and Polar Engineers: Cupertino, CA, USA, 1999; pp. 332–338.
6. Sakai, F.; Takaeda, S.; Tamaki, T. Tuned liquid column damper-new type device for suppression of building vibrations. In Proceedings of the International Conference on Highrise Buildings, Nanjing, China, 25–27 March 1989; pp. 926–931.
7. Lee, H.H.; Wong, S.H.; Lee, R.S. Response mitigation on the offshore floating platform system with tuned liquid column damper. *Ocean Eng.* **2006**, *33*, 1118–1142.
8. Lee, H.H.; Juang, H.H. Experimental study on the vibration mitigation of offshore tension leg platform system with UWTLCD. *Smart Struct. Syst.* **2012**, *9*, 71–104.
9. Taflanidis, A.A.; Angelides, D.C.; Scruggs, J.T. Simulation-based robust design of mass dampers for response mitigation of tension leg platforms. *Eng. Struct.* **2009**, *31*, 847–857.
10. Lee, S.K.; Lee, H.R.; Min, K.W. Experimental verification on nonlinear dynamic characteristic of a tuned liquid column damper subjected to various excitation amplitudes. *Struct. Des. Tall Spec. Build.* **2012**, *21*, 374–388.
11. Chatterjee, T.; Chakraborty, S. Vibration mitigation of structures subjected to random wave forces by liquid column dampers. *Ocean Eng.* **2014**, *87*, 151–161.
12. Zeng, X.; Yu, Y.; Zhang, L.; Liu, Q.; Wu, H. A New Energy-Absorbing Device for Motion Suppression in Deep-Sea Floating Platforms. *Energies* **2014**, *8*, 111–132.
13. Rijken, O.; Spillane, M.; Leverette, S.J. Vibration Absorber Technology and Conceptual Design of Vibration Absorber for TLP in Ultradeep Water. In Proceedings of the ASME 29th International Conference on Ocean, Offshore and Arctic Engineering, Shanghai, China, 6–11 June 2010; ASME: New York, NY, USA, 2010; pp. 629–638.
14. Bian, X.S.; Leverette, S.J.; Rijken, O.R. A TLP solution for 8000 Ft water depth. In Proceedings of the ASME 29th International Conference on Ocean, Offshore and Arctic Engineering, Shanghai, China, 6–11 June 2010; ASME: New York, NY, USA, 2010; pp. 255–262.
15. Spillane, M.W.; Rijken, O.R.; Leverette, S.J. Vibration absorbers for deepwater TLP's. In Proceedings of the 17th International Offshore and Polar Engineering Conference, Lisbon, Portugal, 1–6 July 2007.
16. Bachrach, B.I.; Rivin, E. Analysis of a damped pneumatic spring. *J. Sound Vib.* **1983**, *86*, 191–197.

17. Lee, J.H.; Kim, K.J. Modeling of nonlinear complex stiffness of dual-chamber pneumatic spring for precision vibration isolations. *J. Sound Vib.* **2007**, *301*, 909–926.

18. Jing, X.; Lang, Z. *Frequency Domain Analysis and Design of Nonlinear Systems based on Volterra Series Expansion: A Parametric Characteristic Approach*; Springer: Berlin, Germany; 2015.

19. Xiao, Z.; Jing, X. An SIMO Nonlinear System Approach to Analysis and Design of Vehicle Suspensions, IEEE/ASME trans. *Mechatronics* **2015**, *20*, 3098–3111.

20. Jing, X. Nonlinear characteristic output spectrum for nonlinear analysis and design. *IEEE/ASME Trans. Mechatron.* **2014**, *19*, 171–183.

21. Shin, Y.H.; Kim, K.J. Performance enhancement of pneumatic vibration isolation tables in low frequency range by time delay control. *J. Sound Vib.* **2009**, *321*, 537–553.

22. Moon, J.H.; Lee, B.G. Modeling and sensitivity analysis of a pneumatic vibration isolation system with two air chambers. *Mech. Mach. Theory* **2010**, *45*, 1828–1850.

23. Li, Z.; Li, X.; Chen, X. Generic vibration criteria-based dual-chamber pneumatic spring vibration isolation table design. *Proc. Inst. Mech. Eng. B J. Eng. Manuf.* **2014**, *228*, 1621–1629.

24. Liu, H.; Lee, J.C. Model development and experimental research on an air spring with auxiliary reservoir. *Int. J. Automot. Technol.* **2011**, *12*, 839–847.

25. Zhu, S.; Wang, J.; Zhang, Y. Research on theoretical calculation model for dynamic stiffness of air spring with auxiliary chamber. In Proceedings of the IEEE Vehicle Power and Propulsion Conference, Harbin, China, 3–5 September 2008.

26. Pu, H.; Luo, X.; Chen, X. Modeling and analysis of dual-chamber pneumatic spring with adjustable damping for precision vibration isolation. *J. Sound Vib.* **2011**, *330*, 3578–3590.

27. Erin, C.; Wilson, B.; Zapfe, J. An improved model of a pneumatic vibration isolator: Theory and experiment. *J. Sound Vib.* **1998**, *218*, 81–101.

28. Heertjes, M.; van de Wouw, N. Nonlinear dynamics and control of a pneumatic vibration isolator. *J. Vib. Acoust.* **2006**, *128*, 439–448.

29. Lee, J.H.; Kim, K.J. A method of transmissibility design for dual-chamber pneumatic vibration isolator. *J. Sound Vib.* **2009**, *323*, 67–92.

30. Toyofuku, K.; Yamada, C.; Kagawa, T.; Fujita, T. Study on dynamic characteristic analysis of air spring with auxiliary chamber. *JSAE Rev.* **1999**, *20*, 349–355.

31. Rao, C.R. Factorial experiments derivable from combinatorial arrangements of arrays. *Suppl. J. R. Stat. Soc.* **1947**, *9*, 128–139.

32. Taguchi, G. Performance analysis design. *Int. J. Prod. Res.* **1978**, *16*, 521–530.

33. Azouzi, R.; Guillot, M. On-line prediction of surface finish and dimensional deviation in turning using neural network based sensor fusion. *Int. J. Mach. Tools Manuf.* **1997**, *37*, 1201–1217.

34. Green, P.E.; Krieger, A.M.; Wind, Y. Thirty years of conjoint analysis: Reflections and prospects. *Interfaces* **2001**, *31*, S56–S73.

35. Kutner, M.H.; Nachtsheim, C.; Neter, J. *Applied Linear Regression Models*; McGraw-Hill/Irwin: New York, NY, USA, 2004.

Correlation of Plastic Strain Energy and Acoustic Emission Energy in Reinforced Concrete Structures

Francisco Sagasta, Amadeo Benavent-Climent, Andrés Roldán and
Antolino Gallego

Abstract: This paper presents a comparison of the acoustic emission (AE) energy and the plastic strain energy released by some reinforced concrete (RC) specimens subjected to cyclic or seismic loadings. AE energy is calculated, after proper filtering procedures, using the signals recorded by several AE low frequency sensors (25–100 kHz) attached on the specimens. Plastic strain energy is obtained by integrating the load displacement curves drawn from the measurements recorded during the test. Presented are the results obtained for: (i) two beams (with and without an artificial notch) and a beam-column connection subjected to several cycles of imposed flexural deformations; (ii) a reinforced concrete slab supported by four steel columns, and a reinforced concrete frame structure, both of the latter are subjected to seismic simulations with a uniaxial shaking table. The main contribution of this paper, which is a review of some papers previously published by the authors, is to highlight that, in all cases, a very good correlation is found between AE energy and plastic strain energy, until the onset of yielding in the reinforcing steel. After yielding, the AE energy is consistently lower than the plastic strain energy. The reason is that the plastic strain energy is the sum of the contribution of concrete and steel, while the AE energy acquired with thresholds higher than 35 dB_{AE} captures only the contribution of the concrete cracking, not the steel plastic deformation. This good correlation between the two energies before the yielding point also lends credibility to the use of AE energy as a parameter for concrete damage evaluation in the context of structural health monitoring.

Reprinted from *Appl. Sci.* Cite as: Sagasta, F.; Benavent-Climent, A.; Roldán, A.; Gallego, A. Correlation of Plastic Strain Energy and Acoustic Emission Energy in Reinforced Concrete Structures. *Appl. Sci.* **2016**, *6*, 84.

1. Introduction

Reinforced concrete (RC) structural elements develop several damage mechanisms when subjected to static/dynamic, monotonic/cyclic loadings. Before the onset of yielding of the reinforcing steel, concrete damage is associated with: (i) the opening of new cracks or the extension of existing ones when the tension stress exceeds the tensile strength of the concrete; (ii) the friction between the planes of fracture after cracking; or (iii) the concrete cracking under compressive stress. When cyclic loading is caused for example by earthquakes, concrete degradation due to cracking eventually results in cumulative damage (low-cycle fatigue damage) to the structural RC components and leads to a state in which repair becomes necessary. One serious consequence of concrete degradation under cyclic loading is the slip between the reinforcing steel bars and the surrounding concrete.

Evaluating this damage in concrete is not easy because RC structures are commonly covered up by non-structural elements (brick veneers, casings, cement plasters, stuccos, *etc.*). Such elements make simple visual inspection complicated. Moreover, visual inspection provides only qualitative information on the "apparent" damage, but no information on the "cumulative" damage resulting from numerous cycles of imposed deformations. It is here where non-destructive techniques play an important role. Among them, the measurement, recording and analysis of Acoustic Emission (AE) signals prove very effective [1–4]. When applying AE techniques to damage evaluation, it is of paramount importance [5–11]: (i) to discriminate relevant signals from spurious AE records; and (ii) to correlate AE with a well-established index of mechanical damage. The AE may be characterized in terms of the so-called MARSE (Mean Measured Area under the Rectified Signal Envelope) energy [2]. Meanwhile, the energy dissipated by a structure through plastic strain deformations (irreversible deformations) is commonly recognized as a good indicator of damage. This paper summarizes several research initiatives that prove there is a very good correlation between AE energy calculated from the AE signals, and the plastic strain energy dissipated by concrete. Based on this correlation, the level of damage in an RC structure can be assessed from the AE records. The correlation is demonstrated under relatively simple loading conditions, *i.e.*, static and monotonic loadings [12,13], yet also under extremely cumbersome loads such as random dynamic cyclic loading caused by earthquakes [13–19]. The AE signals recorded during a seismic event are extraordinarily complex; and unveiling their relation with the damage accumulated on the structure requires considerable post-processing work [13,16,18–20].

2. Cyclic Loading

2.1. RC Beams

Firstly, a $100 \times 100 \times 1000$ mm^3 RC beam (Figure 1) was built and subjected to three-point bending tests [13]. Concrete compressive strength was 25 MPa. Reinforcement consisted of four longitudinal bars of 4 mm thickness and 76×76 mm^2 square stirrups. Reinforcement steel yield strength was 500 MPa.

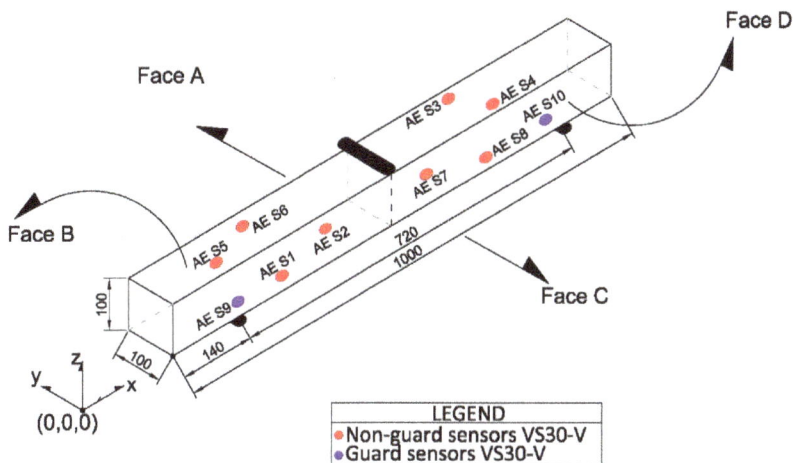

Figure 1. Undamaged Reinforced Concrete (RC) beam and ten Acoustic Emission (AE) sensors. S1 and S2 on face C; S3 and S4 on face B; S5 and S6 on face A; S7 and S8 on face D. Distances in mm.

Applied force, Q, and the corresponding vertical displacement, δ, were measured with a load cell and a displacement transducer. AE signals were acquired with ten channels of AMSY-5 equipment using ten VS30-V sensors (Vallen Systeme, Icking, Germany) with an almost flat bandwidth response in low frequency (25–100 kHz). Sampling frequency was set at 5 MHz, and the number of samples per signal was 4096. The threshold used was 35.1 dB_{AE}. Sensors S9 and S10 were configured as guard sensors to eliminate noise generated by friction at the supports. After recording, only signals passing the guard sensors and a filter based on the Root Mean Square (RMS) (explained in more detail in [13]) were taken into account during the analysis.

The bending test was carried out by applying a history of incremental load cycles controlled by force (Figure 2). Each cycle comprises three branches: loading, holding load and unloading, which are represented in Figure 2 by sets of three consecutive segments separated by a segment of null load.

Final failure occurred at $Q = 30.72$ kN. Figure 3 shows the load-displacement and load-curvature diagrams. Residual plastic deformations are clearly observed from the initial cycles, indicating that damage took place even in the earlier cycles. This damage likewise results in a degradation of stiffness (Figure 4). This figure shows that for cycles 1, 2 and 3, the variation of stiffness between the loading and unloading branches was not significant. However, from cycle 4 onward, this variation becomes more noticeable. Thus, from Figures 3 and 4 it follows that the reinforcement steel began to plastify during cycle 4. The stiffness degradation became more pronounced as the load increased from cycle 5 onward.

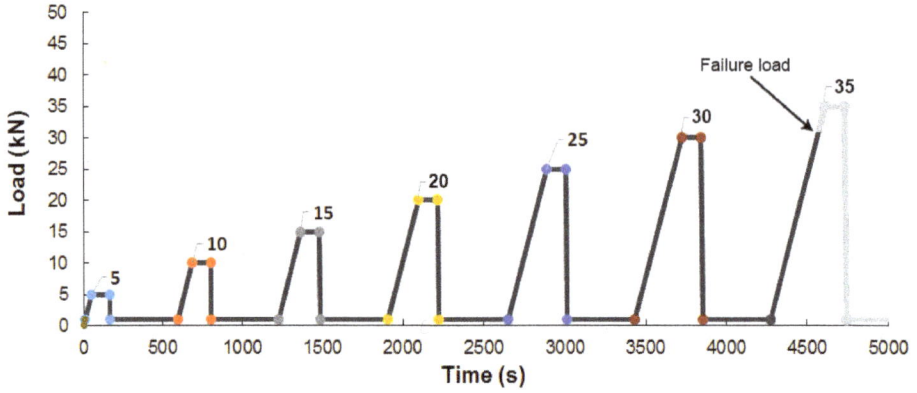

Figure 2. Cycle loading applied to the beam without notch.

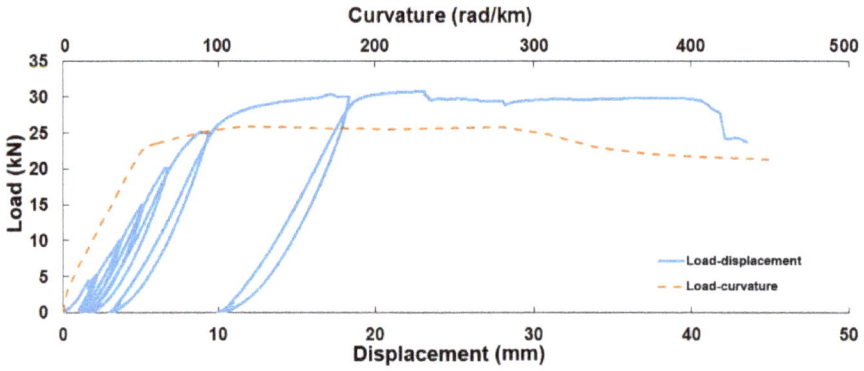

Figure 3. Load-curvature (dashed line) and load-displacement diagrams (solid line).

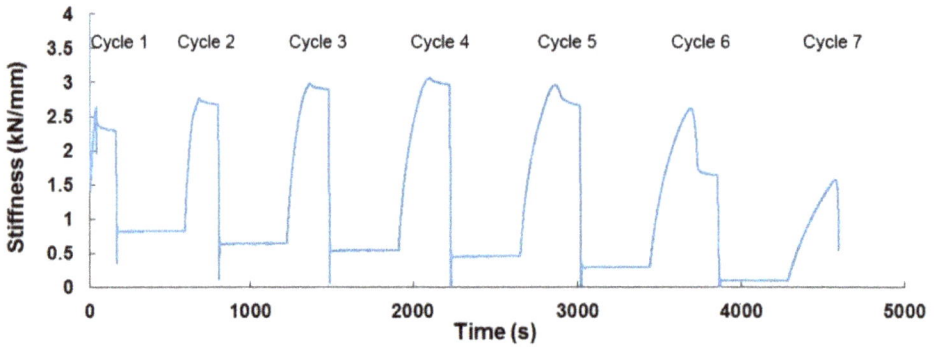

Figure 4. Stiffness degradation of the beam without notch.

336

Accumulated AE MARSE energy, E^{AE}, was calculated for the first hit of the events constructed with sensors S1–S8 [13] (Figure 5). In this figure, the plastic strain energy W_P dissipated by the specimen is also included. Both E^{AE} and W_P were normalized by their respective values E_0^{AE} and $W_{P,0}$ at the onset of yielding in the steel reinforcement, referred to hereafter as t_0. W_P was obtained by integrating the load-displacement curve (Q-δ). W_P is the sum of the energy dissipated by the concrete, W_{PC}, and the energy dissipated by the steel, W_{PS}, as

$$
\begin{aligned}
W_P &= W_{Pc} \; if \; t < t_0 \\
W_P &= W_{Pc} + W_{Ps} \; if \; t \geq t_0.
\end{aligned}
\tag{1}
$$

AE energy is also a double contribution of concrete and steel. However, with the AE threshold used (35.1 dB) during acquisition, the very low amplitude AE proceeding from the plastic strain of steel cannot be recorded, *i.e.*,

$$
E^{AE} = E_c^{AE} + E_s^{AE} \cong E_c^{AE}.
\tag{2}
$$

As observed in Figure 5, before the yielding point, a very good correlation exists between E^{AE}/E_0^{AE} and $W_P/W_{P,0}$. However, after this point, E^{AE} is always lower than W_P. Moreover, the decreasing of AE energy—and thus its separation from the plastic strain energy—is in good agreement with the results reported in [21], *i.e.*, AE energy decreases as the energy absorbed by the reinforcement through plastic deformations increases. This result also supports the increase of the b-value at the onset of yielding in the steel reinforcement [13]. In sum, it is held that

$$
\begin{aligned}
W_P &\sim E^{AE} \; if \; t < t_0 \\
W_P &> E^{AE} \; if \; t \geq t_0,
\end{aligned}
\tag{3}
$$

which is in very good agreement with Equations (1) and (2).

Secondly, a $100 \times 100 \times 1600$ mm^3 beam with an artificial notch in the middle was built and subjected to a three-point bending test (see Figure 6) [13]. Concrete, steel and reinforcement were similar to those of the previous beam. The 5×10 mm^2 notch was made according to [22]. Three VS30-V AE sensors were placed at distances of 20, 40 and 60 cm from the middle of the notch. Sensors 4 and 5 were configured as guards and only signals that passed the RMS filter were taken into account during analysis [13].

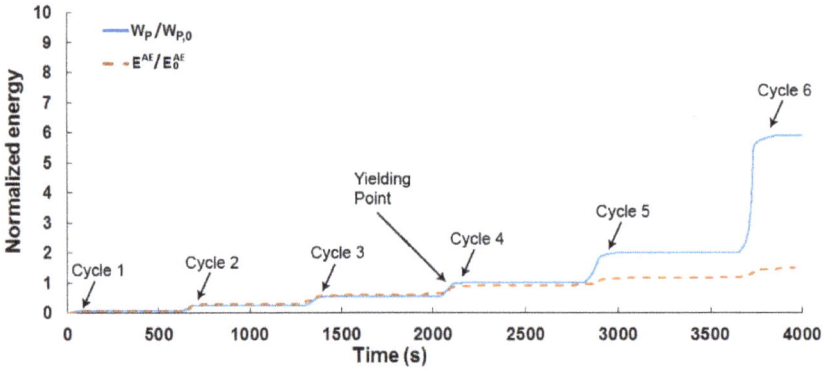

Figure 5. Accumulated AE energy (dashed line) and accumulated hysteretic strain energy (solid line). RC beam without notch.

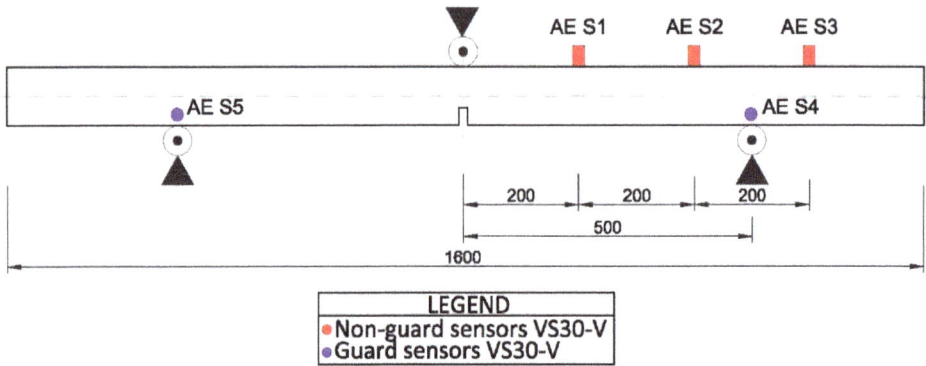

Figure 6. Notched RC beam tested and AE sensors. Distances in mm.

The bending test entailed applying incremental load cycles by force control. Cycles were proportional to each other, with a variation of the maximum load of 4 kN per cycle (see Figure 7). Final failure occurred at $Q = 20.3$ kN. Figure 8 shows the load-displacement and load-curvature diagrams, while degradation of stiffness in each cycle is plotted in Figure 9. Figures 8 and 9 suggest that the yielding point (t_0) was probably reached near the end of cycle 3.

Accumulated AE MARSE energy, E^{AE}, was also calculated for the first-hit of the events constructed with sensors S1 to S3 [13]. Figure 10 shows both energies, E^{AE} and W_P, normalized by their respective values E_0^{AE} and $W_{P,0}$ at the onset of yielding of the steel reinforcement, t_0. The result observed coincides with that of the unnotched beam, $i.e.$, there is a very good correlation between E^{AE} and W_P until t_0 and a divergence of the two energies after this point.

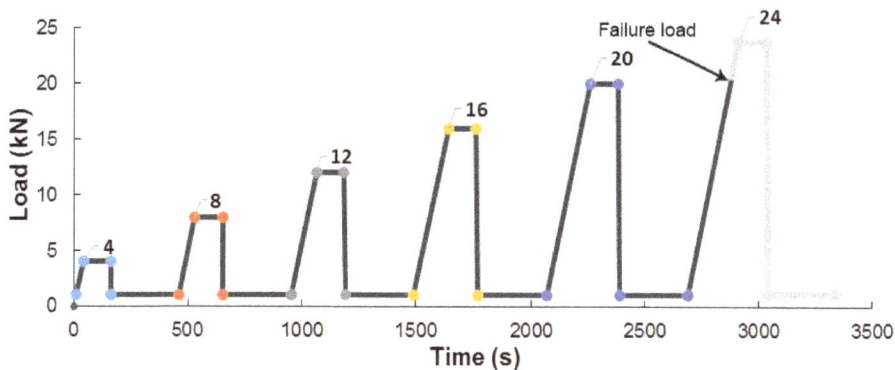

Figure 7. Cycle loading applied to beam with notch.

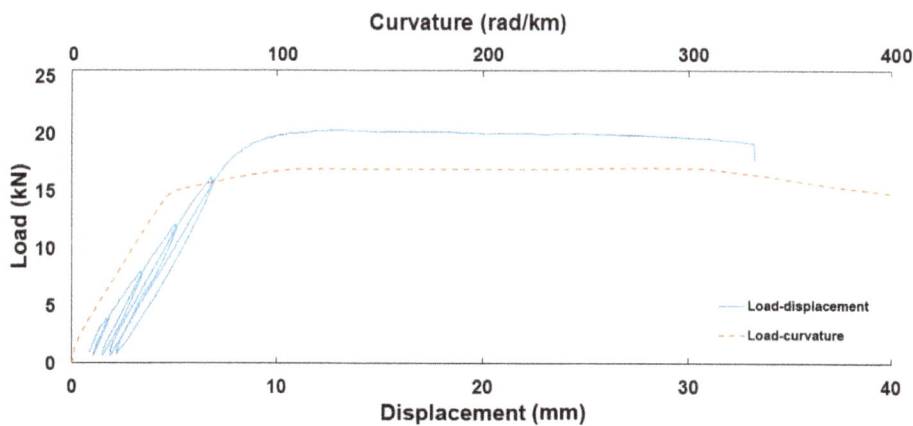

Figure 8. Load-curvature (dashed line) and load-displacement diagrams (solid line).

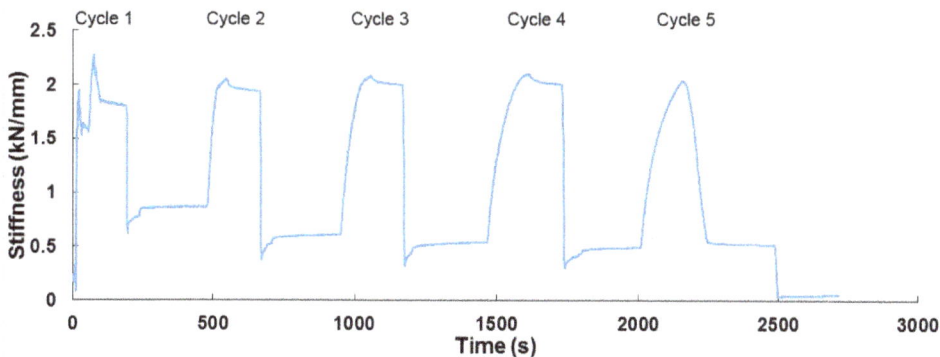

Figure 9. Stiffness degradation of the notched beam.

339

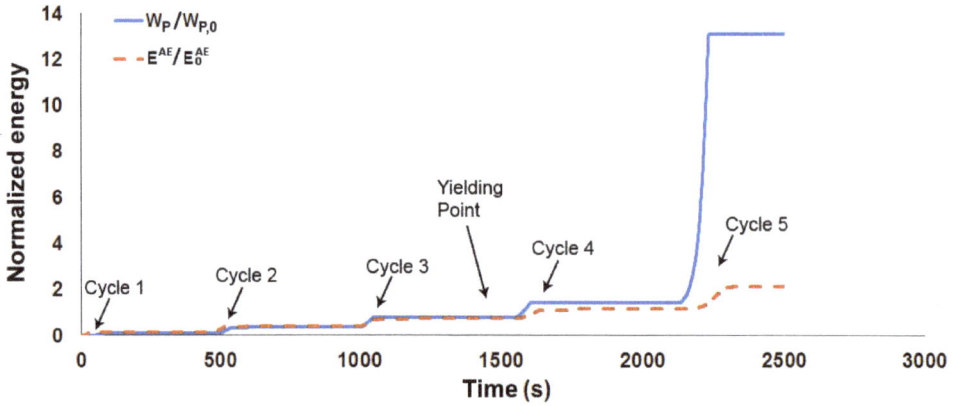

Figure 10. Accumulated AE energy (dashed line) and accumulated hysteretic strain energy (solid line). RC beam with notch.

2.2. Beam-Column Connection

A 3/5 scale specimen was constructed to represent an exterior wide beam-column connection in a prototype RC moment-resisting frame with six stories and four bays. The prototype building was designed assuming a location in the earthquake-prone southern part of Spain. The specimen corresponds to the second story of the building. The steel yield stress was 404 MPa, and the concrete compressive strength 24.9 MPa [12]. The test model was installed in the loading machine shown in Figure 11, reproducing the actual boundary conditions of the subassemblage within the RC frame under lateral loads. Gravity loading was simulated by the combination of wide beam self-weight and sand bags of total weight 40 kN placed on the beam, plus an axial force applied to the column of 214 kN by means of two post-tensioned rods.

Subsequently, cyclic horizontal forced displacements were applied automatically by the actuator following the scheme shown in Figure 12a, which is based on the ATC-24 loading protocol [23], until failure. The amplitude of the cycles was made constant within each set, but it increased with every consecutive set of cycles, following the sequence $0.5\Delta y$, $0.75\Delta y$, $1.0\Delta y$, $2\Delta y$, $3\Delta y$, $4\Delta y$ and so on, where Δy is the quotient between the predicted lateral displacement at the first yielding of the reinforcing steel and the total height of the column (180 cm) expressed as a percentage (drift-ratio).

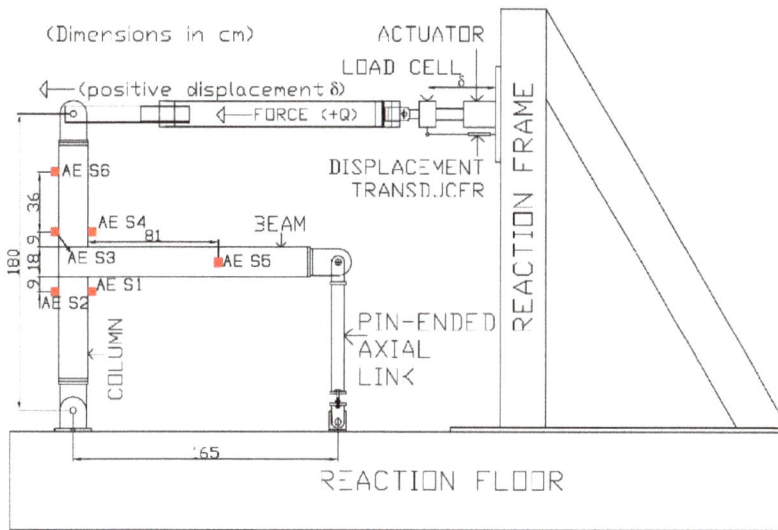

Figure 11. Experimental set up for the beam-column connection subjected to cycling loads.

A load cell and a displacement transducer installed on the actuator measured the overall horizontal force, Q, and the corresponding horizontal displacement, δ, while strain gauges attached to the reinforcing steel measured the corresponding strains (Figure 12b). Six AE low-frequency VS30-V sensors (Figure 11) were placed on each specimen. In all the channels, 25–180 kHz pass filters were used; signals were preamplified 34 dB$_{AE}$, and 40 dB$_{AE}$ was set as the acquisition threshold.

Figure 12b shows the strains recorded by the gauges fixed to the longitudinal reinforcing bars of the beam at the column face (*i.e.*, at the section of maximum bending moment of the beam). The solid horizontal line indicates the strain corresponding to yielding. It is observed that the steel reinforcement remained elastic (*i.e.*, undamaged) during the first two sets of cycles. Figure 13 shows the curves of overall load-displacement, Q-δ, obtained from the tests. The solid circle indicates the onset of plastic deformations in the longitudinal reinforcement, which occurred before the maximum lateral strength was attained. Failure was assumed to occur when the load reached at the peak displacement of a given cycle was lesser than 75% of the maximum load attained by the specimen in previous cycles. This percentage may be considered as the limit of the "usable" capacity of the member [24].

341

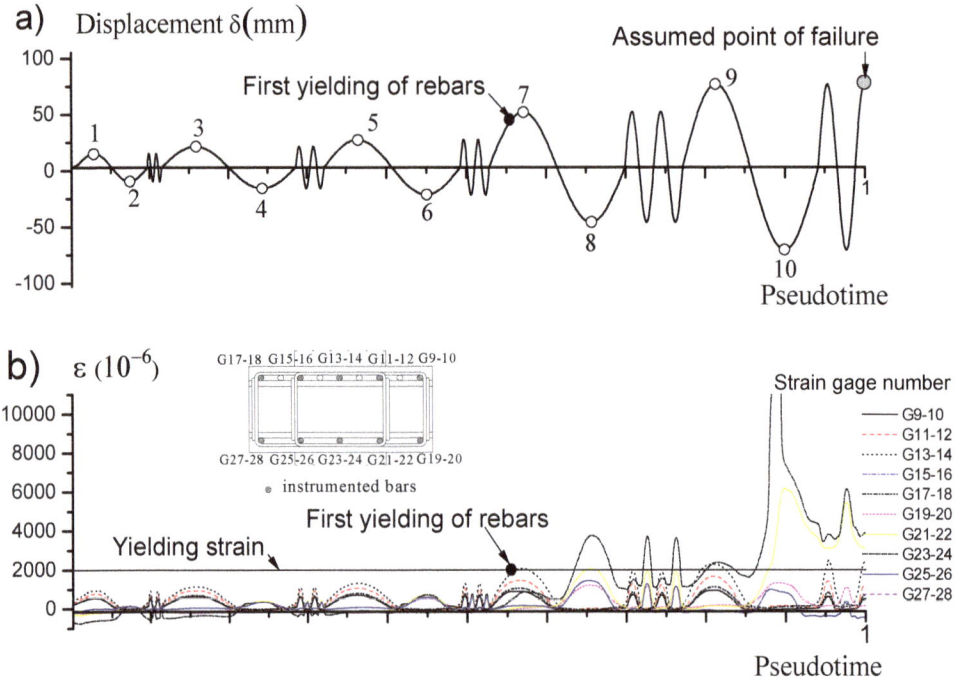

Figure 12. Displacement history applied to beam-column connection (**a**) and corresponding strains in gauges (**b**).

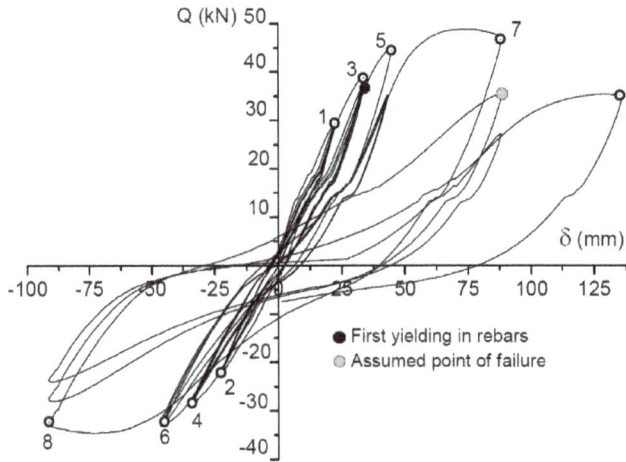

Figure 13. Load-displacement curves of beam-column connection.

The accumulated AE MARSE energy, E^{AE}, was calculated for all sensors. Figure 14 displays with dashed lines the E^{AE} calculated from the signals recorded by sensors one (on the column) and five (on the beam) [12]. As seen in this figure, the overall plastic strain energy W_P dissipated is superimposed by solid lines. Both E^{AE} and W_P were normalized by their respective values E_0^{AE} and $W_{P,0}$ at the onset of yielding in the steel reinforcement, timed at t_0. Once again, there is a very good correlation between E^{AE} and W_P until t_0. After this point, however, E^{AE} is always lower than W_P, which is again in very good agreement with Equations (1) and (2).

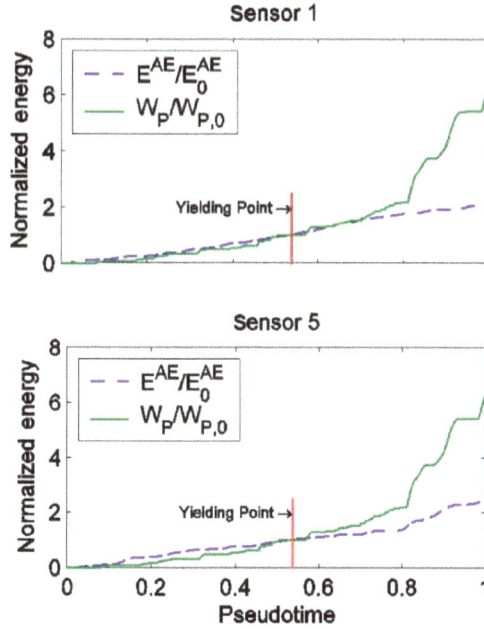

Figure 14. Accumulated AE energy (dashed line) and accumulated hysteretic strain energy (solid line). Beam-column connection.

Since, in specimens subjected to cyclic loadings, the load was applied quasi-statically, the low speed of loading made it possible to follow (in most cases by the naked eye) the evolution of different cracking modes. Initially, microcracking started to develop in the tensile side of beams and columns, followed by macroscopic cracks that formed when the tensile strength of concrete was reached. Both flexural and shear cracks were identified, and the former type predominated over the latter during the loading process. In the case of the RC beams reported in Subsection 2.1, flexural cracks (perpendicular to the axis of the member) started at the midspan of the beam. These cracks enlarged as the level of applied force increased and new flexural cracks appeared, extending progressively to both ends. Shear cracks (inclined approximately 45 degrees) were also observed near the points where the load or the reaction forces were applied. In the case of the beam-column connection, flexural cracks developed at the tension sides of beam and columns, and shear cracks were visible in the region where beam and column intersect (commonly known as beam-column joint).

3. Seismic Loading

3.1. RC Slab

A prototype one-story, one-bay structure (2.8 m high and 4.8 m long) consisting of an RC slab supported on four box-type steel columns was designed following current Spanish codes NCSE-02 and EHE-08. From this prototype structure, the corresponding test model was derived by applying the similarity laws described in [15]. The depth of the slab was 125 mm. It was reinforced with steel meshes: one on the top made with 6 mm diameter bars spaced 100 mm, and another on the bottom consisting of 10 mm diameter bars spaced 75 mm. The average yield stress of the reinforcing steel was 467 MPa, and the average concrete strength was 23.5 MPa. The model was tested with a uniaxial MTS 3 × 3 m² shaking table as shown in Figure 15.

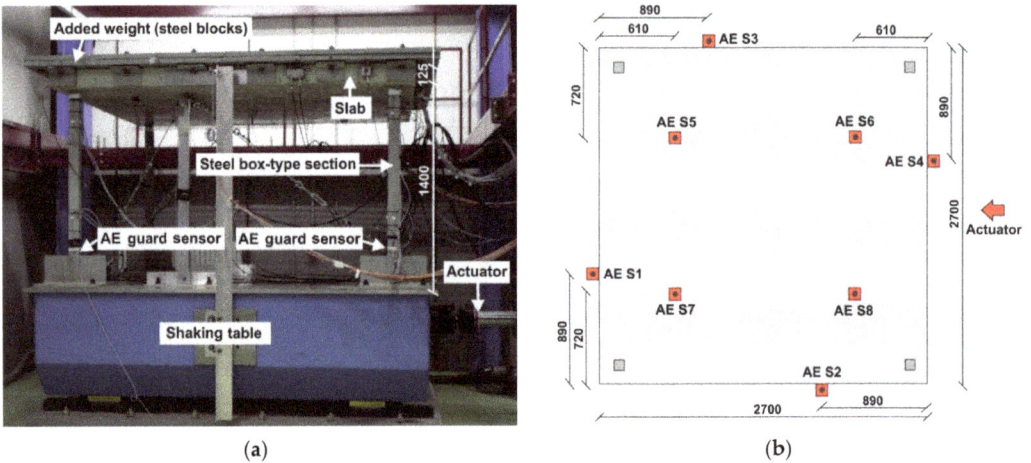

(a)   (b)

Figure 15. RC slab with AE sensors (S1 to S8) on the shaking table. (**a**) Elevation. (**b**) Plan (bottom view). Distances in mm.

The acceleration record used for the tests reproduced the North-South (NS) component of the 1980 Campano-Lucano earthquake recorded at Calitri (Italy). Two series of seismic simulations were applied to the test model. The same accelerogram was used in all simulations, the scaling factor of the peak accelerations (PA) being the only difference (Figure 16). Displacements, strains and accelerations were acquired simultaneously during each seismic simulation.

Eight VS30-V AE sensors were placed on the specimen at the eight positions indicated in Figure 15. AE signal acquisition was carried out using a sampling period of 1.6 μs, 45 dB$_{AE}$ as the detection threshold and amplifiers of 34 dB$_{AE}$. An AE signal discrimination procedure was applied to prevent spurious and friction sources using guard sensors and an RMS filter [18]. The accumulated AE MARSE energy, E^{AE}, was calculated for the first-hit of the events constructed with the eight channels [15]. As an example, Figure 17 shows E^{AE} with dashed lines for simulations D1 (before the sliding point of the reinforcement steel inside the concrete) and D2 (after sliding point). In this figure, the overall plastic strain energy W_P dissipated is superimposed by solid lines. Both E^{AE} and W_P were normalized by their values at the end of each simulation. As seen in Figure 17, for simulation D1, the two energy curves are very well correlated. However, in simulation D2, this correlation is not maintained, the accumulated AE energy being consistently lower than the cumulative strain energy.

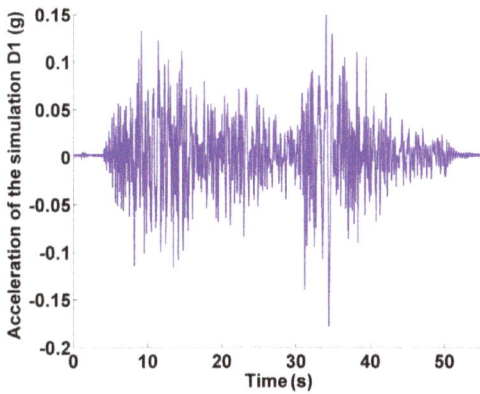

TEST SERIES		PA (g)
1	2	
Simulation		
A1		0.08
B1		0.10
C1		0.12
D1		0.19
E1		0.29
F1		0.38
G1		0.44
H1		0.58
	A2	0.19
	B2	0.38
	C2	0.58
	D2	0.66
	E2	0.74
	F2	0.95

(a)   (b)

Figure 16. (a) Acceleration of Campano-Lucano earthquake scaled at 100% (Simulation D1); (b) Seismic simulations (name and Peak Acceleration (PA)).

Figure 18 represents the final values of both energies obtained in each simulation, normalized by their respective values E_0^{AE} and $W_{P,0}$ at the onset of the sliding point of the steel reinforcement, timed at t_0. At that moment, the steel columns also began to yield. The AE threshold used (45 dB$_{AE}$) during acquisition could not record the very low amplitude AE proceeding from the plastic strain of steel. This fact justifies

345

the very good correlation between E^{AE} and W_P up to t_0 and the deviation of the two energies after this point, again in very good agreement with Equations (1) and (2).

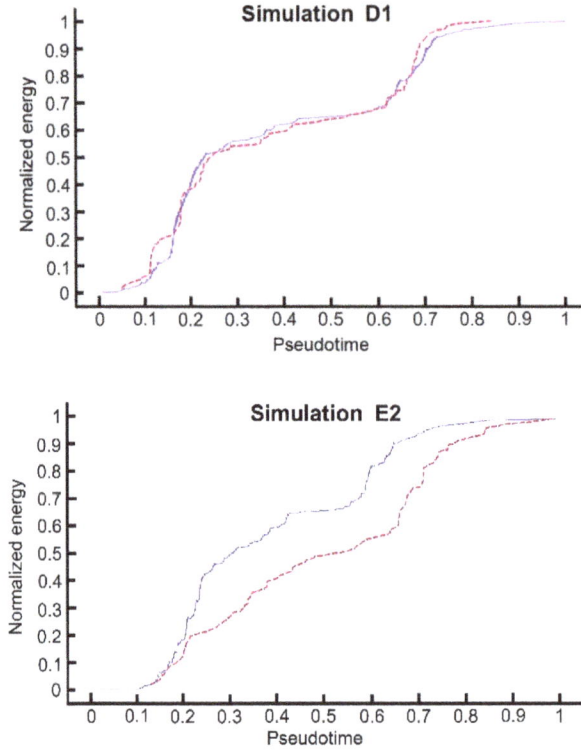

Figure 17. Accumulated AE energy (dashed line) and accumulated plastic strain energy (solid line) obtained for the RC slab. Both were normalized at the end of each simulation.

3.2. RC Frame

A prototype RC building was designed following the current Spanish seismic code NCSE-02, assuming that it was located in Granada (Spain), the most earthquake-prone region in all of Spain. From this prototype, and by applying scale factors, the test model displayed in Figure 19 was projected and constructed [14,19,20]. Basically, the test specimen consisted of four RC columns connected by RC beams (Figure 19).

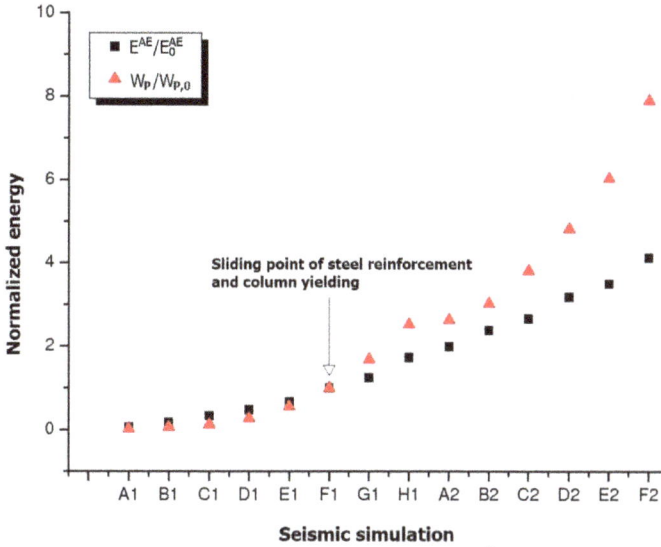

Figure 18. Accumulated AE energy (black square) and accumulated plastic strain energy (red triangle). RC slab.

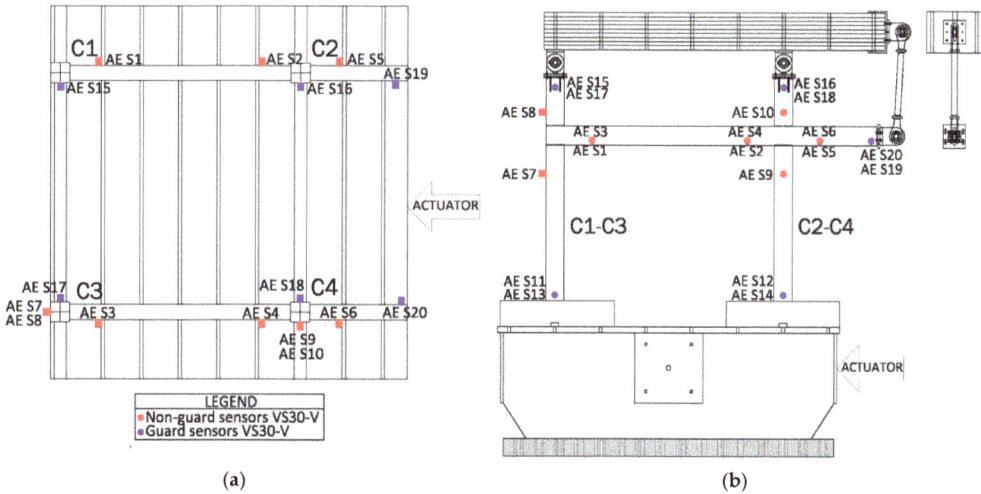

Figure 19. RC frame with AE sensors (S1 to S20). (**a**) Plan (bottom view); (**b**) Elevation.

AE signals were acquired with twenty VS30-V AE sensors (Figure 19), recording the waveform with a sampling frequency of 2.5 MHz and a detection threshold of 50 dB_{AE} and preamplifiers of 34 dB_{AE}. In addition, the steel reinforcement was instrumented with 192 strain gauges and 10 uniaxial accelerometers, and nine displacement transducers were installed on the RC specimen in order to measure the in-plane translations and the inter-story drifts in the direction of the seismic loading.

The RC frame was subjected to the same earthquake pattern as the RC slab. In this case, five seismic uniaxial tests were consecutively applied to the specimen with steadily increasing values of *PA*, *i.e.*, 50%, 50%; 100%; 200% and 300%. Figure 20 shows the corresponding accelerograms.

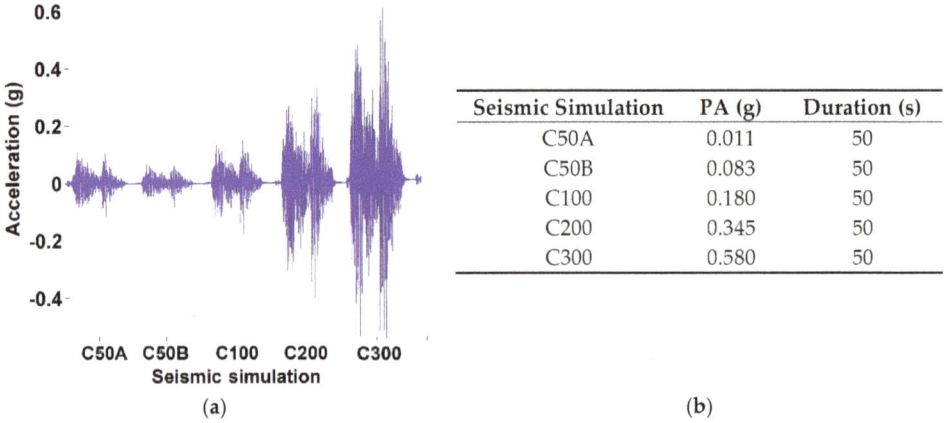

Seismic Simulation	PA (g)	Duration (s)
C50A	0.011	50
C50B	0.083	50
C100	0.180	50
C200	0.345	50
C300	0.580	50

(a)   (b)

Figure 20. (a)Acceleration of the five seismic simulations applied to the RC frame with the shaking table; (b) Seismic simulations (name, Peak Acceleration (PA) and duration).

A local damage study in the C3 and C4 beam-column connections was based on the AE energy generated at the zones around these connections, which are, in fact, the most critical points of the specimen. For interior connection C4, the sensors S2, S6, S9 and S10 were configured to construct the AE events, and the rest were considered as guard sensors. Likewise, for the exterior column-beam connection C3, only the sensors S3, S7 and S8 were configured to construct the AE events, the rest being considered as guards [19].

The Morlet Continuous Wavelet Transform (CWT) was applied in order to reconstruct AE signals using only scales 18 to 20, which correspond to the 45–64 kHz frequency band, tentatively assigned to the concrete cracking [16,17]. The accumulated AE MARSE energy, E^{AE}, was calculated for the reconstructed AE signals corresponding to the first-hit of events. Results are shown in Figure 21 along with the overall plastic strain energy (W_P) dissipated by the specimen [13,14]. Both E^{AE} and W_P were normalized by their respective values E_0^{AE} and $W_{P,0}$ at the onset of yielding of the steel reinforcement. Figure 21 reflects the same result as in previous tests, namely a very good correlation between E^{AE} and W_P until t_0, and a divergence of the energies after this point, in very good agreement with

Equations (1) and (2). It is worth noting that, besides the good correlation between E_0^{AE} and $W_{P,0}$, it is consistently observed that the development of cracks in concrete reduces the fundamental frequency of vibration of the specimen and increments the fraction of damping [14,25,26]. The reduction of the fundamental frequency is due to the decrease of stiffness. The increase of damping is associated with the energy dissipated through friction between the planes of fracture after cracking. Moreover, the accumulated damage (in terms of plastic strain energy) always increases monotonically over time, and the rate of this increase tends to augment with the amplitude of the applied load.

This satisfactory correlation is consistent with the results of other researchers [27–30]. Since in the specimens subjected to seismic loading, the forces were applied dynamically in successive seismic simulations, only macroscopic cracks were observed at the end of each one. These cracks were evident after the first seismic simulation. In subsequent simulations, macroscopic cracks extended and new ones developed. In the case of the RC slab, all the cracks identified were of the flexural type. As for the RC frame, flexural cracks developed at beam and column ends (*i.e.*, near the beam-column joint), and shear cracks were visible only within the beam-column joint.

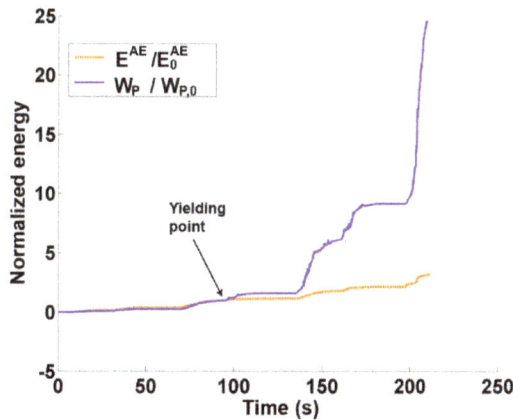

Figure 21. Accumulated AE energy (dashed line) and accumulated plastic strain energy (solid line). RC frame.

4. Conclusions

The Acoustic Emission (AE) technique was used to assess damage in several reinforced concrete (RC) elements and structures subjected to different types of loading. Simple beams, beam-column subassemblages, flat-slab structures supported on steel columns, and RC frame structures were tested. The RC elements and structures were subjected to static cyclic loads and randomly applied dynamic loadings. During each test, the AE was recorded by several low-frequency sensors. In addition, forces, accelerations and displacements were measured with load cells, accelerometers and displacement transducers. Analysis of the AE signals served to calculate the AE MARSE energy accumulated throughout the test. By integrating the load-displacement curves, the energy dissipated through plastic deformations was calculated. Plastic strain energy is commonly recognized as a good indicator of damage; and it was found that there is a good correlation between the AE MARSE energy calculated with the AE measurements and the cumulative damage on the concrete measured in terms of plastic strain energy. Such a correlation points to the utility of AE signals for developing damage indices based on AE MARSE energy, as they are capable of assessing the damage in RC structures subjected to different types of loadings. An example of the damage index based on the correlation of the strain and acoustic energies is the Sentry function, proposed in [31], in terms of the logarithmic ratio of both energies. The reason why AE MARSE energy can be utilized for the establishment of damage indices in reinforced concrete (RC) elements is that the failure of RC structural elements is typically due to the degradation (damage) of concrete, not of the reinforcing steel. That is, what typically limits the ultimate capacity of a RC element is the amount of plastic strain energy accumulated in the concrete, not in the reinforcing steel. It is also worth noting that, in a real experiment or in a real structure, calculating the damage index in terms of AE MARSE energy requires an estimation of the final value of the AE MARSE energy (*i.e.*, the value associated with failure). Past research [15,32] ascertained that this estimation can be made on the basis of the information collected from previous experiments and the volume of concrete of the specimen or of the real structure. A change in specimen size does not have appreciable effects on damage patterns, but the volume V of damaged concrete affects the number of AE events as well as the AE MARSE energy; simple relationships that relate AE MARSE energy with V have been proposed [33–36].

Acknowledgments: This research received financial support from the local government of Spain, Consejería de Innovación, Ciencia y Tecnología, projects P07-TEP-02610 and PE12-TEP-02429, from the Spanish National Plan for Scientific Research, Development and Technological Innovation (projects BIA 2005-00591 and DPI 2006-02970), and from the Formación del Profesorado Universitario (FPU) Program of Spain's Ministry of Education, Culture and Sports (AP2010-2880).

Author Contributions: All authors contributed to the work presented in this paper, and to the writing of the final manuscript. F.S. and A.R were responsible for the Acoustic Emission and the analysis of the data. A.G. and A.B-C. provided discussion on the testing procedure and the results, and were responsible for the mechanical testing and AE monitoring.

Conflicts of Interest: The authors declare no conflict of interest.

Abbreviations

The following abbreviations are used in this manuscript:

AE   Acoustic Emission
RC   Reinforced Concrete

References

1. Grosse, C.U.; Ohtsu, M. *Acoustic Emission Testing*; Springer: Berlin, Germany, 2008.
2. Miller, R.K.; Hill, E.K.; Moore, P.O. *Acoustic emission testing, NDT Handbook*, 3rd ed.; American Society for Nondestructive Testing: Columbus, OH, USA, 2005.
3. Ono, K. Application of acoustic emission for structure diagnosis. *Diagnostyka* **2011**, 2, 3–18.
4. Ohtsu, M. The history and development of acoustic emission in concrete engineering. *Mag. Concr. Res.* **1996**, *48*, 321–330.
5. ElBatanouny, M.; Larosche, A.; Mazzoleni, P.; Ziehl, P.; Matta, F.; Zappa, E. Identification of cracking mechanisms in scaled FRP reinforced concrete beams using acoustic emission. *Exp. Mech.* **2014**, *54*, 69–82.
6. Calabrese, L.; Campanella, G.; Proverbio, E. Identification of corrosion mechanisms by univariate and multivariate statistical analysis during long term acoustic emission monitoring on a pre-stressed concrete beam. *Corros. Sci.* **2013**, *73*, 161–171.
7. Mangual, J.; ElBatanouny, M.; Ziehl, P.; Matta, F. Acoustic-emission-based characterization of corrosion damage in cracked concrete with prestressing strand. *ACI Mater. J.* **2013**, *110*, 89–98.
8. ElBatanouny, M.; Ziehl, P.; Larosche, A.; Mangual, J.; Matta, F.; Nanni, A. Acoustic emission monitoring for assessment of prestressed concrete beams. *Constr. Build. Mater.* **2014**, *58*, 46–53.
9. Abdelrahman, A.; ElBatanouny, M.; Ziehl, P.; Fasl, J.; Larosche, C.J.; Fraczek, J. Classification of alkali-silica reaction damage using acoustic emission: A proof-of-concept study. *Constr. Build. Mater.* **2015**, *95*, 406–413.
10. Laroche, A.; Ziehl, P.; Mangual, J. Damage evaluation of prestressed piles to cast in pace bent cap connections with acoustic emission. *Eng. Struct.* **2015**, *84*, 184–194.
11. Abdelrahman, M.; ElBatanouny, M.; Rose, J.; Ziehl, P. Signal processing techniques for filtering acoustic emission data in prestressed concrete. *Struct. Health Monit.*, in press.
12. Benavent-Climent, A.; Castro, E.; Gallego, A. Evaluation of low cycle fatigue damage in RC exterior beam-column subassemblages by acoustic emission. *Constr. Build. Mater.* **2010**, *24*, 1830–1842.
13. Sagasta, F. Evaluación De Daño En Estructuras De Hormigón Armado Sometidas a Cargas Sísmicas Mediante El Método De Emisión Acústica. Ph.D. Thesis, Universidad de Granada, Granada, Spain, 29 January 2016. (In German)
14. Benavent-Climent, A.; Escolano-Margarit, D.; Morillas, L. Shake-table tests of a reinforced concrete frame designed following modern codes: Seismic performance and damage evaluation. *Earthq. Eng. Struct. Dyn.* **2014**, *43*, 791–810.

15. Benavent-Climent, A.; Gallego, A.; Vico, J.M. An acoustic emission energy index for damage evaluation of reinforced concrete slabs under seismic loads. *Struct. Health Monit.* **2011**, *11*, 69–81.

16. Zitto, M.E.; Piotrkowski, R.; Gallego, A.; Sagasta, F.; Benavent-Climent, A. Damage assessed by Wavelet scale bands and *b*-value in dynamical test of a reinforced concrete slab monitored with acoustic emission. *Mech. Syst. Signal Process.* **2015**, *60*, 75–89.

17. Zitto, M.E.; Piotrkowski, R.; Gallego, A.; Sagasta, F. AE wavelet processing in dynamical tests of a reinforced concrete slab. *J. Acoust. Emiss.* **2012**, *30*, 64–75.

18. Sagasta, F.; Torné, J.L.; Sánchez, A.; Gallego, A. Discrimination of acoustic emission signals for damage assessment in a reinforced concrete slab subjected to seismic simulations. *Arch. Acoust.* **2013**, *38*, 303–310.

19. Sagasta, F.; Morillas, L.; Roldán, A.; Benavent-Climent, A.; Gallego, A. Discrimination of AE signals from friction and concrete cracking in a reinforced concrete frame subjected to seismic trainings. In Proceedings of the 6th ECCOMAS Conference on Smart Structures and Materials SMART2013, Turin, Italy, 24–26 June 2013.

20. Sagasta, F.; Benavent-Climent, A.; Fernández-Quirante, T.; Gallego, A. Modified Gutenberg-Richter coefficient for damage evaluation in reinforced concrete structures subjected to seismic simulations on a shaking table. *J. Nondestruct. Eval.* **2014**, *33*, 616–631.

21. Aggelis, D.G.; Soulioti, D.V.; Getselou, E.A.; Barkoula, N.M.; Matikas, T.E. Monitoring of the mechanical behavior of concrete with chemically treated steel fibers by acoustic emission. *Constr. Build. Mater.* **2013**, *48*, 1255–1260.

22. Japanese Industrial Standards (JIS) A 1132. In *Method of Making and Curing Concrete Specimens*; Japan Concrete Institute Standard: Tokyo, Japan, 2014.

23. Luo, Y.H.; Durrani, A.J. Equivalent beam model for flat-slab buildings—Part II: Exterior connections. *ACI Struct. J.* **1995**, *92*, 250–257.

24. Scribner, C.F.; Wight, J.K. Strength decay in reinforced concrete members under load reversals. *J. Struct. Division ASCE* **1980**, *106*, 861–875.

25. Benavent-Climent, A.; Morillas, L.; Escolano, D. Seismic performance and damage evaluation of a reinforced concrete frame with hysteretic dampers through shake-table tests. *Earthq. Eng. Struct. Dyn.* **2014**, *43*, 2399–2417.

26. Benavent-Climent, A.; Donaire-Ávila, J.; Oliver-Saiz, E. Shaking table tests of a reinforced concrete waffle-flat plat structure designed following modern codes: Seismic performance and damage evaluation. *Earthq. Eng. Struct. Dyn.* **2015**.

27. Niccolini, G.; Borla, O.; Accornero, F.; Lacidogna, G.; Carpinteri, A. Scaling in damage by electrical resistance measurements: An application on the terracotta statues of the Sacred Mountain of Varallo Renaissance Complex (Italy). *Rend. Lincei-Sci. Fis.* **2015**, *26*, 203–209.

28. Niccolini, G.; Borla, O.; Lacidogna, G.; Carpinteri, A. Correlated fracture precursors in rocks and cement based materials under stress. In *Acoustic, Electromagnetic, Neutron Emissions from Fracture and Earthquakes*; Carpinteri, A., Lacidogna, G., Manuello, A., Eds.; Springer International Publishing: Cham, Switzerland, 2016.

29. Carpinteri, A.; Lacidogna, G.; Manuello, A.; Borla, O. Electromagnetic and neutron emissions from brittle rocks failure: Experimental evidence and geological implications. *J. SADHANA* **2012**, *37*, 59–78.

30. Carpinteri, A.; Lacidogna, G.; Manuello, A.; Niccolini, A.; Schiavi, A.; Agosto, A. Mechanical and electromagnetic emissions related to stress-induced cracks. *Exp. Tech.* **2010**, *36*, 53–64.

31. Selman, E.; Ghiami, A.; Alver, N. Study of fracture evolution in FRP-strengthened reinforced concrete beam under cyclic load by acoustic emission technique: An integrated mechanical-acoustic energy approach. *Constr. Build. Mater.* **2015**, *95*, 832–841.

32. Benavent-Climent, A.; Castro, E.; Gallego, A. AE monitoring for damage assessment of RC exterior beam-column subassemblages subjected to cyclic loading. *Struct. Health Monit.* **2009**, *8*, 175–189.

33. Carpinteri, A.; Pugno, N. Fractal fragmentation theory for shape effects on quasi-brittle materials in compression. *Mag. Concr. Res.* **2002**, *54*, 473–480.

34. Carpinteri, A.; Lacidogna, G.; Niccolini, G. Critical behaviour in concrete structures and damage localization by acoustic emission. *Key Eng. Mater.* **2006**, *312*, 305–310.

35. Carpinteri, A.; Lacidogna, G.; Pugno, N. Structural damage diagnosis and life-time assessment by acoustic emission monitoring. *Eng. Fract. Mech.* **2007**, *74*, 273–289.

36. Carpinteri, A.; Lacidogna, G.; Niccolini, G. Fractal analysis of damage detected in concrete structural elements under loading. *Chaos Soliton Fract.* **2009**, *42*, 2047–2056.

Numerical Models for the Assessment of Historical Masonry Structures and Materials, Monitored by Acoustic Emission

Stefano Invernizzi, Giuseppe Lacidogna and Alberto Carpinteri

Abstract: The paper reviews some recent numerical applications for the interpretation and exploitation of acoustic emission (AE) monitoring results obtained from historical masonry structures and materials. Among possible numerical techniques, the finite element method and the distinct method are considered. The analyzed numerical models cover the entire scale range, from microstructure and meso-structure, up to full-size real structures. The micro-modeling includes heterogeneous concrete-like materials, but mainly focuses on the masonry texture meso-structure, where each brick and mortar joint is modeled singularly. The full-size models consider the different typology of historical structures such as masonry towers, cathedrals and chapels. The main difficulties and advantages of the different numerical approaches, depending on the problem typology and scale, are critically analyzed. The main insight we can achieve from micro and meso numerical modeling concerns the scaling of AE as a function of volume and time, since it is also able to simulate the *b*-value temporal evolution as the damage spread into the structure. The finite element modeling of the whole structure provides useful hints for the optimal placement of the AE sensors, while the combination of AE monitoring results is crucial for a reliable assessment of structural safety.

Reprinted from *Appl. Sci.* Cite as: Invernizzi, S.; Lacidogna, G.; Carpinteri, A. Numerical Models for the Assessment of Historical Masonry Structures and Materials, Monitored by Acoustic Emission. *Appl. Sci.* **2016**, *6*, 102.

1. Introduction

Nondestructive experimental techniques and monitoring set-up are increasingly being adopted to acquire and assess the progress of dangerous structural phenomena, such as cracking and damage, and to estimate their successive evolution. The adoption of the better controlling and monitoring technique strictly depends upon the typology of the reinforced concrete or masonry structures under consideration, and on the information to be collected [1]. For historical buildings, non-destructive techniques (NDT) can be exploited for different purposes: (1) revealing hidden structural elements, such as floor beams, or arches and piers which have been incorporated into the walls; (2) assessing the mechanical properties of masonry and mapping the inhomogeneity of the wall components (e.g., adoption of different bricks throughout the life of a building); (3) estimating the extent of cracking in damaged structures; (4) mapping voids and flaws; (5) monitoring moisture content and water rising due to capillary action; (6) surveying surface decay phenomena; and (7) assessing the mechanical and physical properties of brick or stone and mortar.

The acoustic emission (AE) technique has proved particularly effective in the assessment of structural integrity, in that it allows an assessment of the amount of energy emitted due to fracture propagation and helps derive information on the criticality of the undergoing process.

At the present time, the AE technique can be exploited during experimental tests to investigate on the damage advancement in ductile or brittle materials prior to the final collapse [2,3]. Moreover, this non-destructive monitoring technique is successful when studying critical phenomena and to predict the remaining lifetime and durability in full-scale structures [4].

According to this method, it is feasible to acquire the transient elastic waves related to each stress-induced crack advancement event inside a structure or a specimen. These waves can be detected and recorded by transducers applied on the external surface of samples or structural elements. The transducers are piezoelectric sensors that commute the power of the elastic waves into voltaic signals. A proper analysis of the AE waveform parameters (peak amplitude, duration time and frequency) helps provide detailed information about the damage progression, such as the emitted energy, cracking pattern, and fracture mode [5].

The critical conditions that anticipate the collapse can be monitored analyzing the b-value calculated from the Gutenberg-Richter (GR) law. The GR law can fit, with basically the same accuracy, data from both earthquake distributions in seismic areas [6] and AE technique structural monitoring [7–9], even though two different dimensional scales are involved.

The connection between recorded waves and fracture mode depends on various factors like geometry, propagation distances, and relative orientations. Both experimental and numerical results show the crucial influence of heterogeneities in the crack propagation path. Therefore, it should be carefully considered for AE characterization of large structures, while it should not be disregarded even in small-scale sample laboratory studies to improve cracking characterization [10,11].

The energy emitted by the monitored structure is strictly connected to the energy detected by AE sensors. The energy dissipation in correspondence of crack nucleation and propagation in quasi-brittle materials and structures plays a fundamental role and influences their mechanical behavior throughout their entire life. In recent times, an ad hoc method has been implemented for structural monitoring by means of the AE technique according to fractal concepts. The fractal theory inherently accounts for the multiscale nature of energy dissipation and for the resulting strong size effects. This energetic approach let to introduce a useful parameter for structural damage assessment that is based on the correlation between AE monitored in the structure and the corresponding emissions detected in specimens of different sizes.

This methodology is effective in quantifying the effects of the environmental configuration on the final response, since it inherently accounts for the size of structures and the signal attenuation and distortion recorded during the AE monitoring.

The effectiveness of the assessment can be greatly improved when NDT monitoring is combined with proper numerical analysis. Numerical modeling is useful both for more efficient structural assessment and for more extensive comprehension of meso and microstructural phenomena of materials.

Analytical and numerical modeling of acoustic emissions in metals dates back to the eighties of last century [12,13]. The numerical simulation of progressive failure of rock and associated acoustic emissions or seismicity, depending on the scale of interest, was among others addressed by Tang [14]. Acoustic emission monitoring in concrete structures started in the sixties of last century [2], while numerical simulation of the acoustic emission process in concrete, according to the authors' knowledge, are more recent [15,16].

The acoustic emission phenomenon can be modeled numerically following different approaches, depending on the scale and on the aspects of the problem under consideration. Dynamical models are preferred if the acoustic emission source mechanism must be considered in detail. In the literature different source hypotheses have been proposed [17] (point-like, linear or based upon fracture mechanics description of the crack propagation), while the elastic acoustic wave propagation is affected by the attenuation, dispersion and propagation in the guiding media. If a heterogeneous concrete-like material is considered, together with the presence of another source of wave disturbance like tendons, the problem become very challenging [18], and numerical results must be compared with additional very innovative experimental techniques like X-ray tomography [19].

When the geometry of the sample or structure become more complex, quasi-static numerical models are usually preferred, which can provide localization in space and time of the acoustic emission events, based on a detailed description of damage propagation in the media. In this case, the signal acquired by piezoelectric sensors cannot be compared directly with numerical results, and it is necessary to calculate the acoustic emission location with some moment-tensor algorithm. Different numerical strategies can be adapted to this purpose. The so-called lattice-model allows for a detailed discretization of the microstructure, and acoustic emission events can be related to the rupture of lattice elements or bundle of elements [20]. Alternatively, continuous discretization can be coupled with discontinuities described with the cohesive crack model [21], and crack advancements can be directly related to the occurrence of acoustic emissions.

Numerical simulations of acoustic emissions in masonry and masonry structures are quite recent and relatively poorly investigated. For this reason, in the following, a number of applications of combined AE and numerical analysis performed by the authors on masonry and masonry structures are reviewed, from real-size structures down to model structures and meso or microstructural material modeling.

2. Historical Masonry Structures

In the last fifteen years, the authors have investigated a number of relevant existing buildings belonging to the Italian cultural heritage. In this section these structures are briefly described, in order to illustrate the combination of numerical modeling and acoustic emission monitoring.

2.1. Historical Masonry Towers of Alba

These medieval masonry towers are the tallest and most iconic constructions from the XIIIth century that are conserved in Alba up to the present time (Figure 1) [22,23].

Figure 1. View of the Astesiano, Sineo and Bonino towers in Alba.

The three towers have been analyzed by three-dimensional finite element models, set up starting from Autocad® drawings (Autodesk®, San Rafael, CA, USA). The discretization mesh was obtained adopting 20-node isoparametric solid brick elements, and the analysis was performed with the finite element commercial code iDIANA (TNO DIANA BV®, Delft, The Netherlands). The discretization mesh was such that the wall thickness of the towers included five element's nodes minimum.

The models accounted for the dimension and shape of openings and for the variation of the wall thickness at various levels. Conversely, the presence of timber floors has been neglected. Each structure was mainly subjected to its dead load. The three towers present various damage patterns, ranging from smeared cracking to dominant cracks. The acoustic emission (AE) technique was exploited to monitor the crack propagation in the most relevant structural areas of the three towers. Piezoelectric transducers applied to the inner or outer surface of the tower walls were used to acquire the elastic waves emitted as a consequence of crack propagation within the masonry volume.

From the survey it was found that the Sineo tower is not exactly vertical, and therefore the possible evolution of the eccentricity of the tower must be considered. A direct assessment of the tilt evolution based on the mechanical characteristics of the foundation is not very reliable, due to the geotechnical uncertainties of the underlying soil layers. On the other hand, a loading scenario can be analyzed, where an increasing rotation of the base is considered, together with the self-weight of the masonry and wind thrust acting on the higher lateral surface of the tower (Figure 2). The crack pattern scenario allowed for localization of the main regions subjected to cracking in case of tilt evolution. Therefore, the location of AE sensors was optimized [22,23].

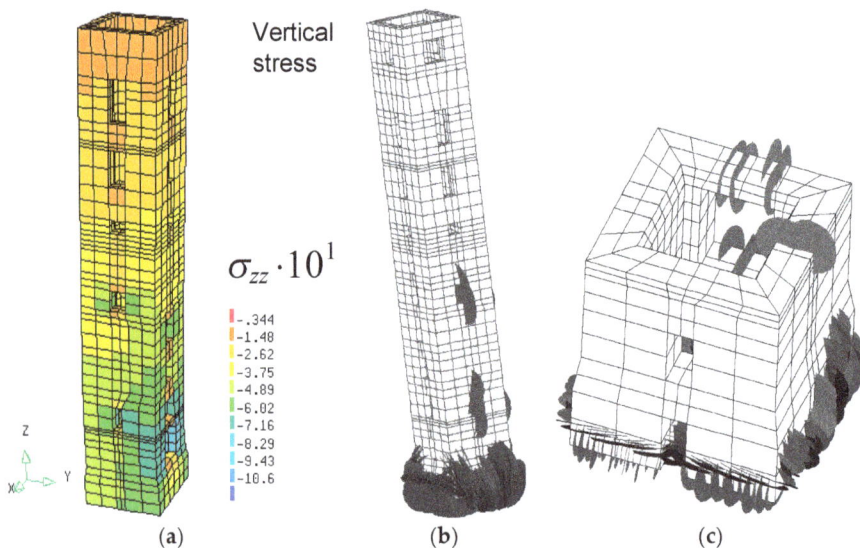

Vertical stress

$\sigma_{zz} \cdot 10^1$

| -.344 |
| -1.48 |
| -2.62 |
| -3.75 |
| -4.89 |
| -6.02 |
| -7.16 |
| -8.29 |
| -9.43 |
| -10.6 |

(a)   (b)   (c)

Figure 2. Torre Sineo: (**a**) vertical stresses (MPa); (**b**) overall crack pattern; (**c**) detail crack pattern of the basement area, corresponding to a 3% deviation from verticality.

The mechanical response of Torre Astesiano was characterized by the presence of a dominant crack, which was uncertainly ascribed to the impact of a cannon ball during a past conflict. The structural analysis (Figure 3) allowed it to be proven that thermal stresses due to seasonal and daily temperature oscillation could justify the opening of the crack. Two piezoelectric transducers were placed in the inner masonry layer of the tower, close to the fourth floor level, and near the tip of the main sub-vertical crack, where dynamic analysis provided the main stress intensification.

The Bonino tower was recently subjected to some restructuring works, which involved the opening of new apertures in the masonry, especially in the correspondence of the first floor level. Another cause of concern was the damage detected in the correspondence of the ornamental layer of the upper part of the tower. The structural analysis (Figure 4) allowed for the localization of stress increases in the correspondence of the new openings, as well as for the determination of tensile stresses due to the higher stiffness of the ornamental layer, which was made with stone whose Young modulus is almost twice that of the surrounding masonry.

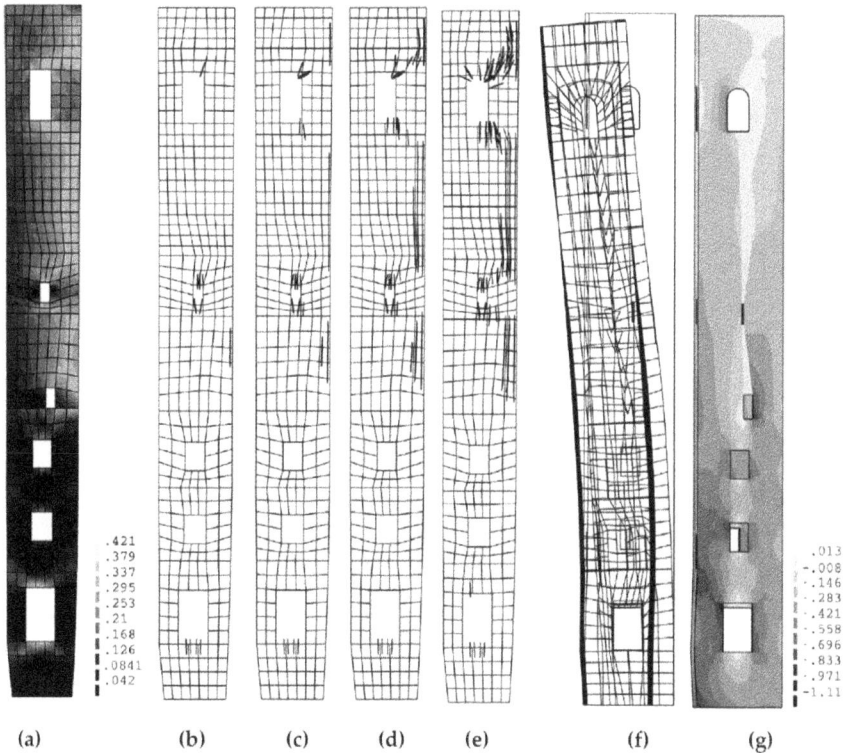

Figure 3. Torre Astesiano: (**a**) internal side of the South wall: principal tensile stress (MPa); (**b**–**e**) thermally induced crack pattern evolution; (**f**–**g**) deformed mesh and vertical stresses (MPa) of the damaged model.

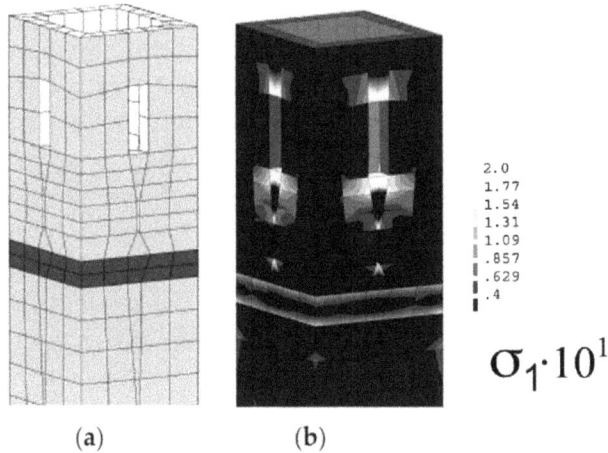

Figure 4. Torre Bonino: (a) detail of the stone layer discretization; (b) principal tensile stresses σ_1 (MPa).

The AE equipment was placed in correspondence of these critical areas, in order to assess the influence on damage propagation. An evolutionary release of energy is recorded under the effect of self-weight, which can be ascribed to pseudo-creep behavior of the masonry.

The rate of propagation of the micro-cracks in the three masonry towers (Figure 5) is correlated to the time dependence of the structural damage observed during the monitoring. The ratio of the cumulative number of AE counts recorded during the monitoring process, N, to the number obtained at the end of the observation period, N_d, as a function of time, t, is equal to:

$$\eta = \frac{E}{E_d} = \frac{N}{N_d} = \left(\frac{t}{t_d}\right)^{\beta_t} \tag{1}$$

The parameter t_d in Equation (1) refers to the whole structure monitoring time, while the E_d and N_d parameters are usually lower than the values attained at critical conditions ($E_d \leqslant E_{max}$; $N_d \leqslant N_{max}$). It is possible to obtain an assessment about the stability condition of the structure deriving the exponent β_t as the linear best-fitting coefficient in the bi-logarithmic diagram (Figure 5) where AE data from monitoring are reported.

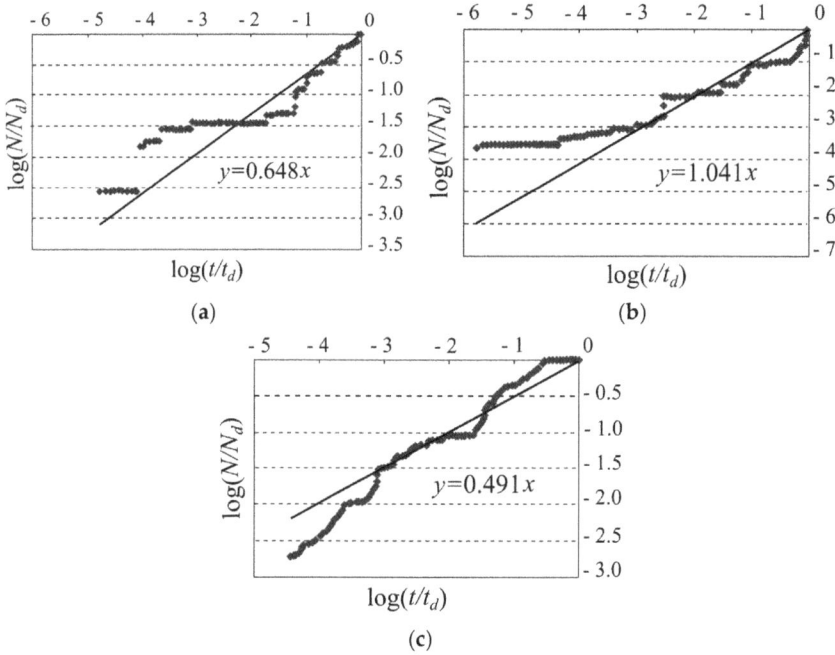

Figure 5. Damage monitoring: (**a**) Sineo tower; (**b**) Astesiano tower; (**c**) Bonino tower.

The exponent $\beta_t < 1$ indicates that the damaging process is slowing down, and that the structure is evolving towards a stable condition. On the contrary, the parameter $\beta_t > 1$ indicates that the process is becoming unstable. Finally, the case $\beta_t \cong 1$ refers to a metastable condition, *i.e.*, although the phenomenon is linearly increasing over time, it could attain either instability or stability conditions. The Sineo tower analysis yielded a slope $\beta_t \cong 0.648$, the Astesiano tower provided $\beta_t \cong 1.041$, and the Bonino tower $\beta_t \cong 0.491$ (Figure 5). During the monitoring period, these results reveal that in the Sineo and Bonino towers the damage process is stable. On the contrary, as far as the Astesiano tower is concerned, the damage is approaching a metastable condition.

2.2. Cathedrals and Vaulted Structures

A monitoring campaign and numerical analysis has been performed on the ancient Cathedral of Syracuse in Sicily [24] (Figure 6a). The acoustic emission (AE) technique is adopted to assess the damage pattern evolution. The localization of the propagating cracks is obtained using six synchronized acoustic emission sensors. A clear correlation between the regional seismic activity and the AE acquisition data has been obtained. In fact the AE count rate presents peaks corresponding to the main seismic events. During the observation period (Figure 6b), the number of AE counts was $\cong 4300$.

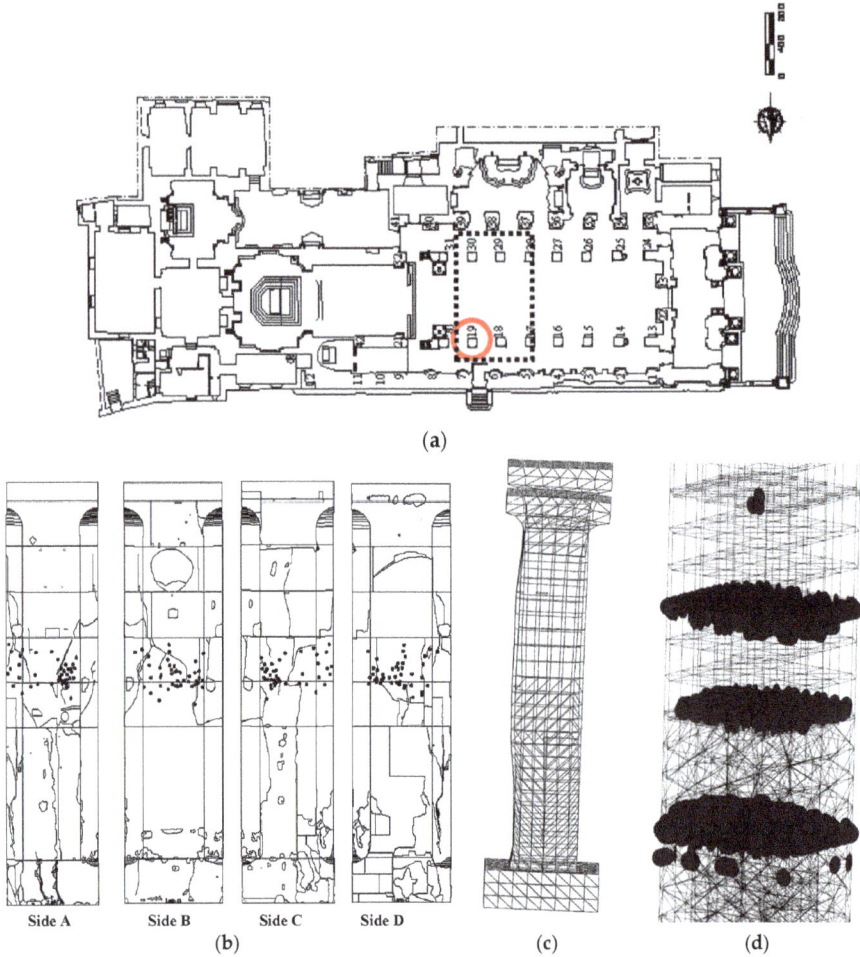

Figure 6. (a) Plan of the Syracuse Cathedral and most critical pillar; (b) cracking pattern and localization of AE sources; (c) deformed mesh due to the seismic load; (d) cracking in the pillar due to to the seismic acceleration.

In order to assess the rate of the damage propagation, as given in Equation (1), the AE technique data were reported in the bilogarithmic plane to obtain the best-fitting slope, providing $\beta_t \cong 0.98$. The quasi-linear progression of damage over time confirms that the process in the pillar is in metastable conditions. A detailed geometrical survey of the most damaged pillar allowed for the definition of an accurate 3D model.

The geometry of each block, as well as the presence of masonry inserts, have been considered. Two main loads were considered: the dead load (of the pillar and of the surrounding structure), and a horizontal seismic load provided as horizontal ground acceleration. Due to the horizontal acceleration, cracking can take place in the pillar. A detail of crack nucleation is shown in Figure 6d. The crack occurrence provided by the analysis agrees quite well with the crack localization provided by the AE recording. Cracking corresponds both to diffuse cracking in the continuum elements of the sandstone blocks and to opening or sliding of the discrete interfaces between blocks.

Another case was the Hospital San Giovanni in Turin (Italy) [25], a masonry building complex initiated in 1680, under the design of the Italian architect Amedeo di Castellamonte (1610–1683) (Figure 7). The ground floor of the complex was recently chosen to host an important fossil collection from the Regional Museum of Natural Science.

Figure 7. Hospital San Giovanni: photograph inside the vault.

Due to this change of use, an assessment of the structural load capacity of the masonry vault beneath the first floor was necessary because the fossil collection involves a significant increase in the vault load. During the *in situ* load test, we recorded the acoustic emissions from the vault, as well as displacements of the vault and strains in the steel rods. We compared the experimental data with the numerical results obtained from finite element modeling of cracking and crushing. After validation, the model allowed us to assess the ultimate load-bearing capacity of the vault. The 3D model provided a slightly better estimate of the displacements close to the abutment (Figure 8). From the direct comparison of the cumulative AE counting distribution with the deformation of the structure, we determined that the two diagrams (Figure 9) match each other in shape quite well. In the present case, if the time scale is plotted without rescaling, the two curves do not match identically. Nevertheless, AE are confirmed as a useful tool to measure damage, specifically in the case of very stiff vaulted structures.

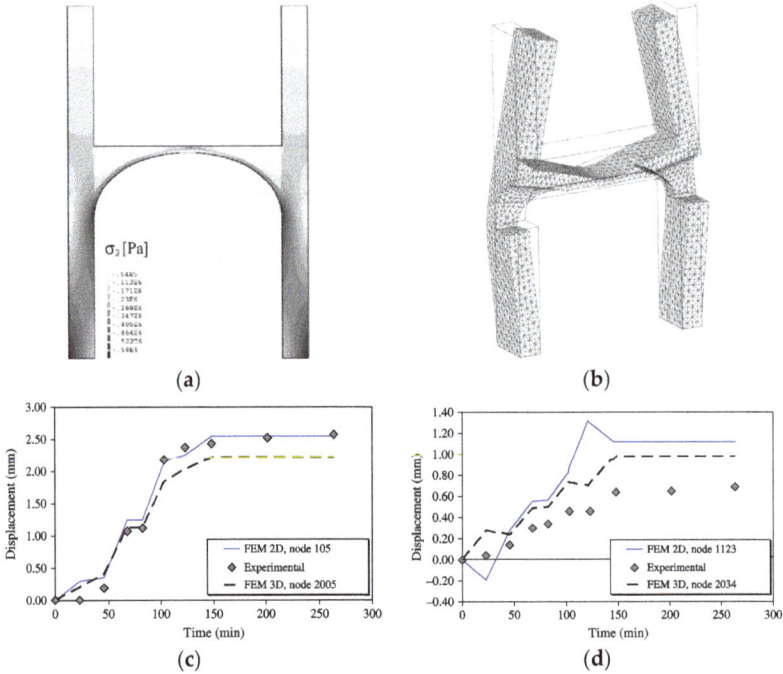

Figure 8. (**a**) Vault subjected to the dead load: principal compression stress contour; (**b**) 3D vault deformed mesh; (**c**) time-displacement diagrams at the springing and (**d**) close to the keystone.

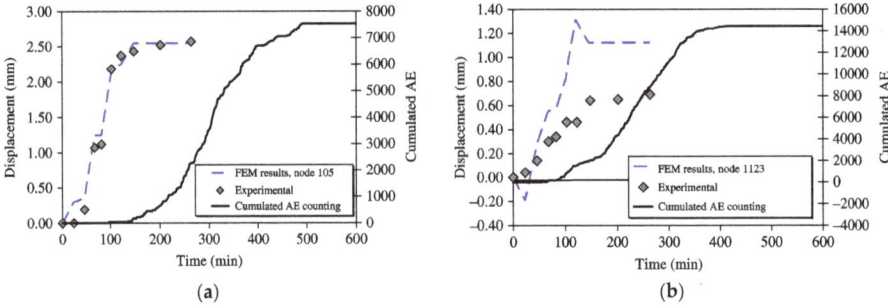

Figure 9. Graph of the comparison between cumulated acoustic emission (AE) counting (right *y*-axis), and the experimental or two-dimensional finite element vertical displacements (left *y*-axis): (**a**) node 105, and (**b**) node 1123.

3. Model Structures

Numerical modeling has also been fruitfully combined with NDT and AE monitoring in the study of scaled model structures in the laboratory.

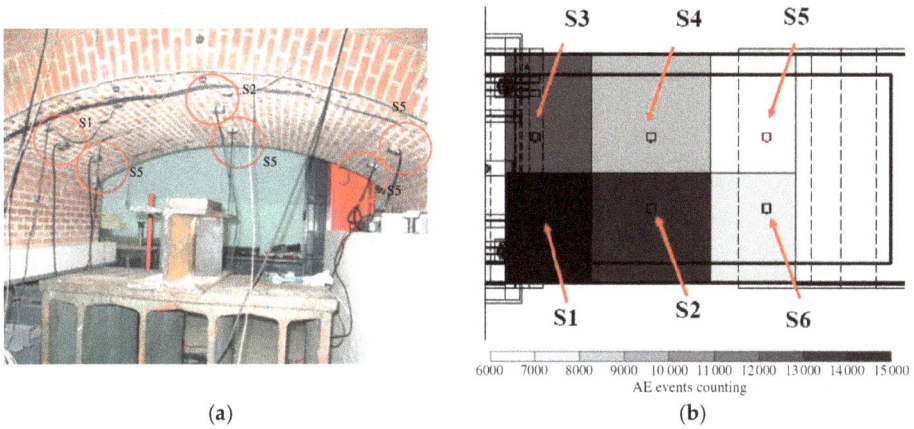

(a)

(b)

Figure 10. (**a**) Placing of the acoustic emissions sensors, and their competence areas; (**b**) number of acoustic events in each competence area.

A scaled model of a two-span masonry arch bridge has been built to assess the consequences of the central pile settlement due to riverbank erosion [26]. The bridge geometry and the structural details, including the masonry bricks and mortar joints, are realized in 1:2 scale (Figure 10). The model bridge has been subjected to incremental settlement of the pile. To this purpose, a devoted mobile support was realized to sustain the central pile.

The AE counting has been recorded during the first stage of the settlement. The criticality of the ongoing process was monitored based on the interpretation of the AE rate acquired during the test. Moreover, it has been possible to localize the main damaged zones on the base of the number of AE recorded by each transducer. The statistical properties AE time series have been analyzed using an estimation of the b-value of the Gutenberg–Richter (GR) law that allows assessment of the damage level reached in the model (Figure 11).

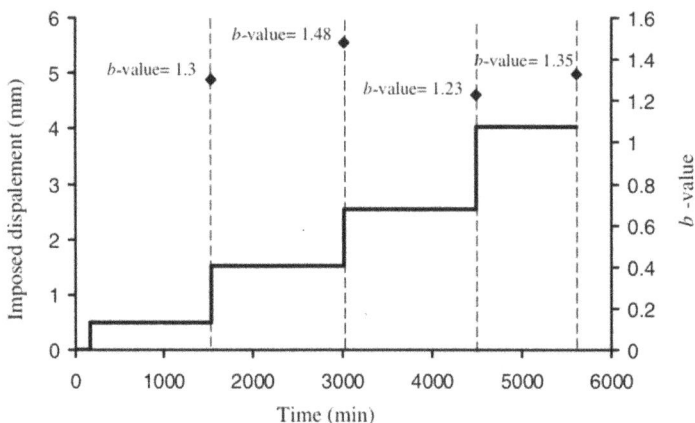

Figure 11. Evolution of the *b*-value compared to pile's incremental settlements.

4. Material Meso and Microstructure

When numerical simulation is combined with AE monitoring at the meso and micro scale, detailed information about the damage evolution are obtained. In particular, the AE statistics and evolution law of the *b*-value with increasing damage can be obtained numerically. The AE statistics is analogous to the well-known Gutenberg–Richter, which express the link between the magnitude and the frequency of seismic events. The *b*-value is related to the slope, in the bilogarithmic diagram, of the Gutenberg–Richter relation, and can be interpreted as an indicator of the criticality of the rupture process. Seismic events and AE share, at very different dimensional scales, the same physical nature, both being energy release phenomena due to damage propagation and rupture.

4.1. Smeared Cracking

Finite element modeling can be performed adopting a smeared cracking constitutive law. In this way the number of crack advance at each gauss point can be put into correspondence with AE events recorded during monitoring.

The double flat jack test has been studied in details [27], combining it with AE monitoring and considering an experimental configuration where the analyzed volume of masonry is not constant.

The numerical model accounts for the detailed mesostructure of the masonry texture (Figure 12). Although it is not possible to obtain an easy direct relation between the acoustic emission and the amount of cracking, nevertheless, it is possible to state that the two quantities are proportional to each other when increasing sizes are considered (Figure 13).

366

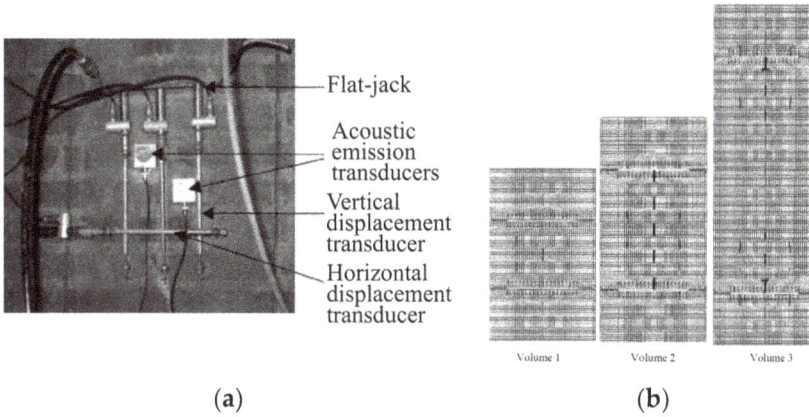

Figure 12. (**a**) Double flat jack experimental set; (**b**) numerical discretization and crack patterns for increasing distance between the flat jacks.

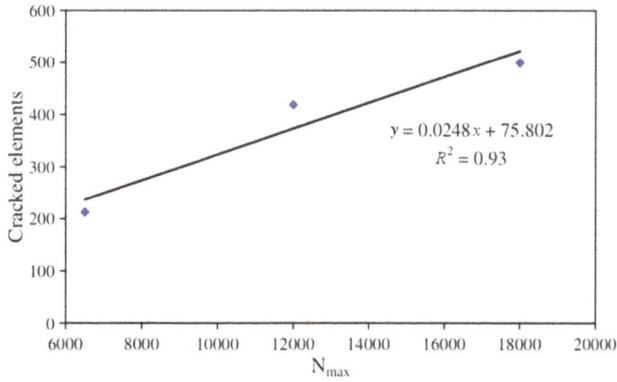

Figure 13. Double flat jack test: Proportionality between N_{max} and the number of cracked finite elements.

Also, concrete-like materials can be studied with a similar strategy, in order to account for the detailed material microstructure composed of matrix and aggregates [28]. In this way the volumetric scaling of AE was obtained numerically (Figure 14).

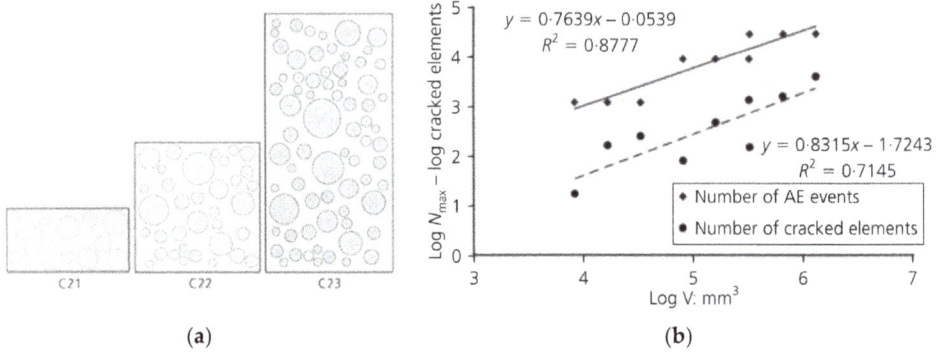

Figure 14. Compression tests: (**a**) sample meshes; (**b**) proportionality between N_{max} and the number of cracked finite elements.

4.2. Particle Strategy

The propagation of micro cracks can be put in direct correspondence with AE events by adopting a particle strategy according to the distinct element method [29]. In this case, every single aggregate is modeled as a particle of appropriate size distribution, while a reciprocal lattice of beams that simulates the matrix (Figure 15a,b). The method is particularly effective in compression crushing simulation (Figure 15c).

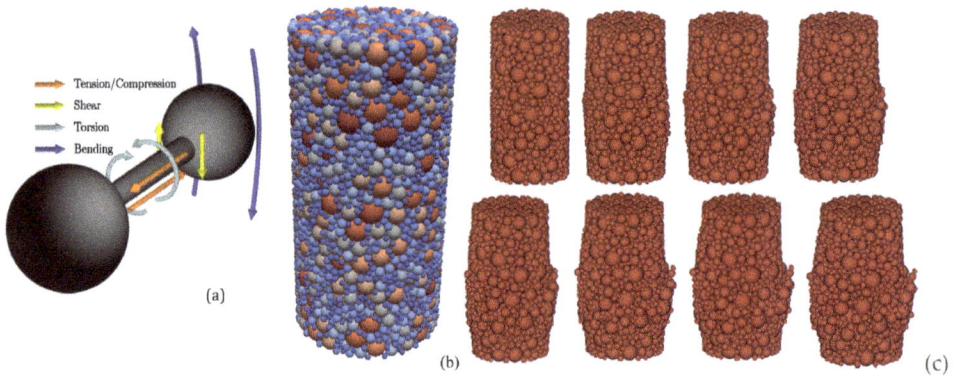

Figure 15. (**a**) Bond interaction scheme between two particles; (**b**) particle size distribution; (**c**) subsequent stages of crushing obtained from a simulated compression test.

The method correctly describes the size effect on compressive strength and the scaling of AE with size. In addition, the b-value can be evaluated numerically, as shown in Figure 16. The evolution of the parameter obtained numerically is in good agreement with theoretical prevision and experimental results.

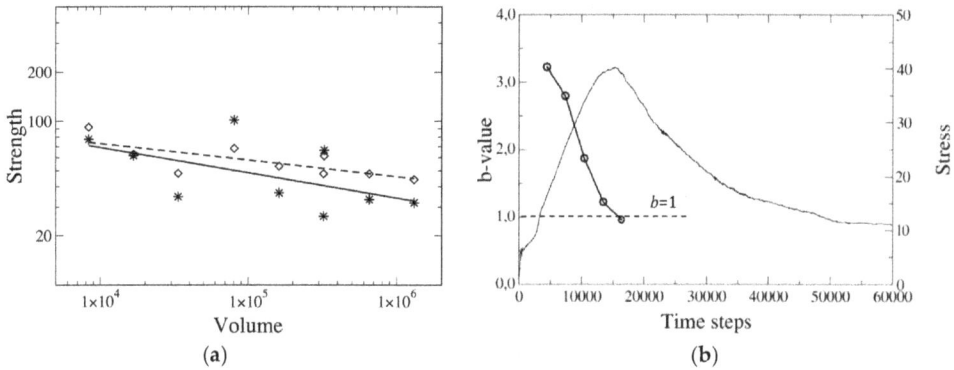

Figure 16. (**a**) Comparison between numerical results (stars and continuous line) and experimental size effect on compression strength (diamonds and dotted line); (**b**) comparison between the stress level and the variation of the simulated b-value (circles) as a function of time.

5. Conclusions

A number of case studies, analyzed by the authors in the last fifteen years, are summarized and briefly described to emphasize the advantage of combined numerical methods and AE monitoring. This analysis shows how proper numerical modeling can improve the comprehension of involved phenomena and the reliability of structural assessment. The role of the geometrical and temporal scale is emphasized, and the complex statistics of the energy release magnitude obtained from monitoring and numerical simulations are successfully compared.

Author Contributions: A.C. supervised the research; S.I. performed the numerical simulations; G.L. carried out acoustic emission and nondestructive technique tests; S.I. wrote the paper after discussion with all the authors.

Conflicts of Interest: The authors declare no conflict of interest.

References

1. Anzani, A.; Binda, L.; Carpinteri, A.; Invernizzi, S.; Lacidogna, G. A multilevel approach for the damage assessment of historic masonry towers. *J. Cult. Herit.* **2010**, *11*, 459–470.
2. Ohtsu, M. The history and development of acoustic emission in concrete engineering. *Mag. Concr. Res.* **1996**, *48*, 321–330.
3. Grosse, C.; Ohtsu, M. *Acoustic Emission Testing*; Springer: Berlin, Germany, 2008.
4. Carpinteri, A.; Lacidogna, G.; Pugno, N. Structural damage diagnosis and life-time assessment by acoustic emission monitoring. *Eng. Fract. Mech.* **2007**, *74*, 273–289.

5. Aggelis, D.G. Classification of cracking mode in concrete by acoustic emission parameters. *Mech. Res. Commun.* **2011**, *38*, 153–157.

6. Scholz, C.H. The frequency-magnitude relation of microfracturing in rock and its relation to earthquakes. *Bull. Seismol. Soc. Am.* **1968**, *58*, 399–415.

7. Shiotani, T.; Fujii, K.; Aoki, T.; Amou, K. Evaluation of progressive failure using AE sources and improved *b*-value on Slope Model Tests. *Progr. Acoust. Emiss.* **1994**, *7*, 529–534.

8. Kurz, J.H.; Finck, F.; Grosse, C.U.; Reinhardt, H.W. Stress drop and stress redistribution in concrete quantified over time by the *b*-value analysis. *Struct. Health. Monit.* **2006**, *5*, 69–81.

9. Carpinteri, A.; Lacidogna, G.; Puzzi, S. From criticality to final collapse: Evolution of the *b*-value from 1.5 to 1.0. *Chaos Solitons Fractals* **2009**, *41*, 843–853.

10. Aggelis, D.G.; Mpalaskas, A.C.; Ntalakas, D.; Matikas, T.E. Effect of wave distortion on acoustic emission characterization of cementitious materials. *Constr. Build. Mater.* **2012**, *35*, 183–190.

11. Carpinteri, A.; Lacidogna, G.; Accornero, F.; Mpalaskas, A.C.; Matikas, T.E.; Aggelis, D.G. Influence of damage in the acoustic emission parameters. *Cement Concrete Compos.* **2013**, *44*, 9–16.

12. Wadley, H.N.G.; Scruby, C.B.; Speake, J.H. Acoustic emission for physical examination of metals. *Int. Metals Rev.* **1980**, *2*, 41–64.

13. Onoe, M.; Tsao, J.W. Numerical simulation of acoustic emissions. *Jap. J. Appl. Phys.* **1981**, *20*, 177–180.

14. Tang, C. Numerical simulation of progressive rock failure and associated seismicity. *Int. J. Rock Mech. Min. Sci.* **1997**, *34*, 249–261.

15. Ohtsu, M.; Kaminaga, Y.; Munwam, M.C. Experimental and numerical crack analysis of mixed-mode failure in concrete by acoustic emission and boundary element method. *Constr. Build. Mater.* **1999**, *13*, 57–64.

16. Mpalaskas, A.C.; Vasilakos, I.; Matikas, T.E.; Chai, H.K.; Aggelis, D.G. Monitoring of the fracture mechanisms induced by pull-out and compression in concrete. *Eng. Fract. Mech.* **2014**, *128*, 219–230.

17. Sause, M.R.G.; Richler, S. Finite element modelling of cracks as acoustic emission sources. *J. Nondestruct. Eval.* **2015**, *34*.

18. Schechinger, B.; Vogel, T. Acoustic emission for monitoring a reinforced concrete beam subject to four-point-bending. *Constr. Build. Mater.* **2007**, *21*, 483–490.

19. Kocur, G.K.; Saenger, E.H.; Vogel, T. Elastic wave propagation in a segmented X-ray computed tomography model of a concrete specimen. *Constr. Build. Mater.* **2010**, *24*, 2393–2400.

20. Grégoire, D.; Verdon, L.; Lefort, V.; Grassl, P.; Saliba, J.; Regoin, J.P.; Loukili, A.; Pijaudier-Cabot, G. Mesoscale analysis of failure in quasi-brittle materials: Comparison between lattice model and acoustic emission data. *Int. J. Numer. Anal. Meth. Geomech.* **2015**, *39*, 1639–1664.

21. Veselý, V.; Vodák, O.; Trčka, T.; Sobek, J.; Koktavý, P.; Keršner, Z.; Koktavý, B. Acoustic emission from quasi-brittle failure of cementitious composites—Experimental measurements and cohesive crack model simulations. *Key Eng. Mater.* **2014**, *592–593*, 676–679.

22. Carpinteri, A.; Invernizzi, S.; Lacidogna, G. Numerical assessment of three medieval masonry towers subjected to different loading conditions. *Mason. Int.* **2006**, *19*, 65–76.

23. Carpinteri, A.; Invernizzi, S.; Lacidogna, G. *In situ* damage assessment and nonliner modelling of an historical masonry tower. *Eng. Struct.* **2005**, *27*, 387–395.

24. Carpinteri, A.; Lacidogna, G.; Manuello, A.; Invernizzi, S.; Binda, L. Stability of the vertical bearing structures of the Syracuse Cathedral: Experimental and numerical evaluation. *Mater. Struct.* **2009**, *42*, 877–888.

25. Carpinteri, A.; Invernizzi, S.; Lacidogna, G. Structural assessment of a XVIIth century masonry vault with AE and numerical techniques. *Int. J. Archit. Herit.* **2007**, *1*, 214–226.

26. Invernizzi, S.; Lacidogna, G.; Manuello, A.; Carpinteri, A. AE monitoring and numerical simulation of a two-span model masonry arch bridge subjected to pier scour. *Strain* **2011**, *47*, 158–169.

27. Carpinteri, A.; Invernizzi, S.; Lacidogna, G. Historical brick-masonry subjected to double flat-jack test: Acoustic Emissions and scale effects on cracking density. *Constr. Build. Mater.* **2009**, *23*, 2813–2820.

28. Invernizzi, S.; Lacidogna, G.; Carpinteri, A. Scaling of fracture and acoustic emission in concrete. *Mag. Concrete Res.* **2013**, *65*, 529–534.

29. Invernizzi, S.; Lacidogna, G.; Carpinteri, A. Particle-based numerical modeling of AE statistics in disordered materials. *Meccanica* **2013**, *48*, 211–220.

Detecting the Presence of High Water-to-Cement Ratio in Concrete Surfaces Using Highly Nonlinear Solitary Waves

Piervincenzo Rizzo, Amir Nasrollahi, Wen Deng and Julie M. Vandenbossche

Abstract: We describe a nondestructive evaluation (NDE) method based on the propagation of highly nonlinear solitary waves (HNSWs) to determine the excess of water on the surface of existing concrete structures. HNSWs are induced in a one-dimensional granular chain placed in contact with the concrete to be tested. The chain is part of a built-in transducer designed and assembled to exploit the dynamic interaction between the particles and the concrete. The hypothesis is that the interaction depends on the stiffness of the concrete and influences the time-of-flight of the solitary pulse reflected at the transducer/concrete interface. Two sets of experiments were conducted. In the first set, eighteen concrete cylinders with different water-to-cement (w/c) ratios were cast and tested in order to obtain baseline data to link the ratio to the time of flight. Then, sixteen short beams with fixed w/c ratio, but subject to water in excess at one surface, were cast. The novel NDE method was applied along with the conventional ultrasonic pulse velocity technique in order to determine advantages and limitations of the proposed approach. The results show that the time of flight detected the excess of water in the beams. In the future, the proposed method may be employed in the field to evaluate rapidly and reliably the condition of existing concrete structures and, in particular, concrete decks.

Reprinted from *Appl. Sci.* Cite as: Rizzo, P.; Nasrollahi, A.; Deng, W.; Vandenbossche, J.M. Detecting the Presence of High Water-to-Cement Ratio in Concrete Surfaces Using Highly Nonlinear Solitary Waves. *Appl. Sci.* **2016**, *6*, 104.

1. Introduction

In concrete and cement based structures, the early stage of hydration and the conditions at which curing occurs, influence the quality and the durability of the final products. For instance, as a result of the chemical reactions between water and the cement during hydration, the mixture progressively develops mechanical properties. Final set for the mixture is defined as the time that the fresh concrete transforms from plastic into a rigid state. At final set, measurable mechanical properties start to develop in concrete and continue to grow progressively. The durability and the strength of concrete may deviate from design conditions as a result of accidental factors. Some of these factors are water-to-cement (w/c) ratio not controlled well and rainfall that permeates the fresh concrete or dampens the forms prior to casting.

As such, the development of nondestructive evaluation (NDE) methods able to determine anomalous concrete conditions is very much needed, and has been a long-standing challenge in the area of material characterization. To date, many NDE methods for concrete exist, and some of them resulted in commercial products. The interested reader is referred to the excellent monograph [1] to gain a holistic knowledge of such methods. The most common technique is probably the one based on the propagation of bulk ultrasonic waves through concrete. This approach measures the speed of the waves propagating through the thickness of the test object to determine the elastic modulus using an empirical formula [1–10]. This approach is usually referred to as the ultrasonic pulse velocity (UPV) method. If the access to the back wall of the sample is impractical, the ultrasonic testing is conducted in the pulse-echo mode. A popular commercial system is the Schmidt hammer [11–14], which consists of a spring-driven steel hammer that hits the specimen with a defined energy. Part of the impact energy is transmitted to the specimen and absorbed by the plastic deformation of the specimen and the remaining impact energy is rebounded. The rebound distance depends on the hardness of the specimen and the condition of the surface. The harder the surface, the shorter the penetration time (or depth) and therefore the higher is the rebound.

Despite decades of research and developments, much research is still ongoing [13,15–28] and many interesting works covering a wide spectrum of NDE techniques are being investigated. In this paper, we propose an NDE method based on the propagation of highly nonlinear solitary waves (HNSWs) along a 1D chain of spherical particles placed in contact with the concrete to be tested. These waves are compact mechanical waves that can form and travel in highly nonlinear systems, such as a chain of elastically interacting spheres. The interaction between two adjacent beads is governed by the Hertz's law [29,30] $F = A\delta^{3/2}$, where F is the compression force between the granules and δ is the closest approach of their centers. The coefficient A is equal to $E \cdot d^{1/2}/[3 \cdot (1 - v^2)]$ where E, d, and v are the modulus of elasticity, diameter, and Poisson's ratio of the spheres, respectively.

The most common way to induce a solitary pulse, hereinafter indicated as the incident solitary wave (ISW), is by tapping the first particle of the chain with a particle striker having at least the same mass of the individual beads forming the chain. Some researchers have demonstrated that a piezo-actuator [31,32] or even laser pulses [33] can be used as well. When the ISW reaches the interface with the material to be tested, the pulse is partially reflected giving rise to the primary reflected solitary wave (PSW). Both ISW and PSW can be sensed either by a force sensor located at the opposite end of the chain or by embedding a sensor bead along the chain. For a detailed description of the underlying basis of HNSWs, see ref. [34–37].

The study presented here expands on a recent work where HNSWs were used to determine the elastic modulus of concrete cylinders fabricated with three different

w/c ratios, namely 0.42, 0.45, and 0.50 [38]. In the present work, we use some of the findings from [38] to predict water in excess in short concrete beams made with w/c = 0.42 but corrupted with water. Two conditions were simulated. The first one consisted of standing water in formworks prior to pouring concrete, whereas the second condition consisted of sprinkling water above the fresh concrete during casting and surface finishing. These two conditions may reflect adverse weather in the field. The objective of the study was the development of a system that, unlike the UPV method, can predict localized deterioration conditions associated with poor quality w/c ratios.

Three HNSW transducers, referred with the descriptor P1, P2, and P3, were used to quantify the elastic modulus of the beams. Owing to the novelty of the transducers design it was decided to assemble and test three of them in order to demonstrate and quantify the repeatability of the design and to quantify any variation of the results associated with differences in the assembly. The findings were then compared to the results of a conventional UPV test in order to evaluate and prove advantages and eventually limitations of the proposed approach.

With respect to ultrasonic-based NDE, the HNSW-based approach: (1) exploits the propagation of HNSWs confined within the particles of the chain; (2) employs a cost-effective transducer; (3) does not require any knowledge of the sample thickness; (4) does not need access to the back-wall of the sample; (5) provides point-like information rather than information about the average characteristics of the whole sample. The method also differs significantly from the Schmidt hammer. In fact, the hammer can be used to test hardened material but the HNSW approach can also characterize fresh concrete and cement as demonstrated in [39,40]; only one parameter, the rebound value, is used in the Schmidt hammer test, while multiple HNSW features can, in principle, be exploited to assess the condition of the underlying material. The hammer may induce plastic deformation or microcracks into the specimen, while the HNSW approach is purely nondestructive.

The paper is organized as follows. The next section describes the concrete samples, the design of the HNSW transducers, and the test protocols adopted throughout the study. Section 3 presents the experimental results associated with the concrete cylinders and it is largely excerpted from reference [38] in order to provide a comprehensive knowledge in support of the findings of Section 4 that presents the results associated with the short beam. The latter represents the core novelty of the present paper. Finally, Section 5 summarizes the findings of the project and provides some suggestions for future studies.

2. Experimental Setup

2.1. Materials

To set the baseline data, eighteen concrete cylinders were cast and tested: nine were evaluated nondestructively, whereas the remaining nine were tested according to the ASTM C469. The cylinders were 152.4 mm (6 in.) in diameter and 304.8 mm (12 in.) high. Three w/c ratios, namely 0.42, 0.45, and 0.50, were considered.

The materials are listed in Table 1a and the mixture designs are presented in Table 1b . To ease identification, we labeled the samples according to Table 1c .

Table 1a. Material used in the concrete mixtures.

Material	Specific Gravity	Water Absorption Capacity (%)
Cement	3.15	n/a
Coarse aggregate	2.71	0.50
Fine aggregate	2.67	1.24
GGBFS [1]	2.83	n/a

[1] ground-granulated blast-furnace slag.

Table 1b. Ingredients of each concrete batch.

Batch	1	2	3
w/c ratio	0.42	0.45	0.50
Paste vol./concrete vol.	0.30	0.30	0.30
Air content (%)	6.50	5.00	6.25
Coarse agg. (kg/m^3)	1054	1054	1054
Fine agg. (kg/m^3)	666	666	666
Cement (kg/m^3)	303	291	274
GGBFS (kg/m^3)	101	97	91
Water (kg/m^3)	170	175	183
Slump (mm)	133	95	203

Then, sixteen 15.2 cm × 15.2 cm × 30.4 cm (6 in. × 6 in. × 12 in.) beams were fabricated using concrete mix design with w/c = 0.42. The beams were subject to the four different scenarios sketched in Figure 1. Each scenario represented either two surface finishing or two standing water situations in the formworks. Conditions 1 and 2 reflected the field case where water accumulates on the formwork as a result of rainfall prior to the placement of the concrete. To create Conditions 1 and 2, a predetermined volume of water based on the surface area of the beam mold was measured and poured on the sealed molds. Concrete was placed as evenly as possible into the molds and a shaft vibrator was then used to consolidate the concrete mixture before finishing the top surface. During the fabrication, the standing water was seen migrating to the top as shown in Figure 2a. After consolidation, the top surfaces of the beam molds were finished.

Table 1c. Detailed information about the concrete cylinders. The samples evaluated with the HNSW (highly nonlinear solitary wave)-transducers were wet and tested after 28 days of curing. The samples evaluated with the UPV (ultrasonic pulse velocity) were wet and tested after 29 days of curing. The samples subjected to compressive load were tested saturated after 28 days of curing.

w/c Ratio	Number of Cylinders	NDE Sample Labels	ASTM C469 Sample Labels
0.42	6	42A, 42B, 42C	42D, 42E, 42F
0.45	6	45A, 45B, 45C	45D, 45E, 45F
0.50	6	50A, 50B, 50C	50D, 50E, 50F

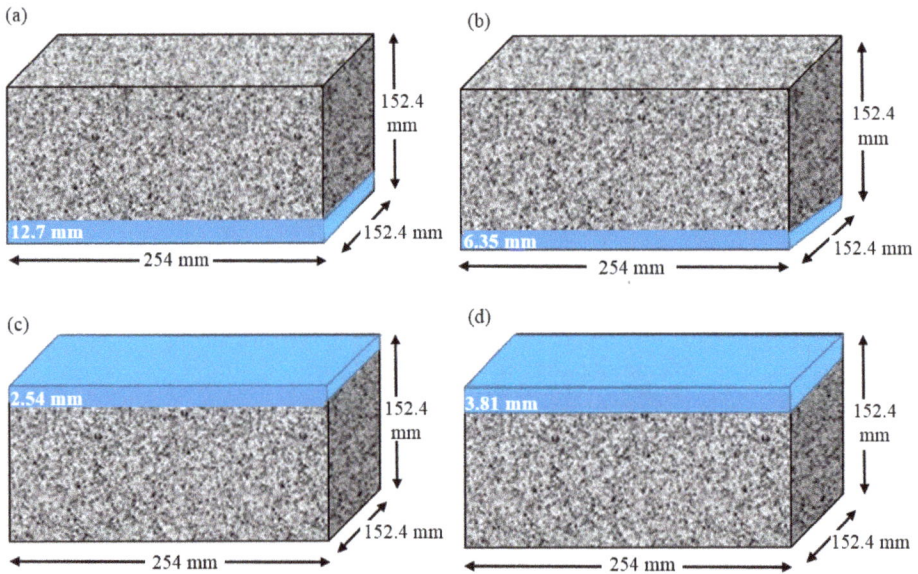

Figure 1. Schematics of the four conditions simulated on the 152.4 × 152.4 × 254 mm³ concrete beams. (**a**) Condition 1 in which 12.7 mm of water sits at the bottom of the form mimicking standing water on the formwork before concreting; (**b**) Condition 2 in which 6.35 mm of water sits at the bottom of the form mimicking standing water on the formwork before concreting; (**c**) Condition 3 in which 2.54 mm of water is sprinkled at the top of the beam mimicking rainfall during concreting; (**d**) Condition 4 in which 3.81 mm of water is sprinkled at the top of the beam mimicking rainfall during concreting.

376

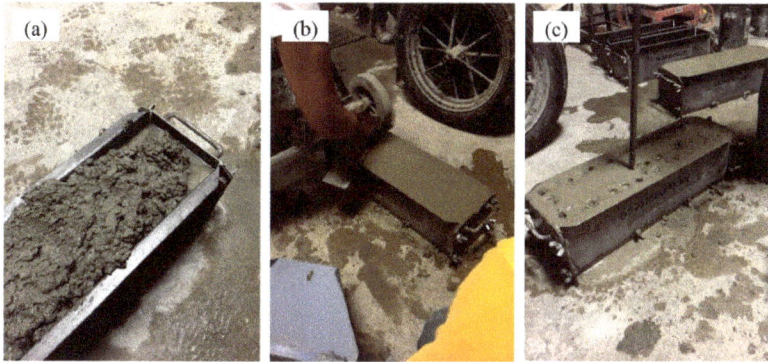

Figure 2. Photos of the preparation of the samples. (**a**) close-up view of one of the samples under Condition 2; standing water from bottom of beam mold migrates to the top; (**b**) preparation of one of the samples under Condition 3: finishing beam surface after second application of water; (**c**) rodding the same sample shown in (**b**) during the third and final application of surface water.

Conditions 3 and 4 simulated instead the occurrence of rainfall during placement and finishing of the concrete. To create these conditions, a specific procedure was developed in an effort to best simulate the finishing of rainfall that would occur on a job site. The procedure began by placing the concrete into the dry beam mold without any consolidation or finishing performed. The predetermined volume of surface water, similarly based on the base area of the mold, was then divided into thirds. The first application of water (one-third of the total surface water) was completed immediately after the concrete was placed into the beam mold. After this first application, a shaft vibrator was used to consolidate the concrete in the mold. The top surface of the beam mold was then struck off and rodded with the rod only penetrating into the concrete approximately 25 mm (1 in.). The second application of surface water was then completed. Following this second application of surface water, the top surface was again finished (Figure 2b) and rodded (Figure 2c). The third and final application of surface water was then applied before the top surface was finished for the last time. This surface finishing process was found to be the best way in controlling the application of surface water and simulating what actually happens on a bridge project. This modified amount of surface water was applied in three separate stages (one-third volume per application), as described above.

Two beams per condition per day were cast. We note here that the amount of water added in the four conditions raised the true w/c ratio to 0.627, 0.524, 0.462, and 0.483, respectively for Conditions 1 to 4. The calculations assumed that the water standing on the formworks or sprinkled above the fresh concrete was uniformly distributed across the entire volume of the specimens.

2.2. HNSW Transducers

Three transducers were assembled. Each transducer contained sixteen AISI 302 steel particles (McMaster-Carr, Aurora, IL, USA) as schematized in Figure 3. The second particle from the top was nonferromagnetic, whereas the others were ferromagnetic. The properties of the particles were: diameter d = 19.05 mm, density ρ = 7800 kg/m^3, mass m = 27.8 g, modulus of elasticity E = 193 GPa, and Poisson's ratio ν = 0.29. Each chain was held by a Delrin tube (McMaster-Carr, Aurora, IL, USA) with outer diameter D_0 = 22.30 mm and inner diameter slightly larger than d in order to minimize the friction between the striker and the inner wall of the tube and to minimize acoustic leakage from the chain to the tube. The striker was driven by an electromagnet (made in the lab) built in our lab and powered by a (DC) power supply (B & K Precision, Melrose, MN, USA).

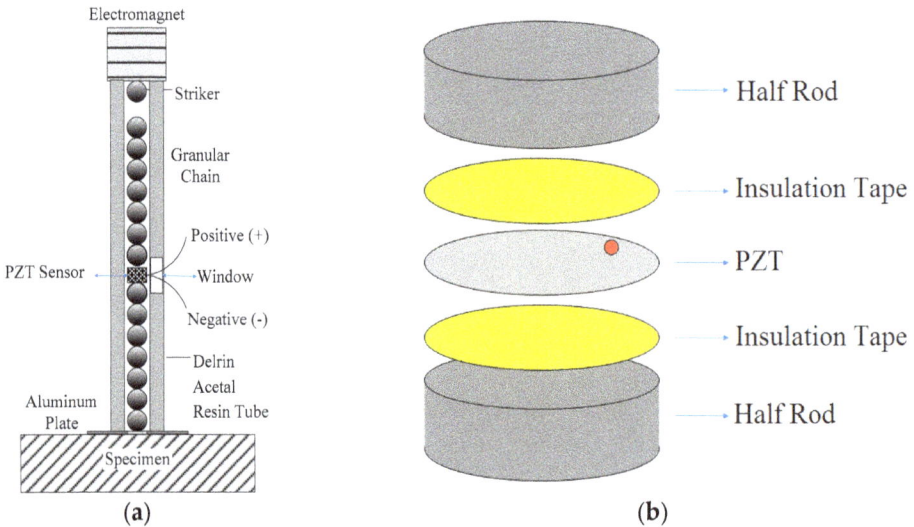

Figure 3. (a) scheme of the solitary waves based transducers used in this study (b) assembly of the sensor-rod inserted in the transducers.

The sensing system consisted of a built-in sensor-rod schematized in Figure 3b and located in lieu of the 9th particle. The sensor-rod was made of a piezo ceramic (Steminc Piezo, Miami, AZ, USA) embedded between two half rods (McMaster-Carr, Aurora, IL, USA). The disc (Steminc Piezo, Miami, AZ, USA) was 19 mm diameter and 0.3 mm thick and insulated with Kapton tape (McMaster-Carr, Aurora, IL, USA). The rod was made from the same material as the beads, and has a mass m_r = 27.8 g, a height h_r = 13.3 mm, and a diameter D_r = 19.05 mm. The sensor-rod had approximately the same mass of the individual particles in order to minimize any impurity in the chain that may generate spurious HNSWs. Finally, the falling height of the striker was 5 mm.

At the bottom of the transducers a 0.254 mm (0.1 in.) thick aluminum sheet (McMaster-Carr, Aurora, IL, USA) was glued to the plastic tube in order to prevent the free fall of the particles. A through-thickness hole was devised to allow for direct contact between the last particle of the chain and the concrete material. The transducers were driven simultaneously by a National Instruments-PXI unit running in LabVIEW (9.0, National Instruments, Austin, TX, USA, 2009), and a DC power supply and a Matrix Terminal Block (NI TB-2643) (National Instruments, Austin, TX, USA) to branch the PXI output into three switch circuits to trigger the action of the electromagnets. The lead-zirconate-titanite (PZT) sensors (Steminc Piezo, Miami, AZ, USA) were connected to the same PXI, and the signals were digitized at 400 kHz sampling frequency, *i.e.*, 2.5 μs sampling period.

2.3. Test Protocol

All the HNSW-transducers were used to test all the specimens, *i.e.*, the experiments were conducted in a round-robin fashion, in order to prevent any bias in the results that may have stemmed from differences during the assembly. The cylinders were tested using the HNSWs immediately after curing the samples at 21 °C (70 °F) 95% relative humidity for 28 days, and completed in a day. The experiments were conducted in a single day. For each test, 50 measurements were taken.

Figure 4. Photo of the three transducers during the experiment.

The beams were tested at room conditions after 28 days of curing at 21 °C (70 °F) and at relative humidity of 95%. The UPV method was employed the day after testing with the solitary waves. Both top and bottom surfaces of the beams were

379

tested by removing them from the mold and eventually rotated. All the transducers were placed on the surface of the beam simultaneously, and fifty measurements were recorded by each transducer. Figure 4 shows the setups relative to the solitary wave measurements. It can be seen that each sample was tested simultaneously with three transducers and at three different locations. This translates in time and cost-savings. Posts (Techspec, Midland, TX, USA) and clamps (Techspec, Midland, TX, USA) were used to hold the transducers.

2.4. Model to Extract the Elastic Modulus

In order to extract the elastic modulus of the concrete in contact with the chain, we modelled numerically the dynamic interaction between a chain of particles identical to the one embedded in the transducer and an elastic material that mimics the concrete. The partial differential equation of the motion associated with the propagation of an HNSW in the chain can be determined using Lagrangian description of particle dynamics:

$$m\partial_{tt}^2 \mu_n = A[\mu_{n-1} - \mu_n]_+^{3/2} - A[\mu_n - \mu_{n+1}]_+^{3/2}. \tag{1}$$

In Equation (1), u_i is the ith particle displacement, $[x]_+$ denotes $max(x, 0)$, and A the stiffness constant present in the Hertz's law. By solving this equation, the time history of the nth particle's oscillation is obtained and then the displacements of the particles can be used to compute the approach $\delta = u_{i+1} - u_i$ and then to apply in the Hertzian relationship. More details about the models are available at [38,41,42].

Figure 5 shows the time-of-flight (TOF) of HNSWs as a function of the Young's modulus for a Poisson's ratio equal to 0.20, which is a typical value for concrete. The graph shows that the Young's modulus affects significantly the wave feature when $E < 100$ GPa; moreover, when the modulus of elasticity of the test sample is higher than 25 GPa, small differences in the measurement of the TOF, let say 3%, yields about 60% change in the estimated modulus. More details about the procedure to generate Figure 5 are available at [43].

In this study, we used Figure 5 to estimate the modulus of the samples by intersecting the experimental TOF to the numerical curve. For illustrative purposes, the figure shows the modulus corresponding to two experimental data, namely 0.5881 ms and 0.5692 ms.

The dynamic modulus of elasticity E_d was then converted into the static modulus of elasticity E_s using the empirical formula proposed by Lydon and Balendran [44,45]:

$$E_s = 0.83E_d. \tag{2}$$

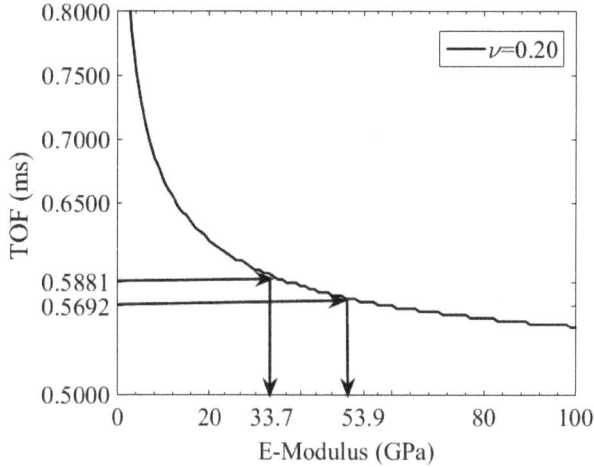

Figure 5. Numerical results: time of flight (TOF) as a function of the dynamic modulus of elasticity when the material in contact with the chain of spherical particles has Poisson's ratio $v = 0.20$.

2.5. Ultrasonic Pulse Velocity (UPV) Method

The conventional UPV method was employed for comparative purposes to determine the dynamic modulus of elasticity E_d of the concrete specimens using the well-known relationship:

$$E_d = V^2/K, \tag{3}$$

where V is the velocity of the bulk wave and K is:

$$K = (1 - \mu)/[(1 + \mu)(1 - 2\mu)]. \tag{4}$$

In Equation (4), μ is the dynamic Poisson's ratio of the sample material. When this ratio is not available, the conventional static Poisson's ratio v can be used [1]. After computing the dynamic modulus, Equation (2) is used to estimate the static modulus.

The UPV test was performed by measuring the velocity of the wave propagating along the axial direction of the cylinders, and the through-thickness (top-bottom) direction of the beams. Two commercial transducers (Olympus X1020, 100 kHz, Center Valley, PA, USA), two pre-amplifiers, a function generator (Tektronix AFG 3022, Fort Worth, TX, USA), and an oscilloscope (LeCroy 44 Xi, Chestnut Ridge, NJ, USA) were utilized. The average of 300 measurements was recorded for each cylinder and for each beam.

381

3. Experimental Results: Baseline Data

3.1. HNSW Transducers and UPV Test

Figure 6a shows one of the fifty waveforms recorded by the transducer P1 when it was placed above the samples 42A, 45A, and 50A. To ease the readability of the time waveforms, the amplitudes were offset. Figure 6b is a close-up view of Figure 6a and it shows that the TOF of the primary reflected wave increases with an increase in w/c ratio. Not presented here, but detailed in [38], the relative standard deviation (RSD), *i.e.*, the ratio of the standard deviation to the mean, ranged from 0.2% to 0.9%. This proved the high repeatability of the setup. The figure reveals that the TOF increases with the increase of the w/c ratio.

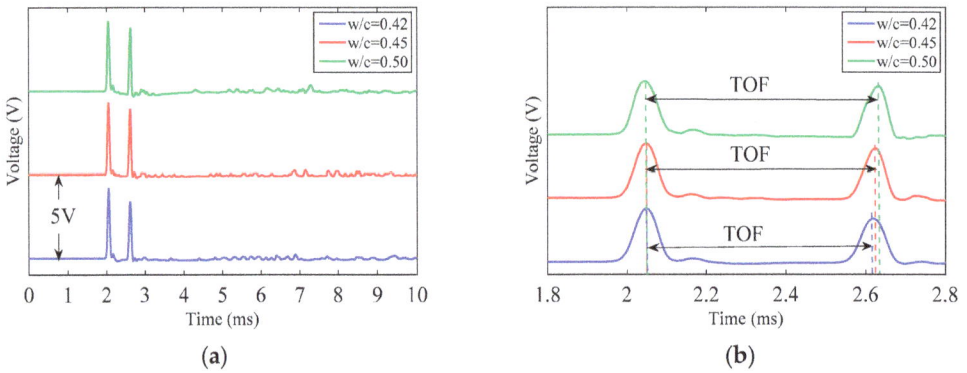

Figure 6. HNSW (highly nonlinear solitary wave)-based measurement using the transducers. (**a**) time waveform relative to P1 for the cylinders with different w/c ratio; (**b**) close-up view of panel (**a**).

The static moduli estimated with Equation (4) are listed in Table 2. They are based on the values of the TOFs averaged from each of the fifty waveforms acquired for every transducer. The values from each transducer and from each cylinder with the same w/c ratio were then averaged and reported on the rightmost column together with the corresponding RSD. The data show that an increase in the w/c ratio decreases the static modulus.

Table 3 presents instead the results associated with the UPV. When compared to the HNSW-based data, the ultrasonic test gives lower modulus and smaller RSD. The latter is due to the better repeatability of the commercial transducer, but it is also the effect of the fact that the ultrasonic test averages the effect of the through-thickness concrete. Finally, it must be noticed that the presence of an ultrasonic gel couplant between the transducer and the concrete mitigates any effect associated with the surface texture.

Table 2. Elastic modulus estimated for the cylinders and based on the time of flight of the reflected wave.

| Cylinder | Elastic Modulus (GPa) | | | | |
	P1	P2	P3	Mean E (GPa) (Per Sample)	Mean (Per w/c Ratio) $\pm$ RSD
42A	41.9	41.9	44.7	42.8	
42B	41.9	37.0	44.7	41.2	41.1% $\pm$ 3.38%
42C	41.9	37.0	39.3	39.4	
45A	39.3	39.3	41.9	40.2	
45B	39.3	34.9	34.9	36.4	37.6% $\pm$ 4.82%
45C	37.0	37.0	34.9	36.3	
50A	28.0	33.0	33.0	31.3	
50B	28.0	33.0	28.0	29.7	31.8% $\pm$ 6.00%
50C	37.0	33.0	33.0	34.3	

Table 3. Modulus of elasticity estimated for each cylinder and based on the value of the ultrasonic velocity of the longitudinal wave propagating along the cylinder axial direction.

Cylinder	Wave Speed (m/s)	Elastic Modulus (GPa)	Mean Elastic Modulus (GPa) (Per w/c Ratio) $\pm$ RSD
42A	4692	34.4	
42B	4730	34.8	34.9% $\pm$ 1.43%
42C	4740	35.6	
45A	4340	31.8	
45B	4456	32.1	31.8% $\pm$ 0.90%
45C	4407	31.4	
50A	4367	29.6	
50B	4344	29.6	29.5% $\pm$ 0.64%
50C	4330	29.2	

3.2. Destructive Testing

As is said in Section 2.1, three specimens from each batch were tested according to C469. This standard is traditionally followed to estimate the static modulus of elasticity. The results are presented in Table 4. Interestingly, the values of the elastic moduli are closer to those predicted with the HNSWs whereas the UPV data underestimated the elastic properties of the cylinders. Nonetheless, the destructive testing confirmed that the modulus of elasticity decreased with an increase in the w/c ratio. On average, the moduli relative to the $w/c = 0.45$ samples and the $w/c = 0.50$ samples were 7.5% and 9.2%, respectively, smaller than the $w/c = 0.42$ sample.

Table 4. Modulus of elasticity estimated using ASTM C469.

Sample	Elastic Modulus (GPa)	Mean (Per *w/c* Ratio) ± RSD
C42D	40.1	
C42E	37.0	38.6% ± 3.29%
C42F	38.8	
C45D	36.2	
C45E	32.4	35.7% ± 4.51%
C45F	35.2	
C50D	32.6	
C50E	31.8	32.4% ± 1.33%
C50F	32.8	

4. Experimental Results: Short Beams

Tables 2–4 constitute the reference data to which compare the findings from the corrupted beams. The results are presented in what follows.

4.1. HNSW-Transducer Measurements

Figure 7a–d show one of the fifty waveforms recorded by the transducer P1 when it was placed above the beams with water in excess. To ease the readability of the time waveforms, the amplitudes were offset. Figure 7e–h are the close-up view of Figure 7a–d, and they reveal some variation in the TOF between the pristine and the conditioned samples. In what follows, the time waveforms are analyzed to validate the hypothesis that the differences are due to the presence of unwanted water.

Table 5 summarizes the results associated with the beam subject to Condition 1, *i.e.*, with the presence of 12.7 mm (0.5 in.) of standing water at the bottom of the formwork. The average value of the fifty TOF measurements per transducer is listed along with the corresponding elastic modulus. The data refer to the beam both top and bottom face. The latter was exposed to the presence of water. However, as is said earlier and displayed in Figure 2a, water migrated toward the top due to vibration and the weight of the concrete. The mean of the twelve moduli is presented in the shaded row along with the corresponding standard deviation and RSD. Surprisingly, the modulus estimated at the top surface is higher than the modulus at the bottom where the standing water was originally located. The results are therefore consistent with the empirical evidence that water migrated to the top and the experimental procedure that water was mixed with the concrete throughout the specimen. The RSD associated with the measurement at the top is almost double the corresponding value at the bottom. This is the result of point-by-point variations due to the presence of water whereas the conditions at the bottom were more uniform due to the self-weight of the concrete and the smoothness of the mold. We will observe that this consideration about the RSD applies to all four conditions.

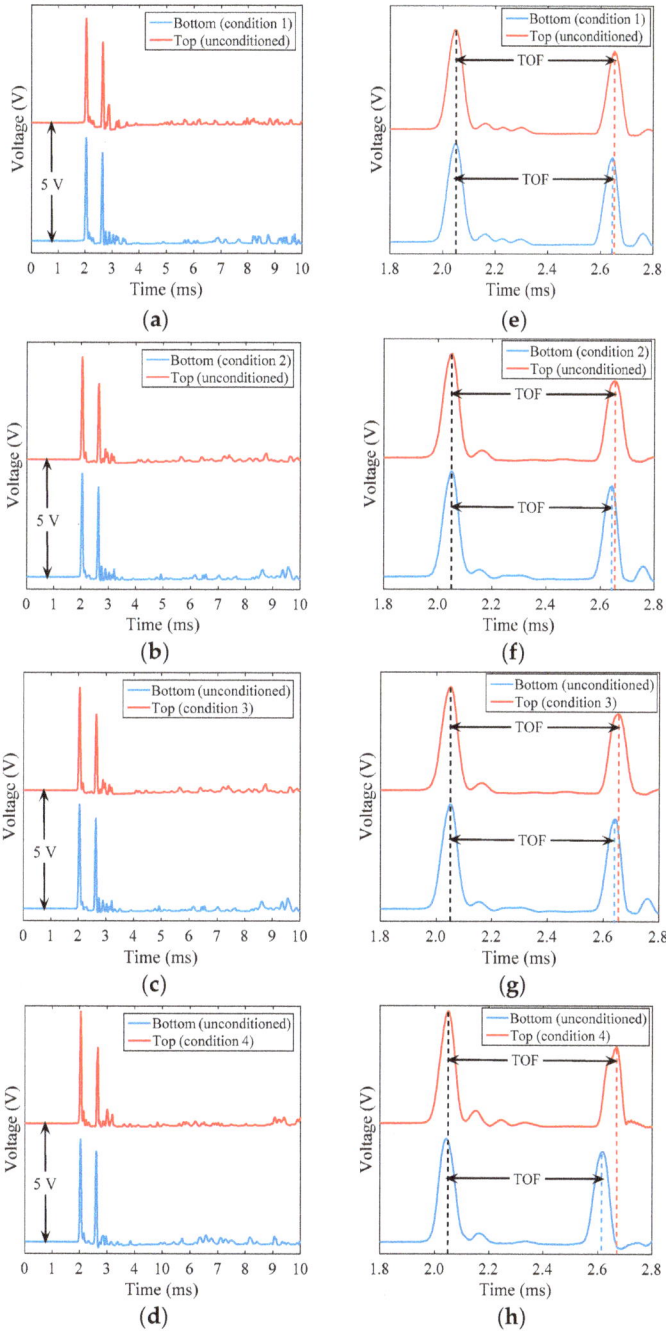

Figure 7. The time waveforms and their close up views obtained from the short beam tests: (**a**) condition 1; (**b**) condition 2; (**c**) condition 3; and (**d**) condition 4; (**e**), (**f**), (**g**), and (**h**) are the close views of (**a**), (**b**), (**c**), and (**d**), respectively.

Table 5. Time of flight and modulus of elasticity of the beams in which Condition 1 was imposed.

Cast Day	Sample	Top		Bottom	
		TOF (ms)	E (GPa)	TOF (ms)	E (GPa)
Day 1	1	0.59224	27	0.57640	37
		0.60296	21	0.58092	33
		0.60428	21	0.57044	42
	2	0.59316	25	0.58244	33
		0.58156	33	0.58072	33
		0.59200	27	0.57508	37
Day 2	3	0.58536	30	0.58364	31
		0.57764	35	0.57880	35
		0.57624	37	0.57728	37
	4	0.57892	35	0.56896	45
		0.60128	22	0.58384	31
		0.56924	45	0.58648	30
Average E (GPa) Short beam $w/c = 0.42$		29.83 ± 7.08 (23.73%)		35.33 ± 4.37 (12.35%)	
Cylinder $w/c = 0.42$		41.1 ± 1.389 (3.38%)			
Cylinder $w/c = 0.45$		37.6 ± 1.815 (4.82%)			
Cylinder $w/c = 0.50$		31.8 ± 1.907 (6.00%)			

The last three rows of Table 5 list the elastic modulus of the baseline cylinders predicted by the same transducers. They are presented here again to ease the prediction of the resulting w/c ratio at the two surfaces of the beams. By comparing the values relative to the beams and to the cylinders, the solitary wave based technique estimates a ratio higher than 0.50 at the top and around 0.46–0.47 at the bottom of the short beams.

Similar to Table 5, Table 6 presents the results relative to Condition 2 where the amount of standing water was half the amount in Condition 1. By looking at the mean of the elastic modulus, we observe that the value relative to the top surface is very close to the value reported in Table 5, whereas the w/c ratio estimated at the bottom is 0.45. The results suggest that the difference in the amount of standing water between Conditions 1 and 2 was not relevant to change the w/c ratio of the overall beam. By looking at the three bottommost rows of Table 6, we can reach the same conclusions for both conditions: the HNSW transducers were able to capture the circumstance that the short beams were corrupted by water in excess.

Table 6. Time of flight and modulus of elasticity of the beams in which Condition 2 was imposed.

Cast Day	Sample	Top		Bottom	
		TOF (ms)	E (GPa)	TOF (ms)	E (GPa)
Day 1	1	0.58332	31	0.57592	37
		0.60476	21	0.57420	39
		0.58260	31	0.56976	45
	2	0.60224	22	0.56684	48
		0.58900	28	0.58128	33
		0.60460	21	0.57644	37
Day 2	3	0.58320	31	0.59280	25
		0.59096	27	0.58152	33
		0.57112	42	0.57180	42
	4	0.57972	35	0.57800	35
		0.57908	35	0.58512	30
		0.58492	31	0.57236	42
Average E (GPa) Short beam $w/c = 0.42$		29.58 ± 6.02 (20.35%)		37.17 ± 6.24 (16.80%)	
Cylinder $w/c = 0.42$		41.1 ± 1.389 (3.38%)			
Cylinder $w/c = 0.45$		37.6 ± 1.815 (4.82%)			
Cylinder $w/c = 0.50$		31.8 ± 1.907 (6.00%)			

The TOF and the static modulus at the surfaces of the beams experiencing Condition 3 are summarized in Table 7. Under this scenario, 2.54 mm (0.1 in.) of water was sprinkled above the samples, whereas the bottom surface was pristine with $w/c = 0.42$ (see scheme of Figure 1c). It is observed that the estimated static modulus for the top surface is 31.8 GPa and it matches the cylinders with $w/c = 0.50$. When compared to the corresponding values presented in Tables 5 and 6 the modulus is slightly higher, and this is likely because the volume of water sprinkled on the samples was lower. Moreover, the value is much smaller than the modulus of elasticity estimated for the cylinders with $w/c = 0.42$. The elastic modulus measured at the bottom surface was higher than the previous two scenarios; this is expected since the specimens with Conditions 3 and 4 were not vibrated, and, therefore, it is unlikely that the water sprinkled on the top reached the bottom of the specimen.

Table 7. Time of flight and modulus of elasticity of the beams in which Condition 3 was imposed.

Cast Day	Sample	Top		Bottom	
		TOF (ms)	E (GPa)	TOF (ms)	E (GPa)
Day 1	1	0.57672	37	0.57420	39
		0.58312	31	0.57420	39
		0.57660	37	0.57888	35
	2	0.57164	42	0.56576	48
		0.59256	25	0.57420	39
		0.57924	35	0.57292	39
Day 2	3	0.59048	27	0.57600	37
		0.57988	35	0.57200	42
		0.58828	28	0.57400	39
	4	0.58448	31	0.57872	35
		0.59096	27	0.57000	42
		0.59000	27	0.57600	37
Average E (GPa) Short beam $w/c = 0.42$		31.83 ± 5.07 (15.96%)		39.25 ± 3.394 (8.650%)	
Cylinder $w/c = 0.42$		41.1 ± 1.389 (3.38%)			
Cylinder $w/c = 0.45$		37.6 ± 1.815 (4.82%)			
Cylinder $w/c = 0.50$		31.8 ± 1.907 (6.00%)			

Finally, the TOF and the modulus of elasticity of the beams subjected to Condition 4 are summarized in Table 8. Under this condition, more water was added onto the fresh concrete. Consistently with the larger volume of liquid, the predicted modulus at the top was lower than what estimated under scenario 3. The table demonstrates that the HNSW-based measurement estimated that the w/c of the unconditioned surface of the concrete beam was close to 0.42, as it is expected, and above 0.50 for the surface of the beam that was sprinkled with 3.8 mm (0.15 in) of water. It is noted here that sampling period used in this study and the design of the metamaterial are such that the potential error in the estimate of the concrete modulus is on the order of 5%. The accuracy can be improved by simply increasing the sapling frequency of the digitizer. Nonetheless, the outcomes of the results are in line with the prediction of water in excess in the concrete beams.

Table 8. Time of flight and modulus of elasticity of the beams in which Condition 4 was imposed.

Cast Day	Sample	Top		Bottom	
		TOF (ms)	E (GPa)	TOF (ms)	E (GPa)
Day 1	1	0.59608	24	0.57420	39
		0.58260	31	0.57420	39
		0.57944	35	0.57768	35
	2	0.59276	25	0.57644	37
		0.58484	31	0.57668	37
		0.59608	24	0.57420	39
Day 2	3	0.57892	35	0.57401	39
		0.57768	35	0.57200	42
		0.58492	31	0.57800	35
	4	0.58932	28	0.57600	37
		0.58512	30	0.56740	48
		0.57172	42	0.57600	37
Average E (GPa) Short beam $w/c = 0.42$		30.92 ± 5.107 (16.52%)		38.67 ± 3.37 (8.73%)	
Cylinder $w/c = 0.42$		41.1 ± 1.389 (3.38%)			
Cylinder $w/c = 0.45$		37.6 ± 1.815 (4.82%)			
Cylinder $w/c = 0.50$		31.8 ± 1.907 (6.00%)			

4.2. UPV Test

UPV was adopted to test the beams as well. The results are summarized in Table 9. The means of the modulus are much closer to each other than what was found with the HNSW method. This remarks upon the fact that the novel NDE method is capable of capturing surface conditions that may have been altered by the presence of water. Moreover, the RSD is smaller, suggesting more homogeneous conditions throughout the four samples. Finally, if we compare the UPV data from Tables 4 and 9 we notice that the UPV method estimates an amount of w/c ratio above 50% for all samples, without any ability at discriminating poor from good surface conditions.

Table 9. The modulus of elasticity of the beams in which different conditions are measured by UPV (ultrasonic pulse velocity) test.

Condition	E (GPa) Mean ± Standard Deviation	RSD (%)
1	31.44 ± 0.966	3.07
2	31.13 ± 1.558	5.00
3	32.37 ± 1.354	4.18
4	29.42 ± 2.660	9.04

5. Discussion and Conclusions

In this article, we showed the principles of a novel NDE method for concrete based on the propagation of highly nonlinear solitary waves along a metamaterial in contact with the concrete to be evaluated. The method aimed at determining the modulus of hardened concrete, in particular to estimate the water-to-cement ratio in a concrete volume close to an HNSW transducer. We demonstrated that the transducers designed and assembled to exploit the principles offer sufficient repeatability and reliability to identify the differences in the amount of water purposely added to the beams in order to mimic rainfall situations.

Owing to the nature of concrete material, it is acknowledged that the w/c ratio is likely not to be the only concrete parameter affecting the amplitude and the time of flight of the reflected solitary wave. Future studies should look at the effect of the aggregate size and overall any other factor that is known to influence the Young's modulus of concrete. Nonetheless, the study presented in this paper is the first attempt to prove that solitary waves can be used to measure water in excess in concrete surfaces.

The advantages of the proposed HNSW-based method are the easy and fast implementation, the possibility to carry out a large number of tests simultaneously, and independence upon internal damage and/or the presence of reinforcing steels inside the concrete. Finally, being a local and contact method, the approach can be successfully applied to characterizing effects of finishing and curing conditions.

Figure 5 showed that the measured values of E are on a part of the curve that is not extremely favorable for sensitivity. Future studies should redesign the metamaterial, such as different sized or different modulus balls, in order to shift the region of influence in a range where a large change in the TOF gives rise to small variation of the Young modulus. Such a region would consent to discriminate small differences in the Young modulus due to changes in the w/c ratio.

Acknowledgments: The study presented in this paper is supported by partial technical results drawn from the currently active work order PITT WO 008 (University of Pittsburgh Work Order 008), funded by the Pennsylvania Department of Transportation and the Federal Highway Administration under contract 4400011482. The third author conducted this research as a visiting scholar at the University of Pittsburgh under a sponsorship from the China Scholarship Council. The use of the results or reliance on the material presented is the responsibility of the reader. The contents of this document are not meant to represent standards and are not intended for use as a reference in specifications, contracts, regulations, statutes, or any other legal document. The opinions and interpretations expressed are those of the authors and other duly referenced sources. The views and findings reported herein are solely those of the writers and not necessarily those of the US Department of Transportation, Federal Highway Administration, or the Commonwealth of Pennsylvania. This paper does not constitute a standard, a specification, or regulations.

Author Contributions: P.R. and J.M.V. conceived and designed the experiments; W.D. and A.N. performed the experiments; A.N. and W.D. analyzed the data; P.R. and J.M.V. contributed reagents/materials/analysis tools; P.R. and A.N. wrote the paper.

Conflicts of Interest: The authors declare no conflict of interest.

Abbreviations

The following abbreviations are used in this manuscript:

AISI	American iron and steel institute
ASTM	American society for testing and materials
DC	Direct current
GGBFS	Ground-granulated blast-furnace slag
HNSW	Highly nonlinear solitary wave
ISW	Incident solitary wave
NDE	Nondestructive evaluation
PSW	Primary reflected solitary wave
PZT	lead-zirconate-titanate
TOF	Time of flight
UPV	Ultrasonic pulse velocity
w/c	Water-to-cement

References

1. Malhotra, V.M.; Carino, N.J. *Handbook on Nondestructive Testing of Concrete Second Edition*; CRC Press: Boca Raton, FL, USA, 2003.
2. Karaiskos, G.; Deraemaeker, A.; Aggelis, D.; van Hemelrijck, D. Monitoring of concrete structures using the ultrasonic pulse velocity method. *Smart Mater. Struct.* **2015**, *24*.
3. Shah, A.; Ribakov, Y.; Zhang, C. Efficiency and sensitivity of linear and non-linear ultrasonics to identifying micro and macro-scale defects in concrete. *Mater. Des.* **2013**, *50*, 905–916.
4. Ye, G.; Lura, P.; van Breugel, K.; Fraaij, A. Study on the development of the microstructure in cement-based materials by means of numerical simulation and ultrasonic pulse velocity measurement. *Cem. Concr. Compos.* **2004**, *26*, 491–497.
5. Lin, Y.; Lai, C.-P.; Yen, T. Prediction of ultrasonic pulse velocity (UPV) in concrete. *ACI Mater. J.* **2003**, *100*, 21–28.
6. Shariq, M.; Prasad, J.; Masood, A. Studies in ultrasonic pulse velocity of concrete containing GGBFS. *Constr. Build. Mater.* **2013**, *40*, 944–950.
7. Komlos, K.; Popovics, S.; Nürnbergerova, T.; Babal, B.; Popovics, J. Ultrasonic pulse velocity test of concrete properties as specified in various standards. *Cem. Concr. Compos.* **1996**, *18*, 357–364.
8. Popovics, S. Analysis of the concrete strength *versus* ultrasonic pulse velocity relationship. *Mater. Eval.* **2001**, *59*, 123–130.

9. Hong, S.; Cho, Y.; Kim, S.; Lee, Y. Estimation of compressive strength of concrete structures using the ultrasonic pulse velocity method and spectral analysis of surface wave method. *Mater. Res. Innov.* **2015**, *19*.

10. Huang, Q.; Gardoni, P.; Hurlebaus, S. Predicting concrete compressive strength using ultrasonic pulse velocity and rebound number. *ACI Mater. J.* **2011**, *108*, 403–412.

11. ASTM (American Society for Testing Material) C805/C805m-08. *Standard Test Method for Rebound Number of Hardened Concrete*; ASTM: West Conshohocken, PA, USA, 2008.

12. Borosnyói, A. NDT assessment of existing concrete structures: Spatial analysis of rebound hammer results recorded *in-situ*. *E. Struct. Technol.* **2015**, *7*, 1–12.

13. Cano-Barrita, P.D.J.; Castellanos, F.; Ramírez-Arellanes, S.; Cosmes-López, M.; Reyes-Estevez, L.; Hernández-Arrazola, S.; Ramírez-Ortíz, A. Monitoring compressive strength of concrete by nuclear magnetic resonance, ultrasound, and rebound hammer. *ACI Mater. J.* **2015**, *112*, 147–154.

14. Proceq. Available online: http://www.Proceq.Com/products/concrete-testing/concrete-test-hammer/ original-schmidt (accessed on 10 January 2016).

15. Choi, H.; Popovics, J.S. Nde application of ultrasonic tomography to a full-scale concrete structure. *IEEE Trans. Ultrason. Ferroelectr. Freq. Control* **2015**, *62*, 1076–1085.

16. Völker, C.; Shokouhi, P. Multi sensor data fusion approach for automatic honeycomb detection in concrete. *NDT E Int.* **2015**, *71*, 54–60.

17. Kim, G.; In, C.-W.; Kim, J.-Y.; Kurtis, K.E.; Jacobs, L.J. Air-coupled detection of nonlinear rayleigh surface waves in concrete—Application to microcracking detection. *NDT E Int.* **2014**, *67*, 64–70.

18. Saravanan, T.J.; Balamonica, K.; Priya, C.B.; Reddy, A.L.; Gopalakrishnan, N. Comparative performance of various smart aggregates during strength gain and damage states of concrete. *Smart Mater. Struct.* **2015**, *24*.

19. Clayton, D.A.; Barker, A.M.; Santos-Villalobos, H.J.; Albright, A.P.; Hoegh, D.K.; Khazanovich, D.L. *Nondestructive Evaluation of Thick Concrete Using Advanced Signal Processing Techniques*; Oak Ridge National Laboratory (ORNL): Oak Ridge, TN, USA, 2015.

20. Shih, Y.-F.; Wang, Y.-R.; Lin, K.-L.; Chen, C.-W. Improving non-destructive concrete strength tests using support vector machines. *Materials* **2015**, *8*, 7169–7178.

21. Amini, K.; Jalalpour, M.; Delatte, N. Advancing concrete strength prediction using non-destructive testing: Development and verification of a generalizable model. *Constr. Build. Mater.* **2016**, *102*, 762–768.

22. Saleem, M.; Al-Kutti, W.A.; Al-Akhras, N.M.; Haider, H. Nondestructive testing procedure to evaluate the load-carrying capacity of concrete anchors. *J. Constr. E. Mang.* **2015**.

23. Pucinotti, R. Reinforced concrete structure: Non destructive *in situ* strength assessment of concrete. *Constr. Build. Mater.* **2015**, *75*, 331–341.

24. Azari, H.; Nazarian, S. Optimization of acoustic methods for condition assessment of concrete structures. *ACI Mater. J.* **2015**, *112*.

25. Ham, S.; Popovics, J.S. Application of micro-electro-mechanical sensors contactless NDT of concrete structures. *Sensors* **2015**, *15*, 9078–9096.

26. Hoegh, K.; Khazanovich, L.; Dai, S.; Yu, T. Evaluating asphalt concrete air void variation via GPR antenna array data. *Case Stud. Nondestruct. Test. Eval.* **2015**, *3*, 27–33.

27. Iliopoulos, S.N.; Aggelis, D.G.; Polyzos, D. Wave dispersion in fresh and hardened concrete through the prism of gradient elasticity. *Int. J. Solids Struct.* **2016**, *78*, 149–159.

28. Dimter, S.; Rukavina, T.; Minažek, K. Estimation of elastic properties of fly ash–stabilized mixes using nondestructive evaluation methods. *Constr. Build. Mater.* **2016**, *102*, 505–514.

29. Hertz, H. On the contact of elastic solids. *J. Reine Angew. Math.* **1881**, *92*, 156–171.

30. Nesterenko, V.F. *Dynamics of Heterogeneous Materials*; Springer Science & Business Media: Berlin, Germany, 2013.

31. Li, F.; Zhao, L.; Tian, Z.; Yu, L.; Yang, J. Visualization of solitary waves via laser doppler vibrometry for heavy impurity identification in a granular chain. *Smart Mater. Struct.* **2013**, *22*.

32. Boechler, N.; Yang, J.; Theocharis, G.; Kevrekidis, P.; Daraio, C. Tunable vibrational band gaps in one-dimensional diatomic granular crystals with three-particle unit cells. *J. Appl. Phys.* **2011**, *109*.

33. Ni, X.; Rizzo, P.; Daraio, C. Laser-based excitation of nonlinear solitary waves in a chain of particles. *Phys. Rev. E* **2011**, *84*.

34. Herbold, E.B. Optimization of The Dynamic Behavior of Strongly Nonlinear Heterogeneous Materials. Ph.D. Thesis, University of California, San Diego, CA, USA, 2008.

35. Daraio, C.; Nesterenko, V.F.; Herbold, E.; Jin, S. Tunability of solitary wave properties in one-dimensional strongly nonlinear phononic crystals. *Phys. Rev. E* **2006**, *73*.

36. Carretero-González, R.; Khatri, D.; Porter, M.A.; Kevrekidis, P.; Daraio, C. Dissipative solitary waves in granular crystals. *Phys. Rev. Let.* **2009**, *102*.

37. Porter, M.A.; Daraio, C.; Szelengowicz, I.; Herbold, E.B.; Kevrekidis, P. Highly nonlinear solitary waves in heterogeneous periodic granular media. *Phys. D Nonlinear Phenom.* **2009**, *238*, 666–676.

38. Nasrollahi, A.; Deng, W.; Rizzo, P.; Vuotto, A.; Vandenbossche, J.M.; Li, K. University of Pittsburgh, Pittsburgh, PA, USA, Unpublished work. 2016.

39. Ni, X.; Rizzo, P.; Yang, J.; Katri, D.; Daraio, C. Monitoring the hydration of cement using highly nonlinear solitary waves. *NDT E Int.* **2012**, *52*, 76–85.

40. Rizzo, P.; Ni, X.; Nassiri, S.; Vandenbossche, J. A solitary wave-based sensor to monitor the setting of fresh concrete. *Sensors* **2014**, *14*, 12568–12584.

41. Herbold, E.; Kim, J.; Nesterenko, V.; Wang, S.; Daraio, C. Pulse propagation in a linear and nonlinear diatomic periodic chain: Effects of acoustic frequency band-gap. *Acta Mech.* **2009**, *205*, 85–103.

42. Yang, J.; Silvestro, C.; Khatri, D.; de Nardo, L.; Daraio, C. Interaction of highly nonlinear solitary waves with linear elastic media. *Phys. Rev. E* **2011**, *83*.

43. Bagheri, A.; Rizzo, P.; Al-Nazer, L. A numerical study on the optimization of a granular medium to infer the axial stress in slender structures. *Mech. Adv. Mat. Struct.* **2016**, *23*, 1131–1143.

44. Lydon, F.; Balendran, R. Some observations on elastic properties of plain concrete. *Cem. Concr. Res.* **1986**, *16*, 314–324.

45. Popovics, J.; Zemajtis, J.; Shkolnik, I. ACI-CRC Final Report. In *A Study of Static and Dynamic Modulus of Elasticity of Concrete*; Civil and Environmental Engineering, University of Illinois: Urbana, IL, USA, 2008.

Opto-Acoustic Method for the Characterization of Thin-Film Adhesion

Sanichiro Yoshida, David R. Didie, Daniel Didie, Tomohiro Sasaki, Hae-Sung Park, Ik-Keun Park and David Gurney

Abstract: The elastic property of the film-substrate interface of thin-film systems is characterized with an opto-acoustic method. The thin-film specimens are oscillated with an acoustic transducer at audible frequencies, and the resultant harmonic response of the film surface is analyzed with optical interferometry. Polystyrene, Ti, Ti-Au and Ti-Pt films coated on the same silicon substrate are tested. For each film material, a pair of specimens is prepared; one is coated on a silicon substrate after the surface is treated with plasma bombardment, and the other is coated on an identical silicon substrate without a treatment. Experiments indicate that both the surface-treated and untreated specimens of all film materials have resonance in the audible frequency range tested. The elastic constant of the interface corresponding to the observed resonance is found to be orders of magnitude lower than that of the film or substrate material. Observations of these resonance-like behaviors and the associated stiffness of the interface are discussed.

Reprinted from *Appl. Sci.* Cite as: Yoshida, S.; Didie, D.R.; Didie, D.; Sasaki, T.; Park, H.-S.; Park, I.-K.; Gurney, D. Opto-Acoustic Method for the Characterization of Thin-Film Adhesion. *Appl. Sci.* **2016**, *6*, 163.

1. Introduction

Thin-film coating is used in a wide variety of applications ranging from micro-electro-mechanical systems (MEMS) to surface treatment of mechanical systems for wear-resistance enhancement. The recent trend indicates that reduced film thickness in a number of applications leads to better performance. In some cases, films of a few tens of nanometers in thickness are exposed to vigorous rubbing motion. With this trend, the quality control of the film-adhesion strength and endurance has become more important than ever. The reduction in the film thickness often means that a minute defect or imperfection at the film-substrate interface can cause significant coating damage, ending the life of the entire system.

A number of destructive and nondestructive methods have been developed to characterize the film-substrate adhesion. Those classified as destructive methods [1–4] assess the ultimate strength of the adhesion via measurement of the critical force that the coating can tolerate. Prevailing techniques classified as nondestructive methods use ultrasonic waves to probe the film-substrate interface [5–9]. An acoustic signal from an ultrasonic transducer is transmitted through the interface for the detection

395

of abnormalities, such as delamination or defects. By measuring the velocity of the acoustic wave, it is possible to characterize the elasticity of the the interface.

The destructive methods are useful for a number of applications. However, they are not suitable for the evaluation of the endurance or fatigue characteristics. It is rare that the surface of a thin-film system is subject to external force comparable to the ultimate stress. The coating is usually damaged by repetitive, low-level stresses.

Similarly, the ultrasonic methods have limitations in practicability as the acoustic frequency used by most methods is unrealistically higher than the frequency range that a thin-film system normally experiences. As will be elaborated in the next section, the film-substrate interface can respond harmonically when driven at certain frequency ranges. Harmonic systems break most easily by repetitive use at or near the natural (resonant) frequency. This is true for a macroscopic system, as represented by the famous collapse of the Tacoma Narrows Bridge, or a micro-, nano-scale system, as represented by the fracture of a cantilever in MEMS. It is important to evaluate the adhesion strength using probing frequencies within a frequency range that thin-film systems are subject to and to examine if the interface exhibits resonance-like behaviors. The use of a proper frequency is important for a good signal-to-noise ratio, as well. If the probing frequency is orders of magnitude higher than the resonant frequency, the diagnostic system has poor sensitivity. The reduction in film thickness raises the ultrasonic frequency even higher. It is possible that the diagnostic system overlooks the most important frequency range.

Considering the above issues, we have devised an opto-acoustic method to characterize the film-substrate interface in a range of audible frequencies. In this method, the specimen is oscillated with an acoustic transducer perpendicular to the film surface, and the resultant differential displacement of the film surface to the substrate is read out optically. By sweeping the driving frequency of the transducer, it is possible to explore for resonance-like behavior of the interface.

Using this method, we studied the effect of a pre-coating surface treatment. The treatment, known as the oxygen-plasma bombardment technique, is widely used for thin-film coating on silicon substrates. According to the theory, plasma bombardment knocks off hydro-carbons and thereby strengthens the chemical bonding between the film and substrate materials. Using various film materials coated on the same silicon substrate with and without the surface treatment, we conducted experiments under various conditions and collected a great amount of data. Consequently, we observed several pieces of evidence that indicate that the film-substrate interface has resonance in a frequency range of 2 kHz to 30 kHz. The resonance behaviors were observed both in the specimen with the surface treatment (the treated specimen) and those with no surface treatment (the untreated specimen).

Analysis of the observed harmonic response indicates that the resonance behavior possibly represents the membrane-mode-like oscillation of the film surface [10] associated with the adhesion weakening mechanism, known as blistering [11–13]. The results of this study indicate the usefulness of this opto-acoustic method for practical applications. As mentioned above, harmonic systems most easily break when disturbed near resonant frequency, and audible frequency is abundant in environmental disturbance.

In this paper, we describe the opto-acoustic method, present the experimental study on the effect of the surface treatment for the silicon-based thin-films and discuss the results of the experiments focusing on the resonance behaviors observed in both the untreated and treated specimens. Furthermore, a possible elastic mechanism behind the resonance-like behavior observed in the audible frequency is discussed.

2. Theory

2.1. Elasticity of the Film-Substrate Interface

The main mechanism of film-substrate adhesion involves a collection of chemical bonds. Although the types and orientations of the chemical bonds are complex, each individual bond is associated with a certain elastic constant [14]. Thus, the film-substrate interface can be viewed as a complex combination of a number of springs, each having a different spring constant from one another. In an actual thin-film system, some of the bonds will not be perfect, leading to a lower elastic constant than the better bonds. Figure 1 illustrates the situation schematically where the middle two springs are weaker than the side springs. When this system is oscillated at a frequency lower than the resonance associated with some of the springs from the rear side of the substrate, the weaker springs oscillate more than the stronger springs. Consequently, the film surface can bulge back and forth as the film oscillates, as illustrated in the lower part of Figure 1. This type of membrane-like oscillation can occur locally in a specimen. By measuring the amplitude of the film surface oscillation scanning through the entire film surface, it is possible to characterize the elasticity of the film-substrate interface as a function of the position in the in-plane coordinates.

The ultrasonic technique known as scanning acoustic microscopy (SAM) probes the film-substrate interface by oscillating it at ultrasonic frequency. By measuring the acoustic velocity, SAM evaluates the elastic constant of the interface. Thus, SAM can potentially detect weak adhesions of thin-film systems. However, when the oscillation associated with adhesion weakness has a resonant frequency significantly lower than the ultrasonic frequency at which SAM is operated, the applicability of SAM becomes questionable for the following reason. When a thin-film system is driven acoustically, an oscillator whose resonance is the closest to the driving frequency responds most sensitively. The transmissibility (the transfer function) of a mechanical oscillator decreases with a quadratic dependence on the frequency (f^{-2})

on the high frequency side of the resonance. If the driving frequency is higher than the resonance, the response of the oscillator decreases quadratically. It is possible that bond imperfection reduces the overall stiffness so significantly that the ultrasonic frequency falls in a range where the sensitivity is unrealistically low.

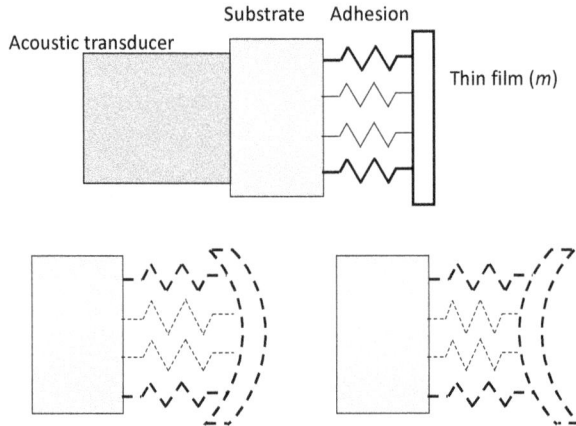

Figure 1. Physical model of thin-film systems. The figures at the bottom show the membrane-like vibration of the film surface due to the weaker elastic constant of the middle part.

2.2. Opto-Acoustic Method

In thin-film adhesion studies, the acoustical characterization of elastic behaviors associated with a resonant frequency lower than 100 MHz is difficult. The wavelength of the probing acoustic wave must be shorter than the width of the thickness of the film and substrate. To generate 10 or more wave numbers in a 100 μm-thick silicon substrate, for instance, the wavelength must be shorter than 10 μm. The acoustic velocity in silicon is approximately 8 km/s. Therefore, the corresponding frequency is of the order of 8 km/s ÷ 10 μm = 800 MHz. For smaller thicknesses, the frequency must be increased further in proportion to the reduction in the wavelength. To overcome this difficulty, we use the opto-acoustic method described here.

2.2.1 Michelson Interferometer

Figure 2 illustrates the optical configuration of this method. The specimen is integrated in a Michelson interferometer as one of the end-mirrors with the film side facing the beam splitter and is oscillated with an acoustic transducer from the rear. The acoustic transducer is driven with a sinusoidal input in a frequency range of 2 kHz to 30 kHz. The resultant oscillatory motion of the film surface is detected as the corresponding optical path difference behind the beam splitter.

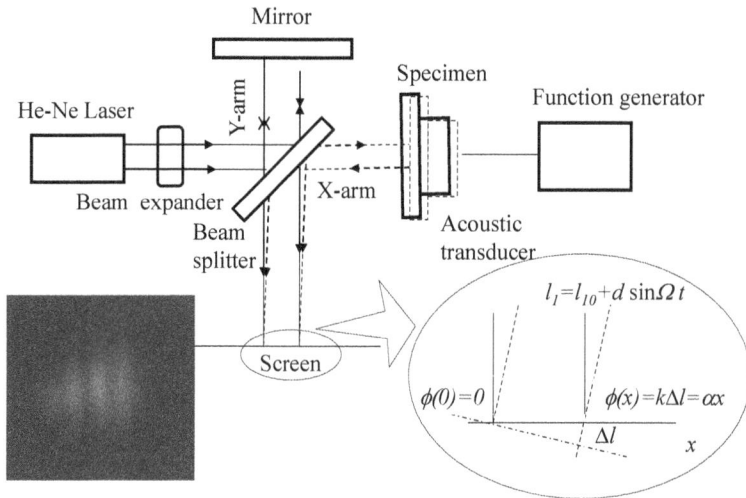

Figure 2. Experimental arrangement with a Michelson-type interferometer. The function generator (Gold Star Model FG 8002) sends sinusoidal input voltage to the acoustic transducer (magnetic type, Design Factory Dokodemo-Speaker 8297).

To visualize the relative optical path length difference between the two arms, one of the end mirrors is slightly tilted. In this way, the relative phase of the optical intensity behind the beam splitter can be detected as parallel interferometric fringes as indicated by the insert in Figure 2. The fringe pattern is projected on a screen, and a digital camera is used to capture the pattern electronically. Here, the dark fringes represent the contour where the relative optical path length difference is totally destructive or the relative phase difference is an odd integer multiple of π. When the specimen is oscillated with the acoustic transducer, the corresponding optical path varies in a sinusoidal fashion, causing the dark fringes to dither around the nominal position where the dark fringes are located when the transducer is off. Since the digital camera's frame rate is much lower than the transducer's oscillation frequency, the dithering lowers the fringe contrast. The present method utilizes the reduction in the fringe contrast to estimate the oscillation amplitude.

In this frequency range, both the film and substrate behave as a rigid body because their elastic constants are so high that the acoustic phase velocities are high enough to make the wavelength at an audible frequency longer than their thicknesses. This means that if the interface has an elastic constant comparable to the film and substrate, the entire specimen should oscillate as a rigid body, as well. In other words, the oscillation amplitude of the rear and the front surface of the specimen is comparable or the transfer function of the rear-surface displacement to the front-surface displacement is unity. However, if the elastic constant of the interface is so low that the substrate-side to the film-side displacement transfer function is higher than unity, the film surface displacement becomes greater, as compared to the transducer's displacement. The preliminary study on the harmonic response of a thin-film specimen demonstrated a case where the low elastic constant of the interface makes the substrate-to-film displacement transfer function greater than unity (see Section 3.1 below).

2.2.2. Principle of Operation

The reduction in the fringe contrast in a given image frame due to the dithering motion of the fringes can be argued quantitatively as follows. The optical intensity on the screen placed behind the beam splitter is given by the following expression.

$$I(t) = 2I_0 + 2I_0 \cos[k(l_{10} - l_{20}) + kd \sin \omega t] \tag{1}$$

Here, I_0 is the intensity of the individual arm; k is the wave number of the laser beam; d is the amplitude of the film surface oscillation; and ω is the driving angular frequency provided by the transducer. Since one of the end-mirrors is tilted, the relative phase difference on the screen can be resolved as a linear function of x with a constant α as $k(l_{10} - l_{20}) = \alpha x$. Here, the x-axis is set perpendicular to the dark fringes. Thus, Equation (1) can be rewritten in the following form.

$$I(t) = 2I_0 + 2I_0[\cos(\alpha x) \cos(\delta \sin \omega t) - \sin(\alpha x) \sin(\delta \sin \omega t)] \tag{2}$$

where $\delta = kd$. Using Bessel functions of the first kind, Equation (2) can further be rewritten as follows.

$$I(t) = 2I_0[1 + \cos(\alpha x)\{J_0(\delta) + 2J_2(\delta) \cos(2\omega t) + \cdots\} - \sin(\alpha x)\{2J_1(\delta) \sin(\omega t) + 2J_3 \sin(3\omega t) \cdots] \tag{3}$$

Therefore, the signal of the digital camera is:

$$
\begin{aligned}
S(\tau) \quad &\propto \quad \int_0^\tau \{1 + \cos(\alpha x)\{J_0(\delta)\}dt \\
&+ \quad 2\cos(\alpha x) \sum_{m=1}^N J_{2m}(\delta) \int_0^\tau \cos(2m\omega t)dt \\
&- \quad 2\sin(\alpha x) \sum_{m=1}^N J_{2m-1}(\delta) \int_0^\tau \sin\{(2m-1)\omega t\}dt
\end{aligned} \tag{4}
$$

Since the frame rate of the digital camera is much lower than the driving frequency, the exposure time τ is much longer than the period of the film surface oscillation. Consequently, the values of the oscillatory terms in Equation (4) $\int_0^\tau \cos(2m\omega t)dt$ and $\int_0^\tau \sin\{(2m-1)\omega t\}dt$ sweep through a number of periods as

t is varied from zero to τ in the integration. On the other hand, the values of the first two terms that do not contain an oscillatory term increase in proportion to τ. Therefore, the signal from the digital camera is dominated by the first non-oscillatory terms of Equation (4). Thus, the signal $S(\tau)$ can be approximately expressed as follows (The concept is similar to the technique known as the amplitude-fluctuation electronic speckle pattern interferometry (AF-ESPI) [15,16]. The difference is that AF-ESPI involves image subtraction from a reference image, and the present method extracts the oscillation amplitude from a given image frame of the interferometric fringe pattern without subtracting a reference image.):

$$S(\tau) \propto \tau + J_0(\delta) \cos(\alpha x) a \tau \tag{5}$$

Equation (5) indicates that the peak value of the spatial Fourier spectrum of $S(\tau)$ (the value of the peak at the lowest spatial frequency) is proportional to $J_0(\delta)$. When the film surface is still or the acoustic transducer is turned off, $d = 0$, hence $\delta = 0$. Since $J_0(0) = 1$ and $J_0(\delta)$ monotonically decreases to zero in the range of $0 < \delta < 2.4$, the peak value of the Fourier spectrum decreases as the oscillation amplitude increases in the corresponding range. The Fourier spectrum peak value of a given signal obtained with the film surface oscillation amplitude can be evaluated relative to the still film surface case (called the relative peak value) as $J_0(\delta)/J_0(0) = J_0(\delta)$. By obtaining the relative peak value experimentally, *i.e.*, by comparing the Fourier spectrum peak value of a given signal with that obtained with no film surface oscillation, it is possible to estimate $\delta = kd$ and, hence, d.

2.2.3. Proof of Principle

Figure 3 (1) shows examples [17,18] of fringe patterns and the Fourier spectrum of the intensity profile along a horizontal line around the vertical center. The left fringe pattern (a) is the case where the film was coated without a surface treatment; the middle (b) is where the film was coated after the silicon substrate was treated with the oxygen-plasma bombardment technique; and the right (c) is where the transducer was turned off. Careful observation will indicate that the fringe contrast increases from (a) to (c). The driving frequency was 11.5 kHz for all three cases. The observed fringe contrasts indicate that at this driving frequency, the surface treatment made the amplitude of the film-surface oscillation smaller than the untreated case. Naturally, the contrast is highest when the transducer was turned off.

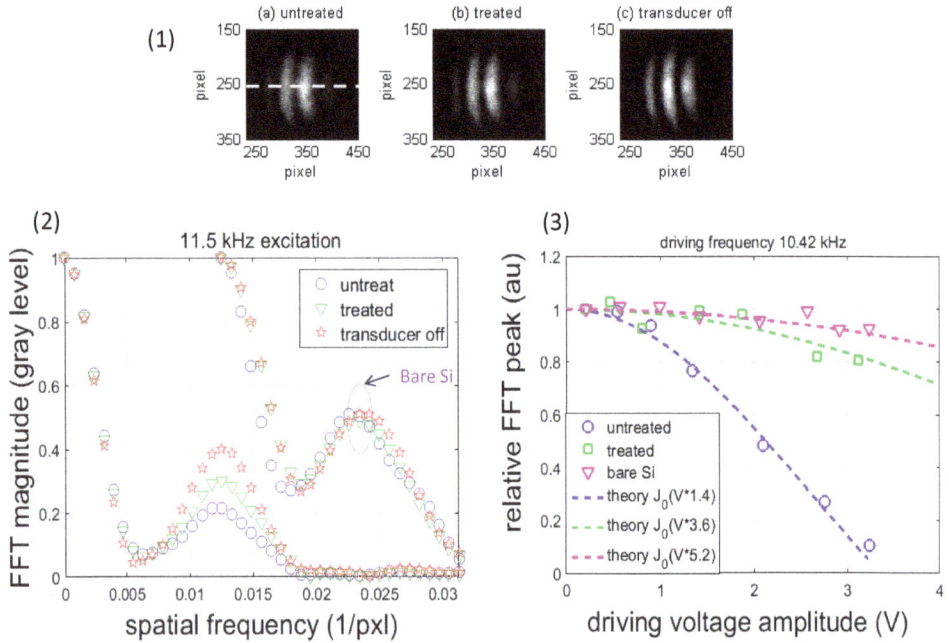

Figure 3. (**1**) Examples of fringe patterns; (**2**) Fourier spectra of intensity along the horizontal line indicated in Fringe Pattern (**a**); and (**3**) driving-voltage dependence of spectral peak value relative to the transducer-off (null driving voltage) case.

The left plots shown under the fringe images, Figure 3 (2), are the Fourier spectra of the optical intensity profile along the horizontal line at the vertical center as indicated in the leftmost fringe image. The unit of the horizontal axis of the spectrum is the inverse of the pixel numbers. The left group of three plots whose first spectrum peaks appear approximately at the spatial frequency of 0.0125 1/pxl result from the three fringe images (a) to (c). The peak values of the three spectra indicate the fringe contrast quantitatively; the spectrum peak decreases as the fringe contrast decreases, being consistent with Equation (5).

The three plots shown on the right part of Figure 3 (2) are the Fourier spectra for the non-coated specimen (the bare silicon specimen) obtained in the same fashion as the left three plots. The three plots are the cases where the transducer oscillated one end-mirror only, oscillated the other end-mirror only and oscillated neither end-mirror, respectively (the use of the same plot marker as the left group indicates that the same mirror was oscillated). The peak values for the three cases are practically the same, indicating that the silicon surface oscillation amplitude was below the sensitivity of the interferometer.

In the frequency domain, the amplitude of the film-surface displacement is the product of the transfer functions of the layers between the transducer surface and the film surface and the Fourier transform of the transducer-surface displacement. The overall transfer function can be expressed as the product of the transfer functions of (a) the transducer surface to the rear surface of the silicon substrate, (b) the rear to the front surface of the silicon substrate, (c) the silicon to the film surface, (d) the rear to the front surface of the Ti film and (e) the transfer functions of the interface between the first and second metal layers (for the Ti-Au and Ti-Pt specimens only). Here: (a) is the transfer function of the glue (epoxy) used to bond the specimen to the transducer; (b) is the silicon substrate itself; (c) is the substrate-film interface; (d) is the film itself; and (e) is the product of the transfer functions of the metal-metal interface and the second metal film itself.

Of these transfer functions, those of the silicon substrate and the metal film itself are considered unity due to the high stiffness as mentioned above. The metal-metal interface in (e) is also considered unity, as it is metallic bonding. As for the epoxy, a theoretical estimation indicates that the resonant frequency of the epoxy is over 30 MHz under the present condition and thereby can be assumed unity in the frequency range used in this experiment. Thus, of (a) through (e), all of the transfer functions, except (c), are considered unity under the driving frequency ranging from 5 kHz to 30 kHz. After all, the harmonic response evaluated by the Fourier spectrum, like Figure 3 (2), represents the behavior of the substrate-film interface.

According to Equation (5), the peak value of a Fourier spectrum, like Figure 3 (2), is proportional to $J_0(\delta) = J_0(kd)$. To validate this proportionality, an additional experiment was conducted. In this experiment, we varied the oscillation amplitude of the transducer surface by changing the amplitude of the electric input and measured the peak value of the resultant Fourier spectrum. Here, since the square of the voltage is proportional to the electric power and the square of the oscillation amplitude is proportional to the mechanical power, the input voltage and the oscillation amplitude are proportional to each other. Figure 3 (3) plots the spectral peak value as a function of the input voltage to the transducer. The three plots are for the untreated and treated Ti-Au specimens and the bare silicon specimen. The driving frequency is 10.42 kHz. Also plotted in Figure 3 (3) is a theoretical curve computed for each specimen based on Equation (5). Since the transfer function from the amplitude of the input voltage to the amplitude of the film-surface displacement is unknown, the theoretical curve is adjusted in the form of $J_0(aV)$, where V is the input voltage and a is a fitting parameter used for each curve. Physically, the parameter a corresponds to the transfer function from the transducer voltage to the film-surface oscillation. Reasonable agreement is seen between the experiment and theory.

3. Experimental Section

3.1. Preliminary Experiment with Doppler Vibrometry

We first studied the harmonic response of a thin-film specimen using a laser Doppler vibrometer (LDV, Model: OFV 5000, Polytec, Irvine, CA, USA [19] with Displacement Decoder Model: DD300, Polytec, Irvine, CA, USA [20]) [21]. Figure 4 illustrates the experimental arrangement. The operation principle of LDV can be found elsewhere [22]. In short, it works in the following fashion. A laser beam is applied normal to an oscillating surface. The sensor of the LDV compares the optical frequency of the reflected laser beam with the incident beam. Due to the Doppler effect, the frequency of the reflected beam is shifted depending on the velocity of the oscillation. From the detected frequency shift, the oscillation amplitude is estimated.

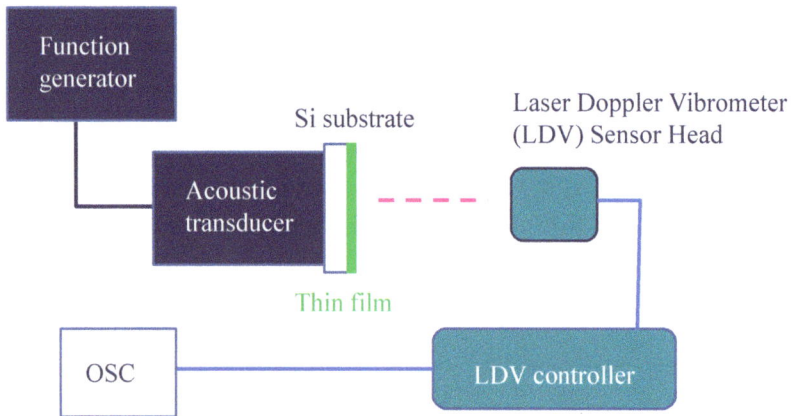

Figure 4. Doppler vibrometry setup.

A pair of polystyrene thin-films were used in this experiment. In the first specimen, 100 nm-thick polystyrene was coated on a 750 μm-thick silicon substrate without pre-coating surface treatment (untreated polystyrene-coated specimen). The silicon substrate was cut along the [1, 0, 0] plane. In the second specimen, polystyrene of the same thickness was coated on the same silicon substrate after the surface was treated with the oxygen-plasma bombardment technique.

An acoustic transducer was attached to the rear surface of the specimen to oscillate it at 50 kHz. The transducer was driven with a function generator (Model 395, Wavetek, San Diego, CA, USA) with a pure sinusoidal output. Figure 5 compares the resultant displacement of the film surface for the untreated and treated specimens along with the oscillation of the transducer surface. (The signals for the untreated and treated specimens are shifted vertically for better visibility. The amplitude of the signals in nm exceeds the maximum value specified by the manufacturer for an unknown reason. However, since the signals do not exhibit any saturation-like behavior, the argument made here is considered legitimate. The absolute amplitude of the oscillation does not affect the gist of the argument.) Here, the middle graph is for the case when the transducer was oscillated without a specimen; the top and bottom graphs are the cases where the treated and untreated specimens were attached, respectively. The top and bottom graphs are shifted vertically for better visibility. As expected, the untreated specimen shows greater oscillation amplitude, indicating that the adhesion is weaker. (Figure 5 indicates that the signal from the treated specimen and the transducer only is much noisier than the signal from the untreated specimen. The reason for this observation is not totally clear. A possible explanation is that: (a) the untreated one is least noisy because the amplitude of the signal is the greatest (among the three signals); (b) the amplitude of the transducer-only signal is not much smaller than the untreated, but the signal is much noisier than the untreated case because the surface of the transducer was not as smooth as the specimen; (c) the treated signal is noisy because the signal was low.) There is a phase difference between the untreated and treated specimens, as will be discussed shortly.

Consider the film surface motion based on the theory of harmonic oscillation. The harmonic response of an oscillatory system of spring constant k, damping coefficient b and mass m driven at angular frequency ω is given as a solution to the following equation of motion.

$$m\ddot{x} + b\dot{x} + kx = fe^{i\omega t} \tag{6}$$

Here, $x(t)$ is the displacement from the equilibrium position of m, and f is the amplitude of the driving force. In the present context, k and b represent the elasticity and viscosity of the adhesion; m is the mass of the thin-film; and ω is the angular frequency of the acoustic transducer. The particular solution to Equation (6) can be put in the following form in general.

$$x(t) = Ae^{i(\omega t - \delta)} \tag{7}$$

Substitution of Equation (7) into Equation (6) leads to the following condition regarding the amplitude and phase of the oscillation of the film displacement:

$$A = \frac{f/m}{\sqrt{(\omega_0^2 - \omega^2)^2 + 4\beta^2\omega^2}} \tag{8}$$

$$\tan\delta = \frac{2\beta\omega}{(\omega_0^2 - \omega^2)} \tag{9}$$

405

Here, $\omega_0 = \sqrt{k/m}$ is the natural frequency of the oscillatory system, and $2\beta = b/m$. f/m in Equation (8) is the acceleration of the transducer surface. The amplitude of the transducer surface A_0 when the transducer is operated at ω can be related to the acceleration as $(f/m) = A_0\omega^2$. Using this relation, we can rewrite Equations (8) and (9) in the following forms.

$$\frac{A}{A_0} = \frac{\xi^2}{\sqrt{(1-\xi^2)^2 + 4\gamma^2\xi^2}} \tag{10}$$

$$\tan\delta = -\frac{2\gamma\xi}{(1-\xi^2)} \tag{11}$$

where the frequency and damping coefficient are normalized by the natural frequency as $\xi = \omega/\omega_0$ and $\gamma = \beta/\omega_0$.

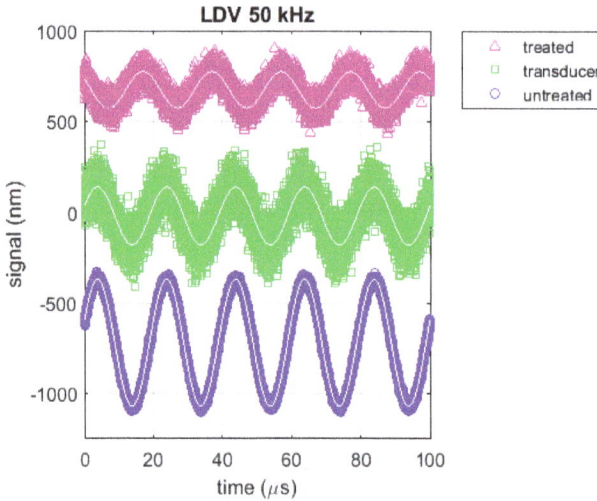

Figure 5. Signal from Doppler vibrometry.

Since the left-hand sides of Equations (10) and (11) can be experimentally determined from Figure 5, we can estimate the natural frequency and decay constant for a given driving frequency by finding the combination of ω_0 and γ that yields A/A_0 and $\tan\delta$ closest to the experimental values. Figure 6 plots theoretical A/A_0 and $\tan\delta$ as a function of ξ (solid lines) and the experimental values determined from Figure 5 (square and circle markers) for the driving frequency of 50 kHz. For the untreated/treated specimen, when ω_0 and γ are as shown in Table 1, the experimental values agree with the theory the best. It is seen that the untreated specimen shows 18% higher resonant frequency than the treated specimen and that the surface treatment makes the elastic dynamics more energy dissipative. This 18%

ratio of the resonant frequency of the untreated specimen to the treated specimen observed in the polystyrene films is approximately the same as the case of the other films tested in the study (see Section 3.2 below for more details). This agreement can be simply a coincidence, but is interesting to note. The higher energy dissipation in the treated specimen can be understood by considering that the oxygen plasma bombardment makes the substrate surface rougher and causes more friction when the layers undergo oscillatory motions.

Table 1. Amplitude and phase of the Doppler vibrometry signal relative to transducer and corresponding natural frequency and decay constant estimated from Equations (10) and (11).

Specimen	Untreated	Treated
A/A_0	2.2	0.62
δ (rad)	1.63	−0.52
$\omega_0/2\pi$ (kHz)	58.5	49.5
γ (1/s)	0.115	0.82

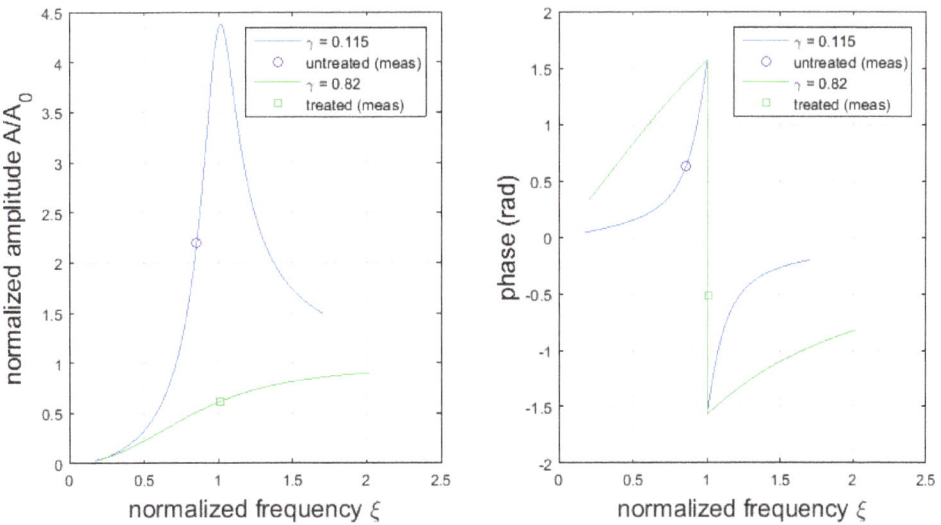

Figure 6. Assimilation of observed relative amplitude and phase with harmonic response function Equations (10) and (11).

3.2.1. Treated and Untreated Specimens

Table 2 summarizes the specimens used in this study. Each film material was coated on a silicon substrate ([1, 0, 0] plane) prepared with the pre-coating plasma treatment (the treated specimen) and on a silicon substrate of the same size without a pre-coating surface treatment (the untreated specimen). The films were initially coated on the silicon substrate as a disk of 75 mm in diameter and cut into an approximately 5 mm × 5 mm square specimen. The bottom row of this table shows the square root of the relative mass to the case of the Ti-only specimen. This quantity is called the mass ratio and will be used later in this section to estimate the resonant frequency of the interface. For dynamics associated with elastic force for a given elastic modulus in general, the resonant frequency is expected to vary inversely proportional to the square root of the mass.

Table 2. Specimens. T: film thickness; ρ: density of the film material; M: total film mass; mass ratio: the ratio of the square root of the mass of the Ti-Au and T-Pt films to that of the Ti film.

Film material	Ti	Ti-Au	Ti-Pt
T (nm)	75	10/100	10/100
ρ (kg/m^3)	4510	4510/19,030	4510/21,450
M (kg)	8.4×10^{-9}	5.0×10^{-8}	5.5×10^{-8}
mass ratio	1	2.4	2.5

3.2.2. Observation of Resonance-Like Behavior

With the use of the interferometer setting shown in Figure 2, experiments were conducted numerous times with the treated and untreated thin-film specimens listed in Table 2. Figure 7 plots the peak value of the Fourier spectrum observed for each specimen as a function of the driving frequency (called the driving frequency sweep). For each specimen, the peak value is normalized to that of the bare silicon evaluated at the same driving frequency (called the peak ratio).

The driving frequency sweep of the peak ratio shows some features. In the Ti-only specimen case, the untreated specimen shows valleys around 14.5 kHz and 27 kHz (called the valley frequency). Similarly, the treated specimen shows valleys at 7 kHz, 13 kHz and 22.5 kHz. The higher valley frequencies observed in the respective specimens are approximately integer multiples of the lowest value. This indicates the possibility that the observed series of valley frequencies represents harmonics of the resonance for the respective specimen, being consistent with the above argument that the film-substrate interfaces have resonance in the swept frequency range. A previous study on Au-coated thin-film specimens clearly shows that the valley observed in the driving frequency sweep around the higher central frequency is the second harmonic of the valley observed around the lower central frequency [17].

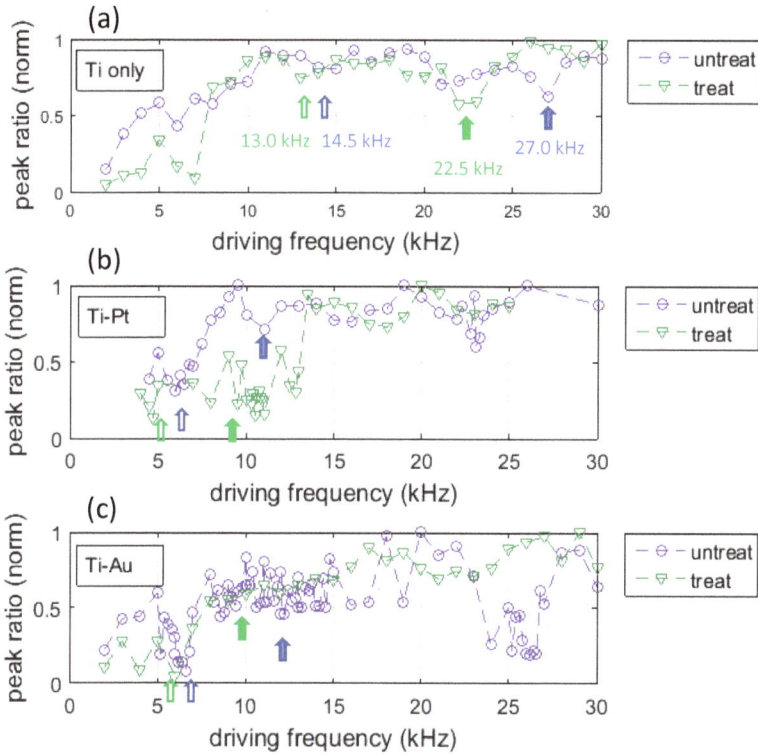

Figure 7. Peak ratio as a function of driving frequency. The valley frequencies indicated in (**a**) the Ti-only specimen are measured values. The valley frequencies pointed to by arrows in (**b**) Ti-Pt and (**c**) Ti-Au indicate valley frequencies estimated from those in (**a**) and the mass ratio defined in Table 2.

Based on the assumption that the elastic property of the Ti-substrate interface is independent of the additional metal layers on top of the Ti layer and on the mass ratio shown in Table 2, the valley frequencies observed in the Ti-only case can be converted to the Ti-Pt and Ti-Au cases. Table 3 lists the valley frequencies estimated in this fashion for the Ti-Pt and Ti-Au specimens. In the driving frequency sweeps shown in Figure 7, the valley frequencies estimated for the Ti-Pt (b) and Ti-Au (c) specimens are indicated with arrows. Valleys are seen approximately at these estimated frequencies.

In Table 3, the valley frequencies are grouped as Valley 1 and Valley 2. Valley 1 refers to the lower frequency pair of valley frequencies observed in the untreated and treated specimens of the same kind. Similarly, Valley 2 refers to the higher frequency pairs. Notice that in each pair, the valley frequency of the untreated specimen is 15% to 20% higher than the treated specimen. This is consistent with the results from the Doppler vibrometry shown in Table 1. These observations support the interpretation that the valleys in the driving frequency sweeps represent the resonance behaviors of the film-substrate interface.

Table 3. Valley frequencies observed in Ti-only specimens and expected valley frequencies in Ti-Pt and Ti-Au specimens based on the mass ratio defined in Table 2 (unit: kHz).

Specimen	Valley 1		Valley 2	
	Treated	Untreated	Treated	Untreated
Ti-only	13.0	14.5	22.5	27.0
Ti-Pt	5.2	5.8	9.0	10.8
Ti-Au	5.4	6.0	9.4	11.3

The reduction in the fringe contrast, *i.e.*, the peak ratio, is very sensitive to the optical alignment and other factors. The longer in the optical path that the interfering laser beams overlap each other, the greater the modulation depth ($J_0(\delta)$ in Equation (4)). The air flow near the interferometer and floor vibration also affect the fringe contrast. The driving frequency sweeps, the Ti-Pt case in particular, indicate that the data fluctuate vigorously near valleys. This is understandable if we consider the nature of the signal detected by the digital camera with the help of Equation (5). The Bessel function J_0 appearing in the second term of Equation (5) is steeper as the Fourier spectrum peak gets lower (Figure 3 (3)). This means that the greater the oscillation amplitude caused by the function generator, the more sensitive the spectrum peak becomes to the fluctuation of the voltage input to the transducer. The output voltage from the function generator used in the present study fluctuated approximately $\pm 0.5\%$ over the period of data taking for each driving frequency. To reduce those errors, we repeated the same measurement a number of times. The plots shown in Figure 7 are averages of over five measurements. The fluctuations in the peak ratio among these measurements were approximately $\pm 5\%$ near a valley frequency and $\pm 2\%$ in the frequency range away from valleys, respectively. From the slopes near valleys in Figure 7, the $\pm 5\%$ error in the peak ratio is translated into $< \pm 1\%$ error in the valley frequency. Thus, the differences in the valley frequencies listed in Table 3 are significant.

In the case of the Ti-Au film, both the untreated and treated specimens show a significant reduction in the peak ratio. It is likely that this is because of poor adhesion between the Au and Ti films, rather than the poor adhesion between Ti and silicon substrate. Poor adhesion between Au and Ti layers was observed elsewhere in a film system similar to the present case [23].

3.2.3. Detailed Analysis on Ti-Pt Resonance

The observation of resonance-like behavior in the audible frequency range was somewhat surprising. In particular, the prominent valleys in the treated Ti-Pt specimen in the lower driving frequency range in Figure 7 was not expected. It was not observed in the previous study [17], where the lowest driving frequency was 10 kHz. We tested the Ti-Pt specimens a number of times using different interferometric setups with a similar interferometric setup in a different laboratory in a different country and still observed resonance-like behavior at the same frequency range. To confirm these observations more precisely, we made two modifications on the experimental arrangement. First, we added a second interferometric beam pair to the interferometer. This allowed us to take the measurement of two specimens simultaneously, thereby removing all temporal fluctuations common to the two interferometric pairs, such as intensity and polarization fluctuation of the laser beam and floor vibration noise, in comparing experimental data from the two specimens. Figure 8 illustrates the dual beam arrangement. The incident beam from the laser was split into two paths before the beam splitter of the interferometer. The two pairs of laser beams were delivered in parallel to each other to the end of the two interferometric arms. At the end of the X-arm (the horizontal arm in Figure 8), the first beam reflected off the first specimen, and the second beam reflected off the X-end mirror. At the end of the Y-arm (the vertical arm), the first beam reflected off the Y-end mirror and the second beam reflected off the second specimen.

Second, we calibrated the acoustic transducer by measuring the transducer's oscillation amplitude as a function of the operating frequency and applied voltage. This ensured that the specimens were oscillated by the transducer with the same oscillation amplitude independent of the driving frequency. The constant transducer oscillation amplitude facilitated the measurement in the lower driving frequency end where the optical alignment was much more difficult than the higher frequency range because the oscillation amplitude was larger for the same applied voltage.

Figure 9a shows the raw data of the driving frequency sweep of the Fourier spectrum peak measured for the untreated and treated Ti-Pt specimens. Also shown in this figure are the Fourier spectrum peaks of a pair of bare Si specimens. One of the bare Si specimens was placed at the X-end of the interferometer, and the other was placed at the Y-end (Figure 8). The signal levels of the raw Fourier spectrum peak values for the two bare Si specimens are different because the first beam splitter did not split the laser beam exactly by 50%. Notice that the treated data converges to the bare Si data of the same arm at the low frequency end (7 kHz) and that the untreated data converges to the bare Si data of the same arm at the low and high frequency ends. This indicates that at these frequencies, the film surface oscillated for the same amount as the bare Si. In Figure 9b, the untreated and treated signals are normalized to the respective bare Si data. The valleys at 8 kHz in the treated specimen and those at 12.5 kHz and 13.5 kHz in the untreated specimen are more clearly seen than in Figure 9a.

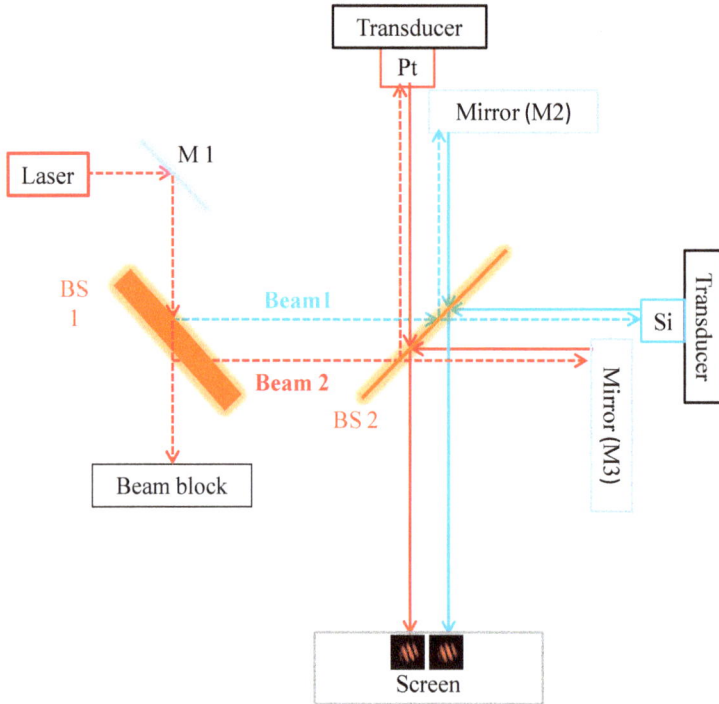

Figure 8. Dual beam Michelson interferometer. Beam Splitter 1 (BS1) splits the incoming laser beam into two passes for two pairs of interferences. Beam 2 is reflected on the rear surface of BS1. Light reflected on the Si/Pt specimen surface interferes with light reflected on M2/M1. Two interferometric fringe patterns are formed on the screen simultaneously. Respective specimens and corresponding mirrors are aligned for fringe formation, as explained in Figure 2. The screen is placed at a distance for better alignment of interfering light beams.

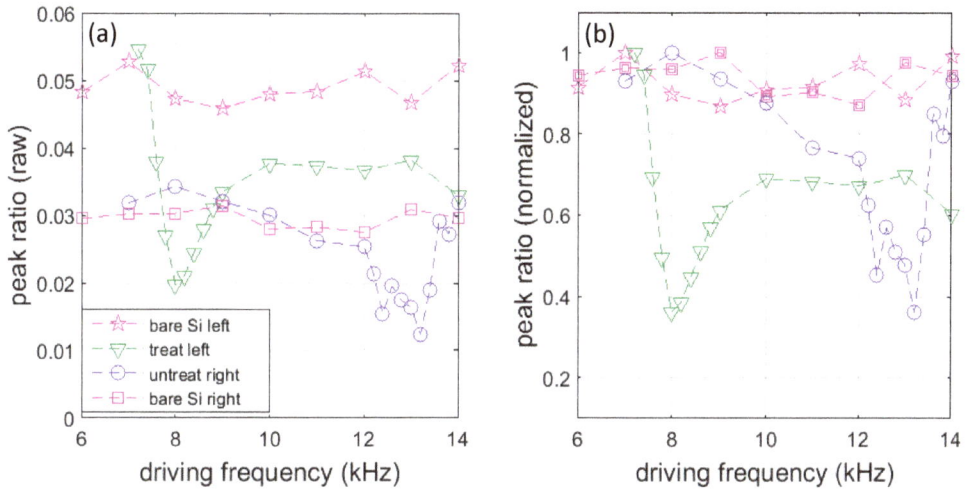

Figure 9. Valley frequencies of untreated, treated Pt-Ti specimens and bare Si specimens. (**a**) Raw data of peak ratio; (**b**) normalized to the bare Si data for the respective mirrors, *i.e.*, the Pt-Ti specimen data obtained with the right/left mirror to bare Si data with the right/left mirror.

3.2.4. Long-Term Temporal Change in the Valley Frequencies

In the course of the driving frequency sweep experiments, we noticed that the valley frequencies tended to shift after the specimen had been oscillated with the acoustic transducer for a certain amount of cumulative time. In some cases, the valley frequency increased and other times decreased. We have not understood this phenomenon on a solid physical basis yet, but it is worth presenting the findings at this time. Compare the lower valley observed in the Pt-Ti treated specimen observed in the driving frequency sweep in Figure 7 and the valley observed in the Pt-Ti treated specimen in Figure 9. This particular specimen was oscillated for approximately one year between the times that the two data were obtained. It is seen that the valley frequency had been shifted from approximately 11 kHz to 8 kHz. A similar phenomenon had been observed in the untreated Au-Ti specimen, as presented in Figure 10. In this figure, the data labeled #1 were taken after the specimen was oscillated for approximately one year, and the data labeled #2 were taken from a new specimen. The valley frequency of 22 kHz is seen to shift to 25 kHz. More investigation is currently being undertaken to explain this phenomenon.

413

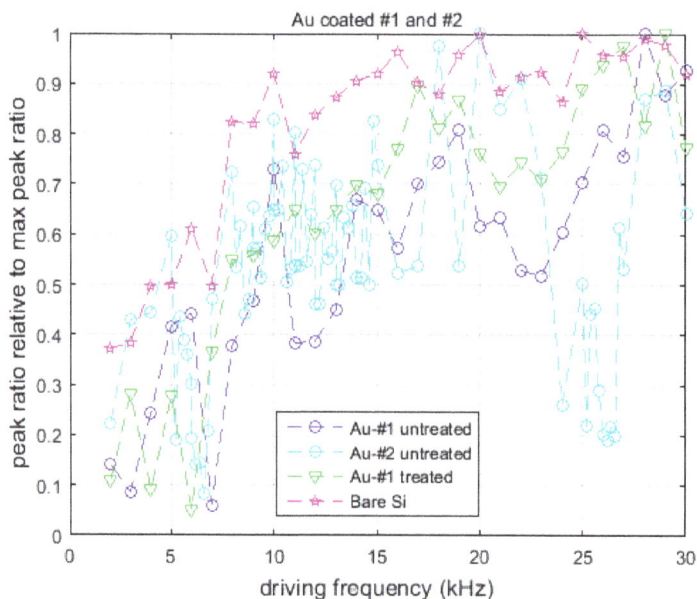

Figure 10. Valley frequency shift observed in the Ti-Au untreated specimen.

3.2.5. Resonant Frequency

Based on the interpretation that the valley frequency represents the resonance of the elastic behavior of the interface and that the resonant frequency can be evaluated by taking the square root of the elastic constant of the film-substrate interface to the mass of the film, it is possible to compare the interfacial elasticity observed in this study with other studies. Table 4 compares the valley frequency of the Ti specimen discussed above and the hypothetical resonant frequency calculated from the elastic constant found in the polystyrene specimen (Table 1) along with the interfacial elastic constant reported by Noijen *et al.* [2] and Ishiyama *et al.* [24]. Here, the hypothetical resonant frequency is calculated under the assumption that a 75 nm-thick, 5 mm × 5 mm Ti film is coated on the respective substrate. Noijen *et al.* [2] measured the interfacial elastic constant for a copper thin-film using the four-point bending test. Ishiyama *et al.* [24] measured the elastic constant of a columnar photoresist adhered to a silicon substrate.

Table 4. Valley frequencies and hypothetical resonant frequencies (unit: kHz). tr: treated specimen; untr: untreated specimen; pol. str: polystyrene specimen.

Ti-tr	Ti-untr	Ref. [2] Min	pol. str-tr	pol. str-untr	Ref. [2] Max	Ref. [24]
22.5	27.0	54.4	75.8	89.6	133.2	13,900

Table 4 indicates that the resonant frequencies observed in this study are of the same order of magnitude as the Noijen's four-point bending test. However, these are three orders of magnitude lower than the value estimated from Ishiyama's measurement. These indicate the possibility that the above-observed resonance-like behaviors represent an elastic mechanism similar to the four-point test and fundamentally different from the column adhesion test. A possible mechanism to explain the elastic behavior observed in this study is discussed in the following section. It is interesting to note that Miyasaka [23] observed that the surface acoustic wave propagates at the interface of the Pt-Ti treated specimen (identical to this study) approximately 1.5% faster than the Pt-Ti untreated specimen (identical to this study) in his SAM experiment using a 400-MHz ultrasonic source. This indicates that the interfacial elastic constant of the treated specimen is higher than the untreated specimen, unlike our observation where the untreated specimen shows higher resonant frequency. It is possible that the 400-MHz acoustic wave probes the elastic mechanism represented by the elastic constant measured by the column test.

3.3. Possible Mechanism of the Observed Elastic Behavior

Although the mechanism behind the interfacial elastic dynamics responsible for the above-discussed resonance-like behaviors in the audible frequency range has not been understood, some arguments in association with the the phenomenon known as the interfacial blistering seem worth making. Thus, an attempt is made in this section to explain the experimental observation according to the following scenario. During the coating process, residual stresses are developed in the Ti layer. The residual stress either compresses or stretches the Ti layer parallel to the interface with the substrate and forms a pattern of compression and tension in the Ti layer. This pattern induces non-uniformity in the interfacial distance between the substrate and the Ti layer, generating a number of local, blister-like weak adhesions. The residual stress is not strong enough to cause delamination, but induces non-uniformity in the elastic strength. Volinsky *et al.* [25] discuss the effect of residual stress on thin aluminum films. Interfacial blistering is a major cause of delamination in the film-substrate interface and is usually discussed in the context of fracture mechanics. However, the dynamics of a blister in a thin-film's interface seems to have resonance in the audible frequency range.

The details of interfacial blistering can be found elsewhere. In short, a blister can be formed in a film-substrate interfacial area where the adhesion is weaker than the surrounding area. Jensen [10] conducted a detailed study on the transition from the plate mode to the membrane mode dynamics in blister tests [12]. In blister tests, pressure is applied to the blistering area from the substrate side to inflate the film surface. When the height of the film-surface inflation reaches a critical value (relative to the film thickness), the elasticity of the film material takes over the longitudinal interfacial elasticity as the dominant restoring force mechanism of the blistering area; *i.e.*, the elastic behavior of the area shifts from the plate mode to the membrane mode. In the present context, no pressure is applied to the film surface from the substrate side, and the structural condition of the interface is different from usual blister tests. However, if the applied dynamic load is at or close to the resonant frequency of the membrane-like elastic behavior, it is possible that the film surface bulges large enough to shift the elastic dynamics to membrane mode.

When a membrane is subject to a dynamic load, the membrane exhibits various oscillation modes, such as the drumhead mode. Each mode has its own resonant frequency that is related to the tension of the membrane, the density of the membrane material and the size of the membrane. In the case of the drumhead mode of a rectangular membrane, the resonant frequency is given by the following equation.

$$f_{mn} = \frac{1}{2}\sqrt{\frac{T}{\rho}[(\frac{m}{a})^2 + (\frac{n}{b})^2]} \tag{12}$$

Here, f_{mn} is the resonant frequency of the drumhead mn mode; T is the membrane tension; ρ is the density of the membrane material; a and b are the sides of the membrane associated with modes m and n, respectively.

Apply the above idea to the Ti film of the present experiment. Since the tension of the film is unknown, Equation (12) cannot be used to estimate the resonant frequency. Therefore, here, we estimate the tension T from the observed valley frequency of 27 kHz (Table 3). The size of the laser beam forming a fringe pattern on the specimen is approximately 2 mm. Substitution of $a = b = 2$ mm, $\rho = 4510$ kg/m^3 (Table 2) and $m = n = 1$ into Equation (12) leads to the tension corresponding to the lowest drumhead mode, 26 MPa. The tensile strength of Ti is approximately 1000 MPa, and the tension of 26 MPa corresponds to 2.6% of the tensile strength. Whether this value is reasonable or not for the tension of the present Ti film specimen cannot be argued from the information available in the present study. However, it is certainly not an impossible number. Residual stress can be tensile or compressive. The scenario of the membrane-mode dynamics in conjunction with residual stress on the Ti film and its relaxation to some extent explains the observation of the long-term effect that, in some cases, the valley frequency is increased and other times decreased after long runs.

To consider the possibility of the above mechanism in terms of the spatial profile of the adhesion weakness, we analyzed the peak ratio around the two valley frequencies observed in Figure 9 two-dimensionally. Figure 11 shows the peak-ratio valleys three-dimensionally for the treated and untreated specimens. Here, the horizontal axes are the driving frequency and the spatial coordinate axis perpendicular to the interferometric fringes. The valley in the untreated specimen appears circular as opposed to that in the treated specimen, which is more cylindrical. This indicates that in the direction parallel to the fringes, the treated specimen shows a wider area of weak adhesion than the untreated specimen. According to Equation (12), the resonant frequency is inversely proportional to the area. This difference in the size of the area possibly explains the fact that the treated valley frequency is lower than untreated. For analysis of the spatial profile of valleys in the orthogonal directions to Figure 11, the fringe systems must be rotated by 90°. This requires a major optical reconfiguration and is a subject of our future study.

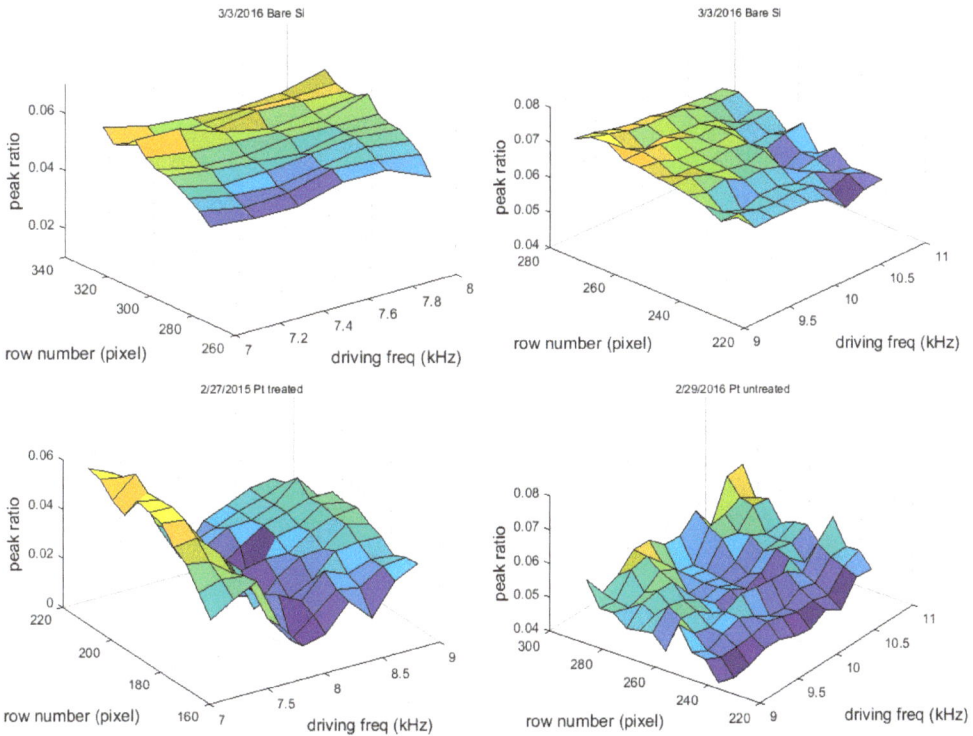

Figure 11. Three-dimensional plots of resonance-like peaks.

4. Conclusions

Opto-acoustic methods to characterize the interface of thin-film systems were described. A thin-film system specimen was oscillated from the rear surface of the substrate in a range of audible frequencies, and the harmonic response of the film surface displacement was measured with a Doppler vibrometer and a Michelson interferometer. The choice of the acoustic frequency allowed us to oscillate the film and substrate as rigid bodies, so that the harmonic response represented the dynamics of the interface. The harmonic response was evaluated in the spatial frequency domain where the film surface oscillation was evaluated by the contrast of the interferometric fringes. Due to the fact that the frame rate of the digital camera that captured the interferometric images was much lower than the acoustic oscillation frequency, the oscillation amplitude decreased the fringe contrast. Theoretical consideration allowed us to relate the reduction in the fringe contrast to the oscillation amplitude using the zeroth order Bessel function of the first kind. Driving frequency sweeps of the fringe contrast measurement provided evidence that the interface had resonance in the tested audible frequency range.

Ti, Ti-Au and Ti-Pt thin-film coated on the same silicon substrate, along with the silicon substrate alone (the bare silicon) were tested with the Michelson interferometer. Each combination of film materials was coated on surface-treated and untreated silicon specimens separately. Comparison with the bare silicon specimen confirmed that the harmonic response of the film surface represented the dynamics of the interface. In each film material, both the untreated and treated specimens show resonance behavior where the resonant frequency of the untreated specimen was 15% to 20% higher than the treated specimen. Some of the observed resonant frequencies were found to shift over long-term use (oscillation) of the specimen.

The elastic constant associated with the observed resonant frequency was orders of magnitude lower than that of the film or substrate material. Neither a clear explanation for this finding nor the finding that the resonant frequency could shift after long-term use have been found. It is possible that the imperfection of chemical bonds in the interface drastically reduced the stiffness and/or that residual stresses induced during the coating process deformed the interface plastically, generating dislocations that reduced the elasticity. The resonant frequency shift seemed to be consistent with the explanation with residual stresses.

Acknowledgments: The present study was supported by the National Research Foundation of Korea (NRF) grants funded by the Korean government MSIP, NRF-2013R1A2A2A05005713, NRF-2013M2A2A9043274.

Author Contributions: S.Y. conceived of the study. D.R.D., D.D., H.-S.P. and S.Y. performed the optical experiments and data analysis. T.S. and S.Y. performed the acoustic transducer calibration. I.-K.P. contributed to the design and analysis of the acoustic transducer arrangement. S.Y. and D.G. wrote the paper.

Conflicts of Interest: The authors declare no conflict of interest.

References

1. Turunen, M.P.K.; Marjamaki, P.; Paajanen, M.; Lahtinen, J.; Kivilahti, J.K. Pull-off test in the assessment of adhesion at printed wiring board metallization/epoxy interface. *Microelectron. Reliab.* **2004**, *44*, 993–1007.

2. Noijenh, S.P.M.; van der Sluis, O.; Timmermans, P.H.M. An Extensive Investigation of the Four Point Bending Test for Interface Characterization. In Proceedings of the 13th International Conference on Thermal, Mechanical and Multi-Physics Simulation and Experiments in Microelectronics and Microsystems, EuroSimE, Cascais, Portugal, 16–18 April 2012.

3. Laugier, M.T. An energy approach to the adhesion of coatings using the scratch test. *Thin Solid Films* **1984**, *117*, 243–249.

4. Bunett, P.J.; Rickerby, D.S. The scratch adhesion test: An elastic-plastic indentation analysis. *Thin Solid Films* **1988**, *157*, 233–244.

5. Lemons, R.A.; Quate, C.F. Acoustic microscopy. In *Physical Acoustics*; Mason, W.P., Thurston, R.N., Eds.; Academic Press: London, UK, 1979; Volume XIV, pp. 1–92.

6. Weglein, R.D. Acoustic microscopy applied to SAW dispersion and film thickness measurement. *IEEE Trans. Sonics* **1980**, *27*, 82–86.

7. Atalar, A. An angular-spectrum approach to contrast in reflection acoustic microscopy. *J. Appl. Phys.* **1978**, *49*, 1530–1539.

8. Atalar, A. A physical model for acoustic signatures. *J. Appl. Phys.* **1979**, *50*, 8237–8239.

9. Telschow, K.L.; Deason, V.A.; Cottle, D.L.; Larson, J.D., III. Full-field imaging of gigahertz film bulk acoustic resonator motion. *IEEE Trans. Ultrason.* **2003** , *50*, 1279–1285.

10. Jensen, H.M. Analysis of mode mixity in blister test. *Int. J. Fract.* **1988**, *94*, 79–88.

11. Bedrossian, J.; Kohn, R.V. Blister patterns and energy minimization in compressed thin films on compliant substrates. *Commun. Pure Appl. Math.* **2015**, *68*, 472–510.

12. Dennenberg, H. Measurement of adhesion by a blister method. *J. Appl. Ploym. Sci.* **1961**, *5*, 125–134.

13. Volinsky, A.A.; Moody, N.R.; Gerberich, W.W. Interfacial toughness measurements for thin films on substrates. *Acta Mater.* **2002**, *50*, 441–466.

14. Liao, Q.; Fu, J.; Jin, X. Single-chain polystyrene particles adsorbed on the silicon surface: A molecular dynamics simulation. *Langmuir* **1999** , *15*, 7795–7801.

15. Huang C.; Ma, C. Vibration characteristics for piezoelectric cylinders using amplitude-fluctuation electronic speckle pattern Interferometry. *AIAA J.* **1988**, *36*, 2262–2268.

16. Wang, W.C.; Hwang, C.H. The Development and Applications of Amplitude Fluctuation Electronic Speckle Pattern Interferometry Method. In *Recent Advances in Mechanics*; Kounadis, A.N., Gdoutos, E.E., Eds.; Springer: New York, NY, USA, 2011; pp. 343–358.

17. Yoshida S.; Adhikari, S.; Gomi, K.; Shrestha, R.; Huggett, D.; Miyasaka, C.; Park, I.K. Opto-acoustic technique to evaluate adhesion strength of thin-film systems. *AIP Adv.* **2012**, *2*, 022126:1–022126:7.

18. Yoshida, S.; Didie, D.R.; Didie, D.; Adhikari, S.; Park, I.K. Opto-Acoustic Technique to Investigate Interface of Thin-Film Systems. In *Advancement of Optical Methods in Experimental Mechanics*; Jin, H., Sciammarella, C., Yoshida, S., Laberti, L., Eds.; Springer: New York, NY, USA, 2014; Volume 3, pp. 117–125.

19. OFV-5000 Vibrometer Controller. Available online: http://www.polytec.com/us/products/vibration-sensors/single-point-vibrometers/modular-systems/ofv-5000-vibrometer-controller/ (accessed on 11 May 2016).

20. DD-300 *24 MHz Dispolacement Decoder*. Available online: http://www.polytec.com/fileadmin/user_uploads/Products/Vibrometers/OFV-Decoder/Displacement_Decoder/Documents/OM_DS_DD-300_2010_07_PDF_E.pdf (accessed on 11 May 2016).

21. Basnet, M.; Yoshida, S.; Tittmann, B.R.; Kalkan, A.K.; Miyasaka, C. Quantitative Nondestructive Evaluation for Adhesive Strength at an Interface of a Thin Film System with Opto-Acoustic Techniques. In Proceedings of the 47th Annual Technical Meeting of Society of Engineering Science, Ames, IA, USA, 4–6 October 2010.

22. Scalise, L.; Paone, N. Laser Doppler vibrometry based on self-mixing effect. *Opt. Lasers Eng.* **2002**, *38*, 173–184.

23. Miyasaka, C. Pennsylvania State University, University Park, PA, USA. Personal communication, 2012.

24. Ishiyama, C.; Tasaki, T.; Tso-Fu Mark Chan, T.M.; Sone, M. Effects of specimen dimensions on adhesive shear strength between a microsized SU-8 column and a silicon substrate. *Jpn. J. Appl. Phys.* **2012**, *51*, doi:10.1143/JJAP.51.06FL19.

25. Volinsky, A.A.; Moody, N.R.; Gerberich, W.W. Superlayer residual stress effect on the indentation adhesion measurement. *Mater. Res. Soc. Proc.* **2000**, *594*, 383–388.

On Site Investigation and Health Monitoring of a Historic Tower in Mantua, Italy

Antonella Saisi, Marco Guidobaldi and Carmelo Gentile

Abstract: The paper describes the strategy adopted to assess the structural condition of the tallest historic tower in Mantua (Italy) after the Italian seismic sequence of May–June 2012 and exemplifies the application of health monitoring using (automated) operational modal analysis. The post-earthquake survey (including extensive visual inspection, historic and documentary research, non-destructive (ND) material testing, and ambient vibration tests) highlighted the poor state of preservation of the upper part of the tower; subsequently, a dynamic monitoring system (consisting of a few accelerometers and one temperature sensor) was installed in the building to address the preservation of the historic structure, and automated modal identification was continuously performed. Despite the low levels of vibration that existed in operational conditions, the analysis of data collected over a period of about 15 months allowed to assess and model the effects of changing temperature on modal frequencies and to detect the occurrence of abnormal behavior and damage under the changing environment. The monitoring results demonstrate the potential key role of vibration-based structural health monitoring, implemented through low-cost hardware solutions and appropriate software tools, in the preventive conservation and the condition-based maintenance of historic towers.

Reprinted from *Appl. Sci.* Cite as: Saisi, A.; Guidobaldi, M.; Gentile, C. On Site Investigation and Health Monitoring of a Historic Tower in Mantua, Italy. *Appl. Sci.* **2016**, *6*, 173.

1. Introduction

The Italian seismic sequence of May–June 2012 [1] highlighted the high vulnerability of the historical architectures, especially in the southern part of the province of Mantua and in the neighboring Emilia-Romagna region, where several brittle collapses of towers, fortification walls, and castles occurred [2,3], despite the supposed low seismicity of the area.

After the earthquake, an extensive research program was carried out to assess the structural condition of the tallest historic tower in Mantua: the *Gabbia* tower [4,5]. The tower (Figure 1), about 54.0 m high, is a symbol of cultural heritage in Mantua so that the fall of masonry pieces from its upper part, reported during the earthquake occurred on 29 May 2012 [1], provided strong motivations for deeply investigating the structural condition of the building. The post-earthquake survey, described in [6], included: (a) historic and documentary research; (b) an on-site survey and visual inspection of the load-bearing walls and structural discontinuities; (c) non-destructive and minor-destructive tests of materials on site; and (d) dynamic tests in operational conditions.

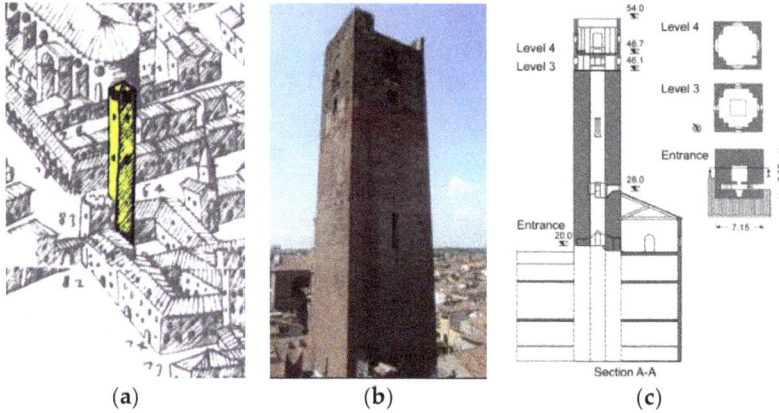

(a)   (b)   (c)

Figure 1. The *Gabbia* tower in Mantua, Italy: (**a**) view from the 17th century [4]; (**b**) recent view from the east; and (**c**) section of the tower (dimensions in m).

As shown in Figure 2, the available historic pictures indicate modifications and successive additions at the top of the tower. Visual inspection [6] of the upper load-bearing walls revealed traces of past structures on all fronts, and ineffective links between the added parts and extended masonry decay. In addition, one local mode—involving the top of the building—was clearly detected in the dynamic tests [6]; the presence of such a local mode highlights the remarkable structural effects of the change in the morphology and quality of masonry observed on top of the tower and, along with the poor structural arrangement of the upper region, suggests the need for repair interventions to be carried out.

While waiting for the retrofit design and funding, a dynamic monitoring system (consisting of three highly-sensitive accelerometers and one temperature sensor) was installed in the tower [7] for structural health monitoring (SHM) purposes. The continuous dynamic monitoring was mainly aimed at: (a) evaluating the response of the tower to the expected sequence of far-field seismic events; (b) assessing the impact of the environmental (*i.e.*, temperature) effects on the natural frequencies of the building [7–13]; and (c) detecting the occurrence of any abnormal change or anomaly in the structural behavior. Further potential objectives are providing the retrofit design with additional information and evaluating the effects of the future strengthening.

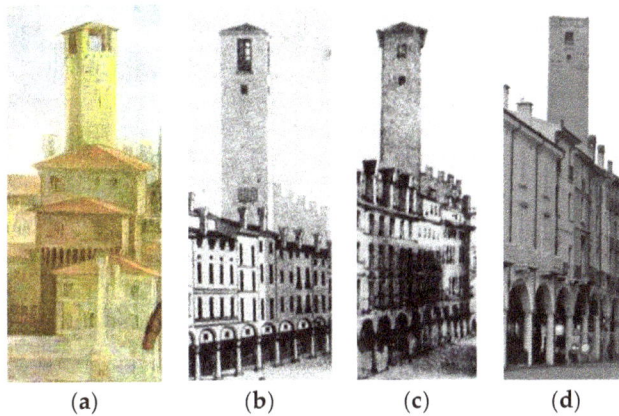

Figure 2. Views of the *Gabbia* tower dating back to: (**a**) 17th century; (**b**) 1830; (**c**) 1852; and (**d**) present days.

Approximately six months after the beginning of the continuous monitoring, the tower was subjected to a far-field earthquake [14] and the maximum measured acceleration exceeded about 50 times the highest response that was observed under normal ambient excitation; hence, the monitoring project provided with a challenging opportunity to identify the occurrence of seismic-induced damage under a changing environment.

After a brief description of the investigated tower and the post-earthquake survey, the paper presents the results of the first 15 months of dynamic monitoring and full details are given on the following main tasks: (1) installation of the continuous dynamic monitoring system with remote control and data transmission via the Internet; (2) development of efficient tools [15] to process the raw data regularly received and to automatically identify the modal parameters [16]; (3) tracking the evolution in time of natural frequencies and modeling the temperature effects on those features; and (4) identification of structural performance anomalies.

2. Description of the *Gabbia* Tower (Mantua, Italy) and On-Site Tests

The *Gabbia* tower [4,5], about 54.0 m high, is the tallest tower in Mantua and recent research dates the end of its construction to 1227. The building (Figure 1) was part of the defensive system of the Bonacolsi family, governing Mantua in the 13th century. As shown in Figure 1, the tower is, nowadays, part of an important palace, whose load-bearing walls seem not to be effectively connected to the tower walls; nevertheless, the several vaults and floors of the palace are directly supported by the tower.

The tower has a nearly square plan and the load-bearing walls, built in solid brick masonry, are about 2.4 m thick except for the upper levels, where the section decreases to about 0.7 m and a two level lodge is hosted (Figure 1c). Over the past centuries, the lodge was used for communication and military purposes and was accessible through a wooden staircase. The staircase has not been practicable since the 1990s and the access to the inside of the tower was re-established only in October 2012, when provisional scaffoldings and a light wooden roof were installed to allow the visual inspection of the inner load-bearing walls.

After the seismic sequence of May–June 2012 [1], an on-site survey of all outer fronts of the tower was performed using a movable platform and visual inspection highlighted two different structural conditions for the tower:

(a) in the main part of the building, until about 46.0 m from the ground level, no evident structural damage was observed and the materials appeared mainly affected by superficial decay;

(b) at about 8.0 m from the top, the brick surface workmanship exhibits a clear change [6] and the upper portion of the tower is characterized by passing-through structural discontinuities on all fronts (Figure 3), corresponding to the successive building phases and modifications (Figure 2), as well as by extensive masonry decay. Moreover, the stratigraphic sequencing of masonry (*i.e.*, the well-known methodology coming from archaeology and aimed at obtaining the relative dating of different parts or stratigraphic units) allowed the recognition of at least six main building phases of the upper region of the tower, as shown in Figure 3.

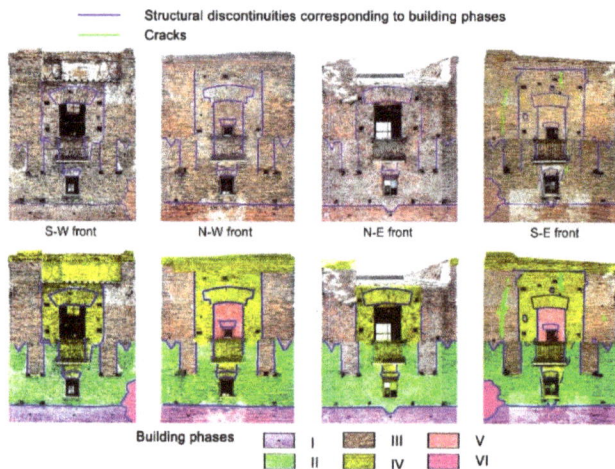

Figure 3. Map of the structural discontinuities observed in the upper part of the *Gabbia* tower and supposed building phases.

It is further noticed that all fronts of the tower includes merlon-shaped discontinuities (building phase II in Figure 3), recalling the defensive purpose of the original medieval architecture.

The mapping of the structural discontinuities (Figure 3) determined concerns on the seismic behavior of the upper part of the tower: due to the lack of effective mechanical links between the added parts, some weakly restrained portions could overturn under the earthquake actions. Based on this investigation, a first evaluation of out-of-plane seismic behavior for each identified masonry portion not effectively linked was carried out, as shown in Figure 4, and safety factors of about 1.15–1.20 were estimated for the earthquake occurred on May–June 2012 [1]. Furthermore, low intensity seismic actions, such as after-shocks or far-field earthquakes, could worsen the weak connection between the different elements by decreasing the adhesion and accentuating the boundaries.

Further evidence of the poor compactness of the materials in the upper region of the structure was attained by performing pulse sonic tests at different levels of the tower: the average value of sonic velocity measured at the upper levels is 600 m/s, whereas the sonic velocity values range between 1100 m/s and 1600 m/s until a height of 46.0 m.

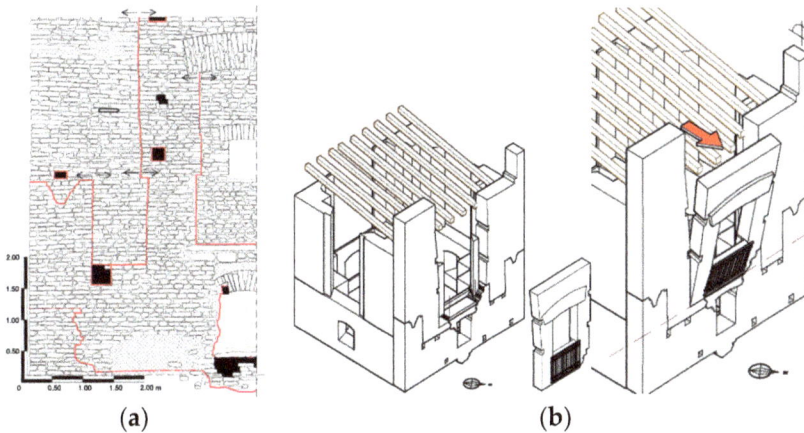

(a)   (b)

Figure 4. (a) Identification of weakly restrained masonry portions and (b) potential overturning mechanisms associated to seismic events.

As previously pointed out, a wooden roof (Figure 5a) was installed in the upper part of the tower when the inner access was re-established through metallic scaffoldings (Figure 5b). The roof, although very light, rests directly on the weakest region of the structure and is slightly inclined (Figure 5a). Hence, the roof slope and the redundant connection with the masonry walls, as well as thermal effects, might cause non-negligible thrusts on the crowning walls, which are very vulnerable due to the presence of several discontinuities, the lack of connection of the different additions (Figure 4), and the extensive decay.

(a)   (b)

Figure 5. Details of (**a**) the added wooden roof and (**b**) the provisional scaffoldings.

3. Preliminary Operational Modal Tests and Dynamic Characteristics of the Tower

Two series of ambient vibration tests (AVTs) were conducted on the tower. The first test was carried out between 31/07/2012 and 02/08/2012 [6] with the objectives of evaluating: (a) the dynamic characteristics of the tower; (b) the possible effects of the poor state of preservation of the upper region on the global behavior; and (c) the impact of temperature changes on the natural frequencies. The second test, preparatory to the continuous dynamic monitoring, was performed on 27/11/2012 with the two-fold objective of evaluating the possible effects of added wooden roof and scaffoldings (Figure 5) and providing reference values of the modal parameters for the subsequent monitoring.

During both tests high-sensitivity accelerometers (WR model 731A, 10 V/g sensitivity and 0.5 g peak acceleration, Wilcoxon Research, Germantown, MD, USA) were used to measure the dynamic response of the tower in 12 points, belonging to four pre-selected cross-sections along the height of the building (Figure 6a). Since access to the inner walls of the tower was not available during the first survey, the accelerometerswere mounted on the outer walls making use of the same movable platform employed for visual inspection. During the second test, performed after the installation of wooden roof and provisional scaffoldings inside the tower (Figure 5), inner mounting of the accelerometers was preferred (Figure 6a).

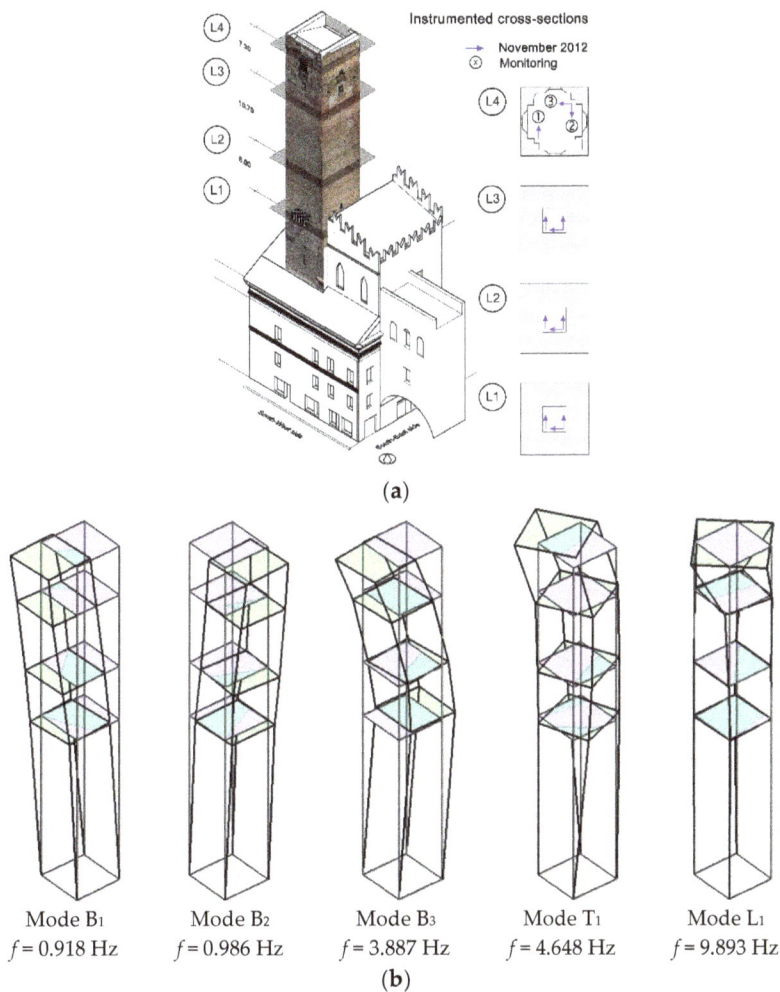

Figure 6. (a) Instrumented cross-sections and layout of the accelerometers during the preliminary tests (November 2012) and the continuous dynamic monitoring; and (b) vibration modes identified applying the data-driven stochastic subspace Identification algorithm (SSI-Data) to the data collected on 27/11/2012.

The modal identification was performed considering time windows of 3600 s and applying the data-driven stochastic subspace Identification algorithm (SSI-Data) [17,18] available in the commercial software ARTeMIS Extractor [19] (Release 5.3, SVS, Aalborg, Denmark, 2012).

Since the second test was aimed at estimating the modal parameters used as reference values in the continuous dynamic monitoring, only the results of this test will be presented herein.

The application of the SSI-Data method to the datasets collected in the test of November 2012 (about two hours of acceleration data recorded with an outdoor temperature of 10 °C–11 °C) allowed the identification of five vibration modes in the frequency range of 0–10 Hz. Figure 6b summarizes the identified dynamic characteristics of the tower and allows the following comments:

(a) two closely-spaced modes (B_1 and B_2) were identified in the frequency range of 0.90–1.00 Hz and are associated to dominant bending in the two main planes of the tower, respectively. The modal deflections of first mode (B_1) involve bending in the N-E/S-W direction, whereas the second mode (B_2) is dominant bending in the orthogonal N-W/S-E direction;

(b) the third mode (B_3) is dominant bending in the N-E/S-W direction with slight components also in the orthogonal direction;

(c) the fourth mode (T_1) is characterized by dominant torsion until the height of 46.0 m and coupled torsion-bending of the top level. The local behavior of the upper region of the tower alters what would otherwise be classified as a pure torsion and was not observed in the previous dynamic test so that it can be conceivably associated to both the poor structural condition of the upper level and the effect exerted by the wooden roof, directly resting on the weakest region of the tower; and

(d) a local mode (L_1) was identified at 9.89 Hz and involved torsion of the top levels of the tower. The presence of a local mode provides further evidence of the structural change in the masonry quality and morphology highlighted in the upper part of the building by the visual inspection.

It is worth mentioning that the fundamental frequencies of about 1 Hz exactly fall in the range of dominant frequencies characterizing the earthquakes recorded in Mantua on May 2012 and, more generally, the earthquakes expected in the Mantua region conceivably explain the fall of small masonry pieces from the upper part of the building, reported during the earthquake of 29 May 2012.

4. Continuous Dynamic Monitoring of the *Gabbia* Tower

A few weeks after the execution of the preliminary AVTs, a simple dynamic monitoring system was installed in the tower. The monitoring system is composed by: (a) three piezoelectric accelerometers (WR model 731A, 10 V/g sensitivity and ±0.50 g peak, Wilcoxon Research, Germantown, MD, USA), mounted on the cross-section at the crowning level of the tower (Figure 6a); (b) one temperature sensor (TRAFAG, Legnano, Italy), installed on the S-W front and measuring the outdoor wall temperature; (c) one four-channel data acquisition system (24-bit resolution, 102 dB dynamic range, and anti-aliasing filters) (National Instruments, Austin, TX, USA); and (d) one industrial PC on site, for the system management and data storage.

A binary file, containing three acceleration time series and the temperature data, is created every hour, stored in the local PC and transmitted to Politecnico di Milano for being processed.

The continuous dynamic monitoring system has been active since 17/12/2012. The data files received from the monitoring system are managed by a LabVIEW toolkit [15] (National Instruments, Austin, TX, USA), including the (on-line or off-line) execution of the following tasks:

(1) creation of a database, with the original data in compact format, for later developments;

(2) data pre-processing, *i.e.*, de-trending, automatic recognition, and extraction of possible seismic events, and creation of one dataset per hour;

(3) statistical analysis of data. This task includes also the evaluation of the hourly-averaged temperature and the computation of the root mean square (RMS) accelerations associated to each 1-h dataset; and

(4) low-pass filtering and decimation of the each dataset and creation of a second database in ASCII format, with essential data records, to be used in the modal identification phase. In more details, the recorded accelerations were low-pass filtered—using a classic seventh-order Butterworth filter with a cut-off frequency of 20 Hz—and decimated five times, reducing the sampling frequency from 200 Hz to 40 Hz.

4.1. Automatic System Identification

As in the preliminary tests, the identification of modal parameters was carried out considering time windows of 3600 s, in order to largely comply with the widely-agreed recommendation [20] of using a minimum duration of the acquired time window (which should be 1000–2000 times the fundamental period of the structure) to obtain accurate estimates from output-only data. In fact, output-only methods assume that the (un-measured) excitation input is a zero mean Gaussian white noise and this assumption is as closely verified as the length of the (measured outputs) time window is longer.

The modal parameters of the building were estimated from the measured signals using an automatic procedure [16], based on the covariance-driven stochastic subspace Identification (SSI-Cov) method [17,18].

In the SSI-Cov algorithm [17,18], the covariance matrices of the m measured outputs are computed, for positive time lags varying from Δt to $(2i-1)\Delta t$, to fill a $(mi \times mi)$ block Toeplitz matrix, that is decomposed to estimate stochastic subspace models—*i.e.*, matrices A, C—of increasing order n. Once matrices A, C of different order n are obtained, $n/2$ sets of modal parameters are extracted from a model of order n: natural frequencies and damping ratios are calculated from the eigenvalues of A, whereas the mode shapes are evaluated from the product of C and the eigenvectors of A.

In the automated procedure [16], each stabilization diagram (where the modes associated to increasing model order are plotted together) was "cleaned" from certainly spurious modes firstly by detecting system poles with non-physical characteristics through a damping ratio check, and subsequently by excluding the poles whose mode shapes exhibit a high modal complexity [21]. Similar modes in the cleaned stabilization diagram are grouped together based on the sensitivity of frequency and mode shape change to the increase of the model order. Finally, the identified modal parameters are compared to the reference results, available from the preliminary test.

In the present application, the time lag parameter i was set equal to 70 and the data was fitted using stochastic subspace models of order n varying between 30 and 120.

4.2. Temperature Effects on Natural Frequencies: Dynamic Regression and ARX Models

As measuring the dynamic response using few accelerometers provides information mainly on the natural frequencies, these parameters have to be necessarily used as features sensitive to abnormal structural changes and damage. However, the natural frequencies of masonry towers are also affected by factors other than damage, such as temperature [9–13]. It should be noticed that, as reported in [9], the humidity might also affect the modal frequencies of masonry towers, with the increase of water absorption (in the rainy season) inducing a temporary decrease of natural frequencies; on the other hand, no appreciable effects of air humidity on natural frequencies have been detected in [11,13].

Since masonry towers are generally subjected to low levels of ambient excitation, it has to be expected that the modal frequencies are mainly affected by changing temperature. Hence, the correlation between natural frequencies and temperature needs to be investigated for early detection of structural anomalies: once the normal response of a structural system to changes in its environmental conditions has been explored and can be filtered out to normalize the response data, any further changes in the sensitive features should rely to changes of the structural condition.

In order to characterize the effects of temperature on the natural frequencies, dynamic regression (DR) models were established for each identified frequency. DR models [22] assume that the value y_k of the dependent variable (*i.e.*, natural frequency) at time instant k is influenced by the values of the model input (*i.e.*, temperature) at current time k, as well as at $(p-1)$ previous time instants:

$$y_k = b_1 \times x_k + b_2 \times x_{k-1} + \ldots + b_p \times x_{k-(p-1)} + \varepsilon_k \tag{1}$$

The linear model (1) can be expressed in matrix form as:

$$y = Xb + \varepsilon \tag{2}$$

where $y \wp R^n$ is the vector containing n observations y_k of the dependent variable, $b \wp R^p$ is the vector formed by the p parameters weighting the contribution of the input, $\varepsilon \wp R^n$ is the vector of the random errors and $X \wp R^{n \times p}$ is the matrix appropriately gathering the values of the input.

Models (1) and (2) can be further generalized so that predictions are computed which also consider previous values of the dependent variables. Among the dynamic methods described in the system identification literature (see e.g., [23]), the Auto-Regressive models with eXogenous input (ARX) are probably the most widely used. ARX models [23] consist of an auto-regressive output and an exogenous input part and can be expressed in the following form:

$$y_k + a_1 \times y_{k-1} + \ldots + a_{n_a} \times y_{k-n_a} = b_1 \times x_{k-n_k} + b_2 \times x_{k-n_k-1} + \ldots + b_{n_b} \times x_{k-n_b+1} + \varepsilon_k \tag{3}$$

Of course, the linear model (3) can be expressed in the matrix form (2). It should be noted that in Equation (3) only one input (e.g., temperature) and one output (e.g., natural frequency) were considered. Equation (3) can be easily generalized to the case of multiple inputs by replacing y_k and x_k with corresponding row and column vectors, respectively. ARX models are characterized by three model orders: the auto-regressive order n_a (corresponding to the number of the considered past measures of the dependent variable), the exogenous order n_b (corresponding to the number of previous model inputs taken into account), and the time delay between input and output n_k. Orders n_a and n_b determine the number of model parameters.

It is worth mentioning that: (a) in the framework of a dynamic monitoring project, different input candidates might be considered in the dynamic regression model (1) and (2), such as temperature, humidity, wind speed, traffic loads on bridges, *etc.* (see e.g., [24–26]); (b) in order to properly describe the influence of changing environmental conditions, the data used to estimate the parameters of the regression models should be collected over a significant period of time, denoted as the training period (or reference period), and including a statistically representative sample of temperature conditions.

Once DR or ARX models have been established for the identified frequencies, the impact of changing environment on each modal frequency should be eliminated (or at least minimized) by defining the following quantities:

(a) The residual error vectors $\varepsilon_i = y_i - X_i b_i$ ($i = 1, 2, \ldots, M$ where M is the number of automatically identified modal frequencies); and

(b)   The cleaned observation vectors $y_i^* = \mathbf{1}\,\mu_i + \varepsilon_i$ ($i = 1, 2, \ldots, M$ where $\mathbf{1}$ is the unit vector and μ_i is the mean value of the identified i-th frequency data in the reference period).

Hence, both the residual error and the cleaned observation vectors might be used to identify the occurrence of abnormal structural changes or damages not observed during the training period.

5. Monitoring Results and Data Analysis

This section presents the main results of the continuous dynamic monitoring for a period of 15 months, from 17/12/2012 to 17/03/2014. During the examined monitoring period, more than 10000 1-h datasets were collected, with the ambient excitation being mainly provided by micro-tremors and wind, and automated modal identification has been performed.

Figure 7 shows the RMS value of the acceleration measured at points 2 and 3 (Figure 6a), computed considering time windows of 1 h. The inspection of Figure 7 highlights that: (a) the ambient excitation is generally very low and the RMS acceleration does not exceed 0.05–0.06 mg (with the corresponding peak accelerations being lower than 0.4 mg); (b) windy days are easily identified through "alignments" of RMS acceleration points in the range 0.06–0.20 mg. Furthermore, the information provided by the weather station in Mantua and the Italian Institute of Geophysics and Volcanology (INGV) allowed the classification of the isolated points exceeding the amplitude of 0.06 mg as generally associated to far-field seismic events or large micro-tremors.

It should be noticed that, until June 2013, the tower's response to different far-field earthquakes was recorded by the monitoring system. The strongest seismic event—corresponding to the earthquake (5.2 Richter magnitude) that occurred in the Garfagnana region (Tuscany) on 21/06/2013 [14] and strongly felt in many regions of Northern Italy—is marked in red in Figure 7 and was characterized by a measured peak acceleration of about 20 mg, exceeding about 40–50 times the highest amplitude of normally observed ambient vibrations.

Figure 8 shows the evolution in time of the outdoor temperature (S-W front) and reveals yearly variations, ranging between $-2\,^\circ$C and $45\,^\circ$C, and daily fluctuations of about $30\,^\circ$C in sunny days. The corresponding variations of the modal frequencies *vs.* time (Figure 9a) and *vs.* temperature (Figure 9b) indicate that the natural frequencies of the global modes (B_1–B_3 and T_1, Figure 6b) are related to the temperature. More specifically, Figure 9b shows that the frequency of global modes tends to increase with increased temperature. This behavior, observed also in other studies of masonry towers [9–13], is conceivably related to the materials thermal expansion that, in turn, gives rise to the closure of minor masonry discontinuities, superficial cracks, and mortar gaps. Hence, the "compacting" of the materials induces a temporary increase of stiffness and modal frequencies, as well.

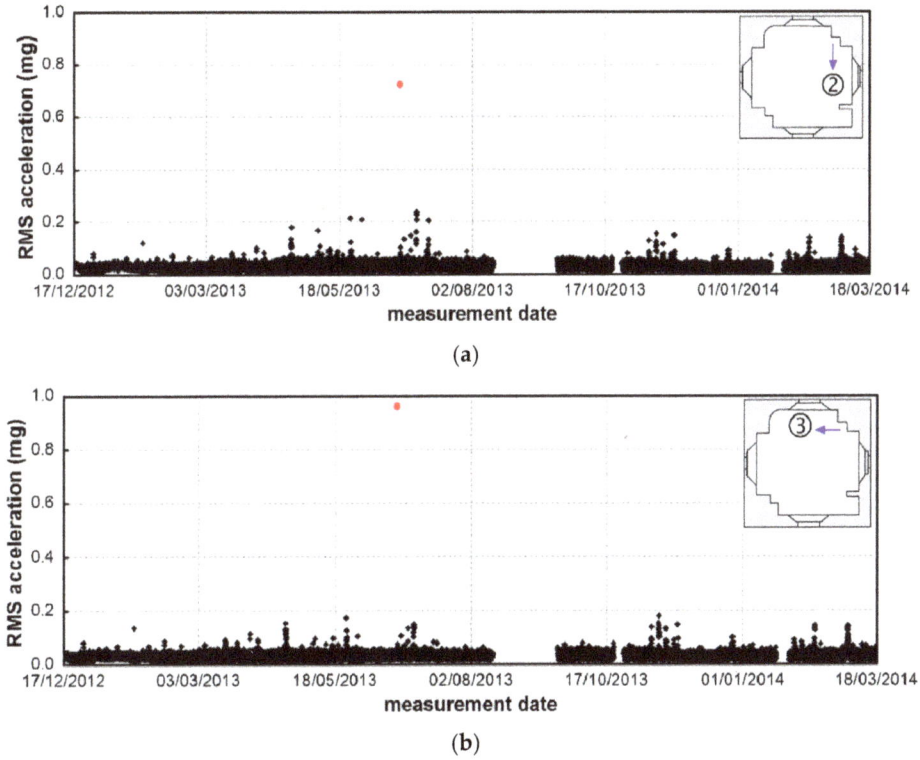

(a)

(b)

Figure 7. The root mean square (RMS) value of the acceleration measured on the tower at points 2 (**a**) and 3 (**b**) from 17/12/2012 to 17/03/2014 (the response to the seismic event of 21/06/2013 is marked in red).

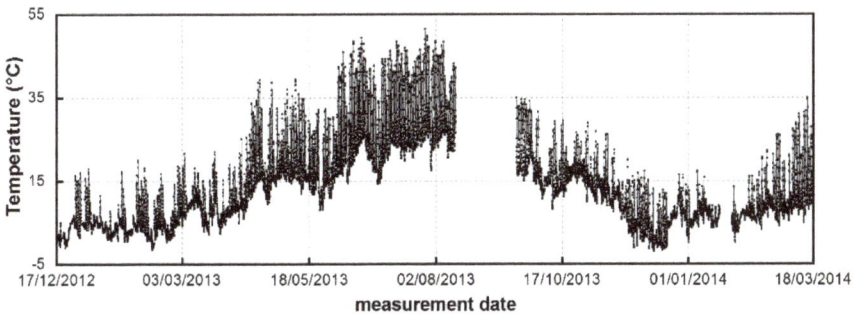

Figure 8. Time evolution of the temperature measured on the S-W front of the tower from 17/12/2012 to 17/03/2014.

433

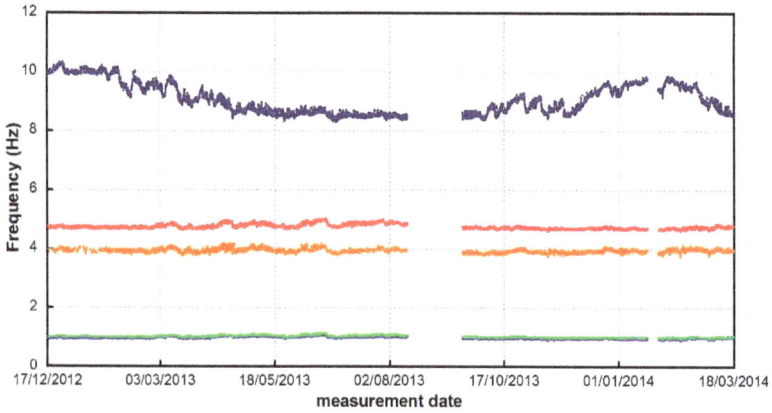

(a)

● Mode B₁   ● Mode B₂   ● Mode B₃   ● Mode T₁

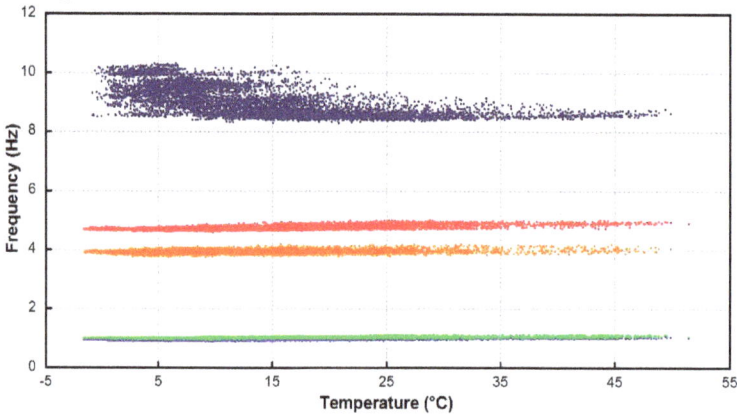

(b)

Figure 9. Natural frequencies, automatically identified from 17/12/2012 to 17/03/2014, plotted (**a**) *vs.* time and (**b**) *vs.* temperature.

The statistics of the natural frequencies identified from 17/12/2012 to 17/03/2014 are summarized in Table 1. This table includes the mean value (f_{mean}), the standard deviation (σ_f), and the extreme values (f_{min}, f_{max}) of each modal frequency. It should be noted that standard deviations are larger than 0.03 Hz for all global modes and especially significant for the local mode. Indeed, the frequency evolution of mode L_1 looks very different from the others (Figure 9) and characterized by remarkable variations, as the natural frequency ranges between 8.33 Hz and 10.33 Hz over the examined time period.

Table 1. Statistics of the natural frequencies identified (SSI-Cov) from 17/12/2012 to 17/03/2014.

Mode	f_{mean} (Hz)	σ_f (Hz)	f_{min} (Hz)	f_{max} (Hz)
1 (B_1)	0.977	0.035	0.903	1.102
2 (B_2)	1.017	0.028	0.960	1.148
3 (B_3)	3.935	0.069	3.742	4.194
4 (T_1)	4.741	0.073	4.600	5.010
5 (L_1)	9.062	0.511	8.332	10.327

B = bending mode; T = torsion mode; L = local mode.

A close inspection of the frequency tracking of mode L_1 allows the recognition of three different phases of the frequency evolution in time: (a) in a first phase (between 17/12/2012 and 14/08/2013) the modal frequency clearly decreases in time, from an initial value of about 10.0 Hz to a final value of about 8.5 Hz; (b) after the summer period (from 19/09/2013 to 19/01/2014) the frequency increases again, even if the new maximum values do not reach those identified one year before in similar environmental conditions; and (c) finally, the natural frequency decreases again from 02/02/2014 to the end of the monitoring period. Furthermore, each of those three phases is characterized by the presence of some discontinuities (drops or increases), which define the general trend of the frequency tracking.

For example, considering the first time period, three drops can be detected at the beginning of February 2013, at mid-March 2013, and at mid-April 2013 [7]. These drops divide the examined time span in four parts, which can be easily identified in the temperature-frequency plot of Figure 10. This figure highlights that the populations of temperature-frequency points, corresponding to the four different periods, exhibiting similar slope of the best fit line, whereas the average frequency value significantly decreases. The subsequent time periods are characterized by similar trends, with the frequency increasing and decreasing, respectively.

It should be noted that the changes exhibited by the frequency of the local mode, illustrated in Figures 9 and 10, can hardly be explained by moisture. In fact, the figures show that the natural frequency of the local mode tends to decrease as: (a) the average daily temperature increases and (b) the temperature daily range includes higher temperatures. On the contrary, it has to be expected [9] that the modal frequency decreases with increased water absorption (*i.e.*, the moisture inside the walls) at the beginning of the raining season and not at the beginning and during the hot seasons, as it happens in the present case (Figures 9a and 10).

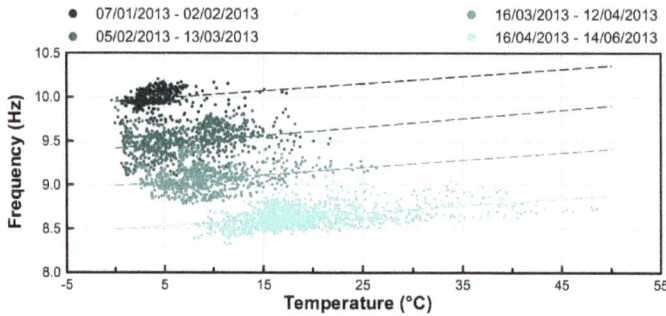

Figure 10. Natural frequency of local mode L_1 plotted *vs.* temperature in the period between 07/01/2013 and 14/06/2013.

The observed behavior suggests the progress of a possible damage mechanism, conceivably related to the effect exerted by the wooden roof with increased temperature, and confirms the poor structural condition of the upper part of the tower, as well as the need of preservation actions. This conclusion seems to be confirmed also by the frequency loss detected after one year of monitoring, with the natural frequency of the local mode being unable to reach the maximum values identified one year before in similar environmental conditions.

As previously pointed out, the earthquake occurring in the Garfagnana region on 21/06/2013 [14] determined accelerations on top of the *Gabbia* tower that were significantly higher (Figure 7) than the usual ambient vibration responses.

Zooming in the time evolution of the natural frequencies of modes B_1, B_2, and T_1—considering three weeks before and three weeks after the earthquake—is shown in Figures 11a, 12a and 13a; the figures show a clear decrease of all modal frequencies on 21/06/2013, corresponding to the occurrence of the seismic event. Furthermore, the temperature-frequency relationships (Figures 11b, 12b and 13b) highlight that the regression lines of all modes exhibited clear variations after the earthquake, with the temperature range being almost unchanged. It is worth mentioning that this trend is confirmed by:

(a) Estimating the distribution of the modal frequencies before and after the seismic event. The estimate—shown in Figure 14 for modes B_1, B_2, and T_1—is based on a normal kernel function, which was evaluated at 100 equally-spaced frequency bins covering the range of variation of each identified frequency. The probability density functions presented in Figure 14 refer to the six months before (from 17/12/2012 to 20/06/2013, solid line) and the six months after the earthquake (dashed line). The inspection of the plots reveals that the frequency distributions change after the seismic event and indicates that the identified frequencies slightly shift towards lower values after the earthquake;

436

(b) The general decrease of the statistics of the natural frequencies (mean value, standard deviation, extreme values) summarized in Table 2 and, again, demonstrating the occurrence of abnormal structural changes induced by the seismic event.

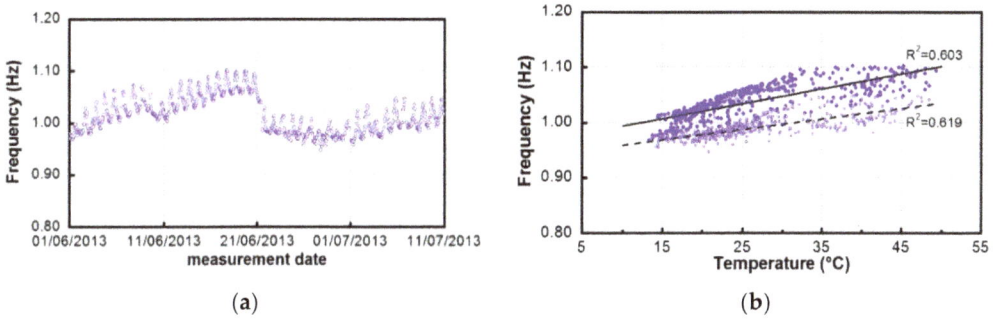

(a)　　　　　　　　　　　　　　　　　　(b)

Figure 11. Mode B_1: (**a**) variation of the natural frequency between 01/06/2013 and 10/07/2013; and (**b**) change in the frequency-temperature correlation induced by the seismic event of 21/06/2013.

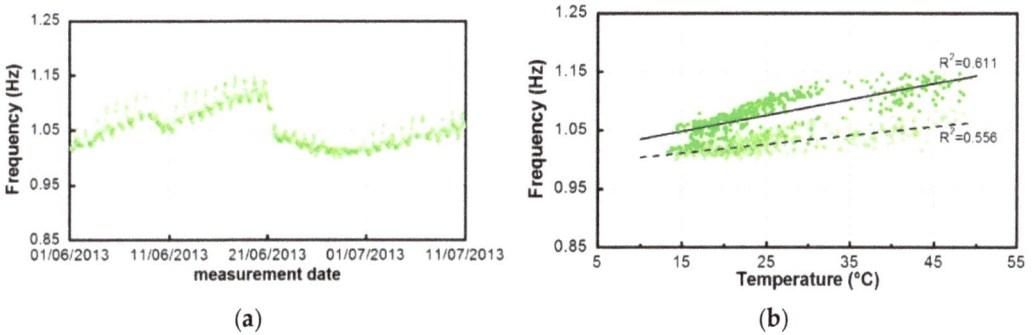

(a)　　　　　　　　　　　　　　　　　　(b)

Figure 12. Mode B_2: (**a**) variation of the natural frequency between 01/06/2013 and 10/07/2013; and (**b**) change in the frequency-temperature correlation induced by the seismic event of 21/06/2013.

Based on the previous evidence, it can be stated that the limited number of accelerometers and temperature sensors installed in the tower and the automatic system identification allowed for the assessment of the effects of changing temperature on modal frequencies and to detect the occurrence of abnormal structural behavior under changing environment.

(a)

(b)

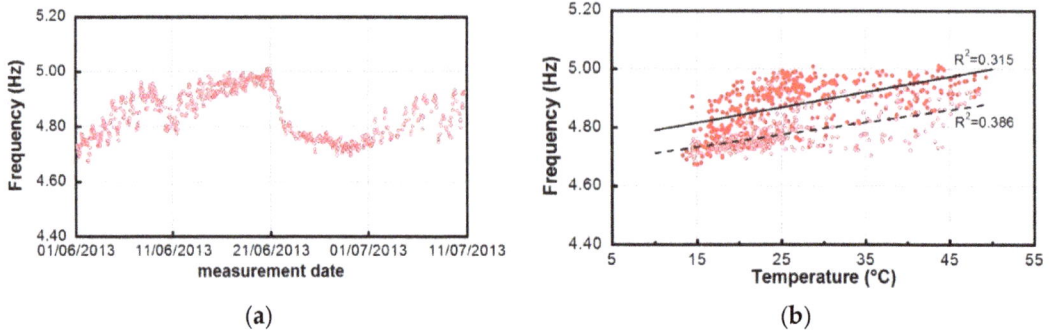

Figure 13. Mode T_1: (**a**) variation of the natural frequency between 01/06/2013 and 10/07/2013; and (**b**) change in the frequency-temperature correlation induced by the seismic event of 21/06/2013.

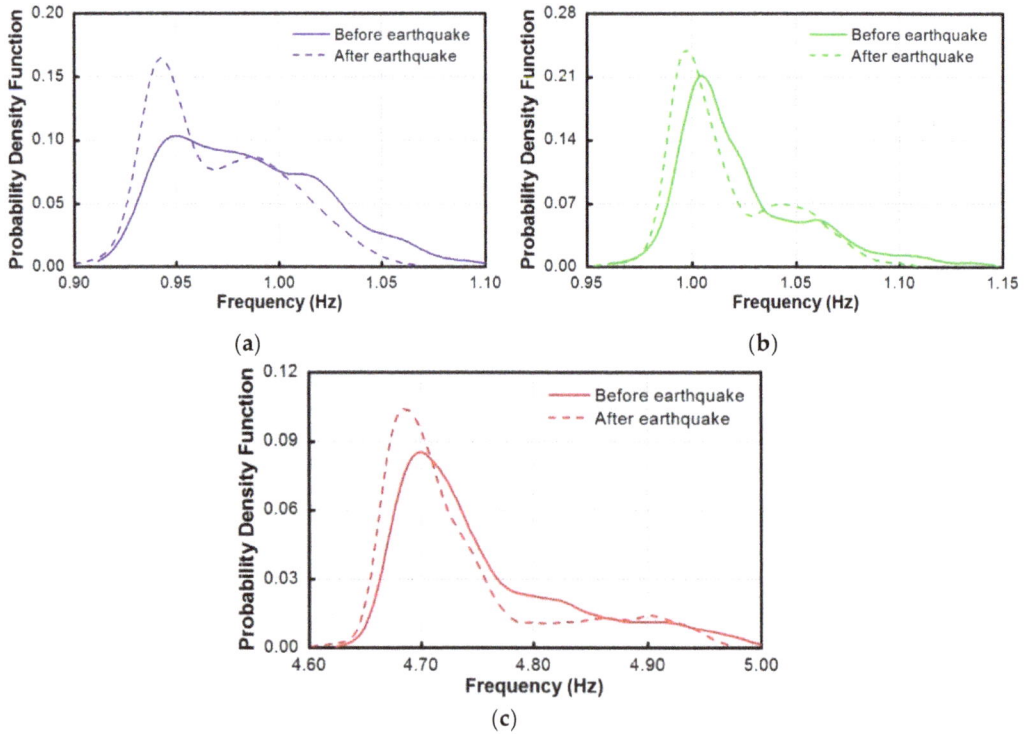

(a)

(b)

(c)

Figure 14. Probability density function of the natural frequency identified from 17/12/2012 to 20/06/2013 (solid line) and after 21/06/2013 (dashed line): (**a**) Mode B_1; (**b**) Mode B_2; and (**c**) Mode T_1.

Table 2. Statistics of the modal frequencies identified before and after the seismic event of 21/06/2013.

Mode	f_{mean} (Hz)		σ_f (Hz)		f_{min} (Hz)		f_{max} (Hz)	
	Before	After	Before	After	Before	After	Before	After
1 (B_1)	0.985	0.968	0.038	0.031	0.910	0.897	1.102	1.070
2 (B_2)	1.024	1.012	0.032	0.025	0.961	0.953	1.148	1.110
3 (B_3)	3.941	3.929	0.075	0.063	3.758	3.742	4.194	4.137
4 (T_1)	4.754	4.727	0.077	0.066	4.621	4.600	5.010	4.982
5 (L_1)	9.222	8.937	0.554	0.433	8.385	8.332	10.327	9.862

B = bending mode; T = torsion mode; L = local mode.

5.1. Damage Detection under Changing Environment

As stated in Section 4.2, the temperature effects on the natural frequency of each global mode have been described by means of DR models, with the outdoor temperature measured on the S-W front as its only input. Based on the classic loss function (LF) and the final prediction error (FPE) criteria [23], regression models depending on the current and 13 previous hourly-averaged values ($p - 1 = 13$) of the external temperature have been established.

As the main goal was to mitigate the environmental effects in order to verify if the structural changes induced by the earthquake of 21/06/2013 are still detected, the models were trained over the six months preceding that seismic event (from 17/12/2012 to 16/06/2013), where the process is supposed to be in control. The dynamic regression models were subsequently used to predict the natural frequencies after the training period and to detect the changes in the structural behavior. The relevant results—in terms of cleaned natural frequencies of modes B_1, B_2 and T_1—are presented in Figure 15 with the confidence limits at plus and minus three times the standard deviation of the frequency estimated obtained over the training period.

It is worth noting that the representation in Figure 15 is centered on the earthquake date (21/06/2013), in order to better highlight the different trends before and after the seismic event. The inspection of the cleaned modal frequencies (Figure 15) allows the following comments:

(a) the natural frequencies still exhibit some correlation, meaning that they are still affected by common factors;

(b) before the seismic events, each frequency oscillates around its mean value and stays within the confidence interval; and

(c) a sudden drop is detected corresponding to the earthquake, along with a clear variation of the previous trend. More specifically, in the months after 21/06/2013, the mean value of the cleaned natural frequency changes and the confidence interval, estimated in the training period, is exceeded.

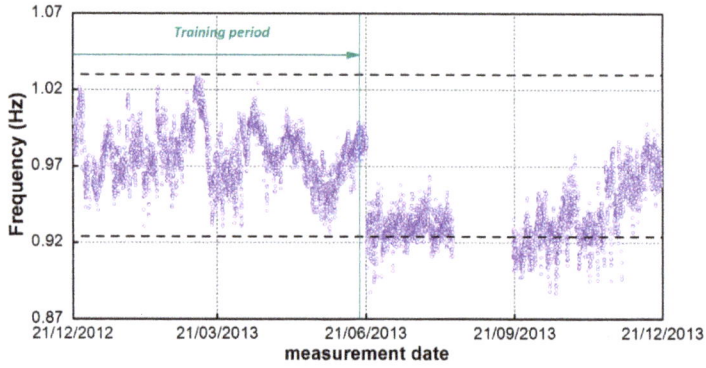

(a)

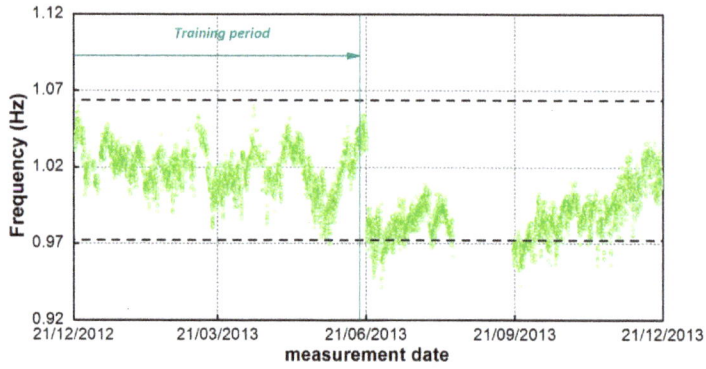

(b)

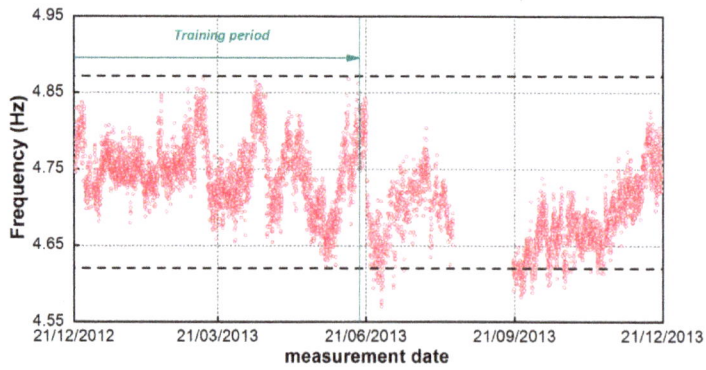

(c)

Figure 15. Change in the cleaned natural frequency (Dynamic Regression model) induced by the seismic event of 21/06/2013: (**a**) Mode B$_1$; (**b**) Mode B$_2$; and (**c**) Mode T$_1$.

Therefore, the correction of the natural frequencies provided by the adopted regression model still indicates that non-reversible structural changes took place after the Garfagnana earthquake.

The DR models, so far discussed, allowed for partially modeling the thermal effects on the natural frequencies of the investigated tower. On one hand, the DR relationship proved to be effective in reproducing the fluctuations of the experimental frequencies due to the daily effect of temperature and in distinguishing the slight damage induced by the Garfagnana earthquake (Figure 15). On the other hand, the model could not as accurately predict the long-term oscillations due to the influence of the average temperature (conceivably combined with the effects of other unobserved factors).

In order to improve the prediction, ARX models have been developed. ARX models assume the value f_k of the natural frequency at time k to be affected by outdoor temperatures at the same time k and at previous time instants, as well as by previous experimental estimates of the input. In particular, the use of past values of the output is expected to indirectly take into account the effects of unmeasured factors that cannot be modeled otherwise (e.g., wind, humidity, *etc.*) leading, in principle, to more accurate predictions. Since several models can be fitted to the experimental data, depending on the amount of previous time instants considered for the input and the output, a comparative study of the performances has been carried out by using the same LF and FPE criteria [23] adopted for establishing the DR models. Eventually, an ARX240 model has been selected, characterized by $n_a = 2$ previous estimates of the experimental frequency, $n_b = 4$ past hourly values of the outdoor temperature, and no delay between input and output ($n_k = 0$). To calibrate the parameters of the ARX240 model, the same training period used for the DR relationship has been considered, *i.e.*, between 17/12/2012 and 16/06/2013, in order to leave out the structural variation due to the earthquake of 21/06/2013.

The evolution in time of the cleaned natural frequencies of modes B_1, B_2, and T_1 is shown in Figure 16: the non-negligible dispersion of the values and the evident fluctuations characterizing the results of the DR are no longer observed, suggesting that the influence of environmental factors has been completely removed. Unfortunately, none of the depurated frequencies exhibits any sudden variation after the Garfagnana earthquake. A possible explanation for this unexpected behavior could lie in the fact that—as shown in Figure 11a, Figure 12a, and Figure 13a—the frequency drop caused by the seismic event has the same order of magnitude as the fluctuations induced by the temperature effects. Therefore, since the ARX model is able to follow the temperature-induced frequency variations with high accuracy, the sudden drop caused by the seismic event is reproduced as well. This, in turn, jeopardizes the possibility of detecting anomalous occurrences from environment-independent features obtained through ARX models.

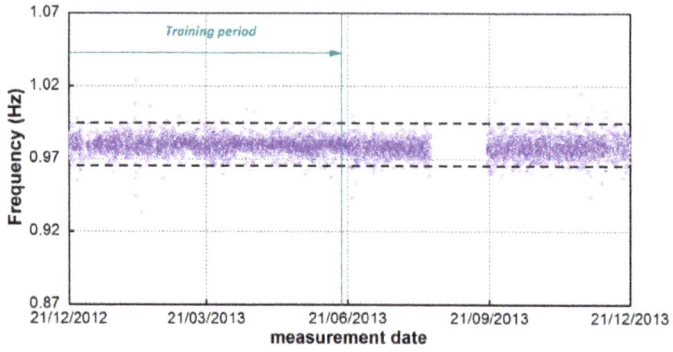

(a)

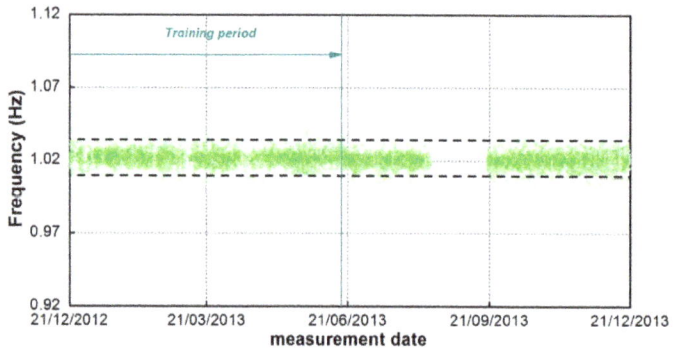

(b)

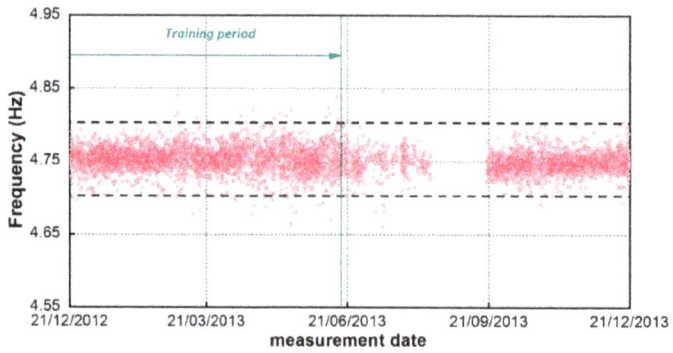

(c)

Figure 16. Change in the cleaned natural frequency (Auto-Regressive model with eXogenous input) induced by the seismic event of 21/06/2013: (**a**) Mode B_1; (**b**) Mode B_2; and (**c**) Mode T_1.

6. Conclusions

The paper focuses on the post-earthquake assessment of a historic masonry tower and summarizes the results of visual inspection, ambient vibration tests, and long-term dynamic monitoring of the building.

Visual inspection and the stratigraphic sequencing of the load-bearing walls clearly indicated that the upper part of the tower is characterized by the presence of several discontinuities due to the historic evolution of the building, local lack of connection, and extensive masonry decay. The poor state of preservation of the same region was confirmed by the observed dynamic characteristics, as one local mode, involving the upper part of the tower, was clearly identified from ambient vibration data.

The results of the SHM exercise performed through the continuous dynamic monitoring of the tower at study highlighted that a limited number of accelerometers and temperature sensors installed in the structure are sufficient to provide meaningful information for the preventive conservation and/or SHM of historic towers. More specifically, the following conclusions can be drawn:

- The application of state-of-art tools for automated operational modal analysis allows tracking of natural frequencies;
- The temperature turned out to significantly affect the (daily) variation of the natural frequencies in masonry towers, with the frequencies of global modes increasing with increased temperature;
- Dynamic regression models are suitable to represent the environmental effects on modal frequencies, even when only one (outdoor) temperature sensor is installed in the structure;
- The application of automated operational modal analysis and dynamic regression models turned out to be effective in the identification of seismic-induced structural anomalies under changing environment;
- ARX models are suitable to completely remove the environmental effects from modal frequencies, but exhibit no damage detection skills (*i.e.*, the frequency drops associated to slight abnormal structural changes are not distinguished from similar variations induced by temperature or other unobserved factors);
- The application of automated operational modal analysis highlighted the possible progress of a damage mechanism, involving the upper part of the tower, remarked by the significant fluctuations of the natural frequency of the identified local mode.

As a concluding remark, this full scale experience demonstrates that low-cost dynamic monitoring systems, combined with state-of-art tools for automated operational modal analysis and statistical tools for the minimization of the environmental effects, are effective and sustainable means for the preservation of historic masonry towers.

Acknowledgments: The research was partially supported by the Mantua Municipality. Marco Antico, Marco Cucchi (VibLab, Politecnico di Milano) and Lorenzo Cantini, PhD are gratefully acknowledged for the assistance during the visual inspection, the field tests, the installation and maintenance of the dynamic monitoring system.

Author Contributions: A.S. supervised the research; A.S. and C.G. conceived and designed the experimental tests as well as the monitoring system; M.G. mainly performed the data analysis.

Conflicts of Interest: The authors declare no conflict of interest.

References

1. Luzi, L.; Pacor, F.; Ameri, G.; Puglia, R.; Burrato, P.; Massa, M.; Augliera, P.; Franceschina, G.; Lovati, S.; Castro, R. Overview on the strong-motion data recorded during the May–June 2012 Emilia seismic sequence. *Seismol. Res. Lett.* **2013**, *84*, 629–644.
2. Sorrentino, L.; Liberatore, L.; Decanini, L.; Liberatore, D. The performance of churches in the 2012 Emilia earthquake. *Bull. Earthquake Eng.* **2014**, *12*, 2299–2331.
3. Cattari, S.; Degli Abbati, S.; Ferretti, D.; Lagomarsino, S.; Ottonelli, D.; Tralli, A. Damage assessment of fortresses after the 2012 Emilia earthquake (Italy). *Bull. Earthquake Eng.* **2014**, *12*, 2333–2365.
4. Bertazzolo, G. *Urbis Mantuae Descriptio*; Biblioteca Teresiana: Mantua, Italy, 1628.
5. Zuccoli, N. *Historic Research on the Gabbia Tower*; Municipality of Mantua Internal Report: Mantua, Italy, 1988. (In Italian)
6. Saisi, A.; Gentile, C. Post-earthquake diagnostic investigation of a historic masonry tower. *J. Cult. Heritage* **2015**, *16*, 602–609.
7. Saisi, A.; Gentile, C.; Guidobaldi, M. Post-earthquake continuous dynamic monitoring of the Gabbia Tower in Mantua, Italy. *Constr. Build. Mater.* **2015**, *81*, 101–112.
8. Sohn, H. Effects of environmental and operational variability on structural health monitoring. *Philos. Trans. R. Soc. A* **2007**, *365*, 539–560. PubMed]
9. Ramos, L.F.; Marques, L.; Lourenço, P.B.; De Roeck, G.; Campos-Costa, A.; Roque, J. Monitoring historical masonry structures with operational modal analysis: Two case studies. *Mech. Syst. Signal Process.* **2010**, *24*, 1291–1305.
10. Cabboi, A. Automatic Operational Modal Analysis: Challenges and Application to Historic Structures and Infrastructures. Ph.D. Thesis, University of Cagliari, Cagliari, Italy, 2013.
11. Cantieni, R. One-year monitoring of a historic bell tower. In Proceedings of the 9th International Conference on Structural Dynamics, EURODYN 2014, Porto, Portugal, 30 June–2 July 2014; pp. 1493–1500.

12. Ubertini, F.; Cavalagli, N.; Comanducci, G. Sensing hardware optimization and automated condition assessment of a monumental masonry bell-tower. In Proceedings of the 1st ECCOMAS Thematic Conference on Uncertainty Quantification in Computational Sciences and Engineering, UNCECOMP 2015, Crete, Greece, 25–27 May 2015; pp. 477–487.

13. Lorenzoni, F.; Casarin, F.; Caldon, M.; Islami, K.; Modena, C. Uncertainty quantification in structural health monitoring: Applications on cultural heritage buildings. *Mech. Syst. Signal Pr.* **2016**, *66–67*, 268–281.

14. Italian Institute of Geophysics and Volcanology (INGV). Available online: http://terremoti.ingv.it/it/ultimi-eventi/921-evento-sismico-tra-le-province-di-lucca-e-massa.html (accessed on 23 June 2013).

15. Busatta, F. Dynamic Monitoring and Automated Modal Analysis of Large Structures: Methodological Aspects and Application to a Historic Iron Bridge. Ph.D. Thesis, Politecnico di Milano, Milano, Italy, 9 November 2012.

16. Cabboi, A.; Magalhães, F.; Gentile, C.; Cunha, À. Automated modal identification and tracking: Application to an iron arch bridge. *Struct. Control Health Monit.* **2016**.

17. Peeters, B.; De Roeck, G. Reference-based stochastic subspace identification for output-only modal analysis. *Mech. Syst. Signal Pr.* **1999**, *13*, 855–878.

18. Peeters, B. System Identification and Damage Detection in Civil Engineering. Ph.D. Thesis, Katholieke Universiteit Leuven, Leuven, Belgium, December 2000.

19. SVS. ARTeMIS Extractor 2011. Available online: http://www.svibs.com/ (accessed on 10 September 2012).

20. Cantieni, R. Experimental methods used in system identification of civil engineering structures. In Proceedings of the 1st International Operational Modal Analysis Conference, IOMAC'05, Copenaghen, Denmark, 26–27 April 2005; pp. 249–260.

21. Reynders, E.; Houbrechts, J.; De Roeck, G. Fully automated (operational) modal analysis. *Mech. Syst. Signal Pr.* **2012**, *29*, 228–250.

22. Hair, J.; Anderson, R.; Tatham, R.; Black, W. *Multivariate Data Analysis*; Prentice Hall: New York, NY, USA, 1998.

23. Ljung, L. *System Identification: Theory for the User*; Prentice Hall: New York, NY, USA, 1999.

24. Magalhães, F.; Cunha, À.; Caetano, E. Vibration based structural health monitoring of an arch bridge: From automated OMA to damage detection. *Mech. Syst. Signal Pr.* **2012**, *28*, 212–228.

25. Cross, E.J.; Koo, K.Y.; Brownjohn, J.M.W.; Worden, K. Long-term monitoring and data analysis of the Tamar Bridge. *Mech. Syst. Signal Pr.* **2013**, *35*, 16–34.

26. Comanducci, G.; Ubertini, F.; Materazzi, A.L. Structural health monitoring of suspension bridges with features affected by changing wind speed. *J. Wind Eng. Ind. Aerodyn.* **2015**, *141*, 12–26.

MDPI AG

St. Alban-Anlage 66

4052 Basel, Switzerland

Tel. +41 61 683 77 34

Fax +41 61 302 89 18

http://www.mdpi.com

Applied Sciences Editorial Office

E-mail: applsci@mdpi.com

http://www.mdpi.com/journal/applsci

www.ingramcontent.com/pod-product-compliance
Lightning Source LLC
Chambersburg PA
CBHW051926190326

41458CB00026B/6417